Biotechnology in Agriculture and Forestry

Springer
*Berlin
Heidelberg
New York
Barcelona
Hong Kong
London
Milan
Paris
Singapore
Tokyo*

Volumes already published
Volume 1: Trees I (1986)
Volume 2: Crops I (1986)
Volume 3: Potato (1987)
Volume 4: Medicinal and Aromatic Plants I (1988)
Volume 5: Trees II (1989)
Volume 6: Crops II (1988)
Volume 7: Medicinal and Aromatic Plants II (1989)
Volume 8: Plant Protoplasts and Genetic Engineering I (1989)
Volume 9: Plant Protoplasts and Genetic Engineering II (1989)
Volume 10: Legumes and Oilseed Crops I (1990)
Volume 11: Somaclonal Variation in Crop Improvement I (1990)
Volume 12: Haploids in Crop Improvement I (1990)
Volume 13: Wheat (1990)
Volume 14: Rice (1991)
Volume 15: Medicinal and Aromatic Plants III (1991)
Volume 16: Trees III (1991)
Volume 17: High-Tech and Micropropagation I (1991)
Volume 18: High-Tech and Micropropagation II (1992)
Volume 19: High-Tech and Micropropagation III (1992)
Volume 20: High-Tech and Micropropagation IV (1992)
Volume 21: Medicinal and Aromatic Plants IV (1993)
Volume 22: Plant Protoplasts and Genetic Engineering III (1993)
Volume 23: Plant Protoplasts and Genetic Engineering IV (1993)
Volume 24: Medicinal and Aromatic Plants V (1993)
Volume 25: Maize (1994)
Volume 26: Medicinal and Aromatic Plants VI (1994)
Volume 27: Somatic Hybridization in Crop Improvement I (1994)
Volume 28: Medicinal and Aromatic Plants VII (1994)
Volume 29: Plant Protoplasts and Genetic Engineering V (1994)
Volume 30: Somatic Embryogenesis and Synthetic Seed I (1995)
Volume 31: Somatic Embryogenesis and Synthetic Seed II (1995)
Volume 32: Cryopreservation of Plant Germplasm I (1995)
Volume 33: Medicinal and Aromatic Plants VIII (1995)
Volume 34: Plant Protoplasts and Genetic Engineering VI (1995)
Volume 35: Trees IV (1996)
Volume 36: Somaclonal Variation in Crop Improvement II (1996)
Volume 37: Medicinal and Aromatic Plants IX (1996)
Volume 38: Plant Protoplasts and Genetic Engineering VII (1996)
Volume 39: High-Tech and Microprogation V (1997)
Volume 40: High-Tech and Microprogation VI (1997)
Volume 41: Medicinal and Aromatic Plants X (1998)
Volume 42: Cotton (1998)
Volume 43: Medicinal and Aromatic Plants XI (1999)
Volume 44: Transgenic Trees (1999)
Volume 45: Transgenic Medicinal Plants (1999)
Volume 46: Transgenic Crops I (1999)
Volume 47: Transgenic Crops II (2001)
Volume 48: Transgenic Crops III (2001)
Volume 49: Somatic Hybridization in Crop Improvement II (2001)

Volumes in preparation
Volume 50: Cryopreservation of Plant Germplasm II (2002)
Volume 51: Medicinal and Aromatic Plants XII (2002)
Volume 52: Brassicas and Legumes: From Genome Structure to Breeding

Biotechnology in Agriculture and Forestry 49

Somatic Hybridization in Crop Improvement II

Edited by
T. Nagata and Y.P.S. Bajaj

With 154 Figures, 4 in Color, and 53 Tables

 Springer

Professor Dr. Toshiyuki Nagata
University of Tokyo
Graduate School of Science
Department of Biological Sciences
7-3-1 Hongo, Bunkyo-ku
Tokyo 113-0033
Japan

Professor Dr. Y.P.S. Bajaj†
New Delhi, India

ISSN 0934-943-X
ISBN 3-540-41112-7 Springer-Verlag Berlin Heidelberg New York

Library of Congress Cataloging-in-Publication Data
Somatic hybridization in crop improvement/edited by Y.P.S. Bajaj. p. cm. – (Biotechnology in agriculture and forestry; 27) Includes bibliographical references and index. ISBN 3-540-57445-X (Berlin: acid-free paper): ISBN 0-387-57445-X (New York: acid-free paper) 1. Crops – Genetic engineering. 2. Somatic hybrids. 3. Crop improvement. I. Bajaj, Y.P.S., 1936– . II. Series. SB123.57.S67 1994 631.5′23 – dc20 94 – 15685. ISBN 3-540-41112-7

This work is subject to copyright. All rights are reserved, whether the whole or part of the material is concerned, specifically the rights of translation, reprinting, reuse of illustrations, recitation, broadcasting, reproduction on microfilms or in any other way, and storage in data banks. Duplication of this publication or parts thereof is permitted only under the provisions of the German Copyright Law of September 9, 1965, in its current version, and permission for use must always be obtained from Springer-Verlag. Violations are liable for prosecution under the German Copyright Law.

Springer-Verlag Berlin Heidelberg New York a member
of BertelsmannSpringer Science & Business Media GmbH

http://www.springer.de

© Springer-Verlag Berlin Heidelberg 2001
Printed in Germany

The use of general descriptive names, registered names, trademarks, etc. in this publication does not imply, even in the absence of a specific statement, that such names are exempt from the relevant protective laws and regulations and therefore free for general use.

Production: PRO EDIT GmbH, Heidelberg

Cover design: *design & production* GmbH, Heidelberg

Typesetting: Best-set Typesetter Ltd., Hong Kong

Printed on acid-free paper   SPIN: 10691023   31/3130/SO 5 4 3 2 1 0

This Volume is dedicated to the late Professor Georg Melchers, Tübingen, Germany, with whom I had intellectual interactions from 1974 to 1998.

Toshiyuki Nagata

Preface

This is the second issue on the "Somatic Hybridization in Crop Improvements". Since the first issue appeared in 1994, further progress has been made, which will be described in this Volume. At the onset, however, I have to explain why I have become the series editor of *Biotechnology in Agriculture and Forestry* (BAF) and thus the editor of this Volume. Towards the end of March, 2000, Dr. Dieter Czeschlik, the executive director for the books and journals program in the life sciences at Springer-Verlag, asked me if I could accept the series editorship of the BAF series, since Professor Y.P.S. Bajaj, the former editor, passed away in May 1999. This was really a surprise, since I am already engaged in various projects, including being Director of the Botanical Gardens of the University of Tokyo, so that I am busy enough. On the other hand, I was aware that this series is both successful and very useful for practical purposes. So it was a great honor for me to have been invited to be the series editor. In the meantime, some of my colleagues encouraged me to do this work and continue this series. So, after consideration and a certain hesitation, finally I dared to accept the proposal on the understanding that some changes in direction of this series could be possible, because plant biotechnology is steadily progressing and its course is changing.

Consequently, my first assignment for the BAF series is to complete Vol. 49, for which most of the manuscripts had accumulated. At this stage, I should thank Prof. Bajaj for designing the rough shape of this Volume; however, I did not ever meet him. Then I asked all the contributors to adjust the shape of their respective chapters.

In this situation, allow me to recall my relationship to this present field. In 1969, I started my studies on culturing tobacco mesophyll protoplasts together with the late Professor Itaru Takebe. After 5 months' efforts, which proved be in vain, I witnessed the first division of tobacco mesophyll protoplasts on November 5, 1969 (described in more detail in the Citation Classic of Current Contents, Feb. 15, 1988). We were convinced at that moment that we could carry out somatic cell genetics by using protoplasts from mesophyll, which until that time had been considered unsuitable for culture. Subsequently, I worked closely with the late Professor Georg Melchers of the Max-Planck Institute for Biology, Tübingen, Germany, on the basic mechanism of protoplast fusion. In close contact with Professor Melchers, I was witness to the production of somatic hybrids between tomato and potato, the so-called pomato and topato. Since then, the idea of using somatic hybridization for crop breeding has become realistic. On this subject, earlier experiments until the year of 1994 are included in the previous volume (Vol. 37 of BAF).

In this present Volume, the further development in this subject is described. Notably, some products are examined in field tests to assess their utility for the purpose of plant breeding; in particular, male-sterile rice produced by this technique has been shown to be useful for the production of hybrid rice. It should also be noted that the transfer of complex genetic traits from some plants to others by asymmetrical hybridization is also described, which cannot be attained even by the advanced transformation technique.

Thus, somatic hybridization is shown to be a useful means to widen the genetic variability of crops. Though the high tide of this subject seems to be passing away, the usefulness of this discipline should be considered.

Tokyo, April 2001   Toshiyuki Nagata

Contents

Section I Cereals

I.1 Somatic Hybridization Between *Hordeum vulgare* L. (Barley)
and *Daucus carota* L. (Carrot)
H. KISAKA, M. KISAKA, A. KANNO, and T. KAMEYA (With 8 Figures)

1 Introduction	3
2 Protoplast Fusion and Culture of Fused Cells	4
3 Analysis of the Three Regenerated Plants	5
4 Characterization of Somatic Hybrids	7
5 Quantification and Effects of Betaine	9
6 Discussion	9
7 Protocol	12
References	14

I.2 Cybridization in *Oryza sativa* L. (Rice)
H. AKAGI (With 8 Figures)

1 Introduction	17
2 Somatic Hybridization	19
3 Charcteriztion of Rice Cybrids	27
4 Selection of CMS Cybrid Plants	29
5 Application of the Cybridization Method in Rice Breeding	29
6 Summary and Conclusion	32
7 Protocol	32
References	34

I.3 Somatic Hybridization Between *Oryza sativa* L. (Rice)
and *Hordeum vulgare* L. (Barley)
H. KISAKA, M. KISAKA, A. KANNO, and T. KAMEYA (With 7 Figures)

1 Introduction	37
2 Protoplast Fusion and Culture	38
3 Analysis of Regenerated Plant	40
4 Assessment of Cold and Salt Tolerance	41
5 Protocol	43
References	46

I.4 Somatic Hybridization Between *Triticum aestivum* L. (Wheat) and *Haynaldia villosa* L.
G.M. XIA, A.F. ZHOU, and H.M. CHEN (With 22 Figures)

1 Introduction	48
2 Preparation and Treatment of Parental Protoplasts	49
3 Protoplast Fusion and Culture	50
4 Verification of Hybridity	53
5 Growth and Development of Hybrid Plants	59
6 Summary and Conclusion	61
7 Protocol	62
References	64

I.5 Asymmetric Somatic Hybridization Between *Triticum aestivum* L. (Wheat) and *Leymus chinensis* (Trin.) Tzvel
G.M. XIA, A.F. ZHOU, F. XIANG, and H.M. CHEN (With 5 Figures)

1 Introduction	65
2 Preparation and Inactivation of Wheat Protoplasts	65
3 Preparation of *L. chinensis* Protoplasts from Irradiated Calli	66
4 Fusion of Parental Protoplasts and Culture of Fusion Products	69
5 Identification of Hybrid Calli and Plants	70
6 Summary and Conclusion	73
7 Protocol	75
References	76

Section II Vegetables and Fruits

II.1 Somatic Hybridization Between *Arabidopsis* and *Brassica*
J. SIEMENS and M.D. SACRISTÁN (With 3 Figures)

1 Introduction	81
2 Somatic Hybridization	82
3 Summary and Conclusion	90
4 Protocol	91
References	92

II.2 Somatic Hybridization in *Asparagus*
H. KUNITAKE and M. MII (With 9 Figures)

1 Introduction	95
2 Somatic Hybridization	97
3 Summary and Conclusion	105
4 Protocol	108
References	110

II.3 Somatic Hybridization in *Cichorium intybus* L. (Chicory)
C. RAMBAUD and J. VASSEUR (With 2 Figures)

1 Introduction	112
2 Somatic Hybridization	113
3 Analysis of the Regenerated Plants	118
4 Summary and Conclusion	120
5 Protocol for the Regenaration of Somatic Hybrids	121
References	122

II.4 Cybridization in *Citrus unshiu* Marc. (Satsuma Mandarin) and *C. sinensis* (L.) Osbeck (Sweet Orange)
M. YAMAMOTO, S. KOBAYASHI, T. YOSHIOKA, and R. MATSUMOTO (With 10 Figures)

1 Introduction	124
2 Somatic Hybridization	126
3 Summary and Conclusion	135
4 Protocol	136
References	137

II.5 Somatic Hybridization in *Cucumis*
C.I. JARL (With 6 Figures)

1 Introduction	139
2 Somatic Hybridization	142
3 Summary and Conclusion	148
4 Protocol	148
References	149

II.6 Somatic Hybridization in *Diospyros* (Persimmon)
M. TAMURA and R. TAO (With 8 Figures)

1 Introduction	152
2 Somatic Hybridization	153
3 Summary and Conclusion	159
4 Protocol	160
References	162

II.7 Somatic Hybridization in *Ipomoea* (Sweet Potato) Species
M.M. BELARMINO and T. SASAHARA (With 8 Figures)

1 Introduction	164
2 Protoplast Isolation	166

3 Protoplast Culture and Plant Regeneration 169
4 Interspecific Hybridization Between Sweet Potato
  and Wild Relatives 172
5 Asymmetric Protoplast Fusion Between Sweet Potato
  (*Ipomoea batatas* Lam.) and African Marigold
  (*Tagetes erecta* L.) 177
6 Summary and Conclusion 180
7 Protocol for Asymmetric Protoplast Fusion Between
  Sweet Potato (*Ipomoea batatas* Lam.)
  and Wild Relative, *I. trifida* Don 183
References 184

II.8 Somatic Hybridization Between *Lycopersicon esculentum*
     Mill. (Tomato) and Wild Nontuberous *Solanum* Species
T. GAVRILENKO (With 4 Figures)

1 Introduction 188
2 Methods of Protoplast Isolation, Protoplast Culture,
  and Plant Regeneration 189
3 Production and Analysis of Intergeneric Somatic Hybrids 191
4 Summary and Conclusion 196
5 Protocol 197
References 198

II.9 Somatic Hybridization Between *Lycopersicon esculentum* Mill.
     (Tomato) and *Solanum melongena* L. (Eggplant)
V.M. SAMOYLOV and K.C. SINK (With 6 Figures)

1 Introduction 199
2 Somatic Hybridization 200
3 Analysis of Somatic Hybrids 202
4 Elimination of the Donor Genome 207
5 Polyploidization and Elimination of the Donor Genome 211
6 Protocol 213
References 214

II.10 Somatic Hybridization Between *Lycopersicon esculentum* Mill.
      (Tomato) and *Solanum ochranthum* Dun.
J.R. STOMMEL, R.S. KOBAYASHI, and S.L. SINDEN (With 8 Figures)

1 Introduction 217
2 Isolation of Protoplasts 218
3 Culture of Protoplasts 219
4 Fusion of Protoplasts and Regeneration of Somatic Hybrids 221

5 Regeneration and Characterization of Somatic Hybrids	222
6 Summary and Conclusion	229
7 Protocol	229
References	230

II.11 Somatic Hybridization Between *Solanum melongena* L. (Eggplant) and *Solanum sanitwongsei* Craib.
H. Asao, S. Arai, and M. Hirai (With 4 Figures)

1 Introduction	233
2 Somatic Hybridization	234
3 Summary and Conclusions	241
4 Protocol	243
References	243

II.12 Somatic Hybridization Between *Solanum commersonii* Dun. and *S. tuberosum* L. (Potato)
T. Cardi (With 4 Figures)

1 Introduction	245
2 Somatic Hybridization	246
3 Summary and Conclusion	258
4 Protocol	259
References	260

II.13 Somatic Hybridization Between *Solanum tuberosum* L. (Potato) and *Solanum phureja*
S. Millam and P. Davie (With 3 Figures)

1 Introduction	264
2 Somatic Hybridization	266
3 Summary and Conclusion	273
References	273

Section III Medicinal and Aromatic Plants

III.1 Somatic Hybridization in *Dianthus* Species
M. Nakano, Y. Hoshino, and M. Mii (With 3 Figures)

1 Introduction	277
2 Protoplast Isolation	278
3 Protoplast Culture	279
4 Protoplast Fusion	280

5 Regeneration of Somatic Hybrids	281
6 Summary and Conclusion	289
7 Protocol	290
References	290

III.2 Somatic Hybridization Between *Nicotiana sylvestris* Speg. & Comes and *N. plumbaginifolia* Viv.
C.C. CHEN, Y.Y. KAO, F.M. LEE, and R.F. LIN (With 4 Figures)

1 Introduction	292
2 Somatic Hybridization	294
3 Applications of Somatic Hybrids	298
4 Summary and Conclusion	300
5 Protocol	301
References	302

III.3 Somatic Hybridization Between *Nicotiana tabacum* L. (Tobacco) and *Atropa belladonna* L. (Deadly Nightshade)
M.K. ZUBKO, E.I. ZUBKO, O.A. KHVEDYNICH, S.V. LOPATO, S.A. LATIPOV, and YU. YU. GLEBA (With 10 Figures)

1 Introduction	304
2 Somatic Hybridization	308
3 Conclusions and Prospects	320
4 Protocol	322
References	324

III.4 Somatic Hybridization and Cell Grafting in *Senecio*
G. WANG and H. BINDING (With 2 Figures)

1 Introduction	328
2 Results and Discussion	328
3 Summary and Conclusion	335
4 Protocol	335
References	336

Section IV Legumes/Pasture Crops

IV.1 Somatic Hybridization Between *Medicago sativa* L. (Alfalfa) and *Lotus corniculatus* L. (Birdsfoot Trefoil)
M. NIIZEKI (With 8 Figures)

| 1 Introduction | 341 |
| 2 Symmetric Somatic Cell Hybridization | 342 |

3 Asymmetric Somatic Cell Hybridization	345
4 Summary and Conclusion	351
5 Protocol	353
References	354

IV.2 Somatic Hybridization Between *Medicago sativa* L. (Alfalfa) and *Medicago falcata* L.
S. ARCIONI, F. DAMIANI, F. PAOLOCCI, and F. PUPILLI (With 2 Figures)

1 Introduction	356
2 Isolation and Fusion of Protoplasts	358
3 Culture and Selection of Fused Protoplasts	360
4 Regeneration of Somatic Hybrids	362
5 Indentification and Characterization of Somatic Hybrid Plants	363
6 Summary and Conclusion	369
7 Protocol	370
References	371

Subject Index	375

List of Contributors

AKAGI, H., Department of Biological Production, Faculty of Bioresource Sciences, Akita Prefectural University, Akita 010-0195, Japan

ARAI, S., Nara Prefectural Agricultural Experiment Station, 88 Shijyo-cho, Kashihara-shi, Nara, 634-0813, Japan

ARCIONI, S., Istituto di Ricerche sul Miglioramento Genetico delle Piante Foraggere del CNR, Via della Madonna Alta 130, 06128 Perugia, Italy

ASAO, H., Nara Prefectural Agricultural Experiment Station, 88 Shijyo-cho, Kashihara-shi, Nara, 634-0813, Japan

BELARMINO, Marilyn M., Tissue Culture Laboratory, Department of Horticulture, Visayas State College of Agriculture, Baybay, Leyte 6521-A, Philippines

BINDING, H., Botanical Institute, Christian-Albrechts-University, Olshausenstr. 40-60, 24098 Kiel, Germany

CARDI, T., CNR-IMOF, Research Institute for Vegetable and Ornamental Plant Breeding, Via Università 133, 80055 Portici, Italy

CHEN, C.-C., Department of Botany, National Taiwan University, Taipei, Taiwan 106, Republic of China

CHEN, H.M., School of Life Science, Shandong University, Jinan 250100, P.R. China

DAMIANI, F., Istituto di Ricerche sul Miglioramento Genetico delle Piante Foraggere del CNR, Via della Madonna Alta 130, 06128 Perugia, Italy

DAVIE, P., Genomics Unit, Scottish Crop Research Institute, Invergowrie, Dundee DD2 5DA, United Kingdom

GAVRILENKO, T., N.I. Vavilov Institute of Plant Industry, B. Morskaya Street 42, 190000 St. Petersburg, Russia

GLEBA, YU. YU., Institute of Cell Biology and Genetic Engineering, National Academy of Sciences of Ukraine, Zabolotnogo Str. 148, 252022 Kiev, Ukraine (Present address: Icon Genetics, Inc., 66 Witherspoon Street, Suite 134, Princeton, New Jersey 08542 USA)

HIRAI, M., National Research Institute of Vegetables, Ornamental Plants and Tea, 360 Ano-cho, Mie, 514-2392, Japan

HOSHINO, Y., Experimental Farms, Faculty of Agriculture, Hokkaido University, Kita 11, Nishi 10, Kita-Ku, Sapporo 060-0811, Japan

JARL, Carin I., Plant Biology, Lund University, P.O. Box 117, 221 00 Lund, Sweden

KAMEYA, T., Institute of Genetic Ecology, Tohoku University, Sendai 980-8577, Japan

KANNO, A., Institute of Genetic Ecology, Tohoku University, Sendai 980-8577, Japan

KAO, Y.-Y., Department of Botany, National Taiwan University, Taipei, Taiwan 106, Republic of China

KHVEDYNICH, O.A., Institute of Cell Biology and Genetic Engineering, National Academy of Sciences of Ukraine, Zabolotnogo Str. 148, 252022 Kiev, Ukraine

KISAKA, H., Institute of Genetic Ecology, Tohoku University, Sendai 980-8577, Japan

KISAKA, M., Institute of Genetic Ecology, Tohoku University, Sendai 980-8577, Japan

KOBAYASHI, Ruth S., US Department of Agriculture, Agricultural Research Service, Plant Sciences Institute, Vegetable Laboratory, Beltsville, Maryland 20705, USA

KOBAYASHI, S., Persimmon and Grape Research Center, National Institute of Fruit Tree Science, Akitsu, Hiroshima 729-2494, Japan

KUNITAKE, H., School of Agriculture, Kyushu Tokai University, Aso, Kumamoto 869-1404, Japan

LATIPOV, S.A., Institute of Genetics, Moldovian Academy of Sciences, Lesnaja Str. 20, 277018 Kishinev, Moldova

LEE, F.-M., Department of Botany, National Taiwan University, Taipei, Taiwan 106, Republic of China

LIN, R.-F., Department of Botany, National Taiwan University, Taipei, Taiwan 106, Republic of China

LOPATO, S.V., Institute of Cell Biology and Genetic Engineering, National Academy of Sciences of Ukraine, Zabolotnogo Str. 148, 252022 Kiev, Ukraine (Present address: University of Vienna, Institute of Biochemistry, Dr Bohrgasse 9, 1030 Vienna, Austria)

MATSUMOTO, R., Department of Citriculture, National Institute of Fruit Tree Science, Kuchinotsu, Nagasaki 859-2501, Japan

MII, M., Laboratory of Plant Cell Technology, Faculty of Horticulture, Chiba University, Matsudo 271-8520, Japan

MILLAM, S., Gene Expression Unit, Scottish Crop Research Institute, Invergowrie, Dundee DD2 5DA, United Kingdom

NAKANO, M., Laboratory of Horticulture, Faculty of Agriculture, Niigata University, 2-8050 Ikarashi, Niigata 950-2181, Japan

NIIZEKI, M., Laboratory of Plant Breeding and Genetics, Faculty of Agriculture and Life Science, Hirosaki University, Hirosaki, Aomori-ken 036-8561, Japan

PAOLOCCI, F., Istituto di Ricerche sul Miglioramento Genetico delle Piante Foraggere del CNR, Via della Madonna Alta 130, 06128 Perugia, Italy

PUPILLI, F., Istituto di Ricerche sul Miglioramento Genetico delle Piante Foraggere del CNR, Via della Madonna Alta 130, 06128 Perugia, Italy

RAMBAUD, C., Laboratoire de Physiologie Cellulaire et Morphogenèse végétales, Bât SN2, Université des Sciences et Technologies de Lille 1, 59655 Villeneuve d'Ascq Cedex, France

SACRISTÁN, M. D., Institute of Biology, Applied Genetics, Free University of Berlin, Albrecht-Thaer-Weg 6, 14195 Berlin, Germany

SAMOYLOV, V.M., Syngenta, P.O. Box 12257, 3054 Cornwallis Rd., Research Triangle Park, North Carolina 27709-2257, USA

SASAHARA, T., Laboratory of Plant Breeding, Faculty of Agriculture, Yamagata University, Tsuruoka-shi 997-8555, Japan

SIEMENS, J., Institute of Biology, Applied Genetics, Free University of Berlin, Albrecht-Thaer-Weg 6, 14195 Berlin, Germany

SINDEN, S.L., US Department of Agriculture, Agricultural Research Service, Plant Sciences Institute, Vegetable Laboratory, Beltsville, Maryland 20705, USA

SINK, K.C., Department of Horticulture, Michigan State University, East Lansing, Michigan 48824-1325, USA

STOMMEL, J.R., US Department of Agriculture, Agricultural Research Service, Plant Sciences Institute, Vegetable Laboratory, Beltsville, Maryland 20705, USA

TAMURA, M., Plant Biotechnology Laboratory, Institute for Fundamental Research, Suntory Ltd., Shimamoto-cho, Osaka 618-8503, Japan

TAO, R., Graduate School of Agriculture, Kyoto University, Kyoto 606-8502, Japan

VASSEUR, J., Laboratoire de Physiologie Cellulaire et Morphogenèse végétales, Bât SN2, Université des Sciences et Technologies de Lille 1, 59655 Villeneuve d'Ascq Cedex, France

WANG, G., Botanical Institute, Christian-Albrechts-University, Olshausenstr. 40-60, 24098 Kiel, Germany (Present address: Department of Cellular and Molecular Physiology, The Pennsylvania State University College of Medicine, 500 University Drive, Hershey, Pennsylvania 17033, USA)

XIA, G.M., School of Life Science, Shandong University, Jinan 250100, P.R. China

XIANG, F., School of Life Science, Shandong University, Jinan 250100, P.R. China

YAMAMOTO, M., Faculty of Agriculture, Kagoshima University, Kagoshima 890-0065, Japan

YOSHIOKA, T., Department of Citriculture, National Institute of Fruit Tree Science, Kuchinotsu, Nagasaki 859-2501, Japan

ZHOU, A.F., School of Life Science, Shandong University, Jinan 250100, P.R. China

ZUBKO, E.I., Institute of Cell Biology and Genetic Engineering, National Academy of Sciences of Ukraine, Zabolotnogo Str. 148, 252022 Kiev, Ukraine (Present address: University of Leeds, LIBA, Leeds LS2 9JT, United Kingdom)

ZUBKO, M.K., Institute of Cell Biology and Genetic Engineering, National Academy of Sciences of Ukraine, Zabolotnogo Str. 148, 252022 Kiev, Ukraine (Present address: School of Biological Sciences, Manchester University, 3.614 Stopford Building, Oxford Road, Manchester M13 9PT, United Kingdom)

Section I
Cereals

I.1 Somatic Hybridization Between *Hordeum vulgare* L. (Barley) and *Daucus carota* L. (Carrot)

H. KISAKA, M. KISAKA, A. KANNO, and T. KAMEYA

1 Introduction

The aim of plant breeding is to construct new genotypes by the introduction and manipulation of genetic variations. The production of somatic hybrid plants by protoplast fusion is a potentially useful method for the combination of genetic materials. Protoplast fusion can sometimes lead to the production of new genetic variants as a consequence of the recombination of nuclear and/or of cytoplasmic genomes. Many intra- and interspecific and several intergeneric somatic hybrid plants have been reported (Melchers et al. 1978; Aviv et al. 1980; Gupta et al. 1982, 1984; Menczel et al. 1983; Negrutiu et al. 1986; Pental et al. 1986; Toriyama et al. 1987; Gleba et al. 1988; Kameya et al. 1989; Toki et al. 1990; Kostenyuk et al. 1991; Perl et al. 1991; Babiychuk et al. 1992). Recently, asymmetric hybrids between remotely related species, for example interfamilial hybrid plants have been remotely exploiting various systems for the selection of hybrids (Somers et al. 1986; Dudits et al. 1987; Kisaka and Kameya 1994; Kisaka et al. 1994).

Intergeneric fusion of *Solanum* and *Lycopersicon* species as a means of introducing tolerance to certain environmental stresses has been reported (Melchers et al. 1978; O'Connell and Hanson 1986). Cold tolerance of plants that were somatic hybrids of potato and tomato was reported by Smillie et al. (1979). Recently, transgenic tobacco plants carrying a bacterial gene for mannitol-1-phosphate dehydrogenase were shown to have enhanced ability to tolerate high salinity as a result of the accumulation of mannitol (Tarczynski et al. 1993). Deping et al. (1996) also produced transgenic rice that introduced a late embryogenesis abundant (LEA) protein gene, the *HVA1* gene, from *H. vulgare*. The transgenic rice plants were also shown to have enhanced ability to tolerant to water deficit and high salinity. Furthermore, transgenic rice plants carrying the gene for betaine aldehyde dehydrogenase or the *codA* gene for choline oxidase were shown to have enhanced ability to tolerate high salinity (Nakamura et al. 1997; Sakamoto et al. 1998).

Barley (*Hordeum vulgare* L.) is a crop plant that tolerates low temperature and salinity. The possibility is examined that these characteristics of barley might be transferable to somatic hybrids of barley and carrot (*Daucus carota*

Institute of Genetic Ecology, Tohoku University, Sendai 980-8577, Japan

L.), utilizing the low-temperature tolerance of barley for selection of hybrids, and that calli induced from the somatic hybrids to determine whether the cold and salt tolerance of *H. vulgare* had been transferred to it by the original protoplast fusion.

2 Protoplast Fusion and Culture of Fused Cells

When cells from 6-month-old suspension cultures of cells isolated from *D. carota* were plated on medium D and incubated at 4°C for various periods and the calli then transferred to 25°C, the number of the regenerated calli decreased with increasing duration of the low-temperature treatment (Fig. 1). On the basis of the result, the low-temperature treatment for selection of hybrid calli consisted of incubation at 4°C for 5 weeks after incubation for 1 month at 25°C of fused cells.

Protoplasts of *D. carota* isolated from 6-month-old suspension culture, and those of *H. vulgare* isolated from young leaves were fused by electrofusion, and fused cells were cultured according to the scheme outlined for selection of hybrids in Fig. 2. After culture for 1 month at 25°C, the fused cells were transferred to medium D and then incubated at low temperature (4°C) for 5 weeks in darkness (Fig. 3a) The resultant calli were transferred to continuous light (4 Wm^{-2}) at 25°C. After visible colonies had developed to about 1–2 mm in diameter, about 2700 colonies were transferred to fresh medium D. Three shoots were regenerated (Fig. 3b) and these were transferred to medium E. The three regenerated plants were potted in soil and designated nos. 1, 2, and 3 (Fig. 3c).

Protoplasts of *H. vulgare* that were isolated from young leaves failed to divide. Protoplasts of *D. carota* that were isolated from 6-month-old suspension cultures proliferated and formed colonies. However, about 1400 colonies

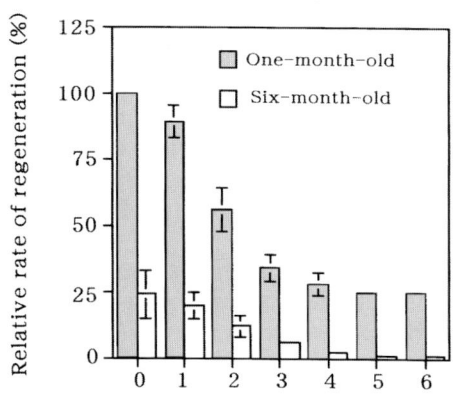

Fig. 1. Effects of low-temperature treatment on cell regeneration. One-month-old and 6-month-old calli of *D. carota* were cultured at 4°C for various periods, and then transferred to regeneration medium and cultured at 25°C for 6 weeks. The regeneration rate in control cells (1 month old) without low-temperature treatment was taken as 100%

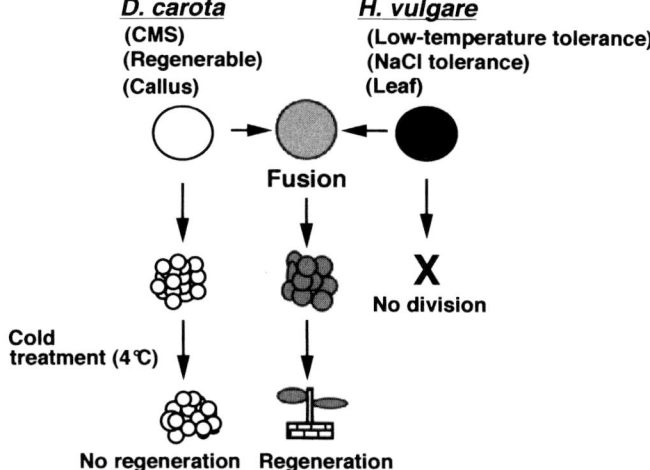

Fig. 2. Scheme for selection of hybrids between *H. vulgare* and *D. carota*

that had been incubated at 4 °C for 5 weeks failed to regenerate any shoots. Furthermore, no plants were obtained from protoplasts of either *H. vulgare* or *D. carota* that were cultured under the same conditions without fusion treatment.

3 Analysis of the Three Regenerated Plants

The somatic hybrid plants closely resembled *D. carota* in morphology (Fig. 3d, e). No. 1 hybrid had variegated green and white leaves and flowers which developed without vernalization (Fig. 3g, h, i). The morphology of roots of the somatic hybrids was similar to that of roots of *D. carota* (Fig. 3f). The flowers exhibited male sterility, as did those of the parent strain of *D. carota*.

Callus cultures induced from leaf segments of the regenerated plants and their parents were analyzed at the cytological and molecular levels. Cytological analysis revealed that the chromosome number of the regenerated plants was about 24 (Table 1), namely, significantly lower than the sum of the chromosome numbers (32) of the parents. Genomic DNA was analyzed by Southern hybridization with a nonradioactively labeled DNA fragment of the *rgp1* gene. The regenerated plants generated both a band specific for *D. carota* (4.4 kbp) and a band specific for *H. vulgare* (3.6 kbp) (Fig. 4). Chloroplast (ct) and mitochondrial (mt) DNAs were also analyzed by Southern hybridization with fragments of ctDNA and mtDNA (Table 1). The results of analysis of ctDNA with a non-radioactively labeled fragment of rice ctDNA of *Bam*HI-8 as probe indicated that the regenerated plants yielded both bands specific for *D. carota* (4.2 and 2.2 kbp) and a band specific for *H. vulgare* (9.0 kbp)

Fig. 3. a Multicellular after 2 weeks of culture. **b** Regeneration of shoots from selected callus. **c** Regenerated plant potted in soil. **d** *Left to right* Plants of *H. vulgare*, a regenerant, and *D. carota*. **e** Plant of somatic hybrid (no. 1). **f** Root of somatic hybrid (no. 1). **g–i** Green and white flowers (**g, h**) and leaves (**i**); *arrow* (**h**) indicates a white flower

Fig. 4. Results of Southern hybridization analysis of genomic DNA. Total DNA was digested with *Hin*dIII, and a fragment of the *rgp1* gene was used as probe. *Arrow* indicates a band specific for *H. vulgare*

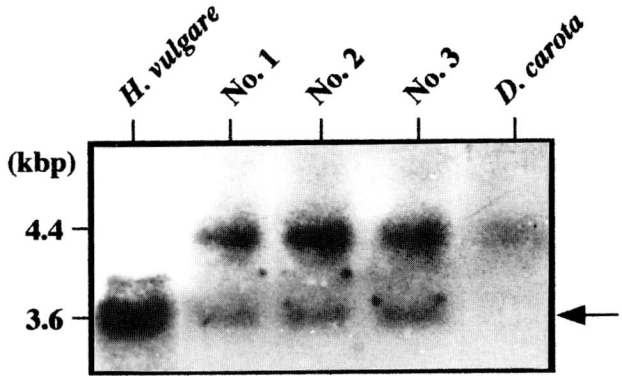

Table 1. Chromosome number and cytoplasmic genotype

Cell line	Chromosome no.	Fragments of mtDNA					Fragments of ctDNA				
		atp6	atp9	cob	18s rRNA	26s rRNA	pSB3	pSB8	pSB10	pSB13	pSB16
No. 1	24.4 (± 1.4)[a]	D + N	DC	DC	DC	DC	H + N	D + H	DC	DC	DC
No. 2	24.9 (± 2.6)	DC	DC	DC	DC	DC	H + N	D + H	DC	DC	DC
No. 3	24.2 (± 1.8)	DC	DC	DC	DC	DC	H + N	D + H	DC	DC	DC

[a] Standard error ($n > 10$). D + N, *D. carota* band plus novel band; H + N, *H. vulgare* band plus novel band; D + H, *D. carota* band plus *H. vulgare* band; DC, *D. carota* type.

(Fig. 5a). The regenerated plants also yielded a band specific for *H. vulgare* (4.4 kbp) and a unique band (8.6 kbp) when the *Bam*HI-3 fragment of rice ctDNA was used as the probe (Fig. 5b). In the analysis of mtDNA, one of the regenerated plants (no. 1) yielded a novel band (9.0 kbp) that was not detected in the analysis of either parent when a fragment of *atp6* was used as the probe (Fig. 6). These results indicated that the regenerated plants were somatic hybrids of *H. vulgare* and *D. carota*.

4 Characterization of Somatic Hybrids

Cells in suspension cultures induced from three somatic hybrid plants and their parents were incubated at 4 °C for various hours after culture for 1 week at 25 °C. From the results of measurement of fresh weight of cold-treatment cells, the growth of *H. vulgare* and no. 2 was more than that of *D. carota* and other somatic hybrids (nos. 1 and 3) in the cell levels (Fig. 7a). TTC-reduction

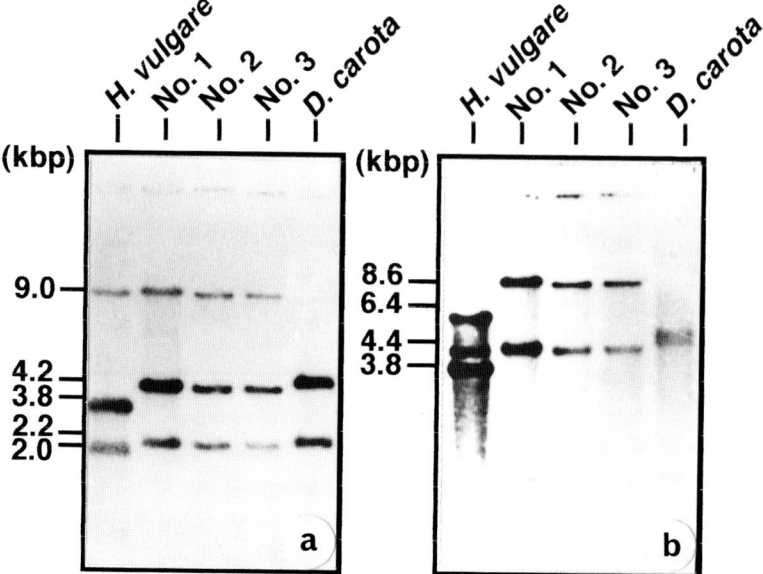

Fig. 5a,b. Results of Southern hybridization of ctDNA. Total cellular DNA was digested with HindIII, and the BamHI-8 fragment (**a**) and the BamHI-3 fragment (**b**) of rice ctDNA were used as probes

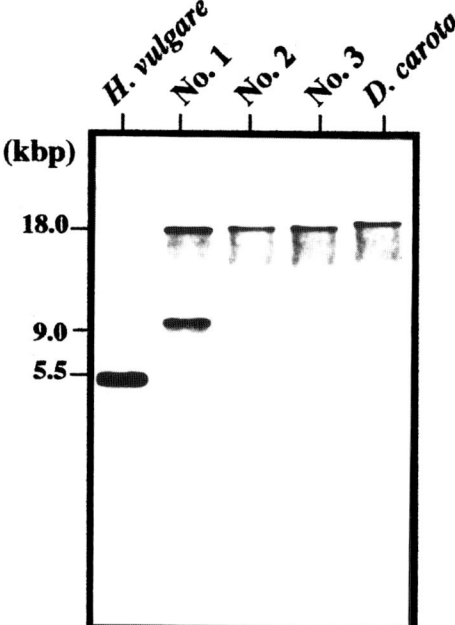

Fig. 6. Results of Southern hybridization of mtDNA. Total cellular DNA was digested with EcoRI, and a fragment of atp6 was used as probe

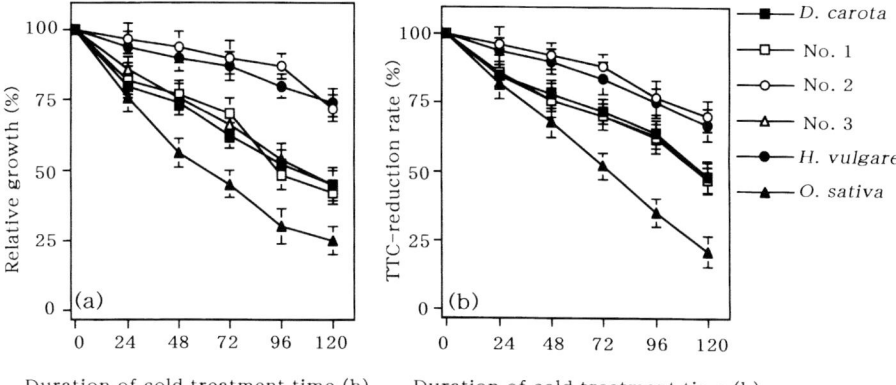

Fig. 7a,b. Effects of cold on cell growth. After culture for 1 week at 25 °C, cells were grown at 4 °C in darkness for the indicated periods. Cell viability was evaluated from measurements of fresh weight (**a**) and TTC-reduction rate (**b**). The fresh weight and TTC-reduction rate for control cells (no cold treatment) were taken as 100%. Each *point* represents the mean ± SE ($n = 3$)

test of cold-treatment cells showed that the cell activity of *H. vulgare* and no. 2 was higher than that of *D. carota* and other somatic hybrids (Fig. 7b).

H. vulgare was more tolerant to NaCl than *D. carota*. The seedlings of *H. vulgare* grew on medium that contained 1.4% (w/v) NaCl, but the growth of *D. carota* seedlings was inhibited by more than 0.8% (w/v) NaCl (Fig. 8a). The inhibition of cell growth of calli that had been induced from *H. vulgare* and *D. carota* was similar to that of the respective seedlings. The results of the NaCl-tolerance test with callus cultures induced from the somatic hybrid plants showed that nos. 2 and 3 were more tolerant to NaCl than *D. carota*. The NaCl tolerance of no. 2 was as high as that of *H. vulgare*, and that of no. 3 was intermediate between that of *H. vulgare* and *D. carota*. However, no. 1 was as sensitive to NaCl as *D. carota* (Fig. 8b).

5 Quantification and Effects of Betaine

Levels of betaine in nos. 2 and 3, which were tolerant to NaCl, were 2.5 and 1.6 times as high as that in *D. carota*, respectively. Furthermore, the levels of betaine in the calli increased upon treatment with NaCl (Table 2).

6 Discussion

To produce somatic hybrids between *H. vulgare* and *D. carota*, we incubated fused cells at a low temperature (4 °C) for 5 weeks, and obtained three somatic

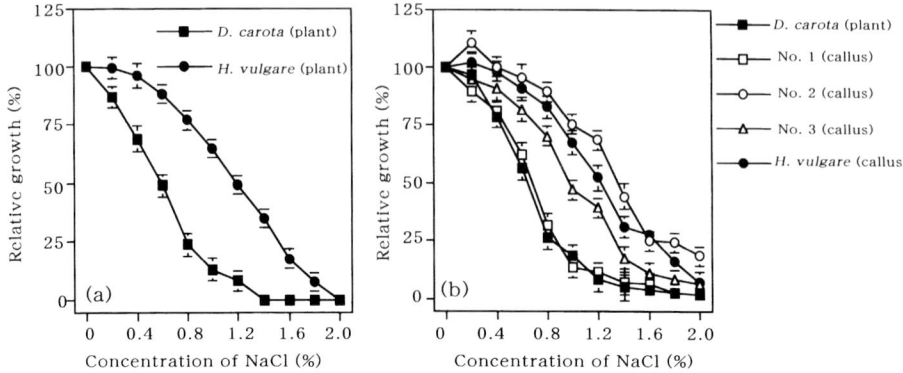

Fig. 8a,b. Effects of NaCl. Plants of *H. vulgare* and *D. carota* (**a**) and cells of the somatic hybrids, *H. vulgare* and *D. carota* (**b**) were exposed to NaCl. After culture for 4 days, cells in suspension cutures were grown on solid medium that contained various concentrations of NaCl. Growth of control cells (no treatment) was taken as 100%. Each *point* represents the mean ± SE (*n* = 3)

Table 2. Accumulation of betaine in the cultured cells from the somatic hybrids and their parents

Cell lines	Betaine (mmol mg^{-1} FW)	
	Nontreatment	NaCl treatment (0.8%)
D. carota	6.51 (± 1.51)[a]	6.68 (± 1.44)
H. vulgare	14.05 (± 2.23)	26.43 (± 1.55)
Somatic hybrids		
No. 1	6.93 (± 1.63)	7.05 (± 0.64)
No. 2	16.53 (± 1.41)	28.79 (± 2.23)
No. 3	10.78 (± 1.52)	16.20 (± 1.61)

[a] Values in parentheses are standard error from three replications.

hybrid plants. In this combination, *H. vulgare* protoplasts failed to divide, while *D. carota* protoplasts did not yield any plants. Thus, selection at the low-temperature condition was effective for the selection of hybrids.

Evidence for the recombination of chloroplast genomes from different genetic lines of higher plants has been reported (Medgyesy et al. 1985; Thanh and Medgyesy 1989). In our experiment, five fragments of ctDNA were used as probes for Southern hybridization to analyze the ctDNA of the regenerated plants. With two fragments of rice ctDNA, *Bam*HI-8 and *Bam*HI-3, as probes the regenerated plants yielded both the parental bands or a band specific to *H. vulgare* and a novel band. Our results indicate that part of the ctDNA was transferred from *H. vulgare* to the regenerated plants. The recombination of mtDNA in interfamilial hybrids has been reported (Smith et al.

1989; Kisaka et al. 1994). In the present study, no. 1 plant yielded a novel band when *atp6* was used as a probe of mtDNA. This is evidence of the possible recombination of the mitochondrial genomes of *D. carota* and *H. vulgare*. When several other fragments of ctDNA and mtDNA were used as probes, the regenerated plants yield the same patterns of bands as *D. carota*. Furthermore, the regenerated plants yield both parental bands with a DNA fragment of the nuclear *rgpl* gene as probe. These results together indicated that the three regenerated plants were somatic hybrids between *H. vulgare* and *D. carota*.

The cold tolerance of no. 2 was more than that of *D. carota*, as was *H. vulgare*. Cells of *O. sativa*, which were used as a weekly control for susceptibility to coldness, was most sensitive to cold among the used materials. In fact, low-temperature incubation (4°C) was successful for the selection of hybrids. However, not all hybrid cells showed cold tolerance. The cold tolerance seemed to be segregated during cell cultures after selection, because the treated cells and regenerated plants were not kept at low temperature after treatment. Smillie et al. (1979) reported that, somatic hybrid plants between tomato and potato were obtained, some of these were tolerant to cold like potato and others had tolerance intermediate between potato and tomato.

In many plant species, salt-tolerant cells were selected on media supplemented with NaCl (Dix and Street 1975; Nabors et al. 1980; Smith and McComb 1983; McHughen and Swartz 1984; Vajrabhaya et al. 1989; Sumaryati et al. 1992). Since the selective media supplemented with NaCl in protoplast cultures were not use in this experiment, the possibility that the salt tolerance of the somatic hybrid calli (nos. 2 and 3) was induced by somaclonal variation was not completely excluded. The synthesis and accumulation of betaine have been best characterized in members the Gramineae and Chenopodiaceae. In barley, the accumulation of betaine was a metabolic response to water or salt stress (Hanson and Nelsen 1978; Wyn Jones and Storey 1978; Hitz et al. 1982). The stress-induced accumulation of betaine was adaptive, with betaine acting as a nontoxic cytoplasmic osmoticum and/or as a protectant against the inactivation of enzyme (Paleg et al. 1981; Wyn Jones 1979). Transgenic rice plants carrying the gene for betaine aldehyde dehydrogenase or the *codA* gene for choline oxidase were shown to have enhanced ability to tolerate high salinity (Nakamura et al. 1997; Sakamoto et al. 1998). In our experiments, the callus of somatic hybrid (no. 2) showed high levels of betaine content, which were increased almost 1.7-fold under high-salt conditions (0.8%), as with *H. vulgare*. The accumulation of betaine is concerned with cell activity because the level of betaine was decreased when the calli treated with 1.2% (w/v) NaCl (data not shown). The calli hardly grew under these conditions.

These results indicated that the cold and salt tolerance might have been transferred to the cells of the somatic hybrid (no. 2) from *H. vulgare* during protoplast fusion. We consider protoplast fusion a useful method to introduce the tolerance of environmental stresses to other plants.

7 Protocol

7.1 Cultivation of Plants and Cells

Seeds of *H. vulgare* (cv. Hashirihadaka) were surface-sterilized in 70% (v/v) ethanol solution for 30s and then in a 2% (v/v) solution of sodium hypochlorite for 15 min. They were washed twice with sterile distilled water, placed on medium A [MS (Murashige and Skoog 1962) medium supplemented with 0.8% (w/v)], and cultured under continuous illumination by fluorescent lamps (4 W m^{-2}) at 25 °C. After 1 week, young leaves were used for isolation of protoplasts. Calli were induced from surface-sterilized leaf segments of a cytoplasmic male sterile (cms) strain of *D. carota* (Kanzaki et al. 1991) cultured on medium B [MS medium supplemented with 1.0 mg l^{-1} 2,4-D, 0.1 mg l^{-1} kinetin and 0.8% (w/v) agar] under continuous illumination by fluorescent lamps (4 W m^{-2}) at 25 °C. The resulting calli were transferred to a liquid medium (medium B minus agar) and subcultured at 2-week intervals.

7.2 Isolation and Fusion of Protoplasts and the Selection of Somatic Hybrids

Protoplasts of *H. vulgare* were isolated from young leaves by incubation in an enzyme solution [1.6% (w/v) Cellulase Onozuka R10 (Yakult Honsha, Tokyo, Japan), 0.3% (w/v) Macerozyme R10 (Yakult Honsha), 8% (w/v) mannitol, and 0.1% (w/v) CaCl$_2$-2H$_2$O (PH 5.5)] for 3 h at 25 °C. Protoplasts of *D. carota* were isolated from cells in 6-month-old suspension cultures by incubation of cells under the same conditions as those described for *H. vulgare*. The protoplasts were filtered through nylon mesh (50-μm mesh) and washed twice with washing solution [8% (w/v) mannitol and 0.1% (w/v) CaCl$_2$-2H$_2$O] with centrifugation at 80 g for 5 min. The protoplasts were purified by floating them in a 25% (w/v) solution of sucrose with subsequent centrifugation at 80 g for 5 min, and then they were washed once with the washing solution. Equal numbers of the two kinds of protoplasts were combined to give a final density of 5×10^3 ml^{-1} and were then subjected to electrofusion using a BTX Electro Cell Manipulator 200.

The fused protoplasts were washed once with washing solution an equal volume of a solution of Gelrite [0.3% (w/v) Gelrite (Sigma), 3% (w/v) sucrose, and 5% (w/v) glucose]. The suspension of protoplasts was added to protoplast culture medium C [MS medium supplemented with 5% (w/v) glucose, 1.0 mg l^{-1} 2,4-dichlorophenoxyacetic acid and 0.5 mg l^{-1} kinetin] in plastic petri dishes. Nonfused protoplasts isolated from *H. vulgare* and *D. carota* were cultured, as controls, under the same conditions as those described for the fused protoplasts. During the first month of culture, the petri dishes were incubated in darkness at 25 °C. The resultant microcalli were subsequently cultured in medium D [MS medium supplemented with 0.1 mg l^{-1} NAA, 1.0 mg l^{-1} N^6-benzylaminopurine and 1.0% (w/v) agar] and incubated in darkness at 4 °C. After 5 weeks at 4 °C, the petri dishes were transferred to continuous light (4 W m^{-2}) at 25 °C for 1 month. The resultant visible colonies of about 1–2 mm in diameter were transferred to the fresh shooting medium D. Regenerated shoots were transferred to root-inducing medium E [MS medium supplemented with 0.8% (w/v) agar].

7.3 Cytological Analysis

Numbers of chromosomes were determined by treating actively growing cells in suspension culture (from the regenerated plants and from *D. carota*) and root tips (*H. vulgare*) with 0.03% (w/v) 8-quinolinol for 3 h at 25 °C, with subsequent fixation in a mixture of ethanol and glacial acetic acid (1:3, v/v) for 16 h at 25 °C. The cells were then treated with 1 N HCl for 7 min at 60 °C and stained with Schiff's reagent for 1 h at 25 °C. Chromosomes were counted in about ten metaphases per regenerated plant.

7.4 Analysis of DNA

Total DNA was prepared from callus (from regenerated plants and *D. carota*) and from the intact plant (*H. vulgare*) by the method described by Honda and Hirai (1990). Southern hybridization was performed with a nonradioactive DNA labeling and detection kit (Boehringer Mannheim, Town, FRG). The cloned fragment of the *rgp1* gene (Sano and Youssefian 1991) was a gift from Dr. H. Sano, (NAIST, Nara, Japan). The clones corresponding to subunit 6 of F_1-F_0 ATPase (*atp6*) and cytochrome b (*cob*) of rice mitochondrial DNA, and fragments of rice ctDNA (*Bam*HI-3, *Bam*HI-8, *Bam*HI-10, *Bam*HI-13, and *Bam*HI-16) were a gift from Dr. A. Hirai (University of Tokyo, Japan). Plasmids containing mitochondrial genes for 26S and 18S ribosomal RNA, for subunit 9 of F_1-F_0 ATPase (*atp9*) were provided by Dr. K. Nakamura (Nagoya University, Nagoya, Japan). The cloned genes were used as probes for Southern hybridization.

7.5 Low-Temperature Treatment and Assessment of Cell Viability

Cells in suspension cultures that had been grown for 1 week at 25 °C were transferred to solid MS medium supplemented with 3% (w/v) sucrose and 0.8% (w/v) agar in plastic petri dishes (9 cm in diameter) and incubated at 4 °C in darkness. Cells in suspension cultures of *Oryza sativa* were used as a weekly control to cold stress. After chilling treatment for 0, 24, 48, 72, or 96 h, cells were collected, washed with distilled water at 0 °C and blotted on filter paper. Nonchilled control cells, which had been incubated at 25 °C in darkness, were collected as described above.

Cell viability after chilling treatment was evaluated from measurements of fresh weight and TTC-reduction test (Steponkus and Lanphear 1967). For the TTC-reduction test, 200 mg (fresh weight) of cells were incubated with 3 ml of a reaction mixture that contained 50 mM potassium phosphate buffer (pH 7.3) and 0.6% (w/v) triphenyltetrazolium chloride (TTC) at 25 °C in darkness under a vacuum. After incubation for 15 h, cells were collected, washed with distilled water, blotted on filter paper, and extracted with 5 ml of 95% ethanol for 30 min. Absorbace of each ethanol extract was measured at 540 nm with 95% ethanol as the blank. The absorbance obtained from treatment of nonchilled, control cells was taken as 100%.

7.6 Assays of Tolerance to NaCl and Other Compounds

Cells in suspension cultures after growth for 4 days at 25 °C were transferred to solid MS medium supplemented with 3% (w/v) sucrose, 0.8% (w/v) agar and various concentrations [0, 0.2, 0.4, 0.6, 0.8, 1.0, 1.2, 1.4, 1.6, 1.8, and 2.0% (w/v)] of NaCl and incubated for 1 month under continuous fluorescent light (4 W m^{-2}) at 25 °C.

Seedlings of *D. carota* and *H. vulgare* were analyzed as controls. Seeds were surface sterilized in 2% (w/v) sodium hypochlorite for 15 min, washed twice in distilled water, placed on solid MS medium supplemented with 3% (w/v) sucrose and 0.8% (w/v) agar and cultured under continuous fluorescent light (4 W m^{-2}) at 25 °C. After culture for 5 days, seedlings that had grown to the same stages were transferred to fresh solid medium supplemented with 3% (w/v) sucrose, 0.8% (w/v) agar, and various concentrations [0, 0.2, 0.4, 0.6, 0.8, 1.0, 1.2, 1.4, 1.6, 1.8, and 2.0% (w/v)] of NaCl, and cultured under the same conditions for 2 weeks.

Growth rates were evaluated from measurements of the fresh weight of calli or the lengths of shoots. The fresh weight or lengths of control calli or plants, cultured without NaCl, were taken as 100%.

7.7 Quantification of Betaine

Cells in suspension cultures of the somatic hybrids *D. carota* and *H. vulgare* were sub-cultured every 2 weeks and cells in 6-day-old suspension cultures were transferred to solid MS medium supplemented with NaCl. The tested concentration of NaCl was 0.8% (w/v). Control calli were cultured without NaCl. The treated calli were collected from 12-day-old material.

Levels of betaine were determined by the method of Gorham et al. (1982). The absorbency of the ester was measured at 262 nm (for quantitation of the *p*-bromophenacyl ester).

References

Aviv D, Fluhr R, Edelman M, Galun E (1980) Progeny analysis of the interspecific somatic hybrids: *Nicotiana tabacum* (CMS) + *Nicotiana sylvestris* with respect to nuclear and chloroplast markers. Theor Appl Genet 56:145–150

Babiychuk E, Kushnir S, Gleba YY (1992) Spontaneous extensive chromosome elimination in somatic hybrids between somatically congruent species *Nicotiana tabacum* L. and *Atropa belladonna* L. Theor Appl Genent 84:87–91

Deping X, Xiaolan D, Baiyang W, Bimei H, Tuan-Hua DH, Ray W (1996) Expression of a late embriogenesis abundant protein gene, *HVA1* from barley confers tolerance to water deficit and salt stress in transgenic rice. Plant Physiol 100:249–257

Dix PJ, Street HE (1975) Sodium chloride-resistant cultured cell lines from *N. sylvestris* and *Capscum annum*. Plant Sci Lett 5:231–237

Dudits D, Maroy E, Praznovszky T, Olah Z, Gyorgyey J, Cella R (1987) Transfer of resistance traits from carrot into tobacco by asymmetric somatic hybridization: regeneration of fertile plants. Proc Natl Acad Sci USA 84:8434–8438

Gleba YY, Hinnisdaels S, Sidorov V, Kaleda van Boryshuk NV, Cherep NN, Negrutiu I, Jacobs M (1988) Intergeneric asymmetric hybrids of *Nicotiana plumbaginifolia* and *Atropa belladonna* obtained by "gamma-fusion". Theor Appl Genet 76:760–766

Gorham J, McDonnell E, Jones RGW (1982) Determination of betaines as ultraviolet-absorbing esters. Anal Chim Acta 138:227–283

Gupta PP, Gupta M, Schieder O (1982) Correction of nitrate reductase defect in auxotrophic plant cells through protoplast-mediated intergeneric gene transfers. Mol Gen Genet 188:-378–383

Gupta PP, Schieder O, Gupta M (1984) Intergeneric nuclear gene transfer between somatically and sexually incompatible plants through asymmertic protoplast fusion. Mol Gen Genet 197:30–35

Hanson AD, Nelsen CD (1978) Betaine accumulation and [^{14}C]formate metabolism in water-stressed barley leaves. Plant Physiol 62:305–312

Hitz WD, Landyman JAR, Hanson AD (1982) Betaine synthesis and accumulation in barley during field water stress. Crop Sci 22:47–54

Honda H, Hirai A (1990) A simple and efficient method for identification of hybrids using non-radioactive rDNA as probe. Jpn J Breed 40:339–348

Kameya T, Kanzaki H, Toki S, Abe T (1989) Transfer of radish (*Raphanus sativus* L.) chloroplasts into cabbage (*Brassica oleracea* L.) by protoplast fusion. Jpn J Genet 64:27–34

Kanzaki H, Takeda M, Kameya T (1991) Sequence analysis of a mitochondrial DNA fragment isolated from cultured cells of carrot cytoplasmic male-sterile strain. Jpn J Genet 66:719–724

Kisaka H, Kameya T (1994) Production of somatic hybrids between *Daucus carota* L. and *Nicotiana tabacum*. Theor Appl Genet 88:75–80

Kisaka H, Lee H, Kisaka M, Kanno A, Kang K, Kameya T (1994) Production and analysis of asymmetric hybrid plants between monocotyledon (*Oryza sativa* L.) and dicotyledon (*Daucus carota* L.). Theor Appl Genet 89:365–371

Kostenyuk I, Lubaretz O, Borisyuk N, Voronin V, Stockigt J, Gleba Y (1991) Isolation and characterization of intergeneric somatic hybrids in the Apocynaaceae family. Theor Appl Genet 82:713–716

McHughen A, Swartz M (1984) A tissue-culture derived salt-tolerant line of Flax (*Linum usitatissimum*). J Plant Physiol 117:109–117

Medgyest P, Fejes E, Maliga P (1985) Interspecific chloroplast DNA recombination in a *Nicotiana* somatic hybrid. Proc Natl Acad Sci USA 82:6960–6964

Melchers G, Sacristan MD, Holder AA (1978) Somatic hybrid plants of potato and tomato regenerated from fused protoplasts. Carlsberg Res Commun 43:203–218

Menczel L, Nagy F, Lazar G, Maliga P (1983) Transfer of cytoplasmic male sterility by selection for streptomycin resistance after protoplast fusion in *Nicotiana*. Mol Gen Genet 189: 395–369

Murashige T, Skoog F (1962) A revised medium for rapid growth and bioassays with tobacco tissue culture. Physiol Plant 15:473–497

Nabors MW, Gibbs SE, Bernstein CS, Meis ME (1980) NaCl-tolerant tobacco plants from cultured cell. Z Pflanzenphysiol 97:13–17

Nakamura T, Yokota S, Muramoto Y, Tsutsui K, Ogura Y, Fukui K, Takabe T (1997) Expression of a betaine aldehyde dyhydrogenase gene in rice, a glycine betaine nonaccumulator, and possible localization of its protein in peroxisomes. Plant J 11:1115–1120

Negrutiu I, Brouwer D, De Watts JW, Sidorov VI, Dirks R, Jacobs M (1986) Fusion of plant protoplasts: a study using auxotrophic mutants of *Nicotiana plumbaginifolia*, Viviani. Theor Appl Genet 72:279–286

O'Connell MA, Hanson MR (1986) Regeneration of somatic hybrid plants formed between *Lycopersicon esculentum* and *Solanum rickii*. Theor Appl Genet 72:59–65

Paleg LG, Douglas TJ, Van Daal A, Keech DB (1981) Proline, betaine and other organic solutes protect enzymes against heat inactivation. Aust J Plant Physiol 8:107–114

Pental D, Hamil JD, Pirrie A, Cocking EC (1986) Somatic hybridization *of Nicotiana tabacum* and *Petunia hybrida*. Mol Gen Genet 202:342–347

Perl A, Aviv D, Galum E (1991) Protoplast fusion-mediated transfer of oligomycin resistance from *Nicotiana sylvestris* to *Solanum tuberosum* by intergeneric cybridization. Mol Gen Genet 225:11–16

Sakamoto A, Alia, Murata N, Murata A (1998) Metabolic engineering of rice leading to biosynthesis of glycine betaine and tolerance to salt and cold. Plant Mol Biol 38:1011–1019

Sano H, Youssefian S (1991) A novel ras-related *rgp1* gene encoding a GTP-binding protein has reduced expression in 5-azacytidine-induced dwarf rice. Mol Gen Genet 228:227–232

Smillie RM, Melchers G, Wettstein D (1979) Chilling resistance of somatic hybrids of tomato and potato. Carlsberg Res Commun 44:127–132

Smith MK, McComb JA (1983) Selection for NaCl tolerance in cell cultured of *Medicago sativa* and recovery of plants from a NaCl-tolerant cell line. Plant Cell Rep 2:126–128

Smith MA, Pay A, Dudits D (1989) Analysis of chloroplast and mitochondrial DNAs in asymmetric hybrids between tobacco and carrot. Theol Appl Genet 77:641–644

Somers DA, Narayanan KR, Kleinhofs A, Cooper-Bland S, Cocking EC (1986) Immunological evidence for transfer of the barley nitrate reductase structural gene to *Nicotiana tabacum* by protoplast fusion. Mol Gen Genet 204:296–301

Steponkus PL, Lanphear FO (1967) Refinement of the triphenyl tetrazolium chloride method of determining cold injury. Plant Physiol 42:1423–1426

Sumaryati S, Negrutiu I, Jacob M (1992) Characterization and regeneration of salt- and water-stress mutants from protoplast culture of *Nicotiana Plumbaginifolia* (Vivani). Theor Appl Genet 83:833–838

Tarczynski MC, Jensen RQ, Bohnert HJ (1993) Stress protection of transgenic tobacco by production of the osmolyte mannitol. Science 259:508–510

Thanh ND, Medgyest P (1989) Limited chloroplast gene transfer via recombination overcomes plastome-genome incompatibility between *Nicotiana tabacum* and *Solanum tuberosum*. Plant Mol Biol 12:87–93

Toki S, Kameya T, Abe T (1990) Production of triple mutant, chlorophyll-deficient, streptomycin-, and kanamycin-resistant *Nicotiana tabacum*, and its use in intergeneric somatic hybrid formation with *Solanum melongena*. Theor Appl Genet 80:588–592

Toriyama K, Kameya T, Hinata K (1987) Selection of a universal hybridizer in *Sinapis turgida* Del. and regeneration of plantlets from somatic hybrids with *Brassica* species. Planta 170:308–313

Vajrabhaya M, Thanapaisal T, Vajrahaya T (1989) Development of salt-tolerant lines KDML and LPT rice cultivas through tissue culture. Plant Cell Rep 8:411–414

Wyn jones RG (1979) An assessment of quaternary ammonium and related compounds as osmotic effectors in crop plants. In: Rains DW, Valentine RC, Hollaender A (eds), Genetic engineering of osmoregulation. Plenum Press, New york, pp 155–170

Wyn jones RG, Storey R (1978) Salt stress and comparative physiology in the Gramineae. II. Glycine betaine and proline accumulation in two salt- and water-stressed barley cultivars. Aust J Plant Physiol 5:817–829

I.2 Cybridization in *Oryza sativa* L. (Rice)

H. AKAGI

1 Introduction

1.1 Need for Somatic Hybridization

Rice is one of the most important crops in the world, and feeds about 40% of the world population. The yield of rice needs to be increased, since a world food crisis has been predicted for the early 21st century. World annual rice production must increase from its current value of 520 to 810 million tons by 2025 (Hossain 1996). Since rice shows heterosis to give a higher yield, hybrid rice may offer a solution to this predicted food crisis. The yield of hybrid rice is about 15% higher than that of conventional pure line varieties (Yuan 1994). In 1991, hybrid rice covered 17.6 million ha, 55% of the total area of rice in China (Yuan 1994).

For practical large-scale seed production of hybrid rice varieties, female strains must be genetically male sterile. Cytoplasmic male sterility (CMS) is a characteristic which can be used to produce stable male sterile lines. CMS occurs widely in higher plants and is due to incompatibility between nuclear and cytoplasmic gene products, which results in a failure to produce mature pollen grains (Newton 1988). Although CMS plants are unable to self-fertilize, their ovules are fertile, and normal seed set can be obtained when they are fertilized with pollen from normal plants (Newton 1988; Hanson et al. 1989). Since CMS eliminates the possibility of self-pollination, it has commercial application in the production of hybrid seed for economically important plants (Newton 1988).

Mitochondrial genomic determination of CMS has been well documented (Leaver and Gray 1982). Reorganization of mitochondrial genomes has been reported in CMS plants, and in many cases this results in the creation of chimeric genes whose products may be responsible for CMS. It is thought that transcripts which have been detected in such chimeric genes may inhibit the expression of the normal gene and that such recombination events result in the CMS trait (Newton 1988; Hanson et al. 1989). In rice, recombination in the downstream from the *atp6* gene in the Chinsurah Boro II cytoplasm results in

---

Department of Biological Production, Faculty of Bioresource Sciences, Akita Prefectural University, Akita 010-0195, Japan

a chimeric sequence configuration (Iwabuchi et al. 1993; Akagi et al. 1994) that includes a novel open reading frame, *orf79* (Akagi et al. 1994), and may inhibit the expression of *atp6*.

In rice, CMS was initially observed by Katsuo and Mizushima (1958). They reported that the CMS trait was expressed when the nuclear genome of *Oryza perennis* M. was replaced with that of the Japonica cultivar Fujisaka5. Several nuclear-cytoplasm combinations which cause CMS in rice have been reported (Li and Zhu 1986; Virmani and Shinjyo 1988). There are two types of CMS in rice, sporophytic and gametophytic, in which pollen grains abort at the uninucleate stage and at the binucleate or trinucleate stage, respectively (Lin and Yuan 1980; Chaudhary et al. 1981).

When the nucleus of Chinsurah Boro II (indica rice) was replaced with that of Taichung65 (japonica rice), the result was CMS (Shinjyo 1969). This CMS system is called the ms-bo type or BT type. BT-type CMS is restored by the nuclear gene, *Rf-1*, which was initially identified in Chinsurah Boro II (Shinjyo 1975). The breeding of hybrid varieties based on CMS is called the three-line method (CMS, maintainer and restore lines). It takes about 3 years to convert fertile cultivars into CMS by the recurrent backcrossing method. Therefore, new methods which would enable the conversion of several fertile cultivars to CMS over a short period are desired.

A donor-recipient protoplast fusion system has been developed in several dicotyledonous species; *Nicotiana* (Zelcer et al. 1978), *Brassica* (Menzel et al. 1987), and *Daucus* (Tanno-Suenaga et al. 1988). X- or γ-Irradiation of protoplasts causes elimination of the chromosomes or nuclei (Dudits et al. 1980; Ichikawa et al. 1987; Sidorov et al. 1987). Iodoacetamide (IOA) inhibits the division of protoplasts (Nehls 1978). Cybrids (cytoplasmic hybrids) can be formed by fusing parental protoplasts which have been inactivated asymmetrically by the two methods (Sidorov et al. 1981; Barsby et al. 1987; Ichikawa et al. 1987). This new system enables CMS traits encoded in the mitochondrial genome to be transferred into a fertile line in a single step.

1.2 Somatic Hybridization in Rice

Somatic hybrid plants of rice and barnyard grass (Terada et al. 1987), interspecies somatic hybrids between cultivated and wild species (Hayashi et al. 1988), and diploid hybrid plants from the cell fusion of haploid cells (Toriyama and Hinata 1988) have been obtained. Cybrid plants of rice have also been created by donor-recipient protoplast fusion between CMS lines and fertile cultivars (Akagi et al. 1989; Kyozuka et al. 1989; Yang et al. 1989). In these studies, the mitochondrial genomes of the CMS lines were introduced into the recipient parents. Gupta et al. (1996) reported the regeneration of putative cybrid plants by asymmetric protoplast fusion between indica subspecies.

Yang et al. (1988) produced cybrid calli between A-58 CMS, which carries the CMS cytoplasm of Chinsurah Boro II, and the fertile Japanese cultivar Fujiminori. After electrofusion between ^{60}Co-irradiated protoplasts of A-58

CMS and IOA-treated protoplasts of Fujiminori, 14 calli were regenerated. Some of these calli were cybrids, as shown by isozyme patterns of peroxidase as markers of nuclei and plasmid-like DNAs as markers of mitochondria. Yang et al. (1988) reported that the γ-ray dose and IOA concentration were the most important factors in cybrid production. Five of seven calli were derived from Fujiminori after inactivation by mild conditions (4.3 krad γ-rays and 2.5 mM IOA), while five of seven calli were cybrids observed under slightly stronger conditions (5.3 krad γ-rays and 4.0 mM IOA). Regenerated plants were also observed from one of these cybrid calli (Yang et al. 1989). Morphological features, isozyme patterns, chromosome numbers, and restriction patterns of the mitochondrial DNA indicated that these plants were cybrids that carried the nucleus from Fujiminori and mitochondria from both Fujiminori and A-58 CMS (Yang et al. 1989). However, the nature of the sterility of these cybrid plants was unclear.

CMS cytoplasm was successfully transferred to a fertile Japanese cultivar directly from Chinsurah Boro II by asymmetric protoplast fusion (Kyozuka et al. 1989). X-ray irradiated (70 krad) protoplasts of Chinsurah Boro II were electrofused with IOA-treated (25 mM) protoplasts of Nipponbare. Sixteen plants with a diploid chromosome set were grown to maturity. Only one of these plants was completely male sterile, and the progeny inherited male sterility maternally. Mitochondrial DNA of Chinsurah Boro II had been introduced to this CMS plant, while the mitochondrial DNA of other fertile plants was identical to that of the recipient parent Nipponbare. The restoration of gametophytic fertility by the *Rf-1* gene suggested that the CMS traits of this cybrid plant were identical to those of Chinsurah Boro II (Kyozuka et al. 1989).

Donor protoplasts of V20A carrying the cytoplasm of *wild abortive*, which includes CMS traits for indica subspecies, were irradiated by 30 krad γ-rays, while recipient protoplasts of three indica cultivars were inactivated with 10 mM IOA. After electrofusion of γ-irradiated and IOA-treated protoplasts, putative cybrid calli were formed, and putative cybrid plants were then regenerated from these calli (Gupta et al. 1996).

In this chapter, we describe a method for producing rice cybrids to transfer CMS traits and discuss the characteristics of cybrid rice plants. We also describe a method for identifying CMS cybrid plants by PCR amplification of the CMS-related mitochondrial region.

2 Somatic Hybridization

2.1 Establishment of Cell Lines

Rice cell lines with high plant-regeneration capacity and with high feasibility for protoplast preparation were established. Calli were initiated from rice seed scutellums on agar-solidified N6 (Chu 1978), R2 (Ohira et al. 1973) or MS (Murashige and Skoog 1962) medium containing $2\,mg\,l^{-1}$ 2,4-D, 0.3% casein hydrolysate, and 3% sucrose. The type of callus initiated varied among rice varieties. The presence of inorganic salts also affected the type of callus. When a friable callus was not observed, we examined which of the basal media would be suitable for production of friable callus.

After 3 to 4 weeks, friable calli were selected from these induced calli. Suspension cultures were then initiated by transferring these friable callis into a liquid medium containing R2 inorganic salts (Ohira et al. 1973), B5 vitamins (Gamborg et al. 1968), 0.3% casein hydrolysate, $1\,mg\,l^{-1}$ 2,4-D, and 3% sucrose (Fujimura et al. 1985). Cells were cultured in this liquid medium at 25 °C under 500 lx of continuous light on a gyratory shaker at 60 rpm, and subcultured every week (Fujimura et al. 1985). The capacity of these cell lines for plant regeneration was monitored.

2.2 Isolation and Culture of Protoplasts

2.2.1 Protoplast Isolation

Protoplasts were isolated according to Fujimura et al. (1985) with minor modifications. The cells were collected 4–5 days after subculturing and incubated in an enzyme solution (pH 5.5) containing 1% Macerozyme R10 (Yakult Honsha Co. Ltd.), 4% Cellulase RS (Yakult Honsha Co. Ltd.), 0.5% $CaCl_2 \cdot 2H_2O$, 0.5% potassium dextransulfate, and 0.4 M mannitol at 27 °C without shaking for 3 h. Under these nonagitated conditions, protoplasts were mainly isolated from surface cells of callus. The protoplast-enzyme mixture was passed through nylon screens to remove undigested cell clumps, and washed three times by centrifugation with modified R2 medium containing 0.4 M glucose.

2.2.2 Protoplast Culture

The protoplasts were cultured according to Fujimura et al. (1985) with some modification. The purified protoplasts were resuspended in the liquid medium described below at $1 \times 10^6\,ml^{-1}$. Aliquots (750 µl) of the protoplast suspension were transferred into 6-cm plastic petri dishes and cultured in the dark at 25 °C.

For vigorous callus growth, fresh medium was added 2 weeks after the beginning of protoplast culture, and colonies were transferred onto agar-solidified medium after another 2 weeks. When calli were larger than 5 mm in diameter, they were transferred to plant-regeneration medium and cultured under 3000 lx of continuous fluorescent light (Fujimura et al. 1985).

2.3 Inactivation of Protoplasts

2.3.1 Inactivation by X-Rays

Donor CMS protoplasts were suspended in 1 ml of enzyme mixture at $1 \times 10^7 \, \text{ml}^{-1}$. This suspension was transferred into a 6-cm plastic petri dish, and then irradiated by X-rays at various dosages ($2 \, \text{krad min}^{-1}$) at 80 kV and 4 mA. Protoplasts were washed three times with 0.4 M glucose after irradiation.

The effect of X-rays on CMS protoplasts was investigated at various dosages. The number of colonies regenerated was determined 1 month after irradiation. At X-ray doses greater than 120 krad, no colony formation was observed (Fig. 1A), even though most of the protoplasts were still alive, as judged by microscopic inspection, and limited cell division was apparent. Protoplasts were routinely X-irradiated at 125 krad to ensure that colony formation was completely inhibited (Akagi et al. 1989).

2.3.2 Inactivation by Iodoacetamide (IOA)

The recipient protoplasts were suspended in 0.4 M glucose. The population density of protoplasts was adjusted to $1 \times 10^7 \, \text{ml}^{-1}$. The recipient protoplasts

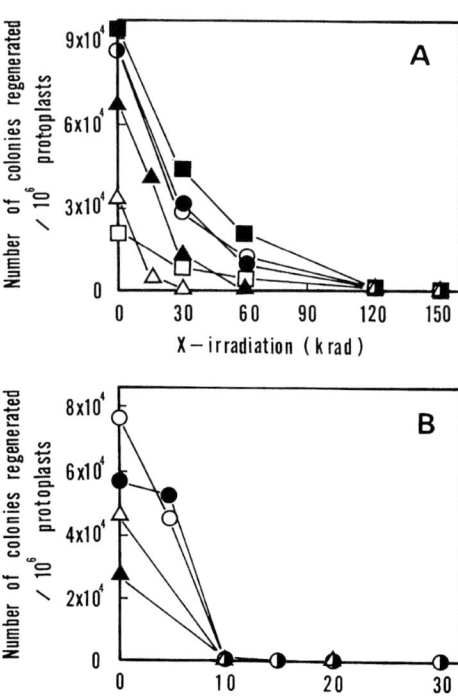

Fig. 1A,B. Effects of X-ray irradiation (**A**) and IOA treatment (**B**) on colony formation by parental protoplasts. (Akagi et al. 1989)

were treated with various concentrations of IOA for 10 min at 27 °C, and then washed four times with 0.4 M glucose.

The concentration of IOA necessary for inactivation of the recipient protoplasts was determined. At 15 mM, IOA completely inhibited cell division (Fig. 1B). In further control experiments, donor and recipient protoplasts were treated with X-rays and IOA, respectively, and then either fused separately and cultured, or mixed and cultured without fusion (Fig. 2). In contrast to the above-mentioned results, many colonies were regenerated from the mixture without fusion when IOA was used at concentrations lower than 25 mM. Therefore, recipient protoplasts must be treated with 30 mM IOA to minimize the possibility of colony formation from non-hybrid cells (Akagi et al. 1989).

Although IOA is widely used in hybrid production, protoplasts that have been inactivated with lower concentrations of IOA sometimes retain the ability to divide. More stringent IOA treatment was required to minimize uniparentally derived callus formation in rice (Akagi et al. 1989). High doses of X-rays were also required to completely eliminate colony formation by parental protoplasts, as compared with other plant species (Sidorov et al. 1981; Ichikawa et al. 1987). Since the inactivation methods primarily affect different targets, we were able to construct a large number of cybrid plants at

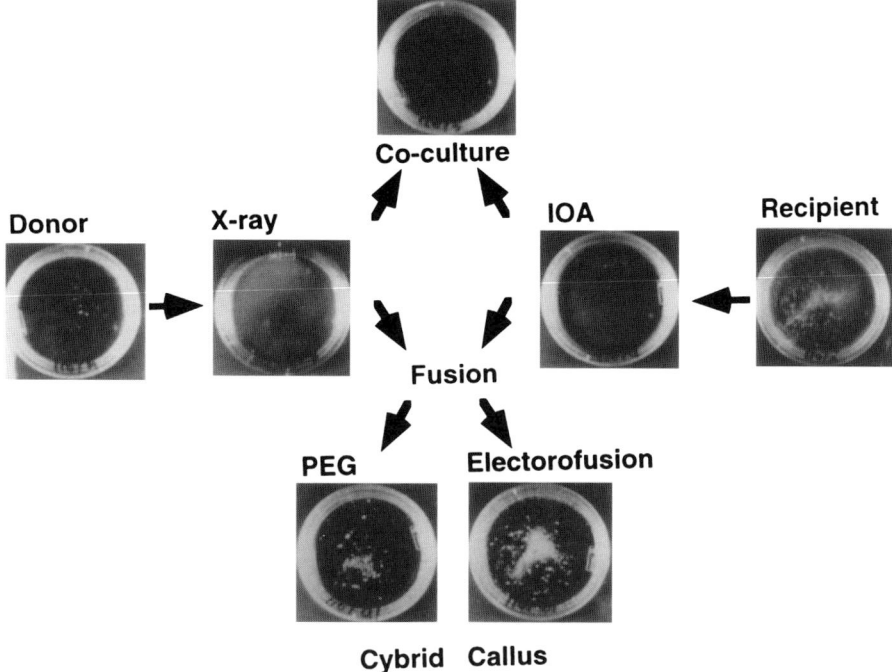

Fig. 2. Restoration of the capacity for cell division by protoplast fusion between X-irradiated and IOA-inactivated protoplasts

high frequency while maintaining a low level of undesired noncybrid plant regeneration.

2.4 Fusion of Protoplasts

2.4.1 Staining of Rice Protoplasts with Fluorescent Dyes

To evaluate the frequency of protoplast fusion, we stained rice protoplasts prior to fusion with one of two lipophilic fluorescent dyes, rhodamine B isothiocyanate (RITC), and fluorescent isothiocyanate (FITC), according to the method described by Galbraith (1984). RITC and FITC were dissolved in ethanol at $5\,mg\,ml^{-1}$. Protoplasts were stained with either of these fluorescent dyes for 3h during protoplast isolation by adding 10μl of RITC or FITC solution to 10ml of enzyme solution.

This method for labeling rice protoplasts with RITC or FITC enabled us to distinguish heterokaryons from parental protoplasts and homokaryons by the presence of both dyes using fluorescence microscopy (Fig. 3). Therefore, we evaluated the frequency of heterokaryon formation by the following fusion experiment.

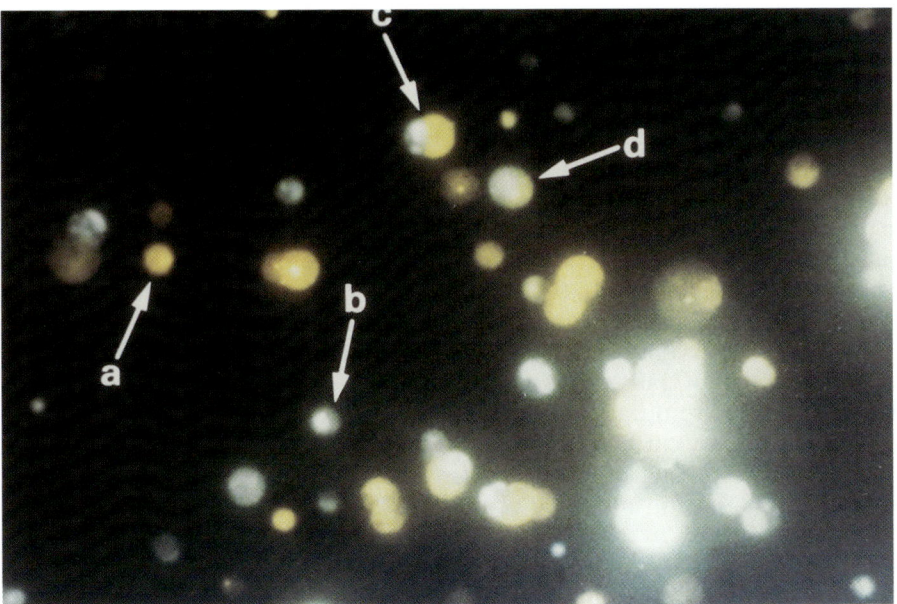

Fig. 3. Polyethylene glycol-mediated protoplast fusion evaluated by fluorescence microscopy. Donor and recipient protoplasts stained with FITC (*a*) and RITC (*b*), respectively, and heterokaryon containing both FITC and RITC (*c,d*)

2.4.2 Fusion with Polyethylene Glycol (PEG)

Stained protoplasts were suspended in 0.4 M glucose at $1 \times 10^7 ml^{-1}$. RITC-stained protoplasts and FITC-stained protoplasts were mixed in a ratio of 1:1. One ml of the mixture was transferred into a centrifuge tube (12 ml), and 30% (w/v) PEG (mw 6000 Pa) solution was added through the tube wall. The PEG solution contained 200 mM $Ca(NO_3)_2$, 50 mM glycine, and 0.1 M glucose (pH 9) (Sala et al. 1985). The concentration of PEG was equalized by rotating the centrifuge tube. After 15 min, 0.4 M glucose was added to dilute PEG. Treated protoplasts were washed twice by centrifugation with 0.4 M glucose.

The number of heterokaryons was counted by fluorescence microscopy. About 20% of the protoplasts contained both RITC and FITC, and were supposed to be heterokaryons. We examined the effect of PEG on the viability of protoplast culture. After 1 week of protoplast culture, the plating efficiency of PEG-treated protoplasts was 4.3%, which was comparable to that of untreated protoplasts (4%). Thus, fusion treatment by PEG solution did not inhibit protoplast division.

2.4.3 Electrofusion of Protoplasts

An efficient and easier method for achieving protoplast fusion in an electric field has been developed (Zimmermann et al. 1981). In this method, protoplasts are aligned like pearls on a string in an AC (alternating current) field. A subsequent DC (direct current) pulse transiently destroys the membrane between the aligned protoplasts, resulting in protoplast fusion. We optimized the electric field strength of both the AC and DC pulses for rice protoplast fusion.

Stained protoplasts were suspended in 0.4 M glucose at $1 \times 10^7 ml^{-1}$. RITC-stained protoplasts and FITC-stained protoplasts were mixed in a ratio of 1:1. The mixture was placed in a fusion chamber. Protoplasts were aligned in an AC field (1 MHz, 150 V cm^{-1}) for a few seconds and then fused by a DC pulse (1.5–2.5 kV cm^{-1}, 50–300 µs). After 15 min, the protoplasts were washed twice by centrifugation with 0.4 M glucose.

A protoplast density of more than $1 \times 10^7 ml^{-1}$ was necessary to observe 10–15% heterokaryons. The efficiency of protoplast fusion increased with prolongation of the DC pulse. However, the viability of rice protoplasts decreased with a longer pulse duration. The frequency of heterokaryon formation with a DC shock (2.5 kV cm^{-1}, 50–100 µs) was estimated to be 15%. Under the above conditions, the plating efficiency of treated protoplasts ranged from 2.6 to 3.3% (3.7 to 4.9% for untreated protoplasts).

2.5 Regeneration of Somatic Hybrids

Many colonies were regenerated after X-irradiated donor protoplasts had been fused with IOA-treated recipient protoplasts, but no colonies were formed without cell fusion (Fig. 2). Metabolic complementation between the nuclear and cytoplasmic compartments restores the ability of fused protoplasts to undergo cell division (Zelcer et al. 1978). The plants that regenerated from these colonies about 3 weeks after being transferred to regeneration medium were putative cybrids (Fig. 4). We randomly selected regenerated plants to analyze the origin of their nuclei and cytoplasms.

Fig. 4a–e. Regeneration of rice cybrids. Division of protoplasts after protoplast fusion (**a**), formation of cybrid colonies (**b**), regeneration of plants from cybrid callus (**c**), regenerated cybrid plants (**d**), and diploid cybrid plants growing in a greenhouse (**e**)

The nuclear characteristics of putative cybrid plants were determined by analyzing their chromosome numbers, a marker enzyme for the recipient strain, and their morphology. Among 38 regenerated plants in which the chromosome number was analyzed, 26 plants had 24 chromosomes, and 12 plants had 48 (Fig. 5). To determine the source of the nucleus, aryl acylamidase I, which is a marker enzyme that is not found in the recipient strain, was assayed in the regenerated plants. All the plants showed enzyme activity as low as that in the recipient parent. Even plants with 48 chromosomes displayed only low levels of enzyme activity (Akagi et al. 1989). Since aryl acylamidase I deficiency is recessive to the nuclear-encoded dominant wild-type allele carried in the donor parent (Matsunaka 1974), we concluded that the nuclei of the regenerated cybrid plants were derived solely from the recipient parent. The plants with 48 chromosomes were presumably fusion products between a CMS cytoplast and two recipient protoplasts (Akagi et al. 1995b). The plants with 24 chromosomes were morphologically quite similar to the recipient parent. These results indicated that no chromosomes had been transferred from the CMS parents (Akagi et al. 1989).

The origin of the mitochondrial genomes of these cybrids was determined based on the restriction patterns of their mitochondrial DNAs (mtDNAs) (Fig. 6). There were three typical restriction patterns of mtDNAs from regenerated plants. Most of the plants were similar to the donor CMS parent, except for fragments specific to the fertile parent. Some of the regenerated plants had restriction patterns that differed from those of both parents. In the case of cybrid plants between N8 (fertile cultivar) and MTC-9A (CMS strain), only 1 of 15 regenerated plants had the same restriction patterns as the recipient parent N8. This plant was presumably regenerated from an unfused N8 protoplast (Akagi et al. 1989).

These electrophoretograms demonstrate that CMS mitochondria were introduced into the fertile cultivar by asymmetric protoplast fusion. Thus, we concluded from these results that only the cytoplasmic traits, and not the nuclei, of CMS lines were introduced into fertile cultivars. The regenerated

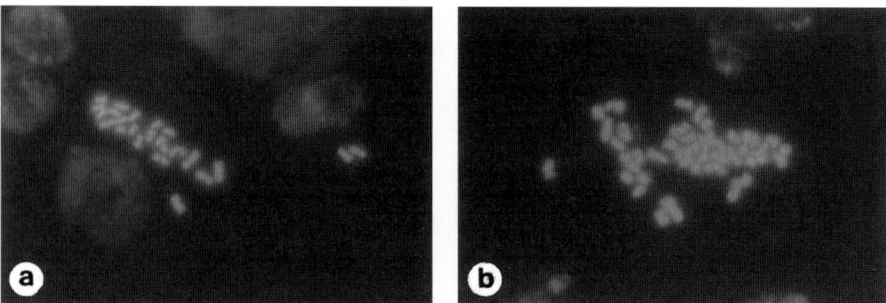

Fig. 5a,b. Root-tip metaphase chromosomes from cybrid plants stained with DAPI (4,6-diamidino-2-phenylindole). **a** CD8-1 (2n = 24). **b** CD8-7 (48 chromosomes). (Akagi et al. 1989)

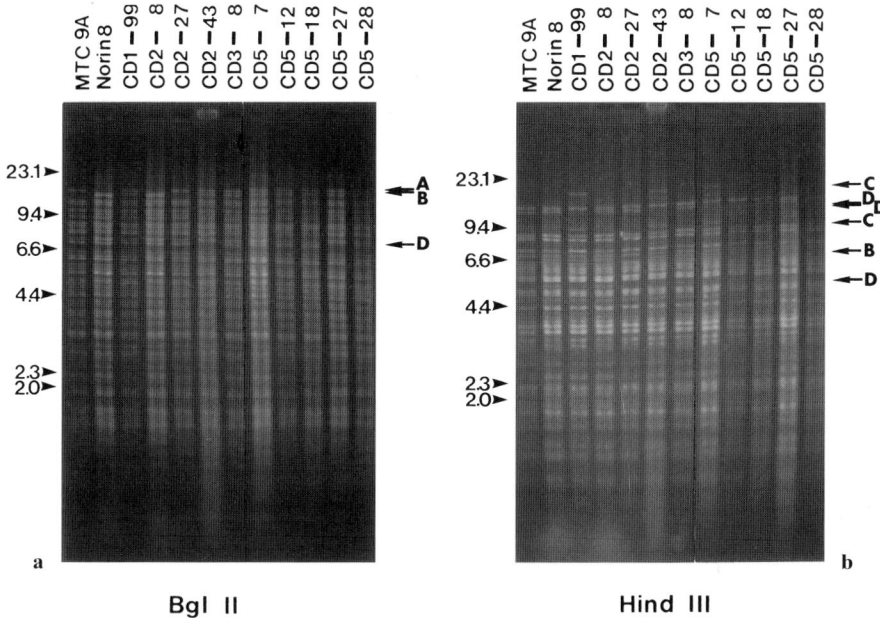

Fig. 6a,b. *Hin*dII (**a**) and *Bgl*II (**b**) restriction patterns of mitochondrial DNA from both parental and cybrid plants. *Arrows* indicate donor CMS parent (*A*)-, recipient Norin 8 (*B*)- and cybrid (*C*)-specific bands. *Arrow D* indicates the band absent from cybrids

plants were cybrids, which had the nucleus of the fertile cultivar and the cytoplasm of both parental strains.

3 Characterization of Rice Cybrids

3.1 Sterility and Restoration of Fertility

More than 80% of the cybrid plants did not set any seeds upon selfing. On the other hand, several sterile plants regenerated from recipient protoplasts without cell fusion (Akagi et al. 1989, 1995a,b). These findings suggest that the sterility of the cybrids was due to the cytoplasmic traits introduced by the CMS parents.

We determined whether the male sterility of cybrid plants was caused by the cytoplasmic or the nuclear genome. Sterile cybrid plants set seeds only when they were fertilized by hand with normal pollen, and yielded only sterile progeny. Thus, male sterile cybrid plants were female fertile. This maternally inherited sterility in cybrid plants showed that these cybrid plants were CMS. The CMS trait of the cybrid plants was stable for at least seven generations with backcrossing with the recipient variety.

To further determine whether or not the CMS of cybrid plants was caused by cytoplasmic traits introduced from CMS lines, we examined the restoration of fertility in CMS cybrids by the *Rf-1* gene, which is the dominant gene for fertility restoration. A high proportion of fully fertile plants was observed in all F_1 progeny among the progeny of CMS cybrids and those of the restore line. Since the CMS of cybrid plants could be restored completely by the single dominant restoring gene, *Rf-1*, CMS of cybrid plants was caused by the mitochondrial genome introduced by protoplast fusion (Akagi et al. 1995b).

The stable transmission to their progeny and the complete restoration of CMS by *Rf-1* demonstrated that cybrids produced by donor-recipient protoplast fusion are useful for creating new CMS rice cultivars for hybrid rice production. It takes about 8 months (from callus induction to seed set by crossing with recipient cultivars) to produce new CMS lines by the donor-recipient protoplast fusion method. However, somaclonal mutation, which occurred during protoplast culture, must be eliminated. Practically, such mutation can be eliminated during two additional backcrossings. Therefore, it takes a total of about 2 years to establish new CMS lines, while it takes about 3 years by the conventional recurrent backcrossing method.

3.2 Mitochondrial Genomes

Use of the protoplast-fusion technique in higher plants has enabled parental mitochondria to coexist in a single cytoplasm. Following such fusion, rearranged mitochondrial DNAs have been observed in several plant species (Galun et al. 1987). These changes may result from interparental recombination between coexisting parental mitochondrial genomes (Temple et al. 1992; Landgren and Glimelius 1994).

In rice cybrids, the occasional modification of mtDNA has been observed (Fig. 6; Akagi et al. 1989, 1995b). To examine whether these parental mitochondrial genomes recombine, the structure around the *atp6* gene of cybrid plants from IR58024A (which carries the cytoplasm from *wild abortive*) and MTC-5A (which carries that from Chinsurah Boro II) were analyzed. By Southern blotting we detected molecules in rice cybrids encompassing the *atp6* gene which contained sequences from both parental mitochondrial genomes. Recombinant molecules that were directly cloned from cybrids confirmed that interparental homologous recombination occurs around the *atp6* gene in the biparental cytoplasm of rice cybrids (Akagi et al. 1995c).

PCR amplification yielded both parental types as well as two recombinant-type molecules, in almost all the callus lines tested after 2 weeks of protoplast culture. Over the course of further cultivation, however, mitochondrial genome diversity decreased as the parental and/or recombinant genomes segregated out. By 4 months, the percentage of lines displaying all four potential types had remarkably decreased (Akagi et al. 1995c). These molecules may either be entirely absent or present at extremely low levels in the cybrid cells.

Thus, interparental recombination occurs with high frequency very soon after protoplast fusion, and most recombinant genomes are eliminated from the cybrid cells by segregation.

About 20% of the cybrid plants between indica and japonica subspecies were fertile (Akagi et al. 1989, 1995a,b). Fertile cybrid plants might be a result of the elimination of CMS traits from the indica subspecies Chinsurah Boro II by segregation (Akagi et al. 1994). By analyzing the recombinant mitochondrial genome of both fertile and sterile cybrids, we successfully identified the CMS trait from Chinsurah Boro II (Akagi et al. 1994), as was done for the CMS traits of *Petunia* (Hanson et al. 1989).

4 Selection of CMS Cybrid Plants

The region downstream from *atp6* in the mitochondrial genome of Chinsurah Boro II is closely related to the appearance of CMS (Iwabuchi et al. 1993; Akagi et al. 1994). This region has been shown to code *orf79*, which is a chimera of another mitochondrial genome region (Akagi et al. 1994). We developed a method for specifically amplifying this region for selection of CMS cybrids. Using a primer pair located in a unique sequence downstream from *atp6* of Chinsurah Boro II, specific amplification was observed in Chinsurah Boro II but not in a fertile Japanese cultivar, Nipponbare (Akagi et al. 1995a).

PCR determination of the region downstream from *atp6* was carried out using both fertile and sterile regenerated cybrid plants. All the progeny of CMS lines had the unique sequence downstream from *atp6*, whereas this sequence was not detected by PCR in fertile cybrids. The progeny of some sterile cybrids also showed recovered fertility. The region downstream from *atp6* was not detected in these progenies. Somaclonal mutation may cause sterility in some cybrid plants (Akagi et al. 1995a). Therefore, the region downstream from *atp6* is closely linked to the CMS phenotype.

We conclude that this selection method based on PCR is very useful for breeding new CMS cultivars by asymmetric protoplast fusion because fertile plants and sterile plants due to somaclonal mutation can be eliminated at an early stage of plant regeneration so that only CMS cybrid plants can be selected (Akagi et al. 1995a).

5 Application of the Cybridization Method in Rice Breeding

5.1 Breeding of New CMS Lines

To convert fertile cultivars to CMS, the cytoplasmic traits of Chinsurah Boro II were transferred to 40 Japanese cultivars by asymmetric protoplast fusion.

About 100 plants were regenerated from calli of each cultivar. However, 30–70% of them showed a tetraploid phenotype, and these plants were sterile with a few exceptions, which set a few selfed seeds. Plants with this phenotype had 48 chromosomes. More than 80% of the diploid cybrid plants were sterile and did not set selfed seeds, except for cybrids that had the nucleus from Hoshiyutaka (Akagi et al. 1995a). The restorer genes for Chinsurah Boro II are widely distributed in the tropics where indica varieties are grown (Shinjyo 1972). Since Hoshiyutaka was bred by crossing japonica and indica rice, it might have the restoration gene. The remaining cultivars were presumed to have no such restorer genes.

We analyzed diploid cybrid plants using the PCR-based method for specifically detecting *orf79*. All the sterile cybrid plants were shown to have this region. On the other hand, no DNA fragment was amplified from fertile cybrids (Akagi et al. 1995a). Interparental recombination between two parental mitochondrial genomes occurs in rice cybrids, and recombinant genomes as well as parental genomes segregate during protoplast culture (Akagi et al. 1995c). Male fertile cybrid plants may have lost the CMS traits by somatic segregation. Thus, the CMS trait was successfully transferred to these sterile cybrids. We concluded that our cybridization method can be applied to most of the cultivars in Japan.

A novel CMS line, called Bio-Mother 1, was bred using this cybridization method from the cultivar Yukigesyo from 1991 to 1993 (Nakamura et al. 1995). The agronomic characteristics of Bio-mother-1 were identical to those of Yukigesyo (Fig. 7), except for pollen fertility (Fig. 8). To date, we have already converted 40 Japanese cultivars to CMS lines as candidates for the female parent of hybrid rice. This large number of new CMS lines may be very useful germ stock for hybrid rice breeding.

We also produced CMS cybrid plants using the WA cytoplasm derived from *wild abortive* which expressed CMS in a sporophytic manner and is widely used for indica subspecies. The cybrid plants between Sasanishiki and IR58024A, which carries WA cytoplasm, were successfully converted to CMS. The timing of the abortion of pollen grains indicated that CMS in these cybrid plants is sporophytic, like the cytoplasmic donor IR58024A.

5.2 Application of New CMS Lines to Hybrid Rice Breeding

We have been developing new hybrid rice varieties. To be accepted in Japan, a novel rice cultivar must (1) give a high yield, (2) be visually appealing, and (3) taste good. We examined these three features in 800 hybrid combinations which were cultured in Chiba, Japan. Tentatively, 10 combinations were selected (Fujimura et al. 1996).

One of the most important features of hybrid rice is a high yield. To precisely evaluate their yield performance, many F_1 seeds are necessary. For this propose, candidates for female parents of hybrid rice were converted to CMS by the cybridization method. Using these new CMS lines, we mass-produced F_1 seeds of several candidate varieties.

Fig. 7. New CMS line Bio-mother 1 created by the cybridization method. A comparison of morphological features in the CMS-converted Bio-mother 1 (*left*) and the fertile parental cultivar Yukigesyo (*right*)

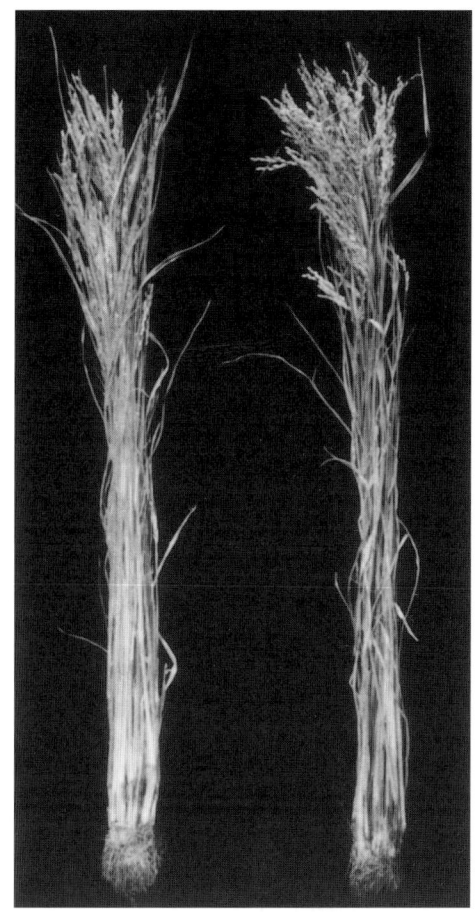

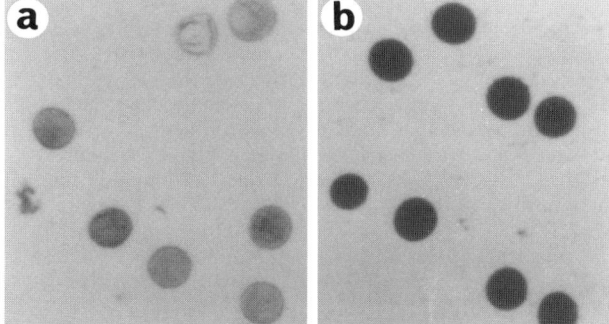

Fig. 8a,b. Sterile pollen from Bio-mother 1 (**a**) and fertile pollen from Yukigesyo (**b**)

These F_1 seeds were then cultured at several experimental stations in Japan, and eventually two combinations, MH2003 and MH2005, were selected. These showed excellent yields that were 30 to 40% higher than those of the leading representative varieties in each region. In addition, their quality was good enough for Japanese consumers (Oka et al. 1995).

6 Summary and Conclusion

We have successfully produced cybrid rice plants based on a donor-recipient protoplast fusion technique. Since only cytoplasmic traits, especially mitochondrial-encoded CMS traits, can be transferred to other cultivars in a single step, we can convert many fertile cultivars to CMS in a short period. It takes about 8 months to produce new CMS lines by the cybridization method. Two additional backcrossings are necessary for the elimination of somaclonal mutation. Therefore, it takes about 2 years for establish the new CMS lines, while it takes about 3 years by the conventional recurrent backcrossing method. Thus, this cybridization method will be especially useful for converting a large number of fertile cultivars into CMS.

The rapid and large-scale production of new CMS lines using the cybridization method enabled us to produce a large number of candidate combinations of new hybrid varieties for evaluation. New hybrid rice varieties that are acceptable in the competitive Japanese market were successfully developed using cybrid-derived CMS lines as female parents.

7 Protocol

7.1 Establishment of Suspension Lines

1. Sterilize husked seeds in 5% sodium chloride for 20min, and then wash three times with sterilized water.
2. Place sterilized seeds on agar-solidified MS or R2 medium containing 2ppm 2,4-D, 0.3% casein hydrolysate, 3% sucrose, and 0.8% agar. Culture seeds at 25°C under 500lx.
3. After 3 to 4 weeks of cultivation, transfer friable calli into 20ml of liquid medium containing R2 inorganic salts, B5 vitamins, 0.3% casein hydrolysate, $1\,mg\,l^{-1}$ 2,4-D, and 3% sucrose. Culture suspension cells at 25°C under 500lx of continuous light on a gyratory shaker at 60rpm.
4. After 3 days, transfer the suspension cells to fresh medium.
5. Subculture every week by transferring 100–200mg of cells without old culture medium to 20 ml of fresh medium.

7.2 Protoplast Isolation

1. Collect about 5g of suspension cells 4–5 days after subculturing, and then add 10ml of enzyme solution (pH 5.5) containing 1% Macerozyme R10, 4% Cellulase RS, 0.5% $CaCl_2 \cdot 2H_2O$, 0.5% potassium dextransulfate, and 0.4M mannitol.

2. Incubate the flask at 27 °C without shaking for 3 h.
3. Remove undigested cell clumps by passing the protoplast-enzyme mixture through a 155-µm nylon screen followed by a 31-µm screen.
4. Remove the enzyme solution by centrifugation ($50\,g$ for 3 min), and then wash three times by centrifugation ($50\,g$ for 3 min) with modified R2 medium containing 0.4 M glucose.

7.3 Inactivation of Protoplasts

1. Adjust the donor protoplast density in the enzyme mixture to $1 \times 10^7\,\text{ml}^{-1}$.
2. Transfer 1 ml of the donor protoplast suspension into a 6-cm plastic petri dish (Falcon no. 3002).
3. Irradiate the petri dish on a turntable by X-rays using a model OM-100R soft X-ray unit (Ohmic, Japan) set at 80 kV and 4 mA for 34 min × 2 (125 krad).
4. Transfer the treated donor protoplasts to a centrifuge tube, and then wash them three times with 0.4 M glucose by centrifugation ($50\,g$ for 3 min).
5. Collect recipient protoplasts by centrifugation ($50\,g$ for 3 min).
6. Add 1 ml of 30 mM IOA solution to the recipient protoplasts. (IOA was dissolved in 0.4 M glucose and sterilized by filtration.)
7. Incubate the recipient protoplasts for 10 min at 27 °C.
8. Add 8 ml of 0.4 M glucose and then wash the recipient protoplasts three times with 0.4 M glucose by centrifugation ($50\,g$ for 3 min).

7.4 Fusion of Protoplasts

1. Adjust the population density of X-irradiated and IOA-treated protoplasts to $2 \times 10^7\,\text{ml}^{-1}$.
2. Mix the X-irradiated donor protoplasts and IOA-treated recipient protoplasts at a ratio of 1:1 or 2:1.
3. Transfer 300 µl of the mixture to a fusion chamber. The fusion chamber was constructed with a pair of stainless steel rods ($2 \times 60\,\text{mm}$) as electrodes and side stop in a glass petri dish. The electrodes were placed in parallel at a distance of 2 mm.
4. Align the protoplasts in an AC field (1 MHz, $150\,\text{V cm}^{-1}$) for 5 s.
5. Fuse the protoplasts by a DC pulse ($2.5\,\text{kV cm}^{-1}$, 50 µs).
6. After 15 min, recover the protoplasts to a centrifugation tube and then wash twice by centrifugation ($50\,g$ for 3 min) with 0.4 M glucose.

7.5 Cultivation of Fused Protoplasts and Regeneration of Cybrid Plants

1. Prepare the liquid culture medium. Remove cells from the suspension culture medium after 3 days of subculture by centrifugation at 12000 rpm for 20 min. Add $2\,\text{mg l}^{-1}$ 2,4-D, 0.4 M sucrose and B5 vitamins (final concentrations). Adjust the pH to 4.25 and then sterilize by filtration.
2. Suspend the fused protoplasts in 0.4 M glucose at $5 \times 10^6\,\text{ml}^{-1}$.
3. Pour 600 µl of this liquid medium into a 6-cm plastic petri dish (Falcon no. 3002) and then add 150 µl of the protoplast suspension (final population density of protoplasts is $1 \times 10^6\,\text{ml}^{-1}$).
4. Culture the protoplasts in the dark at 25 °C.
5. After 2 weeks of culture, add 1 ml of R2 liquid medium supplemented with $2\,\text{mg l}^{-1}$ 2,4-D, and 0.2 M glucose (pH 4.9).
6. After 2 more weeks of cultivation, transfer cell clusters onto agar-solidified medium containing R2 inorganic salts, $2\,\text{mg l}^{-1}$ 2,4-D, and 3% sucrose.
7. When calli are larger than 5 mm in diameter, transfer them to N6 medium without plant growth regulator and culture under 3000 lx of continuous fluorescent light for plant regeneration.
8. After about 4 weeks, transfer regenerated plants to a Magenta box containing 1/2 MS salts, 2% sucrose, and 0.15% Gellan Gum, and then culture under 10000 lx of continuous light.

7.6 Diagnosis of Cybrid Plants (for the CMS Trait of Chinsurah Boro II)

1. Dry leaves from cybrids in an Eppendorf tube at 70°C for 2h.
2. Put small glass beads (ϕ 3mm) in the Eppendorf tube, and then homogenize leaves using a vortex mixer.
3. Add 400µl of an extraction buffer containing 25mM EDTA, 250mM NaCl, 0.5% SDS, and 200mM Tris-HCl (pH 8.0) (Edwards et al. 1991) and incubate for 1h at room temperature to extract crude DNA.
4. Remove cell debris by centrifugation at 12000rpm for 3min.
5. Precipitate DNA by adding an equal volume of 2-propanol to the supernatant.
6. Add 100µl of TE and resolve crude DNA.
7. Add 1µl of DNA solution to 20µl of reaction solution containing 10mM Tris-HCl (pH 8.3), 50mM KCl, 1.5mM $MgCl_2$, 0.5 unit of Taq polymerase (TAKARA), 4nmol dNTP, and 10pmol primers. Nucleotide sequences of the primers are 5'-TCATAGGAGATCTTGGTCCC-3' and 5'-AAGCTTACTTAGGAAAGACTAC-3'.
8. Using a Thermal Cycler 9600 (Perkin-Elmer), perform 35 PCR cycles, each consisting of 30s of denaturation at 94°C, 30s of annealing at 60°C, and 1min of polymerization at 72°C.
9. Detect PCR products about 600bp long by ethidium bromide staining after electrophoresis on 1% agarose gel.

Acknowledgments. This work was performed at the Life Science Laboratory of Mitsui Toatsu Chemicals, Inc. I would like to express my deepest gratitude to Mitsui Toatsu Chemicals for supporting this research. I thank Dr. Atsushi Nakamura for preparing the photograph of Bio-Mother 1. I also gratefully acknowledge Dr. Fujimura for his support.

References

Akagi H, Sakamoto M, Negishi T, Fujimura T (1989) Construction of rice cybrid plants. Mol Gen Genet 215:501–506
Akagi H, Sakamoto M, Shinjyo C, Shimada H, Fujimura T (1994) A unique sequence located downstream from the rice mitochondrial *atp6* may cause male sterility. Curr Genet 25: 52–58
Akagi H, Nakamura A, Sawada R, Oka M, Fujimura T (1995a) Genetic diagnosis of cytoplasmic male sterile cybrid plants of rice. Theor Appl Genet 90:948–951
Akagi H, Taguchi T, Fujimura T (1995b) Stable inheritance of and expression of the CMS traits introduced by asymmetric protoplast fusion. Theor Appl Genet 91:563–567
Akagi H, Shimada H, Fujimura T (1995c) High frequency inter-parental recombination between mitochondrial genomes of rice cybrids. Curr Genet 29:58–65
Barsby TL, Chuong PV, Yarrow SA, Wu S-C, Coumans M, Kemble RJ, Powell AD, Beversdorf WD, Pauls KP (1987) The combination of Polima cms and cytoplasmic triazine resistance in *Brassica napus.* Theor Appl Genet 73:809–814
Chaudhary RC, Virmani SS, Khush GS (1981) Patterns of pollen abortion in some cytoplasmic male sterile lines of rice. Oryza 18:140–142
Chu CC (1978) The N6 medium and its applications to another culture of cereal crops. In: Proc Symp on Plant Tissue Culture, Scientific Press, Peking, pp 45–50
Dudits D, Fejer O, Hadaczly G, Koncz C, Lazar GB, Horvath G (1980) Intergeneric gene transfer mediated by plant protoplast fusion. Mol Gen Genet 179:283–288
Edwards K, Johnstone C, Thompson C (1991) A simple and rapid method for the preparation of plant genome DNA for PCR analysis. Nucleic Acids Res 19:1349
Fujimura T, Sakurai M, Akagi H, Negishi T, Hirose A (1985) Regeneration of rice plants from protoplasts. Plant Tissue Cult Lett 2:74–75

Fujimura T, Akagi H, Oka M, Nakamura A, Sawada R (1996) Establishment of a rice protoplast culture and application of an asymmetric protoplast fusion technique to hybrid rice breeding. Plant Tissue Cult Lett 13:243–247

Galbraith DW (1984) Selection of somatic hybrid cells by fluorescence activated cell sorting. In: Vasil IK (ed) Cell culture and somatic cell genetics of plants, vol 1. Academic Press, New York, pp 433–447

Galun E, Aviv D, Breiman A, Fromm H, Perl A, Vrdri A (1987) Cybrids in *Nicotiana*, *Solanum* and *Citrus*: isolation and characterization of plastome mutants: pre-fusion treatments, selection, and analysis of cybrids. In: von Wettstein D, Chua N-H (eds) Plant molecular biology. Plenum Press, New York, pp 197–207

Gamborg OL, Miller RA, Ojima K (1968) Nutrient requirements of suspension culture of soybean root cells. Exp Cell Res 50:151–158

Gupta HS, Bhattacharjee B, Pattanayak A (1996) Transfer of cytoplasmic male sterility in indica rice through protoplast fusion. IRRN 21:33–34

Hanson MR, Pruitt KD, Nivison HT (1989) Male sterility loci in plant mitochondrial genomes, In: Miflin BJ (ed) Oxford surveys of plant molecular and cell biology, vol 6. Oxford University Press, Oxford, pp 61–85

Hayashi Y, Kyozuka J, Shimamoto K (1988) Hybrids of rice (*Oryza sativa* L.) and wild *Oryza* species obtained by cell fusion. Mol Gen Genet 214:6–10

Hossain M (1996) Economic prosperity in Asia: implications for rice research. In: Khush GS (ed) Rice genetics III. IRRI, Manila, Philippines, pp 3–16

Ichikawa H, Tanno-Suenaga L, Imamura J (1987) Selection of *Daucus* cybrids based on metabolic complementation between X-irradiated *D. capillifolius* and iodoacetamide-treated *D. carota* by somatic cell fusion. Theor Appl Genet 74:746–752

Iwabuchi M, Kyozuka J, Shimamoto K (1993) Processing followed by complete editing of an altered mitochondrial *atp6* RNA restores fertility of cytoplasmic male sterile rice. EMBO J 12:1437–1446

Katsuo K, Mizushima U (1958) Studies on the cytoplasmic difference among rice varieties, *Oryza sativa* L., I. On the fertility of hybrids obtained reciprocally between cultivated and wild varieties. Jpn J Breed 8:1–5

Kyozuka J, Kaneda T, Shimamoto K (1989) Production of cytoplasmic male sterile rice (*Oryza sativa* L.) by cell fusion. Bio/Technology 7:1171–1174

Landgren M, Glimelius K (1994) A high frequency of intergenomic mitochondrial recombination and an overall biased segregation on *B. campestris* or recombined *B. campestris* mitochondria were found in somatic hybrids made within Brassicaceae. Theor Appl Genet 87:854–862

Leaver CJ, Gray MW (1982) Mitochondrial genome organization and expression in higher plants. Annu Rev Plant Physiol 33:373–402

Li Z, Zhu Y (1986) Rice male-sterile cytoplasm and fertility-restoration. In: Hybrid rice. Int Rice Res Inst, Manila, Philippines, pp 85–102

Lin SC, Yuan LP (1980) Hybrid rice breeding in China. In: Innovative approaches to rice breeding. IRRI, Manila, Philippines, pp 35–51

Matsunaka S (1974) Genetic background of specific herbicide tolerance in higher plants. In: Plant growth substances 1973, Hirokawa Publishing, Tokyo, pp 1182–1186

Menzel L, Morgan A, Brown S, Maliga P (1987) Fusion-mediated combination of Ogura-type cytoplasmic male sterility with *Brassica napus* plastid using X-irradiated CMS protoplasts. Plant Cell Rep 6:98–101

Murashige T, Skoog M (1962) A revised medium for rapid growth and bioassays with tobacco tissue culture. Physiol Plant 15:473–497

Nakamura A, Akagi H, Oka M, Arai N, Sawada R, Sano T, Matsumura T, Samoto S, Fujimura T, Tsuchiya T (1995) Breeding of cytoplasmic male sterile rice cultivar Bio-mother 1 as tentatively. Breed Sci 44 (Suppl 1):212 (in Japanese)

Nehls R (1978) The use of metabolic inhibitors for the selection of fusion products in higher plant protoplasts. Mol Gen Genet 166:117–118

Newton KJ (1988) Plant mitochondrial genomes: organization, expression and variation. Annu Rev Plant Physiol Plant Mol Biol 39:503–532

Ohira K, Ojima A, Fujiwara A (1973) Studies on the nutrition of rice cell culture part I. A simple defined medium for rapid growth in suspension culture. Plant Cell Physiol 14:1113–1121

Oka M, Sano T, Matsumura T, Nakamura A, Sawada R, Arai N, Kamiguchi T, Akagi H, Samoto S, Tsuchiya T, Fujimura T (1995) Breeding of a good cooking-quality and high-yield hybrid rice. Breed Sci 44 (Suppl 1):213 (in Japanese)

Sala C, Biasini MG, Morandi C, Nielsen E, Parisi B, Saka F (1985) Selection and nuclear DNA analysis of cell hybrids between *Daucus carota* and *Oryza sativa*. J Plant Physiol 118:409–419

Shinjyo C (1969) Cytoplasmic-genetic male sterility in cultivated rice, *Oryza sativa* L. II. The inheritance of male sterility. Jpn J Genet 44:149–156

Shinjyo C (1972) Distributions of male sterility-inducing cytoplasms and fertility-restoring genes in rice. II. Varieties introduced from 16 countries. Jpn J Breed 22:329–333

Shinjyo C (1975) Genetical studies of cytoplasmic male sterility and fertility restoration in rice, *Oryza sativa* L. Sci Bull Coll Agric Univ Ryukyus 22:1–51

Sidorov VA, Menzel L, Nagy F, Maliga P (1981) Chloroplast transfer in *Nicotiana* based on metabolic complementation between irradiated and iodoacetate-treated protoplasts. Planta 152:341–345

Sidorov VA, Zubko MK, Kuchko AA, Komarnitsky IK, Gleba YY (1987) Somatic hybridization in potato: use of γ-irradiated protoplasts of *Solanum pinnatisectum* in genetic reconstruction. Theor Appl Genet 74:364–368

Tanno-Suenaga L, Ichikawa H, Imamura J (1988) Transfer of the CMS trait in *Daucus carota* L. by donor-recipient protoplast fusion. Theor Appl Genet 76:855–860

Temple M, Makaroff CA, Mutschler MA, Earle ED (1992) Novel mitochondrial genomes in *Brassica napus* somatic hybrids. Curr Genet 22:243–249

Terada R, Kyozuka J, Nishibayashi S, Shimamoto K (1987) Plantlet regeneration from somatic hybrids of rice (*Oryza sativa* L.) and barnyard grass (*Echinochloa oryzicola* Vasing). Mol Gen Genet 210:39–43

Toriyama K, Hinata K (1988) Diploid somatic-hybrid plants regenerated from rice cultivars. Theor Appl Genet 76:665–668

Virmani SS, Shinjyo C (1988) Current status of analysis and symbols for male-sterile cytoplasms and fertility-restoring genes. Rice Genet Newsl 5:9–15

Yang Z-Q, Shikanai T, Yamada Y (1988) Asymmetric hybridization between cytoplasmic male-sterile (CMS) and fertile rice (*Oryza sativa* L.) protoplasts. Theor Appl Genet 76:801–808

Yang Z-Q, Shikanai T, Mori K, Yamada Y (1989) Plant regeneration from cytoplasmic hybrids of rice (*Oryza sativa* L.). Theor Appl Genet 77:305–310

Yuan LP (1994) Increasing yield potential in rice by exploitation of heterosis. In: Virmani SS (ed) Hybrid rice technology. New developments and future prospects. IRRI, Manila, Philippines, pp 1–6

Zelcer A, Aviv D, Galun E (1978) Interspecific transfer of cytoplasmic male sterility by fusion between protoplasts on normal *Nicotiana sylvestris* and X-ray-irradiated protoplasts of male-sterile *N. tabacum*. Z Pflanzenphysiol 90:397–407

Zimmermann U, Scherurich P, Pilwat G, Benz R (1981) Cells with manipulated functions new perspectives for cell biology, medicine and technology. Angew Chem Int Ed 20:325

I.3 Somatic Hybridization Between *Oryza sativa* L. (Rice) and *Hordeum vulgare* L. (Barley)

H. KISAKA, M. KISAKA, A. KANNO, and T. KAMEYA

1 Introduction

Somatic hybridization by protoplast fusion provides plant breeders with the possibility of increasing genetic variability and overcoming sexual cross-incompatibility of plants. Various somatic hybrid plants have been reported. However, within Gramineae, production of somatic hybrids has been hindered by the fact that the mesophyll protoplasts of these species are not totipotent.

Oryza sativa and *Hordeum vulgare* are very important crops and some somatic hybrid cells and hybrid plants have been obtained from these species. For example, Kao and Michayluk (1974) and Kao et al. (1974) reported the fusion of leaf protoplasts from *H. vulgare* with protoplasts isolated from cell suspension cultures of *Glycine max*. Somatic hybridization of protoplasts of *H. vulgare* and *Daucus carota* (Dudits et al. 1976), of *H. vulgare* and *Nicotiana tabacum* (Somers et al. 1986), of *O. sativa* and *D. carota* (Sala et al. 1985; Kisaka et al. 1994), and of *O. sativa* and *G. max* (Niizeki et al. 1985) has also been reported. However, most of the hybrid cells failed to regenerate plants.

Intergeneric fusion of *Solanum* and *Lycopersicon* species as a means of introducing tolerance to certain environmental stresses has been reported (Melchers et al. 1978; O'Connell and Hanson 1986). Cold tolerance of plants that were somatic hybrids of potato and tomato was reported by Smillie et al. (1979).

Tarczynski et al. (1993) reported that transgenic tobacco plants carrying a bacterial gene for mannitol-l-phosphate dehydrogenase had increased tolerance to salinity by virtue of their enhanced ability to accumulate mannitol in their cells. Furthermore, Deping et al. (1996) also produced transgenic rice that introduced a late embryogenesis abundant (LEA) protein gene, *HVA1* gene, from *H. vulgare*. The transgenic rice plants were also shown to have enhanced ability to tolerate water deficit and high salinity.

H. vulgare is a crop plant that tolerates cold and salinity, whereas rice, in particular the japonica type, is very sensitive to these environmental stresses. We attempted to produce somatic hybrids of *O. sativa* and *H. vulgare* that would have the ability to tolerate low temperatures and elevated salinity.

Institute of Genetic Ecology, Tohoku University, Sendai 980-8577, Japan

In this study, we obtained one hybrid plant and analyzed its nuclear and cytoplasmic genomes at the DNA level, and examined the callus induced from the somatic hybrid to determine whether the cold and salt tolerance of *H. vulgare* had been transferred to it by the original protoplast fusion.

2 Protoplast Fusion and Culture

Protoplasts of rice and barley were hybridized by electrofusion and subjected to the selection protocol shown in Fig. 1. The selection of somatic hybrids exploited the low rate of cell division in barley protoplasts and the lack of regenerative ability in rice protoplasts. To obtain somatic hybrid plants between rice and barley, the experiments of fusion and culture were repeated more than 20 times. No regenerated plants were obtained from nonfused protoplasts and fused protoplasts between homologus combinations. Seven shoots were obtained from fused protoplasts (Fig. 2a), but most of them failed to form roots, and stopped growing. Only one regenerated plant was successfully transferred to soil in a greenhouse. The morphology of the regenerated plant closely resembled that of the parental rice plant (Fig. 2b), but the regenerated plant was susceptible to high temperatures in summer and the leaves were wider than those of the parental rice plant. The regenerated plant produced spikes and flowering organs with some anthers (Fig. 2c). After staining with 1% (w/v) acetocarmine, the pollen grains of control plants were stained as shown in Fig. 2d, while in the regenerated plant these grains were not stainable, as shown in Fig. 2e. Moreover, the regenerated plant failed to produce any seeds (Table 1).

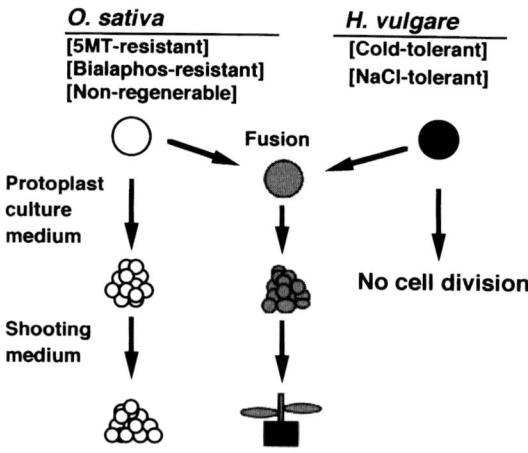

Fig. 1. Scheme for selection of plants that are hybrids of *H. vulgare* and *O. sativa*

Fig. 2. a Regeneration of a shoot from a callus. **b** *Left to right* Plants of *H. vulgare*, somatic hybrid, and *O. sativa*. **c** Spikes and some anthers of the regenerated plant. **d** Pollen of a control rice plant. **e** Pollen of the regenerated plant. Both sets of pollen were stained with acetocarmine

Table 1. Characterization of somatic hybrid and rice

Cell lines	No, of tillers	Fertility rate (%)		Plant height (cm)
		Seeds	Pollens[a]	
O. sativa	11.5	99.6	93.2	97.3
Somatic hybrid	16.0	0	8.3	92.2

[a] Pollen grains of *O. sativa* and of the somatic hybrid were stained with 1% acetocarmine.

3 Analysis of the Regenerated Plant

Cytological analysis indicated that the regenerated plant retained both small chromosomes from rice and large ones from barley (Fig. 3). It appeared that approximately 14 barley ($n = 7$) and 6 rice ($n = 12$) chromosomes were present. Somatic combinations between remote species generally result in the elimination of chromosomes (Kameya et al. 1989; Takamizo et al. 1991; Babiychuk et al. 1992). In this experiment, the regenerated plant had most of the parental chromosomes.

The genomes of the parental plants and the regenerated plant were analyzed at the molecular level. Genomic DNA was analyzed by Southern hybridization using a nonradioactively labeled fragment of the rice *trpB* gene. The regenerated plant yielded bands specific for both rice (9.0 kbp) and barley (3.4 kbp; Fig. 4).

Chloroplast (ct) and mitochondrial (mt) genomes were analyzed by the same method. The regenerated plant yielded novel bands of mtDNA that were not detected in either of the parents when fragments of the *cob* and *atp6* mt

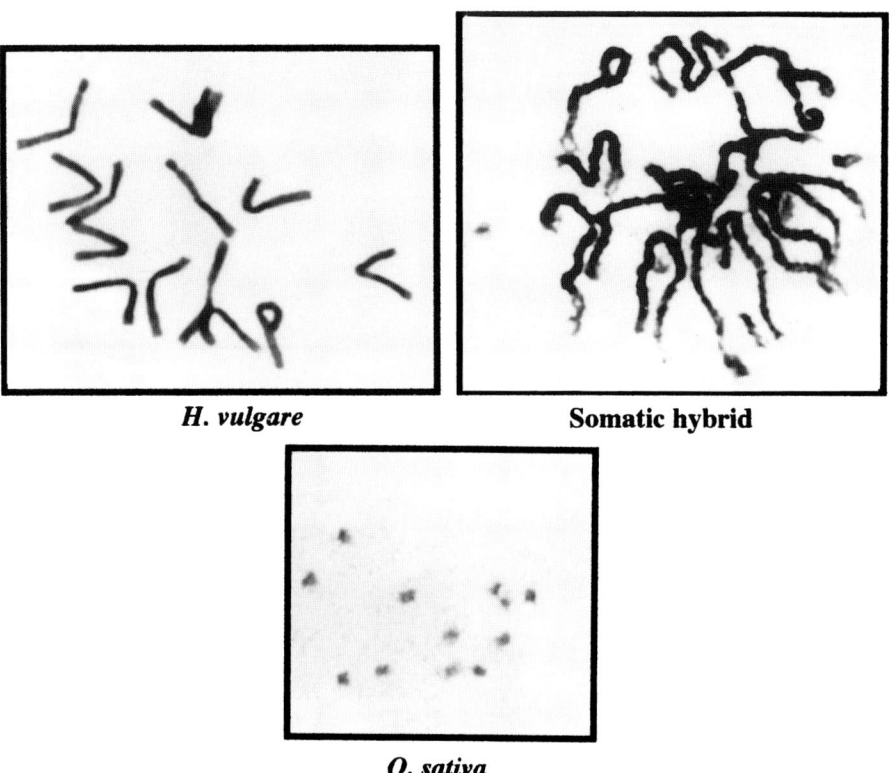

Fig. 3. Chromosomes of *H. vulgare* ($n = 7$), the somatic hybrid, and *O. sativa* ($n = 12$)

Fig. 4. Southern hybridization of genomic DNA. Total DNA was digested with *Bam*HI, and a fragment of the rice *trpB* was used as the probe

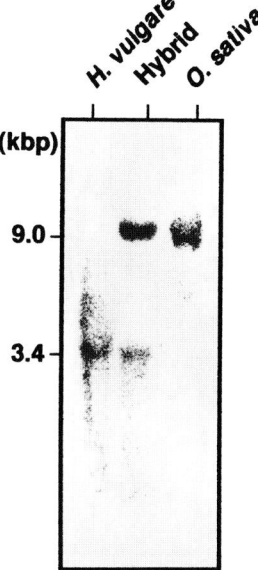

genes were used as probes (Fig. 5a, b). The regenerated plant also yielded a novel band of ctDNA that was not detected in either of the parents when the *Bam*HI-8 fragment of rice ctDNA was used as probe (Fig. 5c). When other fragments of mtDNA and ctDNA (see Protocol, Sect. 5) were used as probes, the regenerated plant yielded the same patterns of bands as rice. As the recombination of mtDNA (Morgan and Maliga 1987; Young and Hanson 1987; Smith et al. 1989; Takamizo et al. 1991; Perl et al. 1991; Kao et al. 1992; Kisaka et al. 1994) and of ctDNA (Medgyesy et al. 1985; Thanh and Medgyesy 1989) has been reported, recombination of chloroplast and of mitochondrial genomes was evident in the regenerated plant.

The results indicated that the regenerated plant was indeed a somatic hybrid plant between rice and barley. However, our single somatic hybrid plant was sterile.

4 Assessment of Cold and Salt Tolerance

Cells in suspension cultures induced from the somatic hybrid plant and from the parent plants were incubated at 4 °C for various times after culture for 1 week at 25 °C. Measurements of the fresh weight of cold-treated cells indicated that the growth of barley cells and hybrid cells was more rapid than that of rice cells (Fig. 6a). The TTC-reduction test showed that the barley cells and hybrid cells were more active than the rice cells (Fig. 6b). These results indicated that the cold tolerance of the somatic hybrid and of barley was greater than that of rice.

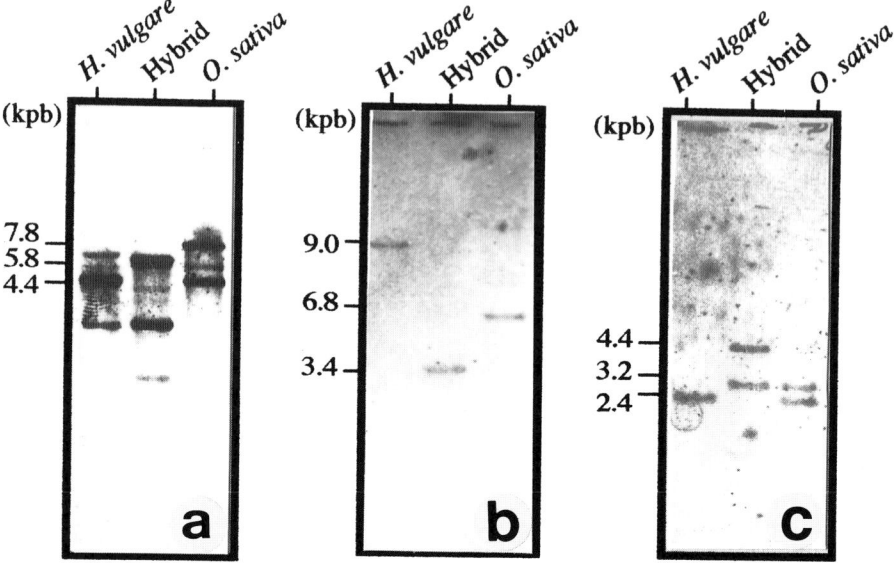

Fig. 5a–c. Southern hybridization of mtDNA and ctDNA. Total DNA was digested with *Eco*RI, and fragments of the *cob* (**a**) and *atp6* (**b**) mtDNA, and the *Bam*HI-8 fragment of rice ctDNA (**c**) were used as probes

The test of the salt tolerance of callus cultures induced from the somatic hybrid and parent plants showed that the hybrid and barley were more tolerant to NaCl than rice. The cells of barley and of the somatic hybrid grew on medium that contained 1.4% (w/v) NaCl, but the growth of rice cells was inhibited by more than NaCl above 0.8% (w/v) (Fig. 7a). In the TTC-reduction test, the activity of barley cells and of somatic hybrid cells was greater than that of rice cells (Fig. 7b). These results indicated that the salt tolerance of the somatic hybrid and of barley was greater than that of rice.

The cells of the somatic hybrid between barley and rice exhibited cold and salt tolerance. Smillie et al. (1979) reported production of somatic hybrid plants from tomato and potato. Some of the somatic hybrids were tolerant to cold, like potato, and others exhibited tolerance intermediate between potato and tomato. In barley, the accumulation of betaine is a metabolic response to water stress or salt stress (Hansen and Nelsen 1978; Wyn Jones and Storey 1978; Hitz et al. 1982). However, rice does not synthesize betaine (Arakawa et al. 1990). Thus, it seems likely that cold and salt tolerance might have been transferred to the somatic hybrid from barley during protoplast fusion. As the somatic hybrid plant was sterile, it is at the present being maintained by vegetative cloning. We think that the somatic hybrid is a useful material for the isolation of genes that might be involved in tolerance to stress.

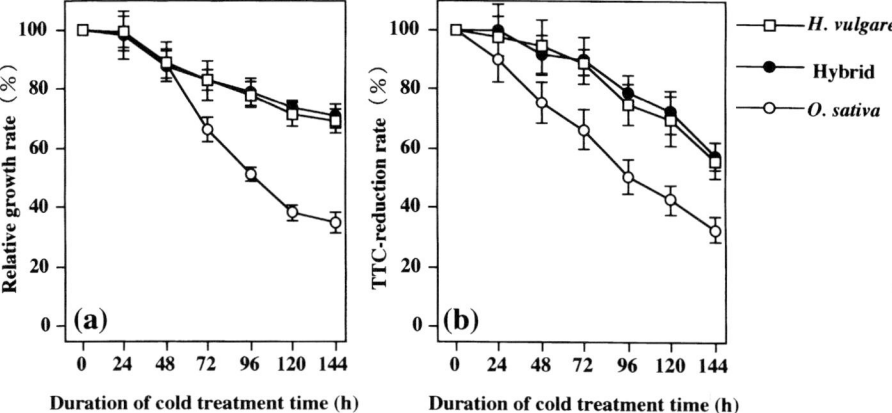

Fig. 6a,b. Effects of cold on cell growth. After culture for 1 week at 25 °C, cells were grown at 4 °C in darkness for the indicated periods. Cell viability was evaluated from measurements of fresh weight (**a**) and TTC-reduction rate (**b**). The fresh weight and TTC-reduction rate for control cells (no cold treatment) were taken as 100%. Each *point* represents the mean ±SE ($n = 3$)

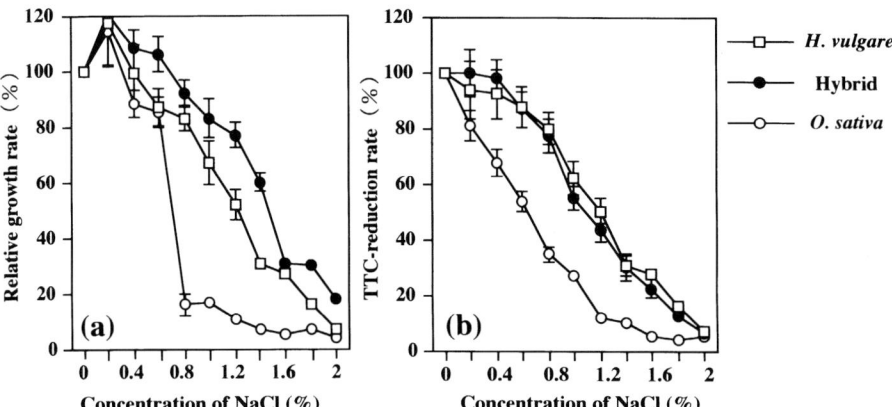

Fig. 7a,b. Effects of NaCl. After culture for 4 days, cells in suspension cultures were grown on solid medium that contained various concentrations of NaCl. Cell viability was evaluated from measurements of fresh weight (**a**) and the TTC-reduction rate (**b**). Results for control cells (no treattment) were taken as 100%. Each *point* represents the mean ± SE ($n = 3$)

5 Protocol

5.1 Plant and Cells

A strain of rice (*Oryza sativa* L.) that was resistant to 5-methyltryptophan and bialaphos was obtained from F_2 plants after crossing a 5-methyltryptophan-resistant mutant of rice cv. Sasanishiki (Lee and Kameya 1991) with rice cv. Yamahoushi (Toki et al. 1992) that had been transformed with a bialaphos-resistance gene. Calluses of 5-methyltryptophan- and bialaphos-resistant rice were initiated from anther cultures. Anthers containing uninucleate pollen grains were plated on B5 medium (Gamborg et al. 1968) supplemented with $2\,mg\,l^{-1}$ 2,4-dichlorophenoxyacetic acid (2,4-D) and 0.8% (w/v) agar to induce formation of calli. The anther-derived calli were transferred to 35ml of liquid AA medium (Toriyama et al. 1985) and cultured with reciprocal shaking in 100-ml flasks at 120rpm to establish cell suspension cultures. The cell suspension cultures were subcultured at 2-week intervals and used for isolation of protoplasts. Seeds of barley (*Hordeum vulgare* L.) were surface-sterilized in 70% (v/v) ethanol for 30s and then in 2% (v/v) sodium hypochlorite solution for 15min, washed twice with sterile distilled water, placed on medium A [MS medium (Murashige and Skoog 1962) supplemented with 0.8% (w/v) agar], and cultured under continuous illumination by fluorescent lamps ($4\,W\,m^{-2}$) at 25°C. After 1 week, young leaves were harvested for isolation of protoplasts.

5.2 Isolation of Protoplasts and Procedures for Fusion, Selection, and Culture

Protoplasts of barley were isolated from young leaves by incubation in a solution of enzyme [1.6% (w/v) Cellulase Onozuka R10 (Yakult Honsha, Tokyo, Japan), 0.3% (w/v) Macerozyme R10 (Yakult Honsha), 8% (w/v) mannitol, and 0.1% (w/v) $CaCl_2$-$2H_2O$ (pH 5.5)] for 3h at 25°C. The protoplasts of rice were isolated from approximately 8-month-old suspension cultures by incubation under the same conditions as those described for barley. The protoplasts were filtered through 50-µm nylon mesh and washed twice with washing solution [8% (w/v) mannitol and 0.1% (w/v) $CaCl_2$-$2H_2O$] with centrifugation at $80\,g$ for 5min. The protoplasts were purified by floating them in a 25% (w/v) solution of sucrose with subsequent centrifugation at $80\,g$ for 5min, and washed once with the washing solution. The two populations of protoplasts were mixed in equal proportions to give a total concentration of 5×10^5 protoplasts ml^{-1} and a 2-ml aliquot of the mixed protoplasts was introduced into the chamber. They were then subjected to electrofusion in an electrocell manipulator (model ECM-200; Biotechnologies and Experimental Research Inc., USA). Alignment condition was 1MHz and 20–$30\,V\,cm^{-1}$ for 30s, and the fusion condition 400–600 $V\,cm^{-1}$ and 20–30µs (pulse width) for two times.

The fused protoplasts were washed once with the washing solution and diluted with an equal volume of a 0.3% (w/v) solution of Gelrite (Sigma Chemical Co., USA) in 3% (w/v) sucrose, and 5% (w/v) glucose, then dropped into protoplast culture medium [medium B {MS medium supplemented with $1.0\,mg\,l^{-1}$ 2,4-D and $0.5\,mg\,l^{-1}$ kinetin and 5% (w/v) glucose} or medium C {MS medium supplemented with $2.0\,mg\,l^{-1}$ 2,4-D and 5% (w/v) glucose}] in plastic petri dishes. Non-fused protoplasts and fused protoplasts between homologous partners of barley and rice were cultured, as controls, under the same conditions. During the first month of culture, the petri dishes were incubated in darkness at 25°C, then transferred to continuous illumination by fluorescent lamps ($4\,W\,m^{-2}$) at 25°C. The resultant microcalli were subsequently cultured in a shooting medium [medium D {MS medium supplemented with $0.1\,mg\,l^{-1}$ naphtaleneacetic acid (NAA), $1.0\,mg\,l^{-1}$ N^6-benzylaminopurine (BAP) and 1% (w/v) agar} or medium E {$1.0\,mg\,l^{-1}$ kinetin and $1.0\,mg\,l^{-1}$ agar}]. After 2 months, regenerated shoots were transferred to a rooting medium [medium F {MS medium supplemented with 1% (w/v) agar}].

5.3 Cytological Analysis

Chromosomes were observed by treating actively growing cells in root tips (of barley and the regenerated plant), and in cell suspension cultures (of rice), with a 0.03% (w/v) solution of 8-hydroxy-quinolinol for 3h at 25°C, with subsequent fixation in a mixture of ethanol and acetic acid (1:3, v/v) for 16h at 25°C. After treatment with 1N HCl for 7min at 60°C, the root tips and cells in suspension culture were stained with Schiff's reagent for 1h at 25°C and then crushed and stained with 1% (w/v) acetocarmine.

Pollen obtained from rice and the regenerated plant was stained with 1% (w/v) acetocarmine. The frequency of stained pollen grains is shown in Table 1 as the fertility rate.

5.4 Analysis of DNA

Total DNA was prepared from rice, from barley, and from the regenerated plant by the method described by Honda and Hirai (1990). Southern hybridizations were performed with a nonradioactive DNA labeling and detection kit (Boehringer Mannheim, FRG). A cloned fragment of the tryptophan B (*trpB*) gene from rice was used as a probe for nuclear genomic sequences. Clones of genes for subunit 6 of F_1–F_0 ATPase (*atp6*) and cytochrome b (*cob*) in the rice mitochondrial (mt) genome and rice chloroplast (ct) DNA fragments (*Bam*H-3, *Bam*H-8, *Bam*H-10, *Bam*H-22) were provided by Dr. A. Hirai, University of Tokyo, Japan. Plasmids containing mitochondrial genes for 26S and 18S ribosomal RNA, and for subunit 9 of F_1–F_0 ATP (*atp9*) were provided by Dr. K. Nakamura (Nagoya University, Nagoya, Japan). The appropriate fragments were used as probes for Southern hybridization.

5.5 Low-Temperature Treatment and Assessment of Cell Viability

Callus in suspension cultures that had been grown for 1 week at 25°C was transferred to solid MS medium supplemented with 3% (w/v) sucrose and 0.8% (w/v) agar in plastic petri dishes (9cm in diameter) and incubated at 4°C in darkness. After chilling treatment at 4°C for 0, 24, 48, 72, or 96h, cells were collected, washed with distilled water at 0°C and blotted on filter paper. Nonchilled control cells, which had been incubated at 25°C in darkness, were collected as described above.

Cell viability after chilling treatment was evaluated from measurements of fresh weight and by the TTC-reduction test (Steponkus and Lanphear 1967). For the TTC-reduction test, 200mg (fresh weight) of cells were incubated with 3ml of a reaction mixture that contained 50mM potassium phosphate buffer (pH 7.3) and 0.6% (w/v) triphenylterazolium chloride (TTC) at 25°C in darkness under a vacuum. After incubation for 15h, cells were collected, washed with distilled water, blotted on filter paper and extracted with 5ml of 95% ethanol for 30min. The absorbance of each ethanol extract was measured at 540nm with 95% ethanol as the blank. The absorbance of a similar extract of nonchilled, control cells was taken as 100%.

5.6 High-Salt Treatment and Assessment of Cell Viability

Cells in suspension culture, after growth for 4 days at 25°C, were transferred to solid MS medium supplemented with 3% (w/v) sucrose, 0.8% (w/v) agar and various concentrations [0, 0.2, 0.4, 0.6, 0.8, 1.0, 1.2, 1.4, 1.6, 1.8, and 2% (w/v)] of NaCl and incubated for 1 month under continuous illumination by fluorescent lamps (4 W m^{-2}) at 25°C.

Cell viability after such treatment was evaluated from measurements of fresh weight and by the TTC-reduction test as described above. Results for control cells that had not been treated with NaCl were taken as 100%.

References

Arakawa K, Katayama M, Takabe T (1990) Levels of betaine and betaine aldehyde dehydrogenase activity in the green leaves, and etiolated leaves and roots of barley. Plant Cell Physiol 31:797–803

Babiychuk E, Kushnir S, Gleba YY (1992) Spontaneous extensive chromosome elimination in somatic hybrids between somatically congruent species *Nicotiana tabacum* L. and *Atropa belladonna* L. Theor Appl Genet 84:87–91

Deping X, Xiaolan D, Baiyang W, Bimei H, Tuan-Hua DH, Ray W (1996) Expression of a late embryogenesis abundant protein gene, *HVA1* from barley confers tolerance to water deficit and salt stress in transgenic rice. Plant Physiol 100:249–257

Dudits D, Kao KN, Constabel F, Gamborg OL (1976) Fusion of carrot and barley protoplasts and division of heterokaryocytes. Can J Genet Cytol 19:263–269

Gamborg OL, Miller RA, Ojima K (1968) Nutrient requirement of suspension cultures of soybean root cell. Exp Cell Res 50:151–158

Hanson AD, Nelson CD (1978) Betaine accumulation and [^{14}C]-formate metabolism in water-stressed barley leaves. Plant Physiol 62:305–312

Hitz WD, Landyman JAR, Hanson AD (1982) Betaine synthesis and accumulation in barley during field water stress. Crop Sci 22:47–54

Honda H, Hirai A (1990) A simple and efficient method for identification of hybrids using non-radioactive rDNA as probe. Jpn J Breed 40:339–348

Kameya T, Kanzaki H, Toki S, Abe T (1989) Transfer of radish (*Raphanus sativus* L.) chloroplasts into cabbage (*Brassica oleracea* L.) by protoplast fusion. Jpn J Genet 64:27–34

Kao HM, Heller WA, Gleddie S, Brown GG (1992) Synthesis of *Brassica oleracea/Brassica napus* somatic hybrid plants with novel organelle DNA compositions. Theor Appl Genet 83:313–320

Kao KN, Michayluk MR (1974) A method for high-frequency intergeneric fusion of plant protoplasts. Planta 115:355–367

Kao KN, Constabel F, Michayluk MR, Gamborg OL (1974) Plant protoplast fusion and growth of intergeneric hybrid cells. Planta 120:215–227

Kisaka H, Lee H, Kisaka M, Kanno, A Kang K, Kameya T (1994) Production and analysis of asymmetric hybrid plants between monocotyledon (*Oryza sativa* L.) and dicotyledon (*Daucus carota* L.) Theor Appl Genet 89:365–371

Lee H, Kameya T (1991) Selection and characterization of a rice mutant resistant to 5-methyltryptophan. Theor Appl Genet 81:405–408

Medgyesy P, Fejes E, Maliga P (1985) Interspecific chloroplast DNA recombination in a *Nicotiana* somatic hybrid. Proc Natl Acad Sci USA 82:6960–6964

Melchers G, Sacristan MD, Holder AA (1978) Somatic hybrid plants of potato and tomato regenerated from fused protoplasts. Carlsberg Res Commun 43:203–218

Morgan A, Maliga P (1987) Rapid chloroplast segregation and recombination of mitochondrial DNA in *Brassica* cybrids. Mol Gen Genet 209:240–246

Murashige T, Skoog F (1962) A revised medium for rapid growth and bioassays with tobacco tissue culture. Physiol Plant 15:473–497

Niizeki M, Tanaka M, Akada S, Hirai A, Saito K (1985) Callus formation of somatic hybrids of rice and soybean and characteristics of the hybrid callus. Jpn J Genet 60:81–92

O'Connell MA, Hanson MR (1986) Regeneration of somatic hybrid plants formed between *Lycopersicon esculentum* and *Solanum rickii*. Theor Appl Genet 72:59–65

Perl A, Aviv D, Galum E (1991) Protoplast fusion-mediated transfer of oligomycin resistance from *Nicotiana sylvestris* to *Solanum tuberosum* by intergenetic cybridization. Mol Gen Genet 225:11–13

Sala C, Biasini MG, Morandi C, Nielsen E, Parisi B, Sala F (1985) Selection and nuclear DNA analysis of cell hybrids between *Daucus carota* and *Oryza sativa*. J Plant Physiol 118:409–419

Smillie RM, Melchers G, von Wettstein D (1979) Chilling resistance of somatic hybrids of tomato and potato. Carlsberg Res Commun 44:127–132

Smith MA, Pay A, Dudits D (1989) Analysis of chloroplast and mitochondrial DNAs in asymmetric somatic hybrids between tobacco and carrot. Theor Appl Genet 77:641–644

Somers DA, Narayanan KR, Kleinhofs A, Cooper-Bland S, Cocking EC (1986) Immunological evidence for transfer of the barley nitrate reductase structural gene to *Nicotiana tabacum* by protoplast fusion. Mol Gen Genet 204:296–301

Steponkus PL, Lanphear FO (1967) Refinement of the triphenyl tetrazolium chloride method of determining cold injury. Plant Physiol 42:1423–1426

Takamizo T, Spangenberg G, Suginobu K, Potrykus I (1991) Intergeneric somatic hybridization in Gramineae: somatic hybrid plants between tall fescus (*Festuca arundinacea* Schreb.) and Italian ryegrass (*Lolium multiflorum* Lam.). Mol Gen Genet 231:1–6

Tarczynski MC, Jensen RQ, Bohnert HJ (1993) Stress protection of transgenic tobacco by production of the osmolyte mannitol. Science 259:508–510

Thanh ND, Medgyesy P (1989) Limited chloroplast gene transfer via recombination overcomes plastome-genome incompatibility between *Nicotiana tabacum* and *Solanum tuberosum*. Plant Mol Biol 12:87–93

Toki S, Takamatu S, Nojiri C, Ooba S, Anzai H, Iwata M, Christensen AH, Quail PH, Uchimiya H (1992) Expression of a maize ubiquitin gene promoter-bar chimeric gene in transgenic rice plants. Plant Physiol 100:1503–1505

Toriyama K, Kameya T, Hinata K (1985) Cell suspension and protoplast culture in rice. Plant Sci 41:179–183

Young EG, Hanson MR (1987) A fused mitochondrial gene associated with cytoplasmic male sterility is developmentally regulated. Cell 50:41–49

Wyn Jones RG, Storey R (1978) Salt stress and comparative physiology in the Gramineae. II. Glycine betaine and proline accumulation in two salt- and water-stressed barley cultivars. Aust J Plant Physiol 5:817–829

I.4 Somatic Hybridization Between *Triticum aestivum* L. (Wheat) and *Haynaldia villosa* L.

G.M. XIA, A.F. ZHOU, and H.M. CHEN

1 Introduction

Wheat is one of the most important cereal crops in the world. Great efforts have been made to enhance its productivity and quality to fullfil the increasing demand for food for man. The genetic approach to arrive at this aim is the most important. However, the genetic pool of wheat is being exhausted through long-term cultivation. It is therefore necessary to introduce into it the desirable agricultural traits from its relatives or remotely related plants, which is hardly completed by conventional breeding and selection alone.

Somatic hybridization is a promising tool for transferring desirable traits across sexual barriers or taxonomic distance. Asymmetric fusion is preferable for introducing only a limited amount of nuclear genome of the donor into the recipient and thus avoiding transferring also too many undesirable traits into the recipient. Much progress, in somatic hybridization for the breeding of economic plants including cereals has been reported (Waara and Glimelius 1995), among which rice is the most successful. Intergeneric (Terada et al. 1987) and interspecific hybrids (Hayashi et al. 1988) as well as cytoplasmic hybrids (Yang et al. 1988; Yang et al. 1989) have been recovered by this technique, but no success has been obtained in wheat. This is because it is difficult to make wheat protoplast regenerable and keep the regeneration capacity of wheat suspension cultures, which remain the most suitable sources for protoplast preparation. This is also the case for many grasses.

Haynaldia villosa ($2n = 14$) belonging to the Haynaldia genera of subtribe the Triticinae, possesses some valuable traits which are absent in wheat, such as high resistance to disease, high content of seed proteins, and strong tillering ability. However, the transfer of these traits into wheat via sexual hybridization is difficult and tedious (Liu et al. 1983). In this work, we succeeded in regenerating plants from fusion products between wheat and *Haynaldia villosa* protoplasts and revealed the phenomenon of complementary effect of regeneration capacity through fusion, which made the regeneration of fusion products possible, although the parents could hardly be regenerated to whole plants.

School of Life Science, Shandong University, Jinan 250100, P.R. China

2 Preparation and Treatment of Parental Protoplasts

2.1 Preparation and Treatment of Wheat Protoplasts

Calli of common wheat cv. Jinan 177 ($2n = 42$) were induced from young embryos on modified MS medium containing $2\,\text{mg}\,l^{-1}$ 2,4-D and subcultured every 2–3 weeks for about 1 year until small granular structures appeared. These calli were selected for suspension preparation as before (Li et al. 1992).

Protoplasts were isolated from 3-year-old suspensions. After treatment with IOA at different concentrations (1.0, 1.5, and $2.5\,\text{mmol}\,l^{-1}$), they were used for fusion experiments. An aliquot of cell suspension was cultured in P_5 solution (Xia et al. 1995) to test the influence of different IOA concentrations upon the growth of the protoplasts.

The protoplasts inactivated by 1 mM IOA retained partial ability to divide; a few of them grew to the four-cell colony stage, but could not divide further. Protoplasts treated with 1.5 mM IOA could not divide, although some of them swelled in volume. Those treated with 2.5 mM IOA turned dark brown and coagulated in 2–5 days. The untreated protoplasts (control) changed their shape after 2 days and began to divide at the 3–4th day of culture. Small calli and some embryo-like structures about 1 mm in size formed after 40 days of culture.

2.2 Origin and Treatment of *H. villosa* Calli and Their Protoplast Preparation

Calli of *H. villosa* were induced from young embryos (as described for wheat). After 1 year of subculture, they formed 1.5-mm granular structures whose color was paler than wheat. They grew rather fast and were subcultured every 12–13 days. A large number of protoplasts could be isolated by the same method as described for wheat, but they could not divide under the culture conditions of this experiment.

H. villosa calli were irradiated with $^{60}\text{Co-}\gamma\text{-ray}$ before protoplast preparation for fusion.

In order to find out the influence of $^{60}\text{Co-}\gamma\text{-ray}$ upon the cell division and chromosome aberration of *H. villosa* calli, a preliminary experiment was carried out. Upon observing the chromosome, particular attention was paid to chromosome fragmentation, because chromosome fragments tend to recombine with the chromosomes of the recipient cells during fusion.

$^{60}\text{Co-}\gamma\text{-ray}$ at doses of 1, 2, 3, 5, 6, 7, 10, and 30 krad (dose rate 130 rad min^{-1}) was applied to the calli after subcultured for 2, 4, 5, 7, and 10 days, respectively. The results showed that 7–30 krad induced lethality of cells (cells died without division) and 3–6 krad division lethality (after one or several divisions the irradiated cells lost the ability of sustained division). Calli subcultured for 4–5 days and 6 krad γ-ray were selected as the optimal condition, because they showed a high frequency (89.9%) of chromosome fragmentation

Table 1. State of chromosome fragmentation after 1–6 krad γ-ray irradiation

Dose (krad)	1	2	3	5	6
Fragmentation frequency at the fragmentation peak after irradiation (%)	26.4	17.3	43.1	71.6	89.9
Time required for the occurrence of fragmentation peak after irradiation (days)	0	5	3	5	3
Time required for the occurrence of cell division peak after irradiation (days)	3	3	5	5	3

(no. of cells possessing chromosome fragments / the total dividing cells observed) and the cells of the calli still retained considerable capacity of division. Some data on chromosome fragmentation after 1–6 krad γ-ray irradiation are listed in Table 1.

The time of fusion after irradiation was also an important factor for the success of hybridization. It was shown that protoplast yield dropped as the time extended after irradiation. Just after or 1–2 days after irradiation was found to be the best time for fusion. This was because not only considerable numbers of vital protoplasts could be obtained, but also this period was shortly before the cell division peak of the irradiated calli (see Table 1). It provided more chance to form heterokaryons. Details on this part of the work have been reported previously (Zhou et al. 1996).

3 Protoplast Fusion and Culture

3.1 Fusion of Parental Protoplasts

The protoplasts were fused with polyethylene glycol according to Lindsey (1991) with a slight modification using PEG6000. Four combinations of fusion were designed:

A: wheat protoplasts untreated (+) *H. villosa* protoplasts derived from irradiated calli,
B: wheat protoplasts treated with 1.0 mM IOA (+) *H. villosa* protoplasts as in A,
C: wheat protoplasts treated with 1.5 mM IOA (+) *H. villosa* protoplasts as in A,
B: wheat protoplasts treated with 2.5 mM IOA (+) *H. villosa* protoplasts as in A.

For each combination, cultures of each parental protoplasts and their mixture were used as control.

At the beginning of fusion, aggregates of heterologous protoplasts could be recognized by the difference in morphology from parental protoplasts. The

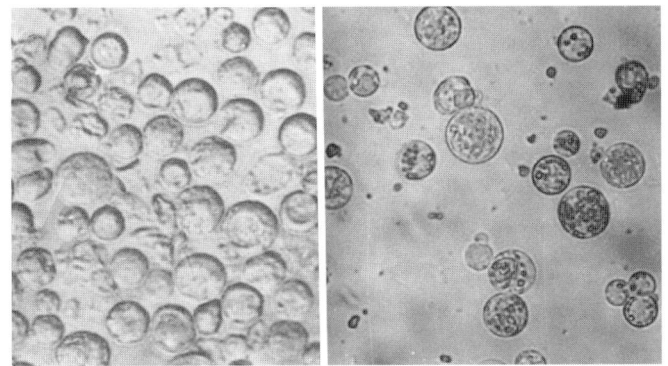

Figs. 1,2. Freshly isolated protoplasts of wheat (**1**) and *H. villosa* (**2**)

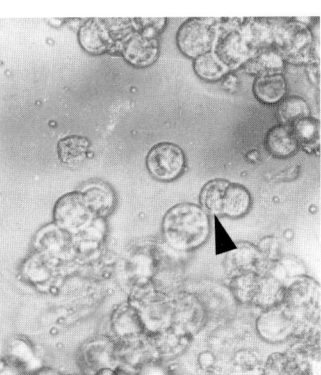

Fig. 3. Aggregate of each protoplast of wheat and *H. villosa* (▶, arrow)

wheat protoplasts were densely cytoplasmic and larger (Fig. 1), while those of *H. villosa* contained many granular inclusions (Fig. 2). The boundary membrane could be distinguished (Fig. 3) after several hours to 1 day. The division and growth of the heterokaryons were different in each combination.

3.2 Growth and Regeneration of the Fusion Products

In combination A, division of the culture began at the 3–4th day after fusion, and colonies were formed after 15 days of culture. Part of the clones grew faster than with the wheat and parental mixtures. They formed embryogenic calli 1.5–2mm in size with a compact structure after a month (Fig. 4). The embryogenic calli were picked up and placed onto the proliferating medium, where some of them became violet red.

The shape of most regenerated calli was intermediate between both parents (M type) and only vary few of them resembled wheat (T type) or *H.*

Fig. 4. Small calli formed from fusion products

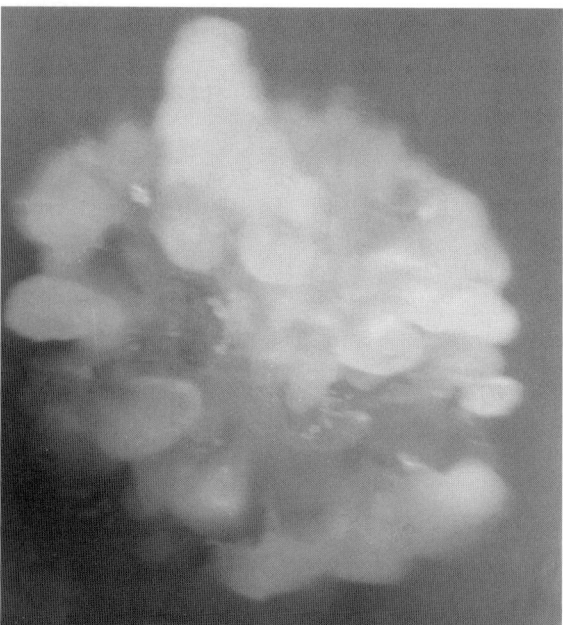

Fig. 5. Embryos and leaves produced from no. 4 hybrid calli

villosa (H type). When they were transferred to a differentiating medium with 12-h illumination of 2000 lx day^{-1}, embryo-like structures and small buds appeared on their surface (Fig. 5), which subsequently regenerated plantlets (Fig. 6). It was notable that only in combination A could embryo formation be

Somatic Hybridization Between *Triticum aestivum* L. (Wheat) and *Haynaldia villosa* L.   53

Fig. 6. Regenerated plants from hybrid no. 4 calli

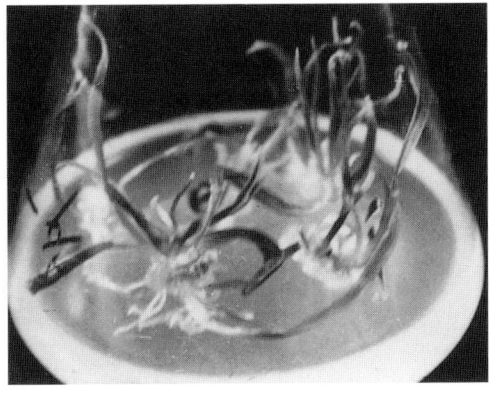

observed. The early-formed clones exhibited a strong differentiating ability compared to the parental calli. More than half of them had regeneration ability and usually differentiated to more than one shoot. From no. 4 clone more than 70 seedlings were obtained.

In combination B, the fusion products grew more slowly. Calli of 1–2 mm in size formed after 40 days and five were large enough to transfer to the solid differentiation medium; but only one differentiated into a few small seedlings, which finally died.

In combination C, the fusion products could grow into small 1-mm calli, most of which were violet red. When transferred to a medium for differentiation, no plantlet could be obtained.

In combination D, the first division of the culture was observed much later (the 6th day of culture) and no visible clones formed.

In control, the untreated wheat protoplasts grew to form many colonies and small calli. However, a few of them differentiated to green spots, small leaves less than 0.5 cm, or roots only. No whole plantlets were regenerated.

The *H. villosa* protoplasts did not divide under the culture conditions of this experiment.

The culture of a mechanical mixture of parental protoplasts gave the same result as that of wheat protoplasts.

4 Verification of Hybridity

4.1 Analysis of Putative Hybrid Calli (see Table 2)

The chromosome numbers of the regenerated calli of wheat protoplasts (control) were mostly (>99%) in the range 31–39, rarely 42, and only a few cells contained chromosome fragments. The chromosome numbers of *H. villosa* calli used for protoplast preparation were mainly 14 (Fig. 7). After irra-

Table 2. Morphology, cytology, isozyme pattern, and plant regeneration of fusion products

Cell clone	Fusion combination	Morphology of calli[a]	No. of chromosomes	Esterase isozyme pattern	Plant regeneration
1	A	M. with violet color	42–56, with fragments	P	✓
2	A	M. with violet color	42–56, with fragments	P	✓
3	A	M. with violet color	45–48, with fragments	P	✓
4	A	M. with violet color	48–56, with fragments	P	✓
5	A	M	43–56 or more, with fragments	P	✓
6	A	M. with violet color	48–56 or more, with fragments	P	✓
7	A	M	about 48, with fragments	P	✓
8	A	M	50–51, with fragments	P	✗
9	A	H	20–42, with fragments	P	✗
10	A	M. with violet color	43–48, with fragments	P	✓
11	A	M. with violet color	–	P	✗
12	A	M	24–39	P	✗
13	A	M	22–37	T	✗
14	A	M	22–25	T	✗
15	A	T	16–29	T	–
16	A	T	–	T	–
17	B	M	28	P N	✗
18	B	M	26–35, with fragments	P	✓
19	C	M. with violet color	28–43, with fragments	P	✗

[a] Morphology of calli: M, middle type; T, wheat type; H, *H. villosa* type. Isozyme pattern: P, containing partial bands of both parent; N, containing new band; T, wheat pattern; ✓, plant regeneration; ✗, no plant was regenerated; –, not examined.

Somatic Hybridization Between *Triticum aestivum* L. (Wheat) and *Haynaldia villosa* L. 55

Fig. 7. Chromosomes of *H. villosa* ($2n = 14$)

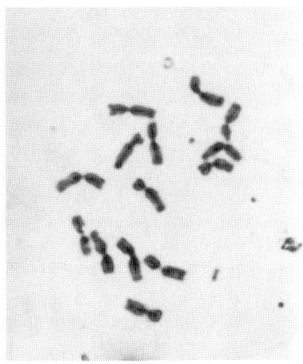

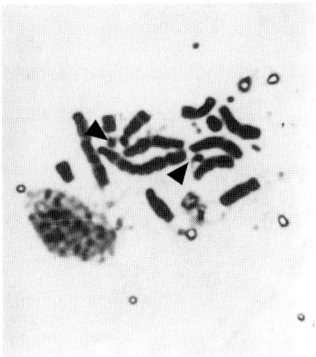

Fig. 8. Chromosomes of *H. villosa* calli irradiated by 6 krad ^{60}Co-γ-ray. ▲ (arrow) Chromosome fragment

diation, the number of intact chromosomes decreased and chromosome fragments appeared (Fig. 8).

Among the regenerated calli, most of the M type contained more chromosomes (42–56) than wheat and chromosome fragments always occurred simultaneously (Figs. 9, 10).

The above observation indicates that the genetic material of *H. villosa* had entered into the wheat cells.

The esterase isozyme pattern of the putative calli was analyzed. Among the 19 calli analyzed, most possessed the characteristic band of both parents except four showing a pattern similar to wheat and one showing an additional new band (Fig. 11). The chromosome numbers of the clones showing bands of both parents were generally higher than those of clones showing wheat bands only or the control wheat protoplasts. This result denoted the coincidence between chromosome and isozyme analyses.

The morphology of most of the regenerated calli was between both parents (M type). The granules of the hybrid calli were 1–1.5 mm in diameter and densely packed, while those of wheat (T) were about 0.5 mm and rather loose in structure. The *H. villosa* calli used for protoplast preparation were composed of larger granules (1.5 mm). It was noticeable that the regenerated

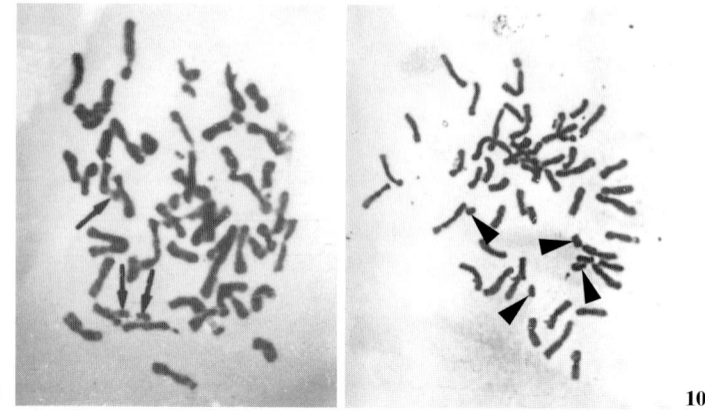

Figs. 9,10. Chromosomes of hybrid no. 2 (**9**) and no. 2 calli (**10**). ▲ Chromosome fragment

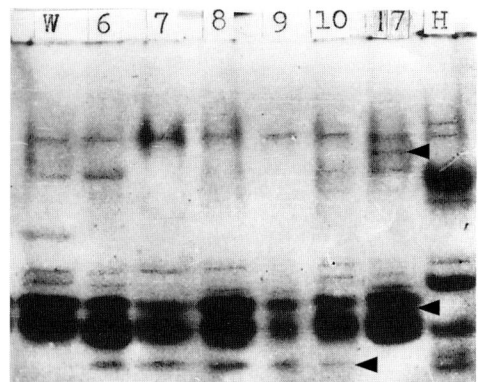

Fig. 11. Esterase isozyme pattern of hybrid calli. ▲ Characteristic band; ↑ new band

calli exhibited two kinds of color – yellowish white or violet red. The latter was remarkably different from both parents. The violet calli were all identified as hybrid by isozyme and chromosome analysis, and therefore the violet color might be considered as a characteristic of hybrid calli.

4.2 Identification of Putative Hybrid Plants

A part of the regenerated plants were used for chromosome counting, isozyme analysis, and morphology inspection. The results for chromosomes and isozymes are listed in Table 3.

As shown above, the regenerated plants had 40–53 chromosomes with concomitance of some fragments in their cells (Figs. 12, 13). The esterase patterns were all P type, showing characteristic bands of each parents (Fig. 14).

Table 3. Cytology and isozyme pattern of partial regenerated plants

Cell clone	No. of plants	Origin and number of chromosome	Isozyme[a]	
			Esterase	amylase
2	2	Young leaf base 43–46, with 3–5 fragments	P	P
4	4-1	Root tip 40–42, with 3–4 fragments	P	P
	4-2	Root tip 46–53, with 2–3 fragments	P	P
	4-3	Young leaf base 40–43, with fragments	P	P
	4-4	–	P	P
	4-5	–	P	P
6	6	–	P	P
8	8	–	P	P
Control	H. villosa	Young leaf base, 14	H	H
	wheat 1	Young leaf base, 42, no fragments (Fig. 15)	T	T
	2	Root tip, 42, no fragments	T	T

[a] P, showing parental bands; H, showing bands of *H. villosa*; T, showing bands of wheat; –, not examined.

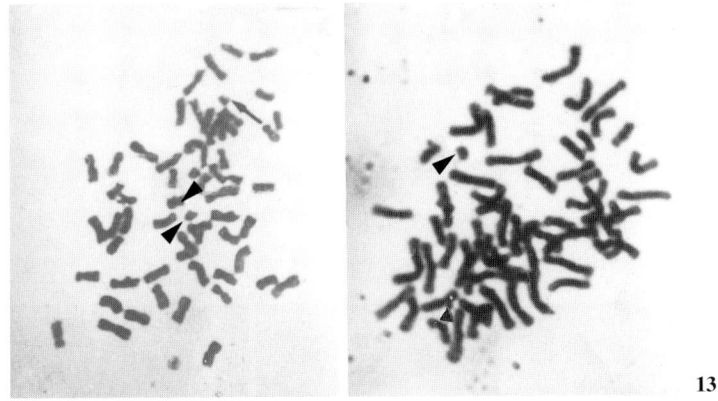

Figs. 12, 13. Chromosomes of no. 2 (**12**) and no. 4-1 (**13**) hybrid plant regenerated from nos. 2 and 4 hybrid calli. ▲ Chromosome fragment

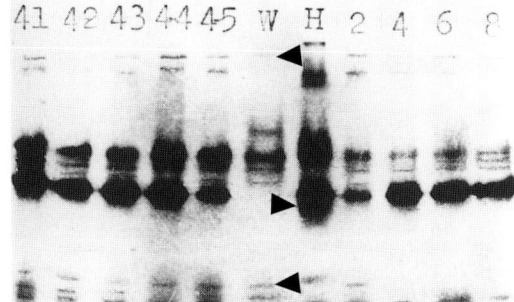

Fig. 14. Esterase of hybrid plants. ▲ Characteristic band

Molecular characterization of the nuclear genome of hybrid plants was analyzed by the RAPD method. The polymorphism of total genome DNA of hybrid no. 4-1 plant (derived from hybrid no. 4 calli) was detected with 4 (OPA-13 CAGCACCCAC, OPA-19 CAAACGTEGG, OPF-01 ACGGATC-CTG, OPF-08 GGGATATCGG) from the 39 primers. The hybrid plant showed characteristic bands or parental genome and new bands (Figs. 16, 17) and conformed the hybrid nature of the analyzed plant.

In the inspection of morphology, we paid attention to the comparison of the ligule and auricle between the hybrids and their parents, because ligule and auricle are the taxonomic characters of the Gramineae. There were remarkable differences between the hybrids and their parents, as shown below (Table 4).

From all aspects of conformation of hybrid nature, we can conclude that the plants regenerated in this experiment were hybrids.

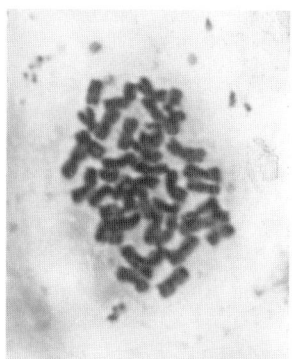

Fig. 15. Chromosomes of wheat plant ($2n = 42$)

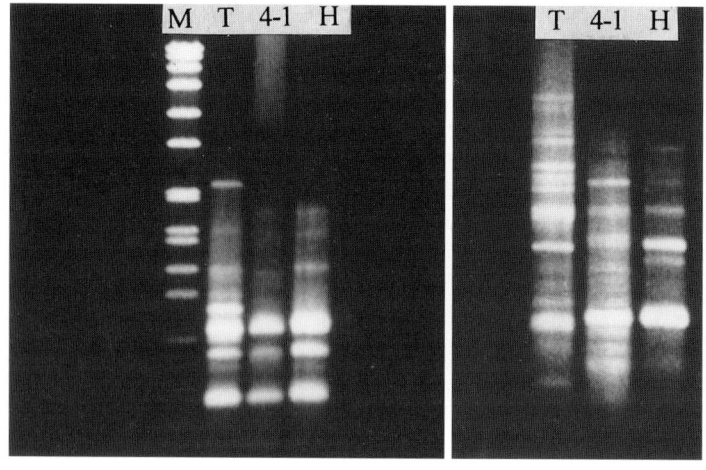

Figs. 16, 17. Electrophoresis of RAPD products using primer OPA-13, OPA-19. *T* Wheat; *Nos. 4-1* hybrid; *H H. villosa*; *M* molecular marker

Table 4. Morphological character of the ligule and auricle at seedling stage

	Ligule	Auricle
Wheat (w) (Fig. 18)	Rather broad and long, a little longer at the middle and saw-shaped at the top	Obvious, bearing scattered hairs
H. villosa (H) (Fig. 19)	Relatively longer than wheat, saw-shaped at the top	Obvious, bearing a lot of long hairs
W + H (Fig. 20)	Relatively short	Partly joined with leaf sheath which has one pointed protruberance on both edges, hairless

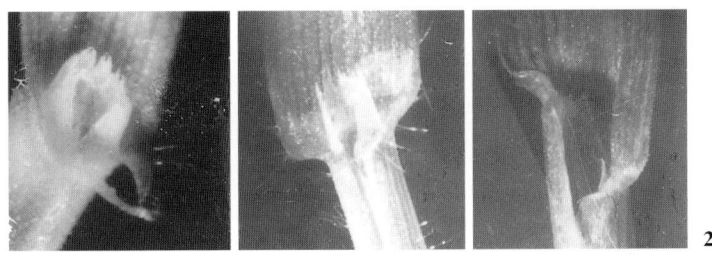

Figs. 18–20. Leaf blades of no. 4-1 hybrid and parents, indicating the ligule and auricle (explained in Table 4); ↑: ligule

5 Growth and Development of Hybrid Plants

In this first experiment more than 170 hybrid seedlings were obtained and about 100 were potted in soil (Fig. 21), but most of them survived only 3 months at most.

In another independent experiment carried out successively, three hybrid plants survived in soil and overwintered. Two of them grew to heading stage in the next year. Their shapes became more and more different from both parents during their growth. They had harder, shorter, and more numerous leaves (generally 8–9, at most 12, while both parents usually had 5–6 or 6–7) on the stem in the shooting stage (Fig. 22). The development of the hybrids was much slower, reaching the heading stage about 1.5 months later than the parents. They produced ears late in summer, which were wrapped up in the sheath of the flag leaves and could not protrude until the plants died.

Recently, we have conducted another hybridization experiment of this same combination. Eighteen hybrid plants were obtained and survived in soil, but they were also sterile. Some of them were backcrossed with wheat and ovaries from some of the others were picked out and cultured in medium with $2\,\text{mg}\,\text{l}^{-1}$ 2,4-D to induce calli and dedifferentiated into "secondary hybrids" in medium containing $0.5\,\text{mg}\,\text{l}^{-1}$ IAA + $0.5\,\text{mg}\,\text{l}^{-1}$ zeatin. Only one plant set seed after backcross, but it was so leptosome that it could not germinate in soil.

Fig. 21. No. 4-1 hybrid plant regenerated from hybrid no. 4 calli

Fig. 22. A mature hybrid plant which has overwintered

6 Summary and Conclusion

As mentioned above, the wheat suspensions used for protoplasts preparation had no regeneration ability and the *H. villosa* calli before irradiation had only 10% regeneration frequency (no. of calli differentiated to plant/no. of total calli tested). However, the fusion products regenerated to plants with rather high frequency. This fact could be explained in that the regeneration capacity of each parent could be complemented through fusion. This complementary effect would be particularly advantageous for somatic hybridization between wheat and grasses, because their cultures were apt to lose the ability for regeneration during subculture. The success of plant production in asymmetric somatic hybridization between wheat and three other closely related intergeneric grasses performed successively (*Leymus chinesis, Agropyron elongatum*, and *Psathyrostachys juncea*; Xia et al. 1996a,b) revealed the common presence of this effect in such fusion combinations. Therefore, this effect is significant for hybrid plant reproduction in the somatic hybridization of wheat, and great efforts should be made to study mutual reactions between both parents, focusing on the cell genetic and molecular level to explain the mechanism of this effect.

The complementary effect played an important role not only in hybrid plant regeneration, but also in their selection, because only hybrid plants could be regenerated after fusion when the parental protoplasts were not regenerated or were low in regeneration capacity.

In this work we found the phenomenon of growth acceleration in hybrid clones. This phenomenon had been observed earlier in experiments (Smith 1976; Shieder 1978; Gleba 1982) and was proposed to be applied as a method for hybrid selection in fusion. This strategy was most successively employed afterwards in somatic hybridization of potato, where hybrid plants could be selected with high efficiency simply by this method (Polgar et al. 1993). In combination A of our experiment the early-formed clones (nos. 1–12) were all hybrids, as confirmed by isozyme and chromosome analysis. Similar results were also observed in subsequent work. This phenomenon of vigorous growth of hybrids offered us a simple method for wheat hybrid selection.

The concentration of IOA played a critical role in the growth and development of the fusion products, as shown in Table 5.

The above result indicates that clone and plant formation was sensitive to the concentration of IOA applied to wheat protoplasts. Since hybrid selection may be accomplished by utilizing the vigorous growth effect of the fused calli, as stated above, in the case of wheat hybridization between a parent of low or no regeneration capacity, fusion without IOA pretreatment on wheat protoplasts followed by "double selection" insured the best yield of regenerated hybrid plants. Moreover, the experimental procedure was quite simple. We adopted this strategy in analogous fusion combinations of wheat (+) *Agropyron elongatum* as well as wheat (+) *Psathyrostachys juncea* (Xia et al. 1996b) and wheat (+) *Leymus chinesis* (Xia et al. 1996a) and obtained satisfisfactory results.

Table 5. The relation between the inactivation degree of wheat protoplast (acceptor) and the growth and development of fusion products

IOA conc. (mmol l^{-1})	Inactive state of wheat protoplasts	Development of fusion products
0	First division at 3–4th day of culture, divided vigorously	Differentiated to strong plants with high frequency (combination A)
1.0	Divided 1–2 times, no sustained division	Differentiated to very weak plants with very low frequency (combination B)
1.5	Some protoplasts swelled, no division	Slow-growing small calli (combination C)
2.5	Died after 2–3 days	Small cell colonies (combination D)

Although this experiment succeeded in hybrid plant regeneration from asymmetric somatic hybridization of wheat and *H. villosa*, many problems need to be resolved before its application to the practical work of wheat breeding such as the fertility of hybrid plants, etc.

7 Protocol

7.1 Preparation and Fusion of Parental Protoplasts

Protoplasts of wheat were prepared from suspension-cultured cells, which had been established for 3 years and were subcultured every 6–7 days, according to the method already described (Li et al. 1992). For isolation of protoplasts, cell clusters (>0.2 mm) were collected from 3-day-old suspension culture and put in an enzyme solution (2% Cellulase Onozuka RS, 0.3% Pectolyase Y-23, 5 mM $CaCl_2$, and 0.6 M mannitol, pH 5.8) in the volume proportion of 1:4. After 4–5 h of incubation in the dark at 25°C with occasional shaking, protoplasts were filtered through a 45-μm stainless mesh and centrifuged at 100 g for 5 min. After being washed with washing solution (5 mM $CaCl_2$ and 0.6 M mannitol, pH 5.8) twice, different amounts of IOA were added to each tube containing protoplasts until the final concentrations reached 1, 1.5, and 2.5 mM, respectively. The protoplasts stayed in IOA for 15 min (including the time of centrifugation) at room temperature (23–25°C). They were washed three times with washing solution and finally adjusted to a density of 10^6 ml^{-1} before use.

H. villosa protoplasts were prepared from calli induced from immature embryos on MB medium containing 2 mg l^{-1} 2,4-D, and subcultured every 12–13 days on the same medium. One-year-old calli with pale yellow granular structures after 4–5 days of subculture were used. After γ-irradiation, they were chopped into small pieces before isolation. The procedure of protoplasts preparation was the same as in wheat.

Wheat protoplasts were mixed with *H. villosa* protoplasts in a 1:1 ratio at a density of 10^6 ml^{-1} and fused according to Lindsey (1991) with slight modification. Fusion was performed in 3-cm petri dishes. Each portion of a 0.15–0.2-ml mixture was placed on a petri dish to form a thin layer in the middle of the dish and left for 15–20 min. Then an equal volume of PEG solution [2% glucose, 1.5% $Ca(NO_3)_2$, 40% PEG (6000), pH 7] was dropped in around the boundary of the layer. After 15 min, a twofold volume of 0.275 M $Ca(NO_3)_2$ (pH 7) solution was dropped in, and 10 min later this step was repeated once for the aggregation of the protoplasts. After removing the liquid in the dish, the protoplast mixture was washed twice with washing solution and once with P_5 liquid medium. Finally, it was cultured in P_5 liquid medium.

7.2 Culture of Fusion Products

The fusion products in petri dishes were placed in the dark at 25 °C for about 1 month until the clones grew to a size of about 1.5–2 mm. They were picked out successively in lots in the order of size and each one was put on solid proliferating medium (MB medium containing 1 mg l^{-1} 2,4-D) in a 50-ml Erlenmeyer flask. Differentiation of the proliferation calli was performed by transferring them to MB medium containing 0.5 mg l^{-1} IAA, and 0.5 mg l^{-1} zeatin under 2000 lx fluorescent light 12 h day^{-1}. When two to three leaflets with a height of 1–2 cm were formed, they were retransferred to the same medium containing 0.5 mg l^{-1} NAA and 0.1–0.2 mg l^{-1} Paclobutrozol for rooting and strengthening. Care must be taken to avoid staying too long in this medium or increasing the Paclobutrozol concentration, otherwise the seedlings aged quickly and hardly grew afterwards. About 3–4 weeks later, they were potted in soil and placed in a growth chamber at about 5–10 °C under 12 h day^{-1} of 1000 lx fluorescent light for 2–3 weeks, followed by transfer to the greenhouse.

7.3 Identification of Hybridity of the Regenerated Calli and Plants

7.3.1 Chromosome Inspection

Calli, root tips, and young leaf bases of the regenerated plants were incubated at 4 °C for 24 h, then fixed at room temperature with acetic alcohol (99% ethanol : acetic acid = 3 : 1). After washing with distilled water, the fixed samples were softened in an enzyme solution containing 1% Cellulase Onozuka RS (YakultHonsla Co. Ltd, Tokyo), 0.2% Pectolyase Y-23 (Seishi Co., Tokyo) for 1–3 h, washed again; and then kept in the water for 0.5–1 h. The samples were refixed with a small amount of acetic acid – alcohol for 30 min and pounded into suspensions with a round-headed glass rod. A few drops of the upper suspensions were spread on a glass slide and dried by flame. Chromosomes were stained with a 5% Giemsa solution for 30 min.

7.3.2 Isozyme Analysis

The samples were ground in 0.1 M Tris-citric acid buffer (pH 8.2) on an ice bath. The homogenate was centrifuged at 5000 g for 10 min and the supernatant was mixed with equal volume of 10% glycerin; 50–100 µl of each mixture was loaded onto 10–4% polyacrylmaid gel and run for 4–5 h at 4 °C and 30 mA. Gels were stained for esterase and amylase following the procedure described previously (Hu 1985).

7.3.3 RAPD Analysis

Total genomic DNA of hybrid and parents was isolated from young leaves according to the CTAB method (Sambrook et al. 1989). PCR reaction was performed in a 15-µl reaction volume containing 20 µM Tris-HCl, pH 8; 50 mM KCl; 3 mM MgCl$_2$; 250 µM each of dATP, dGTP, dCTP, dTTP; 15 ng primer (Operon Technology Inc, USA); 1 U Ampli Taq polymerase (Promega Crop. USA) and 50 ng of genomic DNA. The thermal cycling sequence was 95 °C for 5 min, 35 °C for 2 min, and 72 °C for 3 min followed by 43 cycles of 94 °C for 10 s; 38 °C for 10 s and 72 °C for 50 s. Amplification products were loaded on 1% agarose gels and electrophoresed for 3–4 h. Gels were stained with ethidium bromide and visualized under UV light and photographed.

Acknowledgements. The project was supported by the National Natural Science Foundation of China: No.30070397 and Trans-century Training Program Foundation for the Talents by the Ministry of Education.

References

Gleba YY, Momot VP, Cherep NN, Skarzynskaya (1982) Intertribal hybrid cell lines of *Atropa belladonna* (+) *Nicotiana chinensis* obtained by cloning individual protoplasts fusion products. Theor Appl Genet 62:75–79

Hu NS, Wan GX (1985) Application of isozyme techinique. Hunan Science and Technology Publishers, ChangSha, China, pp 96–104

Li ZY, Xia GM, Chen HM (1992) Somatic embryogenesis and plant regeneration from protoplasts isolated from embryogenic cell suspension of wheat (*Triticum aestivum* L.) Plant Cell Tissue Organ Cult 28:72–83

Lindsey K (1991) Plant tissue culture manual. Section D. Kluwer, Dordrecht, pp 1–17

Liu DJ, Chen PD, Pei GZ (1983) Studies on transfer of genetic material from *Haynaldia* to *Triticum aestivum*. Acta Genet Sin 10(2):103–113

Polgar ZS, Preiszner J, Dudits D, Fehe'r A (1993) Vigorous growth of fusion products allow highly efficient selection of interspecific potato somatic hybrids: molecular proofs. Plant Cell Rep 12:399–402

Preiszoner J, Feher A, Veisz Stuka J, Dudits D (1991) Characterization of morphological variation and cold resistance in interspecific somatic hybrids between potato (*Solanum tuberosum* L.) and *S. brevidens* Phil. Euphytica 57:37–49

Sambrook J, Fritsch EF, Maintis T (1989) Molecular cloning, a laboratory manual (2nd edn). Cold Spring Harbor Laboratory Press, New York, pp 9.16–9.19

Schieder O (1987) Somatic Hybrids of *Datura irnoxia* Mill + *Datura discolor* Bernn and of *Datura innoxia* Mill + *Datura stramonium* L.var tatula L. I. selection and characterization. Mol Gen Genet 162:113–119

Smith HH, Kao KN, Combatti NC (1976) Interspecific hybridization by protoplasts fusion in *Nicotiana* conformation and extension. J Hered 67:123–128

Terada R, Kyozuka J, Nishibagashi S, Shimamoto K (1987) Plantlet regeneration from somatic hybrids of rice (*Oryza sativa* L.) and barnyard grass (*Echinochloa oryzicola* Vasting). Mol Gen Genet 210:39–43

Warra S, Glimlius K (1995) The potential of somatic hybridzation in crop breeding. Euphytica 85:217–233

Xia GM, Chen HM (1996a) Plant regeneration from intergenric somatic hybridization between *Triticum aestivum* L. and *Leymus chinensis* (Trin.) Tzvel. Plant Sci 120:197–203

Xia GM, Li ZY, Zhou AF, Guo GQ, Chen HM (1995) Plant regeneration of wheat protoplasts prepared from different composition of suspension culture. Chin J Biotechnol 11(1):289–294

Xia GM, Wang H, Chen HM (1996b) Plant regeneration from intergeneric somatic hybridization between wheat (*Triticum aestivum* L.) and Russian wild rye (*Psathyrostachys juncea* (Fisch) Nevski.) and couch grass (*Agropyron elongatum* (Host) Nevski.) Chin Sci Bull 41(16):-1382–1386

Yang ZQ, Shikanai T, Yamada Y (1988) Asymmetric hybridization between cytoplasmic male-sterile (CMS) and fertile rice (*Oryza sativa* L.) protoplasts. Theor Appl Genet 76:801–808

Yang ZQ, Shikanai T, Yamada Y (1989) Plant regeneration from cytoplasmic hybrids of rice (*Oryza sativa* L.) protoplasts. Theor Appl Genet 77:305–310

Zhou AF, Xia GM, Chen HM (1996) Effect of ^{60}Co-γ radiation on calli of *Haynaldia villosa*. Chin J Biotechnol 12 (Supple):127–130

I.5 Asymmetric Somatic Hybridization Between *Triticum aestivum* L. (Wheat) and *Leymus chinensis* (Trin.) Tzvel

G.M. XIA, A.F. ZHOU, F. XIANG, and H.M. CHEN

1 Introduction

Wheat (*Triticum aestivum* L.) is a cultivated crop with a narrow genetic base. Due to long-term cultivation it is now difficult to enhance its yield and quality to a great extent by conventional breeding for the increasing need of food in the world. It is necessary to introduce exogenous beneficial traits from related plants to overcome this difficulty.

Somatic hybridization by protoplast fusion has offered the possibility for increasing genetic variability and creating hybrids across sexual barriers. Many intra- and interspecific hybrids, as well as some intergeneric somatic hybrid plants, have been created by this technique. Recently, the production of an interfamilial somatic hybrid plant between barley and carrot was reported (Kisaka et al. 1997). Among the most important cereal crops, intergeneric hybrids have been recovered by this technique for rice, but then died young (Terada et al. 1987).

Leymus chinensis is a forage grass of high quality and resistant to cold, drought, salinity, and many diseases. In this work we try to transfer some of its favorable traits to wheat via asymmetric somatic hybridization.

2 Preparation and Inactivation of Wheat Protoplasts

Protoplasts of wheat cv. Jinan 177 were isolated from the young embryo-derived cell suspensions as described previously (Li et al. 1992). Due to long-term subculture, the suspensions were capable of regenerating to proembryos and globular embryos, which could not develop further. Portions of protoplasts were treated separately with iodoacetamine (IOA) at concentrations of 1.25 and 2.5 mM.

School of Life Science, Shandong University, Jinan 250100, P.R. China

3 Preparation of *L. chinensis* Protoplasts from Irradiated Calli

L. chinensis ($2n = 28$, Fig. 1A) protoplasts were used as donor in the experiment. They were isolated from nonregenerable 2-year-old calli which had been irradiated by ^{60}Co-γ-ray for inducing chromosome elimination and variation in their cells. The optimal irradiation condition was determined by a preliminary test, in which different doses at 30, 50, and 100 Gy with dose rate $1.3\,\text{Gy}\,\text{min}^{-1}$ and $3.9\,\text{Gy}\,\text{min}^{-1}$ were applied to the calli, which had been subcultured for 2, 5, and 8 days. Chromosome variation as well as the

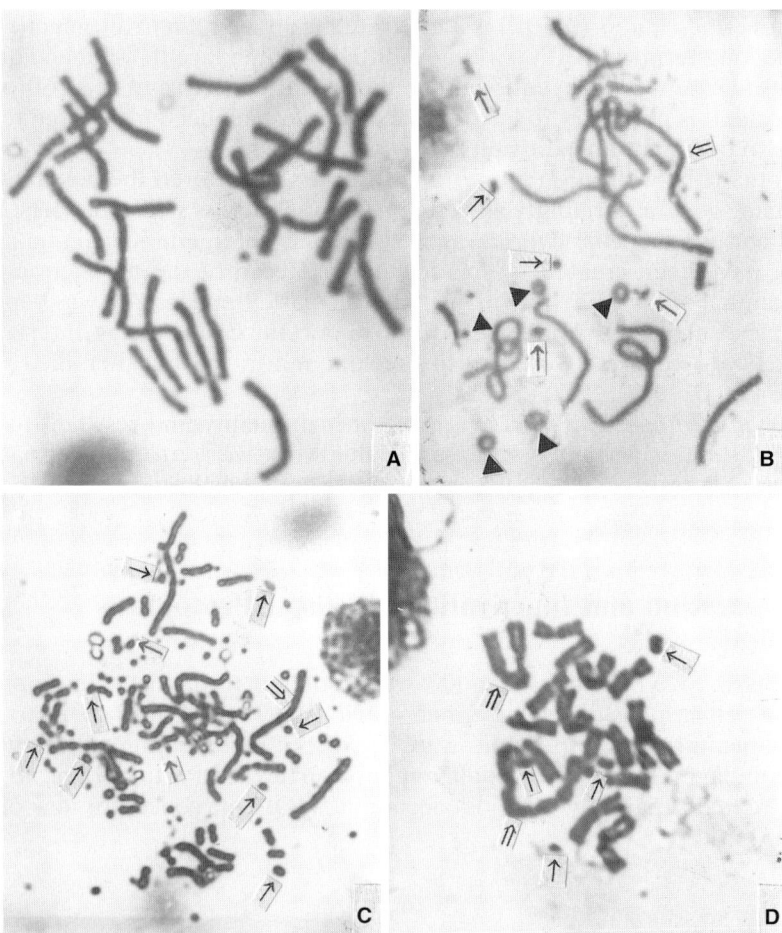

Fig. 1. A Normal chromosome set of *L. chinensis*, $2n = 28$. **B, C** Chromosomes of *L. chinensis* cell irradiated by ^{60}Co γ-ray at a dose of 50 Gy and dose rate of $3.9\,\text{Gy}\,\text{min}^{-1}$. **D** Chromosomes of *L. chinensis* cell irradiated by ^{60}Co γ-ray at a dose of 50 Gy and dose rate of $1.3\,\text{Gy}\,\text{min}^{-1}$. ↑: chromosome fragment; ⇑: dicentric or polycentric chromosome; △: chromosome ring

inhibition of division were investigated. The results are described in Sections 3.1 and 3.2.

3.1 The Structural Aberration of Chromosomes Induced by ^{60}Co-γ-Ray

All the doses and dose rates of ^{60}Co-γ-ray in the experiments induced different sorts of aberration of chromosomes in the culture cells, such as chromosome fragmentation (Fig. 1B,C,D), chromosomal ring (Fig. 1B), and dicentric or polycentric chromosomes (Fig. 1D). As shown in Tables 1 and 2, the frequency of aberration was related to dose and dose rate, and the most frequently observed aberration was chromosome fragmentation, which was beneficial to asymmetric hybridization because it led to intact chromosome elimination and the chromosome fragments thus formed could be easier to recombine with the chromosomes of the acceptor cells during or after fusion.

3.2 Effect of γ-Ray on the Cell Division

Calli subcultured for 5 days were irradiated by γ-ray at a dose rate 1.3 Gy min^{-1} and total doses were 30, 50, and 100 Gy, respectively. Frequencies of cell division at 1–6 days after irradiation are listed in Table 3.

Different calli subcultured for 2, 5, and 8 days were irradiated at 3.9 Gy min^{-1} and 50 Gy γ-ray, respectively. Division frequencies at 1–4 days after irradiation are shown in Table 4.

The above experiments showed that low frequency of chromosome aberration was induced by 30 Gy γ-ray (Table 1); on the other hand, 100 Gy γ-ray injured the cells so much that only a few of them could divide (Table 3). Both doses were not suitable for asymmetric fusion.

γ-Ray at 50 Gy (3.9 Gy min^{-1}) could initiate fragmentation in 100% of the cells (Table 2) without too much injury (they retained a division frequency of about 50% of the control) (Tables 3 and 4). It was also found that cells

Table 1. Frequency of chromosomal structural aberration in dividing cells after γ-ray irradiation at different doses and dose rates

Dose	No. of dividing cells observed	No. of cells with chromosomal structural aberration	Variation frequency (%)
30 Gy (1.3 Gy min^{-1})	98	25	25.6
50 Gy (1.3 Gy min^{-1})	105	48	45.6
100 Gy (1.3 Gy min^{-1})	102	102	100
50 Gy (3.9 Gy min^{-1})	101	101	100

Note: Calli subcultured for 5 days were used as material for irradiation. Variation frequency (no. of cells with variation/no. of observed cells) was calculated on the 4th day after irradiation.

Table 2. Major type and frequency of chromosomal structural variation at different doses of γ-ray irradiation

Dose of irradiation	No. of dividing cells observed	No. of intact chromosomes	Chromosome fragmentation			Dicentric chromosomes		Chromosome ring	
			No. of fragments	No. of cells	Freq. (%)	No. of cells	Freq. (%)	No. of cells	Freq. (%)
50 Gy (1.3 Gy min^{-1})	30	14–24	2–18	27	90.0	5	16.7	2	6.67
100 Gy (1.3 Gy min^{-1})	30	10–17	10–45	30	100	10	33.3	3	10.0
50 Gy (3.9 Gy min^{-1})	30	11–20	8–27	30	100	6	20.0	8	26.7

Note: Calli subcultured for 2 days were used as materials for irradiation. Frequencies were calculated on the 4th day after irradiation.

Table 3. The effect of different irradiation doses on cell division (dose rate 1.3 Gy min^{-1})

Days after irradiation	No. of observed cells	No. of dividing cells			
		0 Gy	30 Gy	50 Gy	100 Gy
1	3000	78	0	0	0
2	3000	17	14	7	0
3	3000	5	29	25	2
4	3000	9	48	45	19
5	3000	7	52	31	15
6	3000	8	7	5	2

Note: The calli used had been subcultured for 5 days when the cell division peak occurred.

Table 4. Effect of γ-ray irradiation (50 Gy, 3.9 Gy min^{-1}) on cell division in calli subcultured for different days

Days after irradiation	No. of observed cells	No. of dividing cells in calli subcultured for:		
		2 days	5 days	8 days
1	3000	0	0	9
2	3000	0	0	3
3	3000	0	10	0
4	3000	37	40	35

subcultured for 2–5 days were more sensitive to irradiation on cell division than those subcultured for 8 days. Therefore, calli subcultured for 2–5 days were used in this experiment for γ-ray irradiation at a dose of 50 Gy (3.9 Gy min^{-1}). Protoplast isolation and fusion were carried out 1–3 days after irradiation.

4 Fusion of Parental Protoplasts and Culture of Fusion Products

The recipient and donor protoplasts were fused by a modified PEG method (Lindsey 1991) and three combinations were carried out:

A) Wheat protoplasts (T) (+) *L. chinensis* protoplasts (L),
B) Wheat protoplasts treated with 1.25 mM IOA ($T_{1.25}$) (+) L,
C) Wheat protoplasts treated with 2.5 mM IOA ($T_{2.5}$) (+) L.

The controls included the cultures of T, $T_{1.25}$, $T_{2.5}$, T + L (mixture), $T_{1.25}$ + L, $T_{2.5}$ + L, T (+) T, and L (+) L.

All the fusion products and controls were cultured in P_5 medium (Xia et al. 1996b). The fusion products in combination (A) grew to small calli after 40–50 days of culture, and were transferred successively according to their size to the medium for proliferation and then for differentiation. Among the 16 early-formed cell clones, 10 exhibited phenotypes similar to that originating from wheat protoplasts (T type) and 6 were middle type (M type) with pale yellow color (like the wheat calli) and dense structure (like *L. chinensis* calli). The T-type clones could not differentiate, while the M type were capable of regeneration. However, the time required for shooting was in each case quite different. In the earliest-formed clone, no. 1, the first leaf began to emerge about 28 days after transfer to the differentiation medium, while later clones required 52–58 days. A difference also existed in regeneration capacity among the clones. More seedlings were differentiated from the early-formed clones than the later. For example, clone no. 1 produced 17 regenerated plants while no. 6 produced only one. In combination (B), the fusion products grew much more slowly and only one differentiated small leaflets, indicating that IOA had a harmful effect, as seen in wheat (+) *Haynaldia villosa* (Zhou et al. 1996). In combination (C), no calli of the size of 1–2mm could be obtained.

All controls produced no regenerated plants.

5 Identification of Hybrid Calli and Plants

The chromosome numbers, isozyme patterns of esterase and peroxidase, as well as the morphology of the regenerated calli and plants were analyzed and the results are summarized in Table 5.

The chromosome numbers of wheat protoplast-derived calli were mainly 31–38, and chromosome fragments were rarely observed. In the regenerated calli of M type, the number of chromosomes (40–68) (Fig. 2B,C) were more than in wheat (Fig. 2A) and chromosome fragment(s) could be observed in nearly all the dividing cells. This indicated that genetic material of *L. chinenses*

Table 5. Morphology, cytology, and isozyme patterns of fusion products

Fusion combination	Cell clone	Morphology	No. of chromosome	Esterase isozyme	Peroxidase isozyme
A	1 (Calli)	M	43–62	P,N	P,N
A	2	M	40–60	P	P
A	3	M	42–58	–	–
A	4	M	43–48	–	–
A	5	M	50–56	–	–
A	10	M	31–38	T	T
B	17	M	50–68	–	–
A	1–3 (Plant)	See text	44–58	P	P
A	2–1		43–56	P	P
A	3–1		46–49	P	P,N
A	4–1		–	P	P
A	5–1		–	P	P

–, Not examined; M, middle type; T, wheat type; P, containing partial band(s) of both parent; N, containing new band(s).

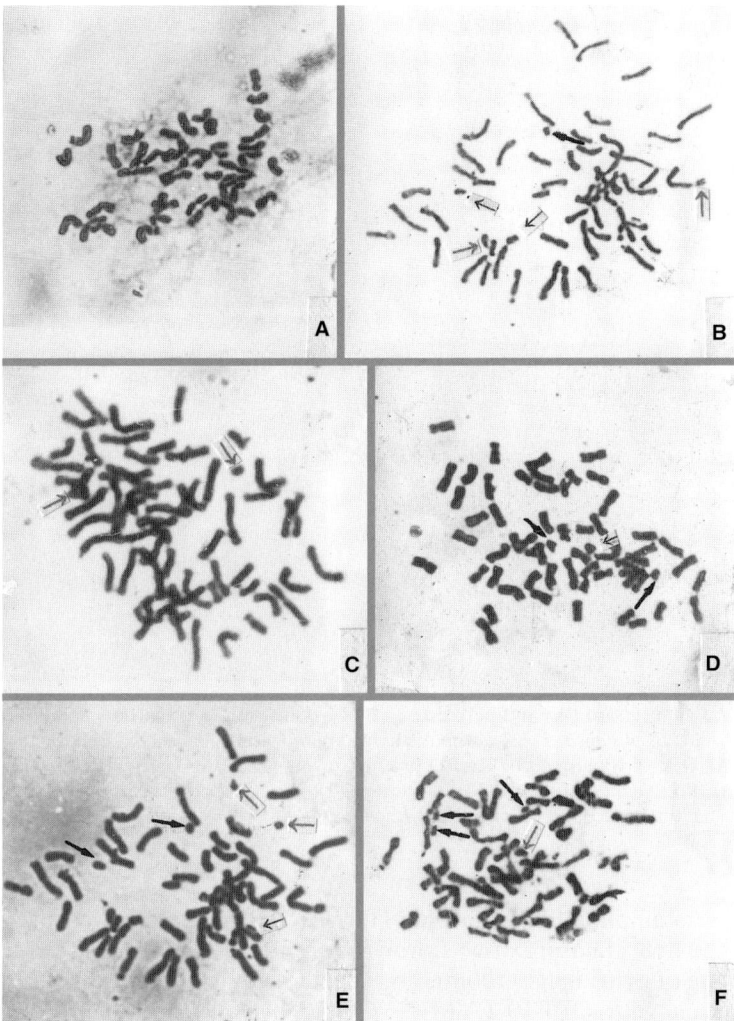

Fig. 2. A Chromosomes of *T. aestivum*. **B, C** Chromosomes in no. 1 and no. 2 cell line. ↑: chromosome fragment. **D, E, F** Chromosomes in the root tip cell of nos. 1-3 (**D**); 2-1 (**E**); 3-1 (**F**), ($2n > 42$); ↑: chromosome fragment

had entered into wheat cells. The analysis of isozyme patterns showed that M-type calli had bands characteristic of both parents.

As for the regenerated plants, the chromosome numbers examined were 43–58 (Fig. 2D,E,F) and chromosomal fragments also existed. This revealed that the foreign genetic material was still retained in the fused cells through differentiation. Both esterase and peroxidase isozyme analyses of regenerated plants gave results similar to those of calli (Fig. 3A,B).

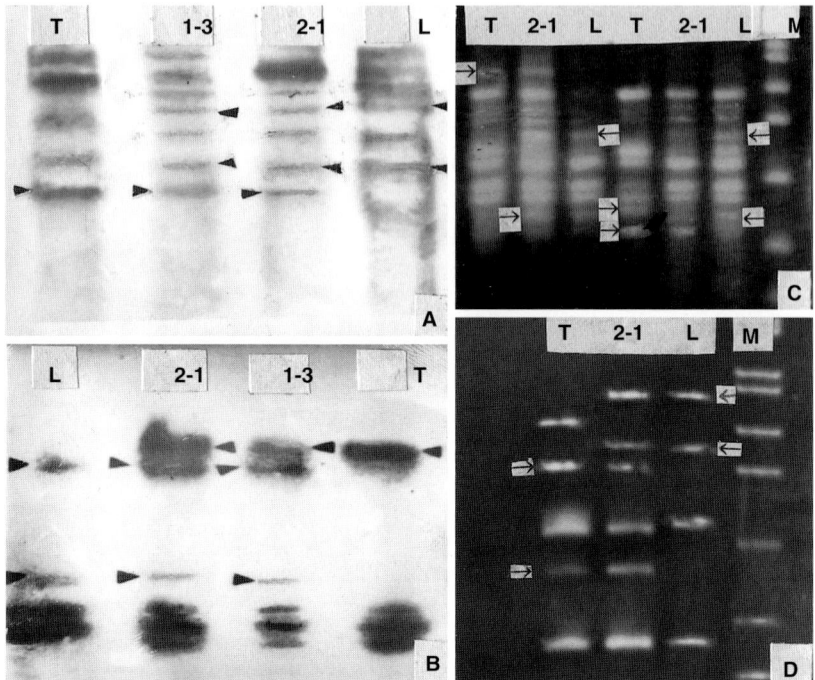

Fig. 3A,B. Leaf esterase (**A**) and peroxidase (**B**) isozyme pattern of parents and hybrid plants; **T** *T. aestivum*; **L** *L. chinensis*; *1-3, 2-1* somatic hybrid plants nos. 1-3 and 2-1; △ characteristic band of parent. **C, D** Electrophoresis of RAPD products using OPJ-10 (**C** *left*), OPJ-11 (**C** *right*), and OPJ-13 (**D**); **T** *Triticum aestivum*; *2-1* W (+) L no. 2-1 hybrid; **L** *L. chinensis*; *M* molecular weight marker; ↑ characteristic band of wheat and *L. chinensis*

Random amplified polymorphic DNA (RAPD) analysis was also used to identify the hybrid nature. It was shown from the electrophoresis profiles that the genome of hybrid plants contained specific sequences of both parents after amplification with OPJ-10 (AAGCCCGAGG), OPJ-11 (ACTCCTGCGA) and OPJ-13 (CTGGGGCTGA) among 18 arbitrary primers used in this experiment (Fig. 3C,D).

More than 20 seedlings were potted in soil. Three of them remained alive and two overwintered outdoors. They matured, but did not flower. Their morphology differed from each other and from their parents (Fig. 4A,B). One hybrid had profuse, crowded shoots (Fig. 4C), while the other bore leaves decanting downward from the base of the leaf blade (Fig. 4D); however, they grew hardly at all in the hot summer and finally died.

Fig. 4A–D. Mature parents and hybrid plants. **A** Wheat plant regenerated from subcultured calli of cv. Jinan 177. **B** *L. chinensis* plant growing in soil. **C, D** no. 1-3 (**C**) and no. 2-1 (**D**) plants regenerated from no. 1 and no. 2 hybrid cell lines

6 Summary and Conclusion

Hybrid plants between wheat and *L. chinensis* were recovered via fusion of γ-ray–irradiated protoplasts in this experiment, with methods and results similar to those of wheat (+) *Hynaldia villosa* (Zhou et al. 1996). Complementary effects in regenerating capacity between parental protoplasts as well as a vigorous growth effect of the hybrids were all observed in both cases; the role of these effects has been discussed in the former papers (Xia et al. 1996a,b; Zhou et al. 1996).

Calli instead of protoplasts originating from suspension cells were irradiated by γ-ray in both experiments. This procedure has the advantage of avoiding the establishment of a grass suspension, which is always difficult to achieve. Besides, when donor protoplasts are irradiated directly, it is difficult to ascertain the optimum irradiation conditions if they cannot divide as in these two experiments. Moreover, calli once subjected to radiation may provide several successive lots of donor protoplasts over a period of time, which is very useful in the study of chromosome behavior in fusion of γ-ray–irradiated protoplasts.

Although γ-ray is a good radiation source and usually used in asymmetric somatic hybridization, no regularity has been found for the relationship between irradiation dose and chromosome elimination. One reason for this irregularity may be the different stability of chromosomes against irradiation in different plants. In the above two fusion experiments of γ-ray–irradiated protoplasts, we found many chromosome fragments (Fig. 1C) in the cells of *L. chinensis* ($2n = 28$) when irradiated at 50 Gy (1.3 Gy min^{-1} or 3.9 Gy min^{-1}) but only a few fragments formed in *H. villosa* ($2n = 14$) at the similar dose (50–60 Gy, 1.3 Gy min^{-1}). When the irradiation dose was raised (>70 Gy), the cells of *H. villosa* died without division. In another work (unpubl. obs.), we also found that numerous fragments existed in the dividing cells of *Agropyron elongatum* ($2n = 70$) irradiated by 50–100 Gy (3.9 Gy min^{-1}) (Fig. 5). From the above three instances, it seems that the more chromosome numbers the cell has, the more fragments may be induced by γ-ray. This is interesting and worth further examination with different groups of closely related plants.

Our experiment denotes that although somatic hybridization makes direct mixing of two sets of the parental genome possible, normal hybrid plant development as well as normal seed production are still the major problems hindering the progress of this technique. Many factors influence the success of somatic hybridization, such as the phylogeneical distance of the parents, the cell cycle of the cells at fusion, the chromosome numbers and behavior of the

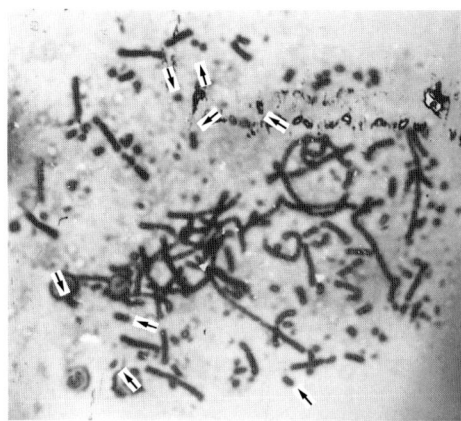

Fig. 5. The chromosomes of *Agropyron elongatum* cell irradiated by ^{60}Co-ray at dose 50 Gy and dose rate 3.9 Gy min^{-1}, ↑: chromosome fragment, ↑: chromosome fragment

fusion partners, as well as the effect of physical or chemical agents on the protoplasts and culture conditions, etc. In short, a normal development of the hybrids requires the hormonic cooperation of both genetic materials. Therefore, theoretical investigations aimed at the interaction of two sets of genomes in the course of hybrid development at cytological and molecular levels are urgently needed, which will greatly promote the practical application of somatic hybridization in plant breeding.

7 Protocol

7.1 Origin and Inactivation of Wheat Protoplasts

Suspensions of wheat cv. Jinan 177 were derived from granular embryogenic calli, which originated from young embryos according to the method described by Li et al. (1992). The protoplasts were isolated from the suspension cells with enzyme mixture containing Onozuka RS 2%, Pectolyse Y-23 0.3%, $CaCl_2$ 5mM, mannitol 0.6M. After purification (Li et al. 1992), portions were treated with 1.25 and 2.5mM IOA, respectively, at room temperature (~25 °C) for 15 min. After washing with washing solution ($CaCl_2$ 5 mM, mannitol 0.6 M) three times, they were resuspended in the same solution and ready for use.

7.2 Origin and Irradiation of *L. chinensis* Calli and Their Protoplast Preparation

Young embryo-derived calli of *L. chinensis* were initiated and subcultured on MB medium (Xia et al. 1996b) containing 2,4-D 2 mg l^{-1} as for the method in wheat (Li et al. 1992). In a preliminary test, 2-year-old calli after being subcultured for different days were irradiated by ^{60}Co γ-ray at different doses and dose rates.
According to the result of the preliminary test, *L. chinensis* calli were irradiated at 50 Gy (3.9 Gy min^{-1}) ^{60}Co γ-ray, and then transferred to a fresh medium in the dark and isolated for 1–3 days after irradiation by the same method as in wheat.

7.3 Fusion and Culture of Protoplasts

Both parental protoplasts were adjusted to 10^6 ml^{-1} and mixed in equal volume, a portion of each 0.15–0.2-ml mixture was placed on a 3.5–4.5 cm petri dish to form a thin layer and remained for 10 min. The procedure described in Lindsey (1991) was carried out.
All fusion products and controls were cultured in the dark at 25 °C in P_5 medium (Xia et al. 1996b). When the fusion products grew to small calli, they were transferred to MB medium (Xia et al. 1996b) containing 0.5–1 mg l^{-1} 2,4-D for proliferation, then transferred to MB medium with 0.5 mg l^{-1} IAA and 0.5 mg l^{-1} zeatin for regeneration. They were put under 12 h fluorescent light at 2000 lx. Seedlings of about 2 cm in height were transferred to MB medium with 0.5 mg l^{-1} NAA and 0.1–0.2 mg l^{-1} Paclobutrozol for strengthening and rooting, and potted in soil about 3–4 weeks later.

7.4 Identification of Hybrid Nature

7.4.1 Chromosome Counting

The calli and root tips or young leaf bases of regenerated plants derived from different cell clones were used for chromosome analysis according to the method described by Xia et al. (1996b).

7.4.2 Isozyme Analysis

The regenerated calli and young leaves of putative hybrids as well as the parents were used for esterase and peroxidase isozyme analysis according to the procedure already described (Xia et al. 1996b).

7.4.3 RAPD Analysis

Total genomic DNA was isolated from young leaves of putative hybrids and parents according to the CTAB method (Sambrook et al. 1989). PCR reaction was performed in 15 µl reaction volume containing 20 mM Tris-HCl, pH 8; 50 mM KCl; 3 mM $MgCl_2$; 250 µM each of dATP, dTTP, dCTP, and dGTP; 15 ng primer (10-mer primer, Operon Technology, USA); 1 U Taq polymerase (Promega Crop, USA), and 50 ng genomic DNA. The thermal cycling was performed as follows: 95 °C for 5 min, 35 °C for 2 min, and 72 °C for 50 s followed by 43 cycles at 94 °C for 10 s; 38 °C for 10 s, and 72 °C for 50 s. Amplified products were loaded on 1% agarose gels and electrophoresed for 3–4 h. Gels were stained with ethidium bromide and observed under UV light. Eighteen arbitrary primers were used in this experiment: OPA-01, OPA-09, OPA-11~OPA-15, OPA-19, OPA-20, OPJ-01, OPJ-03, OPJ-05, OPJ-07, OPJ-09~OPJ-13.

Acknowledgements. The project was supported by the National Natural Science foundation of China: No. 30070397 and Trans-century Training Program Foundation for the Talents by the Ministry of Education.

References

Kameya T (1989) The complementary restoration of plant regeneration, and flower initiation ability through cell fusion. Cytologia 54:385–388
Kisaka H, Kisaka M, Kanno A, Kameya T (1997) Production and analysis of plants that are somatic hybrids of barley (*Hordeum vulgar* L.) and carrot (*Daucus carrot* L.). Theor Appl Genet 94:221–226
Li ZY, Xia GM, Chen HM (1992) Somatic embryogenesis and plant regeneration from protoplasts isolated from embryogenic cell suspension of wheat (*Triticum aestivum* L.). Plant Cell Tissue Organ Cult 28:73–85
Lindsey K (1991) Plant tissue culture manual, Section D. Kluwer, Dordrecht, pp 1–17
Sambrook J, Fritsch EF, Maintis T (1989) Molecular cloning, a laboratory manual, 2nd end. Cold Spring Harbor Laboratory Press, New York, pp 9.16–9.19.
Terada R, Kyaruka J, Nishbayashi V, Simamoto K (1987) Plantlets regeneration from somatic hybrids of rice (*Oryza sativa* L.) and barnyard grass (*Echinochloa oryzicola* vasting). Mol Gen Genet 210:39–43

Vasil IK (1993) Molecular genetic improvement cereal and grass crops. Newsl IAPTC (Int Assoc Plant Tissue Culture) 724:2–10

Xia GM, Chen HM, Wang H (1995) Somatic hybridization and regeneration capacity complementation between common wheat (*Triticumn aestivum* L.) and *Agropyron elongatum*. J Shandong Univ 30(3):325–330

Xia GM, Wang H, Chen HM (1996a) Plant regeneration from intergeneric asymmetric somatic hybridization between wheat (*Triticumn aestivum* L.) and Russian wild rye (*Psathyrostichs juncea* (Fisch) Neweski) and couch grass (*Agropyron elongatumn* Host Neviski). Chin Sci Bull 41(160):1382–1386

Xia GM, Chen HM (1996b) Plant regeneration from intergeneric somatic hybridization between *Triticumn aestivum* L. and *Leymus chinensis* (Trin.) Tzvel. Plant Sci 120:197–203

Zhou AF, Xia GM, Chen HM (1996) Asymmetric somatic hybridization between *Triticumn aestivum* L. and *Haynaldia villosa* Schur. Sci China (Ser C) 39(6):617–626

Section II
Vegetables and Fruits

II.1 Somatic Hybridization Between *Arabidopsis* and *Brassica*

J. SIEMENS and M.D. SACRISTÁN

1 Introduction

1.1 Importance and Distribution

A large set of vegetables cultivated worldwide belong to the genus *Brassica*. Furthermore in the temperate climate zone the most important oil plant is oilseed rape (*B. napus* L.). Therefore, the agronomical relevance of the genus *Brassica* is undoubtedly very high. In contrast, mouse-ear cress (*Arabidopsis thaliana*) is just a weed, which is distributed worldwide in genetics laboratories and only useful in terms of a model plant. However, this tiny plant has extraordinary advantages, e.g., small genome, fast generation cycle, high seed yield per plant, a tight molecular map, a large set of available molecular probes as well as transgenic lines (for review see: Meyerowitz 1992; Rédei 1992; Kunkel 1996). For these reasons, this useful weed has provided insights into numerous fields of plant research.

1.2 Significance of Somatic Hybridization

In addition to sexual crosses combined with ovule or embryo culture, somatic hybridization by protoplasts fusion has become a tool in *Brassica* breeding. Many intrageneric and intergeneric somatic hybrid combinations have been examined for the transfer of agronomically important traits, such as resistance traits, which could be found in allies of the *Brassica* crops (Glimelius et al. 1991). For example, resistance genes of *B. nigra* for *L. maculans* and *P. brassicae* have been succesfully transferred into *B. napus* (Sjödin and Glimelius 1989; Gerdemann-Knörck et al. 1994). Somatic hybridization between *Sinapis alba* (*B. hirta*) and *B. napus* (Primard et al. 1988) as well as *B. oleracea* (Hansen and Earle 1997) has been used to transfer resistance genes for *Alternaria brassicae*. Even intertribal hybridization combinations with *B. napus* were realized to induce cold tolerance (Fahleson et al. 1994a) and to modify oil quality (Fahleson et al. 1994b; Skarzhinskaya et al. 1996). Fertile hybrid plants *B. napus* (+) *Thlaspi perfoliatum* could be selected and backcrossed to *B. napus*

Institute of Biology, Applied Genetics, Free University Berlin, Albrecht-Thaer-Weg 6, 14195 Berlin, Germany

(Fahleson et al. 1994a). A summary of all intertribal hybridizations in the family Brassicaceae is given in Table 1.

Using *Arabidopsis thaliana* as one partner in fusion experiments, probably no agronomically important trait can be achieved in cultivated plants, but studies with the model plant might give insights into the basic processes of introgression by somatic hybridization.

1.3 Brief Review of the Work Done

The first intertribal somatic hybrid plants between *A. thaliana* (tribe: Sisymbrieae-Arabidopsidinae) and a species of the genus *Brassica* (tribe: Brassiceae-Brassicinae) were described by Gleba and Hoffmann (1979, 1980). As early as the late 1970s, Gleba and Hoffmann (1978) fused protoplasts of *B. campestris* and *A. thaliana*. They obtained calli with a hybrid genome and later regenerated sterile plants which had intermediate morphology (Gleba and Hoffmann 1979, 1980; Hoffmann and Adachi 1981). More recently, Bauer-Weston et al. (1993) fused *B. napus* protoplasts (recipient) with irradiated *A. thaliana* cells (donor) and also obtained sterile plants with intermediate morphology. Fertile *B. napus* (+) *A. thaliana* plants, which could be backcrossed to *B. napus*, were selected (Forsberg et al. 1994; Forsberg et al. 1998a). In the experiments of Bauer-Weston et al. (1993) and Forsberg et al. (1994, 1998a) a regeneration protocol established for *B. napus* was used for the regeneration of the *B. napus* (+) *A. thaliana* somatic hybrids. A fusion and regeneration protocol using *A. thaliana* as recipient and *B. nigra* lines as irradiated donor was established by Siemens and Sacristan (1995). In this work the transgenic donor lines (*B. nigra* 184+ and 68+) were expressing a hygromycin resistance gene (Sacristan et al. 1989).

2 Somatic Hybridization

Methods for direct gene transfer to protoplasts and regeneration of fertile plants were published for different ecotypes of *A. thaliana* (Damm et al. 1989; Damm and Willmitzer 1991; Masson and Paszkowski 1992; Park and Wernicke 1993; Siemens et al. 1993; Mathur et al. 1995; Torres et al. 1997). The protoplast regeneration protocol of Damm and Willmitzer (1991) has been modified to establish a protocol for somatic hybridization.

2.1 Isolation of Protoplasts

2.1.1 Culture Conditions

Seeds were surface-sterilized by 70% ethanol for 2 min and 2.5% sodium hypochlorite solution (20% of the technical solution) for 30 min. After washing

Table 1. Summary of intertribal somatic hybridizations in the family Brassicaceae

Hybrid combination[a]	Treatment	Genome size[b] Donor/recipient (pg)	Chromosome number Donor/recipient	HF[c] (%)	Lines obtained[d]	Fertile plants obtained	AHyF[e] ($\times 10^{-6}$)	Reference
Symmetric								
B. napus (+) *Barbarea vulgaris*		2.1/2.5	38/16	17 (79)	98	−		Fahlesson et al. (1994a)
B. napus (+) *Thlaspi perfoliatum*		2.1/0.9	38/ca. 70	7 (76)	50	+		Fahlesson et al. (1994b)
B. napus (+) *Lesquerella fendleri*		2.1/nd	38/12		33	+		Skarzhinskaya et al. (1996)
B. campestris (+) *Barbarea stricta*		1.05/nd	10/16		(10/5)	+	0.01	Oikarinen and Ryöppy (1992)
B. campestris (+) *Barbarea vulgaris*		1.05/2.5	10/16		(10/5)	+	0.01	Oikarinen and Ryöppy (1992)
B. campestris (+) *A. thaliana*	Mechanical isolation of hybrid colonies	1.05/1.32 (octoploid)	20/40		3	−		Gleba and Hoffmann (1978, 1979); Hoffmann and Adachi (1981)
B. napus (+) *A. thaliana*	Chlorosulfuron-selection	2.1/0.33	38/10		2	−	0.5	Bauer-Weston et al. (1993)
B. napus (+) *A. thaliana*		2.1/0.33	38/10	8 (70)	25	+		Forsberg et al. (1994)
B. carinata (+) *Camelina sativa*			34/40	6.8	3	−	>0.5	Narasimhulu et al. (1994)
Camelina sativa (+) *B. oleracea*	Iodoacetate (donor)		40/20		14	−		Hansen (1998)
Capsella bursa-pastoris (+) *B. oleracea*	Iodoacetate (donor)	0.94/1.30	32/20			−		Sigareva and Earle (1999)

Table 1. Continued

Hybrid combination[a]	Treatment	Genome size[b] Donor/recipient (pg)	Chromosome number Donor/recipient	HF[c] (%)	Lines obtained[d]	Fertile plants obtained	AHyF[e] (×10⁻⁶)	Reference
Asymmetric								
B. napus (+) Thlaspi perfoliatum	70 Gy (donor)	2.1/0.9	38/ca. 70	7 (87)	10	+		Fahlesson et al. (1994b)
B. napus (+) Lesquerella fendleri	180–200 Gy (donor)	2.1/nd	38/12		43	+		Skarzhinskaya et al. (1996)
B. napus (+) A. thaliana	600–900 Gy (donor) and chlorosulfuron selection	2.1/0.33	38/10		15	–	2	Forsberg et al. (1994)
B. napus (+) A. thaliana	UV-light (donor)	2.1/0.33	38/10			+		Forsberg et al. (1998a)
A. thaliana (+) B. nigra	450–1720 Gy (donor) and hygromycin selection	0.33/3.9 (octoploid)	10/56–60	(ca. 28)	3	–	0.015	Siemens and Sacristán (1995)
A. thaliana (+) B. napus	Iodoacetamid (recipient) 600 kJ UV light (donor)	0.33/2.1	10/38		21	?	ca. 1	O'Neill and Mathias (1994); O'Neill, Norwich (pers. comm.)

[a] The regeneration protocol is always established for the first-mentioned species, which is taken as recipient.
[b] The genome size is given according to Fahleson et al. (1988, 1994b). The genome size of A. thaliana and B. nigra are given according to Marie and Brown (1993).
[c] HF Heterokaryon frequence estimated 1 day after fusion, the number in parentheses is the heterokaryon frequence after enrichment by "flow sorter".
[d] Lines are defined as calli with shoot regenerates.
[e] The absolute hybrid frequence (AHyF) is defined as lines in relation to fused recipient protoplasts. The values are calculated or estimated from the published data.

with sterile water they were completely dried in a lamina flow chamber. Dry seeds were stored at 4 °C. Surface-sterilized seeds were germinated on AM medium (Damm and Willmitzer 1991) under 16h light (60–80 µmol m^{-2} s^{-1}). After 3–4 weeks leaves were harvested for protoplast isolation.

2.1.2 Protoplast Isolation

Sterile leaves (1 g) from *A. thaliana* were cut into stripes in 2 ml of 0.5 M mannitol. For the enzymatic cell wall degradation, 25 ml enzyme solution [0.4% (w/v) cellulase Onozuka R-10 (Yakult Honsha Co. Ltd., Tokyo), 0.2% (w/v) Macerozyme R-10 (Yakult Honsha Co. Ltd., Tokyo), 0.4 M mannitol, 8 mM CaCl$_2$, 2 mM MES, pH 5.6] was added. After 12–16h incubation in the dark the protoplasts were sieved (60-µm gauze), washed twice with seawater, and floated on a 0.6-M sucrose cushion. After washing with seawater, the protoplasts were counted with a hemocytometer and stored on ice in W$_5$ solution (minimally for 30 min, maximally for 4h).

Protoplast yield and quality of *A. thaliana* are dependent on the age of the plants. Bolting plants should be omitted from the starting material. Masson and Paszkowski (1992) described the influence on cultivation conditions mainly by light enriched in blue and UV spectra. Park and Wernicke (1993) enriched division competent protoplasts by a Percoll gradient. However, the described simple conditions also resulted in a yield of 10^6 protoplasts g^{-1} fresh weight, plating efficiencies between 1 and 4%, and shoot regeneration frequencies of 40–60% for several *A. thaliana* ecotypes (Siemens et al. 1993; Siemens and Sacristan 1995; Torres et al. 1997). Furthermore, protoplasts of the donor plant *B. nigra* could be isolated using the same protocol.

2.2 Fusion of Protoplasts

2.2.1 Inactivation of Donor Protoplasts

B. nigra protoplasts in W$_5$ solution were inactivated in 2.2-ml Eppendorf tubes (0.5 ml, 2×10^6 protoplasts ml^{-1}) by X-rays (X-ray tube MCN 101 (Philips, NL), 90 kV, 10 mA, 28 Gy min^{-1}, half-value layer 0.06 mm Al). During irradiation the tubes were rotated at 5 rpm to avoid protoplast sedimentation.

Using this equipment, combined with feeder cell cultivation of irradiated protoplasts, dosages of 1000 and 1800 Gy were estimated for complete inactivation of *B. nigra* and *A. thaliana* protoplasts, respectively (Siemens and Sacristan 1995).

This inactivation dosage is very high compared to other published data for species of the genus *Brassica* (Menczel et al. 1987; Morgan and Maliga 1987; Yamashita et al. 1989; Sjödin and Glimelius 1989) and for *A. thaliana* (Bauer-Weston et al. 1993), which might be explained in part by differences in the cultivation system and also by the quality of X-rays used for inactivation.

2.2.2 Fusion of Protoplasts

Protoplast suspensions of donor and recipient were washed to remove the W_5 solution completely. Afterwards, the 10^6 donor cells were resuspended with 0.6 ml *A. thaliana* protoplast suspension ($2 \times 10^6 \text{ml}^{-1}$) in fusion buffer (15 mM $MgCl_2$, 0.5 M mannitol, 2 mM MES, pH5.6). The same volume of PEG solution [26% (w/v) PEG 6000, 65 mM $Ca(NO_3)_2$, 0.26 M mannitol, pH 7–9] was added to the protoplast suspension and the fusion mixture was incubated for 30 min at room temperature. It was then carefully diluted with at least 10 ml W_5 solution. The protoplasts were then washed to remove the W_5 solution completely. Finally, they were suspended in 0.3 ml 0.5 M mannitol ($2 \times 10^6 \text{ml}^{-1}$ protoplast density) and mixed with the same volume of alginate solution [2.0% (w/v) alginate, 0.4 M mannitol]. This mixture was poured on calcium agar (20 mM $CaCl_2$, 0.4 M mannitol, 1% agar) to produce alginate disks (2–3 cm in diameter). After complete polymerization the alginate disks were transferred to petri dishes (6 cm diameter) with 2 ml of medium.

2.3 Culture of Fused Protoplasts

The media were changed every 10 days. For the first 10 days each alginate disk was cultivated in 2 ml NT medium (Gleba and Hoffmann 1978) and then transferred to MI medium (Li and Kohlenbach 1982). When the first divisions were visible (10th day), the selection by $20 \mu g \text{ml}^{-1}$ hygromycin was started and it continued until the 60th day of culture. Twenty days after embedding, each protoplast disk was transferred to 2 ml MII medium (Damm and Willmitzer 1991) and to dim light (approx. $5 \mu mol\, m^{-2}\, s^{-1}$). When microcalli had a diameter of more than 1 mm (55–60 days of culture) they were picked out of the alginate matrix and transferred to SRM medium (Damm and Willmitzer 1991) and further cultivated in bright light (16 h light, $60–80 \mu mol\, m^{-2}\, s^{-1}$).

Alginate embedding was shown to be necessary for fusion cells. Park and Wernicke (1993) established a regeneration protocol for *A. thaliana* without embedding after enrichment of protoplasts. For initial cultivation of protoplasts without embedding, IMH medium (Park and Wernicke 1993) was used with limited success for PEG-treated protoplasts (Torres et al. 1997). Furthermore the alginate embedding represents a culture system with certain advantages, i.e., the cultures can be easily monitored, the replacement of media is simple and can be done without disturbing protoplast-derived colonies.

The regeneration protocol for *A. thaliana* C24 used by Siemens et al. (1993) was applied for the cultivation of PEG-treated protoplasts. The plating efficiencies of PEG-treated *A. thaliana*, *B. nigra* protoplasts, and mixtures of both, estimated after 25 days of culture, are given in Table 2. According to these experiments, NT medium was used as initial medium for cultivation of PEG-treated cell mixtures of fusion partners, because cell division was achieved even when the plating density was reduced due to severe PEG treatment (Siemens 1994).

Table 2. Plating efficiencies estimated after 25 days (PE_{25}) of *A. thaliana*, *B. nigra* 184 protoplasts, and PEG-treated mixtures of *A. thaliana* (+) *B. nigra* 184 (AB 184) and *A. thaliana* (+) *B. nigra* 68 (AB 68) in different initial media. Mean values and standard deviation of three independent experiments with five replications

Medium	*A. thaliana* (recipient) PEG-treated	*B. nigra* 184 (donor)	AB 68 (fusion)	AB 184 (fusion)
B5	1.51 ± 0.23	–	No. div.	No. div.
Ca/B5[a]	1.76 ± 0.28	–	No. div.	No. div.
MI	1.86 ± 0.25	(0–0.55)	0.6 ± 0.3	0.6 ± 0.3
Ca/MI	1.75 ± 0.22	–	No. div.	No. div.
MI + C[b]	1.81 ± 0.23	–	0.6 ± 0.3	0.6 ± 0.3
Ca/MI + C	No. div.	–	No. div.	No. div.
Kmp8	No. div.	–	No. div.	No. div.
Ca/Kmp8	No. div.	–	No. div.	No. div.
NT	1.6 ± 0.22	(0–0.72)	0.5 ± 0.3	0.5 ± 0.3

[a] Ca indicating an incubation in calcium-mannitol solution (0.4 M mannitol, 10 mM $CaCl_2$) at 4 °C for 2 days.
[b] MI + C: MI medium supplemented with 2% (v/v) coconut milk.

An initial incubation in calcium-mannitol solution increased the division frequency of *A. thaliana* cells (Damm and Willmitzer 1991). In contrast, this treatment was not suitable to stimulate division of PEG-treated cells. The medium MI was established for the initial cultivation of mesophyll protoplasts of *B. napus* (Li and Kohlenbach 1982) and has been used for several *A. thaliana* ecotypes (Siemens et al. 1993). Fusogen-treated cells of *B. napus* and *B. nigra* were successfully cultivated within this medium (Sacristán et al. 1989; Gerdemann-Knörck et al. 1994). NT medium was successfully used for the cultivation of hybrid cells of *A. thaliana* and *B. campestris* (Gleba and Hoffmann 1978). In contrast to B5 medium, in both media, MI and NT, *B. nigra* cells were also stimulated to divide. For the cultivation of PEG-treated cells this probably weakened the negative effect of dying neighbor cells.

The maximal fusion frequency which could be obtained using this protocol was estimated by a fusion between the nearly isogenic *A. thaliana* lines C24 and the transgenic line D4E. The dose for mitotic inactivation of *A. thaliana* D4E in mixed culture together with 600 000 C24 feeder cells per alginate disk was estimated at 1800 Gy. Hybrid colonies could be easily detected due to the expression of the *uidA* gene (glucuronidase activity), which was under the control of the CaMV-35S promotor in the D4E cells.

Using this fusion combination of two *A. thaliana* lines, the putative maximal fusion frequency, estimated from the number of blue colonies, was assessed at a minimum of 28% (Siemens and Sacristan 1995). Fusion frequencies were usually determined as heterokaryon frequencies and ranged from 5 to 23% (Kirti et al. 1991, 1992; Fahleson et al. 1994a, 1994b; Forsberg et al. 1994). Compared to heterokaryon frequencies, the putative maximal fusion frequency is probably underestimated, because not all hybrid cells will develop hybrid colonies, but it is probably overestimated with respect to fusions between taxonomically distant species, in which the development of a hybrid cell can be disturbed by any kind of incompatibility. However, the value is an indication of the efficiency of the fusion protocol and an alternative estimation, if the fused cells cannot be easily detected by morphological markers or by staining of the fusion partners.

2.4 Regeneration of Somatic Hybrids

In the majority of regeneration schemes for somatic hybridizations the protocol established for the recipient species is used for regeneration of hybrids. This strategy was also successful for *A. thaliana* hybrids. Among 13 shoot-regeneration media published for *A. thaliana*, the media ARMIIR, ARMIIC (Marton and Browse 1991), N2 (Chuong et al. 1987, modified according to Sacristán et al. 1989), and SRM (Damm and Willmitzer 1991) proved to be suitable for hybrid calli (Siemens 1994). Using these media, the first shoots of calli of the recipient *A. thaliana* could be observed after 2 weeks. Shoot regenerates of hybrid calli could be observed 2 months after transfer to SRM medium.

The hybrid plantlets revealed strongly curled leaves, which lift the shoot axis of the plants out of the medium. This was prevented by cultivation on artifical stalks (Fig. 1). Regenerated shoots (3–5 mm length) were cut off and placed on the top of cut pipette tips filled with solid AM medium. Developing shoots were transferred to AM or Gl medium to promote rooting. Rooted shoots were transferred to soil.

For the hybrid combination *A. thaliana* (+) *B. nigra*, a total number of 66 hygromycin-resistant calli were obtained and shoots could be regenerated from five calli, which represents an absolute fusion frequency of 3.1×10^{-7} and an absolute hybrid frequency of 1.5×10^{-8} in regard to the total number of PEG-treated *A. thaliana* protoplasts. Relative fusion frequencies (RFF) were estimated by comparing the plating efficiencies under selective conditions with those in control cultures without hygromycin selection. The number of calli

Fig. 1. Cultivation method of small regenerated shoots on artificial stalks

and the number of calli which regenerated shoots compared to PEG-treated viable *A. thaliana* C24 protoplasts were defined as absolute fusion frequency (AFF) and as absolute hybrid frequency (AHyF), respectively. The "absolute hybrid frequency" was introduced as basis for comparison with published data and is defined as the quotient of the number of calli with one or more shoot regenerates and the number of fused recipient protoplasts. The value for the combination *A. thaliana* and *B. nigra* is low. The values calculated from the data given in the various publications range from 10^{-5} and 10^{-6} for intrageneric and intergeneric combinations with close allies of the genus *Brassica* (Kirti et al. 1991, 1992; Gerdemann-Knörck et al. 1994) to 10^{-8} for intertribal combinations (Oikarinen and Ryöppy 1992). For the combination *B. napus* (+) *A. thaliana*, an absolute hybrid frequency of approximately 1×10^{-6} was achieved (Bauer-Weston et al. 1993; Table 1).

Chromosome numbers, Southern analysis, and PCR analysis indicate that the hybrid lines AB22 and AB18 and the subclones of AB18 are nearly symmetrical hybrids in regard to the nuclear genome (Siemens and Sacristan 1995). The two hybrid lines AB22 (Fig. 2) and AB18 showed an intermediate morphology. In vitro shoots of the line AB22 initially developed a rosette. Afterwards the plants developed roots and a few axillary shoots on a very short shoot axis. Sometimes flower stalks with different flower morphologies could be detected on a single plant, but flower development was never observed on plants transferred to soil. No seed set was obtained after pollination with *A. thaliana* or *B. nigra* pollen. These hybrid lines were maintained over 5 years, continuously regenerating shoots on SRM medium comparable to the recipient *A. thaliana*. A similar regeneration capacity was observed in only one hybrid line of *B. campestris* (+) *A. thaliana* after prolonged in vitro culture (Hoffmann and Adachi 1981), whereas only a few shoots could be obtained from a single *B. napus* (+) *A. thaliana* hybrid callus (Bauer-Weston et al. 1993; Forsberg et al. 1994).

Some morphological characteristics of the hybrid plants *A. thaliana* (+) *B. nigra* had been also observed in the hybrids *B. campestris* (+) *A. thaliana*

Fig. 2. The somatic hybrid line AB22 compared to the fusion partners *A. thaliana* and *B. nigra* (*center*)

(Hoffmann and Adachi 1981) and in *B. napus* (+) *A. thaliana* (Bauer-Weston et al. 1993; Forsberg et al. 1994; Forsberg et al. 1998a). All hybrid plants showed shoot habitus and leaf shape intermediate between *A. thaliana* and the different *Brassica* species, and all plants showed deviant and variable flower morphology. However, the hybrids *B. napus* (+) *A. thaliana* could be maintained under greenhouse conditions and some plants have been successfully backcrossed to *B. napus* (Forsberg et al. 1994; Forsberg et al. 1998a; Bohman et al. 1999).

B. napus might be more suitable as recipient in somatic hybridization experiments with *A. thaliana* due to the relationship of the nuclear genome size between donor and recipient. The *A. thaliana* nuclear genome represents approximately 16% of the *B. napus* nuclear DNA content (Marie and Brown 1993). Using *A. thaliana* as recipient and *B. nigra* as donor, thus the genome size relations between donor and recipient were inverted in comparison with the above-mentioned *B. napus* (+) *A. thaliana* hybrids.

2.4.1 Subcloning Experiments

Most initial intertribal hybrids were revealed to be sterile; therefore the success of backcrossing is an exception (Fahlesson et al. 1994a; Forsberg et al. 1994; Skarzhinskaya et al. 1996). Subcloning might be an alternative. The *A. thaliana* (+) *B. nigra* hybrid line AB18 was successfully subcloned by protoplast isolation and cultivation together with feeder protoplasts of *A. thaliana* C24 under hygromycin selection conditions (Siemens and Sacristán 1995). Most of the subclones of AB18 also revealed characteristics similar to the original line AB18, but from a few lines shoots could be cultivated on artificial stalks and maintained as small plantlets with five to ten green leaves for up to 5 weeks. One of these hybrid subclones showed chromosome loss without any chemical or physical treatment (Siemens and Sacristán 1995). Shoots of this subclone, designated AB18d61 (Fig. 3), could be maintained on Gl medium and as rooted plants with an intermediate morphology on AM medium. They could also be transferred to soil, but were highly susceible to changes in environmental conditions. Exceptionally, and not before 4 months of culture on AM medium, plants developed flowers with a reduced number of *Arabidopsis*-like petals and sepals, but with abnormal pistils and anthers.

3 Summary and Conclusion

The possibility of combining the *A. thaliana* genome with the genomes of *B. rapa*, *B. napus*, and *B. nigra* has been proven (Gleba and Hoffmann 1979; Bauer-Weston et al. 1993; Forsberg et al. 1994, 1998a; Siemens and Sacristán 1995). Thereby, the advantages of the model plant *A. thaliana* could also be used for studies in somatic hybridization with the important crop plants of the

Fig. 3. The somatic hybrid line AB18d61 transferred to soil

family Brassicaceae. However, somatic hybridization with *A. thaliana* would only be a useful tool for studying the transfer of genome fractions if the experimental design could be improved by reducing the amount of transferred donor chromosomes, or by induction of chromosome loss in the selected hybrids. After several subculture steps on different media, Hoffmann and Adachi (1981) obtained a hybrid line *B. campestris* (+) *A. thaliana* which had lost several *B. campestris* chromosomes and showed chromosome translocations. Chromosome loss was also achieved for one of the hybrids, *A. thaliana* (+) *B. nigra*, without further chemical or physical treatment by regeneration of protoplast-derived subclones (Siemens and Sacristán 1995). When *A. thaliana* was taken as donor and was hybridized with the recipient *B. napus*, several hybrid lines *B. napus* (+) *A. thaliana* were fertile and could be studied in backcross programs (Forsberg et al. 1994; Forsberg et al. 1998a). The analysis of these progenies has provided insight into introgression at the molecular level (Forsberg et al. 1998b; Bohman et al. 1999).

4 Protocol

The fusion procedure of *A. thaliana* protoplasts with irradiated *B. nigra* protoplasts was based on the method used by Damm et al. (1989) for direct gene transfer to *A. thaliana* protoplasts. The regeneration procedure was optimized for the somatic hybridization of the species *A. thaliana* and *B. nigra*.

4.1 Isolation of Protoplasts

1. Cut 1 g sterile leaves of 3-week-old seedlings in 2 ml 0.5 M mannitol into 0.5-mm stripes.
2. Add an enzyme mixture and incubate the leaves for 12–16 h in the dark.

3. Isolate the protoplasts by sieving the suspension through a 60-μm nylon gaze and floating on a sucrose cushion (0.6 M sucrose) by centrifugation (120 g, 10 min).
4. Wash the protoplast with W_5 solution and store them on ice for 30 min.

4.2 Fusion of Protoplasts

1. Inactivate donor protoplasts by irradiation, thereby preventing sedimentation of protoplasts.
2. Remove the W_5 solution totally by centrifugation (120 g, 5 min), resuspend the donor protoplasts with 0.6 ml *A. thaliana* protoplasts suspension ($2 \times 10^6 \text{ml}^{-1}$) in fusion buffer, and add the same volume of PEG solution to the protoplast suspension.
3. Incubate the fusion mixture for 30 min at room temperature.
4. Dilute the fusion mixture carefully and slowly with at least 10 ml W_5 solution.
5. Centrifuge the protoplast suspension and wash each pellet with a mixture of 0.4 M mannitol and seawater to remove all Ca^{2+} ions.
6. Resuspend the protoplasts in 0.5 M mannitol, mix this protoplast suspension with the same volume of alginate solution, and pour the suspension on Ca^{2+} agar plates for alginate polymerization.

4.3 Regeneration of Hybrids

1. Add 2 ml of NT medium to each alginate disk within a petri dish (6 cm diameter)
2. After 10 days replace the medium with 2 ml of MI medium.
3. Replace the medium with MI medium 10 days later and transfer the colonies to dim light (approx. $5 \mu\text{mol m}^{-2}\text{s}^{-1}$). Change the medium every 10th day.
4. Pick microcalli with a diameter of more than 1 mm (55–60 days of culture) out of the alginate matrix and transfer them to solid SRM medium for further cultivation in bright light (16 h light, $60-80 \mu\text{mol m}^{-2}\text{s}^{-1}$). Change the medium every 20th day.
5. Cut off shoots (3–5 mm length) and place them on the top of cut pipette tips filled with solid AM medium.
6. Transfer developing shoots to AMI or Gl medium to promote rooting.

References

Bauer-Weston B, Keller W, Webb J, Gleddie S (1993) Production and characterization of asymmetric somatic hybrids between *Arabidopsis thaliana* and *Brassica napus*. Theor Appl Genet 86:150–158

Bohman S, Forsberg J, Glimelius K, Dixelius C (1999) Inheritance of *Arabidopsis* DNA in offspring from *Brassica napus* and *A. thaliana* somatic hybrids. Theor Appl Genet 98:99–106

Chuong PV, Pauls KP, Beversdorf WD (1987) Plant regeneration from *Brassica napus* (L.) Koch stem protoplasts. In Vitro Cell Dev Biol 23:449–452

Damm B, Willmitzer L (1991) *Arabidopsis* protoplast transformation and regeneration. In: Lindsey K (ed) Plant tissue culture manual: fundamentals and applications. Kluwer Dordrecht, Boston

Damm B, Schmidt R, Willmitzer L (1989) Efficient transformation of *Arabidopsis thaliana* using direct gene transfer to protoplasts. Mol Gen Genet 217:6–12

Fahleson J, Dixelius J, Sundberg E, Glimelius K (1988) Correlation between flow cytometry determination of nuclear DNA content and chromosome number in somatic hybrids within *Brassicaceae*. Plant Cell Rep 7:74–77

Fahleson J, Eriksson I, Landgren M, Stymne S, Glimelius K (1994a) Intertribal somatic hybrids between *Brassica napus* and *Thlaspi perfoliatum* with high content of the *T. perfoliatum*-specific nervonic acid. Theor Appl Genet 87:795–804

Fahleson J, Eriksson I, Glimelius K (1994b) Intertribal somatic hybrids between *Brassica napus* and *Barbarea vulgaris* production of in vitro plantlets. Plant Cell Rep 13:411–416

Forsberg J, Landgren M, Glimelius K (1994) Fertile somatic hybrids between *Brassica napus* and *Arabidopsis thaliana*. Plant Sci 95:213–223

Forsberg J, Dixelius C, Lagercrantz U, Glimelius K (1998a) UV dose-dependent DNA elimination in asymmetric somatic hybrids between *Brassica napus* and *Arabidopsis thaliana*. Plant Sci 131:65–76

Forsberg J, Lagercrantz U, Glimelius K (1998b) Comparison of UV light, X-ray and restriction enzyme treatment as tools in production of asymmetric somatic hybrids between *Brassica napus* and *Arabidopsis thaliana*. Theor Appl Genet 96:1178–1185

Gerdemann-Knörck M, Sacristán MD, Braatz C, Schieder O (1994) Utilization of asymmetric somatic hybridization for the transfer of disease resistance from *Brassica nigra* to *Brassica napus*. Plant Breed 112:106–113

Gleba YY, Hoffmann F (1978) Hybrid cell lines *Arabidopsis thaliana* + *Brassica campestris*: no evidence for specific chromosome elimination. Mol Gen Genet 165:257–264

Gleba YY, Hoffmann F (1979) *"Arabidobrassica"*: plant genome engineering by protoplast fusion. Naturwissenschaften 66:547–554

Gleba YY, Hoffmann F (1980) *"Arabidobrassica"*: a novel plant obtained by protoplast fusion. Planta 149:112–117

Glimelius K, Fahleson J, Landgren M, Sjödin C, Sundberg E (1991) Gene transfer via somatic hybridization in plants. Trends Biotechnol 9:24–30

Hansen LN (1998) Intertribal somatic hybridization between rapid cycling *Brassica oleracea* L. and *Camelina sativa* (L.) Crantz. Euphytica 104:173–179

Hansen LN, Earle ED (1997) Somatic hybrids between *Brassica oleracea* L. and *Sinapis alba* L. with resistance to *Alternaria brassicae* (Berk) Sacc. Theor Appl Genet 94:1078–1085

Hoffmann F, Adachi T (1981) *Arabidobrassica*: chromosomal recombination and morphogenesis in asymmetric intergeneric hybrid cells. Planta 153:586–593

Kirti PB, Prakash S, Chopra VL (1991) Interspecific hybridization between *Brassica juncea* and *B. spinescens* through protoplast fusion. Plant Cell Rep 9:639–642

Kirti PB, Narasimhulu SB, Prakash S, Chopra VL (1992) Somatic hybridization between *Brassica juncea* and *Moricandia arvensis* by protoplast fusion. Plant Cell Rep 11:318–321

Kunkel B (1996) A useful weed put to work: genetic analysis of disease resistance in *Arabidopsis thaliana*. Trends Genet 12:63–69

Li L, Kohlenbach HW (1982) Somatic embryogenesis in quite a direct way in cultures of mesophyll protoplasts of *Brassica napus* L. Plant Cell Rep 1:209–211

Marie D, Brown SC (1993) A cytometric exercise in plant DNA histograms, with 2C values for 70 species. Biol Cell 78:41–51

Marton L, Browse J (1991) Facile transformation of *Arabidopsis*. Plant Cell Rep 10:235–239

Mathur J, Koncz C, Szabados L (1995). A simple method for isolation, liquid culture, transformation and regeneration of *Arabidopsis thaliana* protoplasts. Plant Cell Rep 14:221–226

Masson J, Paszkowski J (1992) The culture response of *Arabidopsis thaliana* protoplasts is determined by the growth conditions of donor plants. Plant J 2:829–833

Menczel L, Morgan A, Brown S, Maliga P (1987) Fusion-mediated combination of Ogura-type cytoplasmic male sterility with *Brassica napus* plastids using X-irradiated CMS protoplasts. Plant Cell Rep 6:98–101

Meyerowitz EM (1992) Introduction to the *Arabidopsis* genome. In: Koncz C, Chua NH, Schell J (eds) Methods in *Arabidopsis* research. World Scientific Publ, Singapur, pp 100–118

Morgan A, Maliga P (1987) Rapid chloroplast segregation and recombination of mitochondrial DNA in *Brassica* cybrids. Mol Gen Genet 209:240–246

Narasimhulu SB, Kirti PB, Bhatt SR, Prakash S, Chopra VL (1994) Intergeneric protoplast fusion between *Brassica carinata* and *Camelina sativa*. Plant Cell Rep 13:657–660

Oikarinen S, Ryöppy PH (1992) Somatic hybridization of *Brassica campestris* and *Barbarea* species. XIIIth EUCARPIA Congress, Book of Poster Abstracts, Angers, France, July 1991, pp 261–262

O'Neill CM, Mathias RJ (1994) Somatic fusion for the transfer of agronomically important traits into *Arabidopsis thaliana* from *Brassica* ssp. Abstr VIIth Int Congr of Plant Tissue and Cell Culture, Florence, Italy, June 1994, 105 pp

Park HY, Wernicke W (1993) Improved mesophyll protoplast culture and plant regeneration in *Arabidopsis thaliana* (L) Heynh, genotype Landsberg erecta. J Plant Physiol 141:376–379

Primard C, Vedel F, Matthieu C, Pelletier G, Chevre AM (1988) Interspecific somatic hybridization between *Brassica napus* and *Brassica hirta* (*Sinapis alba* L.) Theor Appl Genet 75:546–552

Rédei GP (1992) A heuristic glance at the past of *Arabidopsis* genetics. In: Koncz C, Chua NH, Schell J (eds) Methods in *Arabidopsis* research. World Scientific Publ, Singapur, pp 1–15

Sacristán MD, Gerdemann-Knoerck M, Schieder O (1989) Incorporation of hygromycin resistance in *Brassica nigra* and its transfer to *Brassica napus* through asymmetric protoplast fusion. Theor Appl Genet 78:194–200

Siemens J (1994) Asymmetrische somatische Hybridisierung zwischen *Arabidopsis thaliana* (L.) Heynh. und *Brassica nigra* (L.) Koch. PhD Thesis Free University of Berlin, Berlin

Siemens J, Sacristán MD (1995). Production and characterization of somatic hybrids between *Arabidopsis thaliana* and *Brassica nigra*. Plant Sci 111:95–106

Siemens J, Torres M, Morgner M, Sacristán MD (1993) Plant regeneration from mesophyll protoplasts of four different ecotypes and two marker lines from *Arabidopsis thaliana* using a unique protocol. Plant Cell Rep 12:569–572

Sigareva MA, Earle ED (1999) Regeneration of plants from protoplasts of *Capsella bursa-pastoris* and somatic hybridization with rapid cycling *Brasscia oleracea*. Plant Cell Rep 18:412–417

Sjödin C, Glimelius K (1989a) *Brassica naponigra*, a somatic hybrid resistant to *Phoma lingam*. Theor Appl Genet 77:651–656

Skarzhinskaya M, Landgren M, Glimelius K (1996). Production of intertribal somatic hybrids between *Brassica napus* L. and *Lesquerella fendleri* (Gray) Wats. Theor Appl Genet 93:1242–1250

Torres M, Siemens J, Meixner M, Sacristan MD (1997). An improved method for direct gene transfer and subsequent regeneration of *Arabidopsis thaliana* Landsberg erecta and two marker lines. Plant Cell Tissue Organ Cult 47:111–118

Yamashita Y, Treada R, Nishibayashi S, Shimamoto K (1989) Asymmetric somatic hybrids of *Brassica*: partial transfer of *B. campestris* genome into *B. oleracea* by cell fusion. Theor Appl Genet 77:189–194

II.2 Somatic Hybridization in *Asparagus*

H. Kunitake[1] and M. Mii[2]

1 Introduction

Asparagus (*Asparagus officinalis* L.) ($2n = 20$) is a monocotyledonous perennial species belonging to the Liliaceae family and is cultivated as a vegetable. Plants of asparagus are usually raised from seeds and cultivated continuously in the field for 10 years or more. Young stems, known as spears, emerging through the ground, are harvested as the edible part of the plants from spring to autumn in Japan.

Asparagus originates from the Mediterranean area and is especially well adapted to mild and sunny climates and sandy soil. Recently, its production has increased year by year in several countries as one of the most refined and healthy vegetables. Nowadays, the main producers are USA, Spain, France, and Taiwan. White asparagus production predominates in Europe (France and Spain) and Taiwan, while green asparagus is essential in North America, Australia, and Japan.

In Japan, the annual total yield has increased year by year until 1990. In the southwest region, the main production area, however, production has recently started to decrease due to a disease syndrome known as asparagus decline, primarily caused by *Fusarium oxysporum*, *F. moriforme*, and *Phomopsis asparagi*. Particularly, stem blight caused by *P. asparagi* has spread over the southwest area, and is one of the serious problems in year-round cultivation of asparagus. These diseases have also been reported in every asparagus-growing region surveyed in the world (Cohen and Heald 1941; Grogan and Kimble 1959; Johnston et al. 1979). The yield in France decreased to 2500 ha in 1991 because of these diseases (Delbreil and Jullien 1996).

The most successful strategy for disease control in other vegetable crops has been the development of resistant varieties (Mace et al. 1981). Ohgoshi and Yoshioka (1987) reported that a wild species of asparagus, *A. macowanii* ($2n = 20$; Fig. 1) was resistant to *P. asparagi* which had been sprayed for infection, as compared with several other wild species. Therefore, it could be expected that the resistant trait to *P. asparagi* could be incorporated into asparagus by interspecific hybridization. However, no interspecific hybrid

---

[1] School of Agriculture, Kyushu Tokai University, Aso, Kumamoto 869-1404, Japan
[2] Faculty of Horticulture, Chiba University, Matsudo 271-8520, Japan

Fig. 1. Mature plants of *Asparagus macowanii* Bak

between *A. officinalis* and *A. macowanii* has been produced by conventional breeding methods.

Somatic hybridization by protoplast fusion offers an alternative method to conventional plant breeding for obtaining inter- and intrageneric somatic hybrids between sexually incompatible species. For the production of somatic hybrids, it is indispensable to establish plant regeneration systems from protoplasts. Plant regeneration from asparagus protoplasts was first reported by Bui Dang Ha and Mackenzie (1973), who isolated protoplasts from cladodes and succeeded in regenerating whole plants via somatic embryogenesis. However, they could not obtain consistent results, and the division frequency varied among culture attempts. Although various systems of protoplast culture have since then been reported in *Asparagus* (Kong and Chin 1988; Elmer et al. 1989; Dan and Stephens 1991; Mukhopadhyay and Desjardins 1994; May and Sink 1995), somatic hybridization has not been reported in this genus.

Recently, we have succeeded in establishing an efficient method for protoplast culture from embryogenic calli and subsequent whole-plant regeneration via somatic embryogenesis (Kunitake and Mii 1990; Kunitake et al. 1995). In this chapter, we describe the successful production of a somatic hybrid plant between *Asparagus officinalis* and *A. macowanii* using electrofusion by utilizing the protoplast culture system we established.

2 Somatic Hybridization

2.1 Isolation of Protoplasts

Protoplasts were first isolated from cladode tissues of *Asparagus officinalis* L. cv. Marche de Malines, grown under constant conditions of light, temperature, and mineral supply in the growth cabinet (Bui Dang Ha and Mackenzie 1973). We also tried to isolate protoplasts from cladodes, stems, young spears, and roots of cv. Mary Washington cultivated under the same conditions. However, we could not obtain a yield sufficient for protoplast culture and somatic hybridization. Therefore, callus cultures induced from stem segments were used as the source of protoplasts for both *A. officinalis* and *A. macowanii*.

Embryogenic calli of *A. officinalis* cv. Mary Washington, which had a yellow, friable, and nodular appearance, were induced from stem segments on MS (Murashige and Skoog 1962) medium containing $1\,\text{mg}\,\text{l}^{-1}$ 2,4-dichlorophenoxyacetec acid (2,4-D), $30\,\text{g}\,\text{l}^{-1}$ sucrose, and $2\,\text{g}\,\text{l}^{-1}$ gellan gum. These embryogenic calli were maintained without the loss of embryogenic capacity by subculturing at 1-month intervals on the same medium at $25\,°\text{C}$ under continuous illumination ($38\,\mu\text{mol}\,\text{m}^{-2}\,\text{s}^{-1}$).

In *A. macowanii* Bak., stem segments of in vitro-grown seedlings were cultured for inducing calli on MS medium containing $5\,\text{mg}\,\text{l}^{-1}$ 4-amino-3,5,6-trichloropicolinic acid (Picloram), $1000\,\text{mg}\,\text{l}^{-1}$ glutamine, $30\,\text{g}\,\text{l}^{-1}$ sucrose, and $2\,\text{g}\,\text{l}^{-1}$ gellan gum, and compact, yellow calli induced 2 months after culture were maintained by subculturing on the same medium at 2-month intervals. However, this callus showed no somatic embryo formation or plant regeneration.

Protoplasts of *A. officinalis* and *A. macowanii* were both prepared from these calli. The calli of each species were gently crushed and put in an enzyme solution containing 2% (w/v) Cellulase Onozuka RS (Yakult Honsha Co. Ltd.), 0.5% (w/v) Macerozyme R-10 (Yakult Honsha Co. Ltd.), 0.05% (w/v) Pectolyase Y-23 (Seishin Co. Ltd.), $10\,\text{mM}$ $CaCl_2 \cdot 2H_2O$, $5\,\text{mM}$ MES, and $0.6\,\text{M}$ sorbitol, pH 5.7. The mixture was incubated on a rotary shaker (60 rpm) for 6 h at $25\,°\text{C}$ to liberate protoplasts. Protoplasts were collected by filtration through a nylon sieve ($60\,\mu\text{m}$) and washed twice with $0.6\,\text{M}$ mannitol solution after centrifugation ($100\,g$ for 5 min). Protoplasts of these two species showed the same appearance, with opaque and slightly yellow cytoplasm and small vacuoles. Usually, the yield of protoplasts for each species was $5\sim10 \times 10^6$ cells g^{-1} callus. More than 95% of the protoplasts were viable, as assessed by staining with fluorescein diacetate (FDA). The average size of protoplasts was approximately $20\,\mu\text{m}$ in diameter in both species.

2.2 Fusion of Protoplasts

Electrofusion was carried out using an Electro Cell Manipulator 401A (BTX Inc., San Diego, CA) connected with a fusion chamber with parallel stainless steel electrodes 2 mm apart. The fusion chamber was sterilized by autoclaving before use. Protoplast fusion was observed under an inverted microscope.

Purified protoplasts of each species were resuspended in a fusion solution containing 0.6 M mannitol, 1 mM $CaCl_2 \cdot 2H_2O$, 5 mM MES, pH 5.7. Protoplasts of both species were mixed at a ratio of 1:1, and 1 ml of the protoplast suspension ($5 \times 10^5 ml^{-1}$) was pipetted into a fusion chamber. The protoplasts were aligned into short chains by applying an laternating current of 100 V cm^{-1}, at 1 MHz for 20s. Fusion was induced by a direct current square pulse of 0.25–1.75 kV cm^{-1} for 40 μs. The fusion frequency was observed under a light microscope after staining with lacto-propionic orcein, and the viability of these protoplasts was assessed using fluorescein diacetate (FDA).

When an alternating current field was applied, so-called pearl chains, each consisting of five to ten protoplasts, were formed within 20s. Field strength markedly affected the fusion frequency of the protoplasts aligned. A field strength of 0.25~0.5 kV cm^{-1} did not induce protoplast fusion, and at 0.75 kV cm^{-1} the electrofusion frequency was less than 1% (Fig. 2). The maximum frequency, approximately 9%, was obtained at 1.0~1.5 kV cm^{-1} for 40 μs. However, at 1.25 kV cm^{-1} or more, protoplast viability showed a marked decrease. Therefore, we applied a direct current pulse of 1.0 kV cm^{-1} for 40 μs to induce electrofusion between *A. officinalis* and *A. macowanii*. After the application of direct current square pulse, the fused protoplasts rapidly became round but slightly damaged, as assessed by the FDA staining method.

After applying the fusion pulse, the protolast suspension was kept for 10 min at room temperature. Then the protoplasts were carefully pipetted into a glass tube and collected by centrifugation at 100 g for 2~3 min.

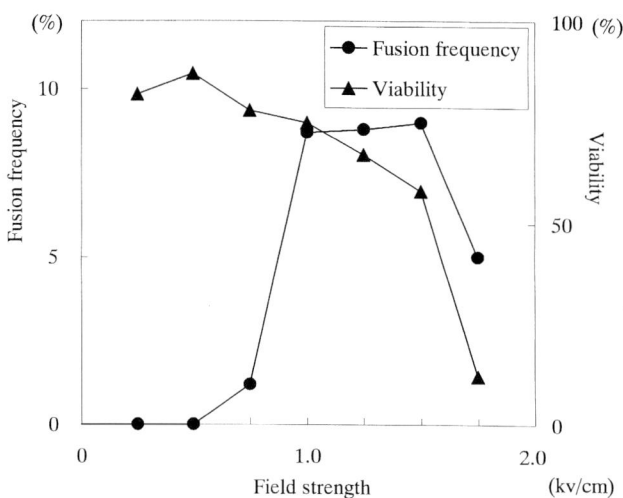

Fig. 2. Effect of field strength on protoplast viability and fusion frequency between *Asparagus officinalis* and *A. macowanii*. (H. Kunitake et al., unpubl.). Fusion frequency was observed under a light microscope after staining with lacto-propionicorcein, and the viability of these protoplasts was assessed using fluorescein diacetate (FDA)

2.3 Culture of Fused Protoplasts

For both preventing the formation of colonies derived from *A. officinalis* protoplasts and selecting only somatic hybrid colonies, protoplasts of *A. officinalis* were inactivated with 10 mM iodoacetamide (IOA) in 0.6 M mannitol solution for 10 min at room temperature. They were subsequently washed twice with 0.6 M mannitol solution and then used for electrofusion experiments.

The fusion-treated protoplasts were cultured in $1\,g\,l^{-1}$ gellan gum-solidified 1/2 MS medium containing $1\,mg\,l^{-1}$ 1-naphthaleneacetic acid (NAA), $0.5\,mg\,l^{-1}$ zeatin, $1\,g\,l^{-1}$ glutamine and 0.6 M glucose (Kunitake and Mii 1990). On this medium, protoplasts of *A. officinalis* formed visible colonies, whereas those of *A. macowanii* stopped growth after a few divisions when they were cultured separately. Division of the protoplasts of *A. officinalis* was completely inhibited by treating with 10 mM IOA for 10 min at room temperature. In the fusion-treated cultures, however, some of the protoplasts divided continuously, and visible colonies were formed after 40 days of culture. These colonies had a compact, yellow appearance with browning, unlike the colonies derived from protoplasts of *A. officinalis*, which had a friable and nodular appearance. All of the seven colonies produced were transferred onto MS medium containing $2\,mg\,l^{-1}$ 2,4-D, $30\,g\,l^{-1}$ sucrose and $2\,g\,l^{-1}$ gellan gum for callus proliferation, and each minicallus obtained was further transferred to liquid MS medium without plant growth regulators for proliferation of calli with removal of endogenous auxin. The suspension cultures thus established were harvested by sieving through a stainless steel mesh of 600 μm pore size. An aliquot of 0.1 ml packed cell volume calli were transferred to $1\,g\,l^{-1}$ gellan gum-solidified MS medium for inducing somatic embryogenesis. Although six calli lines browned and died within 30 days after transfer, the remaining one callus line (SH2) produced somatic embryos.

As the remaining six calli, which died on the regeneration medium, had morphological characters similar to those of *A. macowanii*, they probably derived from "escaped" protoplasts of *A. macowanii*. The formation of "escaped" colonies was also reported in somatic hybridization experiments on *Brassica* (Terada et al. 1987) and *Dianthus* (Nakano and Mii 1993).

2.4 Regeneration of Somatic Hybrids

SH2 callus was nodular and intermediate in color between those of the parents. SH2 callus-derived somatic embryos were yellow, compact and appeared larger as compared with those of *A. officinalis* (Fig. 3). Most of the somatic embryos derived from SH2 callus showed abnormal germination with only root development. Only a few embryos (less than 1%) germinated and developed shoots and roots, but the plantlets thus obtained showed very poor growth and failed to acclimatize.

In asparagus, somatic embryos induced on MS medium containing 0.1 M sucrose and $10\,g\,l^{-1}$ gellan gum most normally germinated after cold treatment at 4 °C for 3~30 days (Nakashima et al. 1993). According to this result, SH2

Fig. 3A,B. Difference in size of somatic embryos between somatic hybrid SH2 (**A**) and *Asparagus officinalis* (**B**). *Bar* 5 mm

callus-derived somatic embryos were transferred to the germination medium after cold treatment at 4°C for 14 days in the dark on $10\,g\,l^{-1}$ gellan gum-solidified MS medium. Then, the somatic embryos were transferred to the germination medium which was modified MS medium consisting of half-strength MS salts, MS vitamins, $30\,g\,l^{-1}$ sucrose, and $2\,g\,l^{-1}$ gellan gum. They were kept at 25°C under continuous illumination of $38\,\mu mol\,m^{-2}\,s^{-1}$ with warm white fluorescent lamps. Using cold treatment, the germination rate increased (12.6%) and the frequency of abnormal germination markedly decreased. Several green roots were visible within 7 days after transfer, and shoots developed 10 days after transfer (Fig. 4a). Crowns were produced at the base of the shoots. The SH2 plantlets thus obtained grew slowly and formed *A. macowanii*-type cladodes.

Cold treatment of SH2-derived somatic embryos increased the germination rate and inhibited abnormal germination. For the microspore-derived embryos of *Brassica napus*, germination rate was significantly increased by cold (Kott and Beversdorf 1990), ABA (Finkelstein et al. 1985) or partial desiccation (Kott and Beversdorf 1990; Senaratna et al. 1991) treatment. Particu-

Fig. 4A,B. Plantlet regeneration from somatic embryos in the somatic hybrid SH2. **A** Normally germinated somatic embryo 10 days after transfer onto the germination medium. *Bar* 1 cm. **B** Slowly growing somatic hybrid plants 20 days after transfer to soil. *Bar* 3 cm

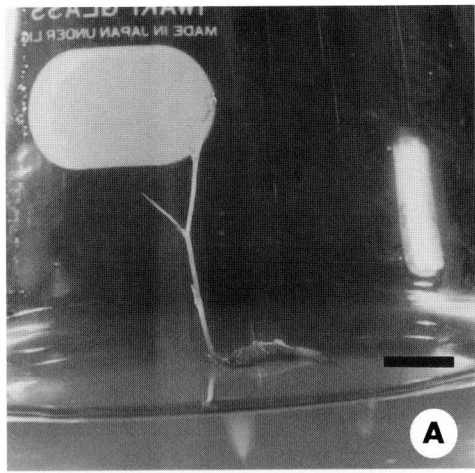

larly, cold treatment of the embryos at 4°C for 9–12 days was more efficient for enhancing germination than other treatments (Kott and Beversdorf 1990). In the present study, somatic embryos of SH2-derived somatic embryos also required approximately the same period of cold treatment as those of *Brassica napus* for normal germination. Although the mechanism involved in the effectiveness of cold treatment for enhancing normal germination of in vitro-produced embryos is still unclear, Kott and Beversdorf (1990) suggested that these treatments may realign the morphological and physiological patterns and thus promote direct, vigorous germination.

More than 50 plantlets of SH2 were potted in soil for plug nursery of foliage vegetables (Kasahara Kogyo Co. Ltd., Japan) and placed in an incubator to acclimatize for 10 days, after which they were transferred to a greenhouse (Fig. 4b). However, these plantlets grew very slowly and have not yet attained the flowering stage.

2.5 Analysis of Somatic Hybrids

2.5.1 Isozyme Analysis

Electrophoretic analysis of leaf isozyme is a simple and efficient method of confirming the hybridity. Because somatic hybridization causes the addition of parental genomes, dominant genes from both parents are expected to be expressed in somatic hybrid plants. In the present study, however, we used calli of the putative somatic hybrid SH2 for isozyme analysis, because a leaf sample of SH2 was not adequate for the analysis. A sample of 2g (FW) calli was collected from each of the parental species and their putative somatic hybrid, SH2. Each sample was washed and homogenized in 1.5 ml buffer (0.2 M Tris-HCl, 1 M sucrose, 56 mM 2-mercaptoethanol, pH 8.5). The homogenates were then centrifuged at 18000g for 20 min at 4°C and the supernatant was collected. Aliquots of 50 μl extract were subjected to isoelectric focusing on polyacrylamide gels for 3h (90–100mA) at 5°C. The gels were stained for esterase (EST) and leucine aminopeptidase (LAP) by the methods outlined by Wetter and Dyck (1983). The results showed that SH2 possessed bands from both parents, and no additional novel bands were observed in the isozyme patterns of these two enzymes.

2.5.2 Endonuclease Analysis of the Nuclear Ribosomal RNA Gene

Restriction endonuclease analysis of the nuclear ribosomal RNA gene (rDNA) has been used extensively to confirm hybridity, instead of electrophoresis of isozymes and electrofocusing of the small subunit of ribulose bisphosphate carboxylase. Uchimiya et al. (1983) first reported that the analysis of rDNA was useful for the identification of *Nicotiana* somatic hybrids. Furthermore, high-resolution rDNA analysis using nonradioactive-labeled rRNA probes has been incorporated to identify somatic hybrids of various plant species (Ohgawara et al. 1985; Grosser et al. 1988; Kisaka et al. 1997; Miranda et al. 1997). Therefore, we carried out the identification of the putative somatic hybrid SH2 using rice rRNA probe.

Total DNAs were extracted from in vitro-growing plantlets of parental species and SH2, according to the method of Rogers and Bendich (1985). The total DNAs were digested with *Eco*R I or *Hind*III endonuclease, separated on agarose gels and blotted onto nylon membrane filters (Hybond-N, Amersham). Analyses of rDNA were performed using DNA labeling with a digoxigenin and detection kit (Boehringer Mannheim). The DNA fragment containing the entire rDNA sequences of rice was prepared from plasmid pRR217 and used as a probe.

The results of the blot-hybridization of digoxigenin-labeled rDNA fragments ro *Eco*R I- or *Hind*III-digested total DNAs of the parental species and the somatic hybrids are shown in Figs. 5 and 6. *Eco*R I digestion gave 6 and 0.5 kbp fragments specific for *A. officinalis*, and 5.6 and 0.9 kbp fragments specific for *A. macowanii* (Fig. 5), whereas *Hind*III digestion gave 30-kbp frag-

Fig. 5. Blot hybridization of digoxigenin-labeled rDNA fragments to EcoR I-digested total DNAs of the parental species and the somatic hybrid. *R* Rice (control); *A A. officinalis*; *H* somatic hybrid SH2; *M A. macowanii*

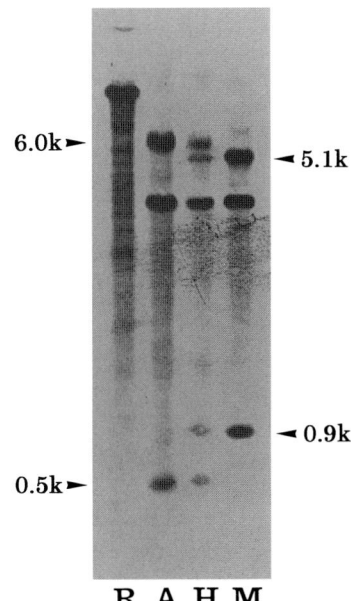

ment specific for *A. officinalis* and a 8.0-kbp one specific for *A. macowanii* (Fig. 6). The putative somatic hybrid plant of SH2 had the fragments specific for both parents.

2.5.3 RAPD Analysis

Most methods for the identification of putative somatic hybrids, including isozyme and rDNA analysis, depend on the use of fully grown plants for either a comprehensive morphological study or mass extraction of DNA or proteins. These identification methods are time-consuming, as a long growth period of the plant to maturity is required. Therefore, a quick and simple method which could detect hybridity of fusion products at either the callus level or the juvenile stage of growth would be of great advantage for preliminary screening in these somatic hybridization programs.

Williams et al. (1990) and Welsh and McCletland (1990) reported the use of a PCR technique by which DNA sequences in total genomic DNA were amplified using arbitrary 10-bp primer to generate randomly amplified polymorphic DNA (RAPD). Recently, this technique has been popular for the identification and screening of somatic hybrids. We also tried to analyze putative somatic hybrids using RAPD marker.

PCR and electrophoresis were performed by the methods described by Williams et al. (1990) with some modifications. The reaction mixtures contained 10mM Tris-HCl, pH 8.3; 50mM KCl; 3.0mM $MgCl_2$; 100μm each of

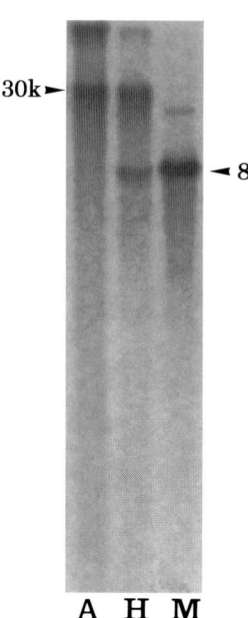

Fig. 6. Blot hybridization of digoxigenin-labeled rDNA fragments to Hind III-digested total DNAs of the parental species and the somatic hybrid. *A A. officinalis*; *H* somatic hybrid SH2; *M A. macowanii*

dATP, dCTP, dTTP, and dGTP, 0.3M primer; 0.5U Ampli Taq polymerase and 10ng of genomic DNA, in a total volume of 18.5μl. Reactions were cycled 45 times, each at 92°C for 1min, 40°C for 2min, and 72°C for 3min in an ASTEC Program Control System PC-700. Primers of ten nucleotides in length were purchased from Operon Technology Inc. (CA, USA). After all the PCR cycles were completed, 5μl of the samples were loaded on 0.8% agarose gels and subjected to electrophoresis at $10\,\text{V}\,\text{cm}^{-1}$ for 15min. Then, the gels were stained with $10\,\text{mg}\,\text{l}^{-1}$ of ethidium bromide and photographed under UV light (360nm). For each combination of sample and primer, the PCR was carried out twice and only stable polymorphism was taken into account.

The profiles of the amplified products from both parents by six primers were clearly different from each other, and the somatic hybrid SH2 had fragments specific for both parents (Fig. 7).

2.5.4 Chromosome Number

Ten putative somatic hybrids, SH2, were randomly selected for chromosome observation. Root tips of in vitro plantlets were pretreated with 2mM 8-hydroxyquinoline for 2h at room temperature and fixed in a mixed solution of ethanol: acetic acid (3:1) for 24h. The root tips were then macerated in 1 N HCl for 3min at 60°C, and stained and crusheded in 1% aceto-orcein. The results showed that the chromosome number of SH2 was 50 (Fig. 8a), which was different from the sum of the chromosome numbers of *A. officinalis* ($2n = 20$, Fig. 8b) and *A. macowanii* ($2n = 20$).

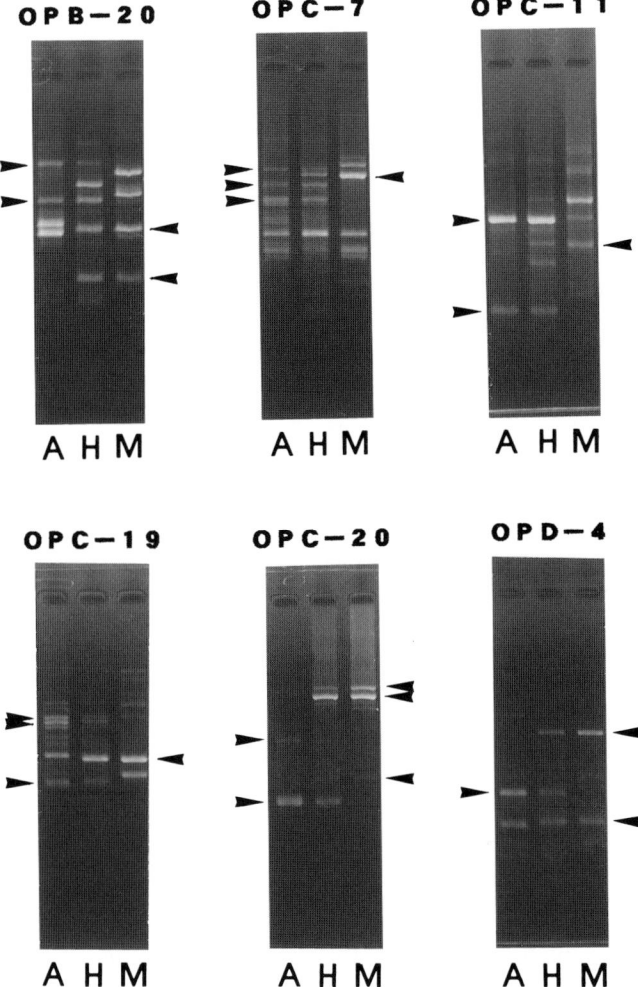

Fig. 7. RAPD analysis of a somatic hybrid. DNAs from somatic hybrid and parental species were used as templates for PCR amplification using OPERON primers, OPA-9, OPA-11, OPA-12, OPA-17, OPA-19, and OPB-14. PCR products were subjected to electrophoresis on a 1.5% agarose gel and stained with 10 mg l^{-1} ethidium bromide. The DNA fragments were revealed by UV light and photographed. *Arrows* indicate the bands specific to the parental species. *A* *A. officinalis*; *H* somatic hybrid SH2; *M* *A. macowanii*

3 Summary and Conclusion

As described above, the present investigation showed that somatic hybrids interspecific between *A. officinalis* and *A. macowanii* can be obtained by using embryogenic calli as the source of protoplasts, by applying the appropriate

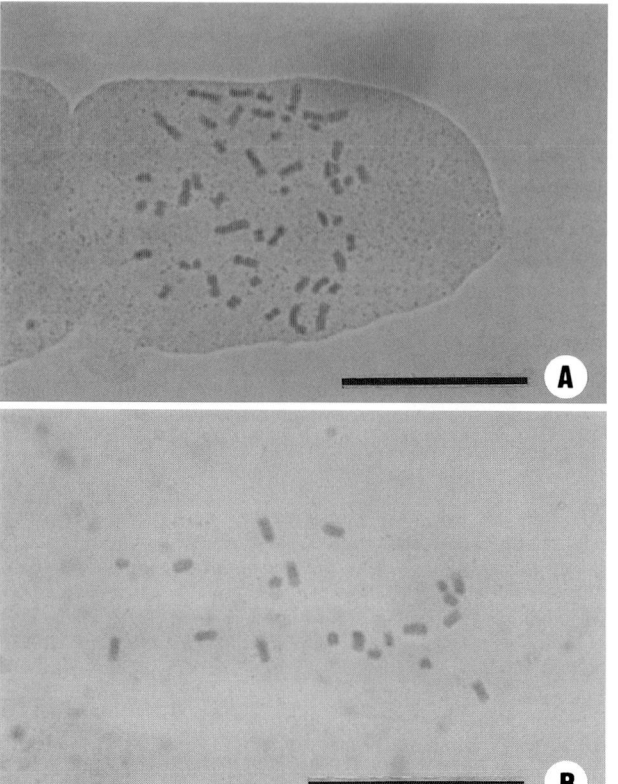

Fig. 8A,B. Chromosomes in a root-tip cell of somatic hybrid SH2 ($2n = 50$) (**A**) and *Asparagus officinalis* ($2n = 20$) (**B**). *Bar* 10 µm

conditions of electrofusion, by selecting somatic hybrid using IOA treatment and the difference of cell division capacity, and by the induction of normal germination in somatic embryos by cold treatment (Fig. 9). Furthermore, the hybridity of the SH2 was confirmed by analyses of isozyme, nuclear rDNA, and RAPD analyses, and by counting the chromosome number. However, the interspecific somatic hybrid, SH2, had a chromosome number of $2n = 50$, which was more than the sum of the parents ($2n = 40$). One of the possible explanations is that this somatic hybrid could be derived from the fusion among three protoplasts, followed by elimination of ten chromosomes. In previous studies, genetic instability in interfamilial and intergeneric hybrids often resulted in incapability of morphogenesis or in the formation of abnormal plants that frequently lacked root organogenesis (Krumbiegel and Schieder 1981; Gleba et al. 1983). The abnormal growth of the somatic hybrid obtained in the present study may also be due to such an aberrant chromosome number or genetic instability. For obtaining hybrid plants with normal growth, further studies are needed to produce symmetric as well as asymmetric hybrids with

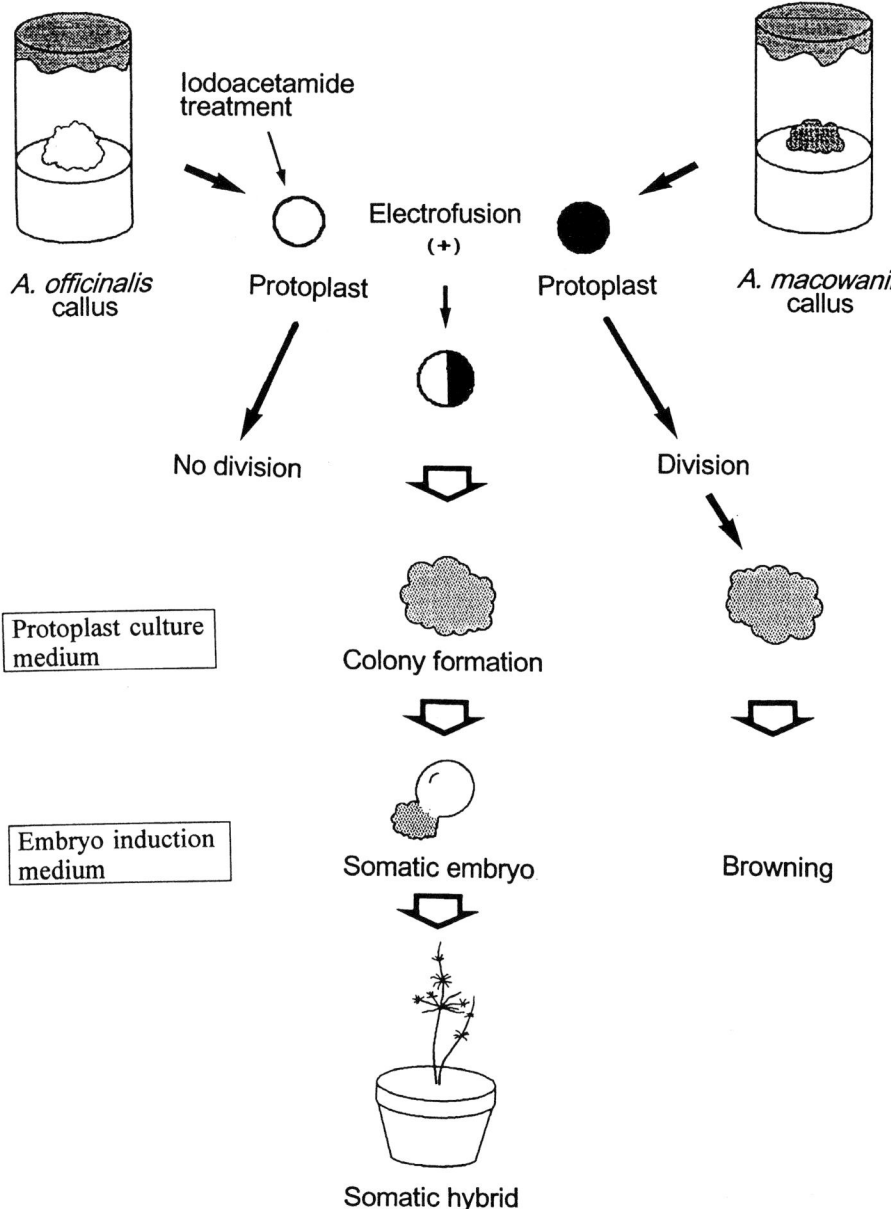

Fig. 9. Screening procedure of somatic hybrids between *Asparagus officinalis* and *A. macowanii*

varied genomic constitution by utilizing various techniques such as irradiation of *A. macowanii* protoplasts.

In the future, the use of genetic engineering techniques to produce varieties of asparagus with viral, fungal, and/or weed resistance will be of significant value in asparagus production. Transgenic asparagus plants have been obtained by using a natural gene transfer system with *Agrobacterium tumefaciens* (Bytebier et al. 1987; Delbreil and Jullien 1993), and by direct gene transfer methods such as electroporation of protoplasts and microprojectile bombardment to the embryogenic calli (Shigemoto et al. 1995; Jose et al. 1997). These, and modifications to other genetic transformation system that are now available due to the development of the appropriate technologies will have significant impact on asparagus improvement in the near future.

4 Protocol

4.1 Preparation of Protoplast Source

4.1.1 Embryogenic Callus Induction of Asparagus officinalis L.

1. Harvest stems (approximately 1-month-old shoots) or young spears from mature asparagus plants.
2. Surface-sterilize stems or young spears successively with 70% ethanol for 30s and 0.2% sodium hypochorite for 7 min, and then wash three times with sterile distilled water.
3. Excise stem segments or meristems from young spears or upper part of stems and inoculte them onto Murashige and Skoog (MS) medium containing $2\,mg\,l^{-1}$ 2,4-D, $30\,g\,l^{-1}$ sucrose and $2\,g\,l^{-1}$ gellan gum.
4. Keep at 25°C under continuous illumination at $38\,\mu mol\,m^{-2}\,s^{-1}$.
5. When somatic embryos or yellow, friable and nodular calli are induced from the explants, transplant these cultures onto the same medium.
6. Subculture the embryogenic calli at 1-month intervals under the same conditions.

4.1.2 Callus Induction of Asparagus macowanii Bak.

1. Sow surface-sterilized seeds of *A. macowanii* into test tubes, each containing 20 ml of $2\,g\,l^{-1}$ gellan gum-solidified MS medium without plant growth regulators.
2. Transfer the stem segments without nodes (1 cm in length) to MS medium containing picloram, $1\,g\,l^{-1}$ glutamine, $30\,g\,l^{-1}$ sucrose and $2\,g\,l^{-1}$ gellan gum.
3. Keep at 25°C under continuous illumination at $38\,\mu mol\,m^{-2}\,s^{-1}$.
4. When compact, yellow calli are induced from the explants, transplant these cultures into the same medium.
5. Subculture the calli at 2-month intervals under the same conditions.

4.2 Isolation of Protoplasts

1. Transfer approximately 1 g of each callus into 10 ml of filter-sterilized (Millipore, 0.45-μm pore size) enzyme solution which contains 2% (w/v) Cellulase Onozuka RS, 0.5% (w/v) Macerozyme R-10, 0.05% (w/v) Pectolyase Y-23, 10 mM $CaCl_2 \cdot 2H_2O$, 5 mM MES and 0.6 M sorbitol, pH 5.7.
2. Incubate the mixture on a rotary shaker (60 rpm) for 6 h at 25 °C to liberate protoplasts.
3. Filter through a nylon sieve (60 μm) and wash twice with 0.6 M mannitol solution after centrifugation ($100\,g$ for 5 min).

4.3 Inactivation of Protoplasts

1. Treat the protoplasts of *A. officinalis* with 10 mM iodoacetamide (IOA) in 0.6 M mannitol solution for 10 min at room temperature to prevent the formation of colonies derived from *A. officinalis* protoplasts.
2. Wash twice with 0.6 M mannitol solution and then use for electrofusion experiments.

4.4 Electrofusion

1. Resuspend the IOA-inactivated *A. officinalis* protoplasts and the *A. macowanii* protoplasts separately in a fusion solution containing 0.6 M mannitol, 1 mM $CaCl_2 \cdot 2H_2O$, 5 mM MES, pH 5.7.
2. Mix the protoplasts of both species at a ratio of 1:1. The density of each protoplasts is $5 \times 10^5 \, ml^{-1}$.
3. Pipette 1 ml of the protoplast suspension into a sterilized fusion chamber connected with an Electro Cell Manipulator 401A (BTX Inc., San Diego, CA).
4. Apply an alternate current of $100 \, V \, cm^{-1}$, at 1 MHz for 20 s and then a direct current square pulse of $1.0 \, kV \, cm^{-1}$ for 40 µs to induce electrofusion.
5. Keep the electrofusion-treated protoplast suspension for 10 min at room temperature.
6. Carefully pipet the protoplast suspension into a gl-1 ass tube and collect by centrifugation ($100 \, g$, 2–3 min).

4.5 Culture of Somatic Hybrid Cells

1. Resuspend the protoplast suspension in $1 \, g \, l^{-1}$ gellan gum-solidified 1/2 MS medium containing $1 \, mg \, l^{-1}$ NAA, $0.5 \, mg \, l^{-1}$ zeatin, $1 \, g \, l^{-1}$ glutamine and 0.6 M glucose.
2. Keep at 25 °C in the dark for 40 days.
3. Transfer protoplast-derived colonies to MS medium containing $2 \, mg \, l^{-1}$ 2,4-D for callus proliferation.

4.6 Optimal Conditions/Medium for Regeneration of Plants

1. Sieve the calli through a stainless steel mesh (pore size 600 µm).
2. Transfer the sieved calli onto $10 \, g \, l^{-1}$ gellan gum-solidified MS medium without growth regulators for inducing somatic embryogenesis.
3. Keep at 25 °C under continuous illumination at $38 \, \mu mol \, m^{-2} \, s^{-1}$ for 30 days.
4. Treat the cultures with induced somatic embryos at 4 °C for 14 days in the dark.
5. Transfer the cold-treated somatic embryos onto $2 \, g \, l^{-1}$ gellan gum-solidified 1/2 MS medium without plant growth regulators for germination.
6. Pot the regenerated plantlets in the soil for plug nursery of foliage vegetable (Kasahara Kogyo Co. Ltd., Japan) 2 months after culture and place in incubators to acclimatize for 10 days.
7. Transfer the plants to a greenhouse.

Acknowledgments. The authors are grateful to Messrs. Masanobu Tanaka, Kinya Mori, and Toshiki Nakashima of The Saga Prefectural Agricultural Research Center for their advice on tissue culture of *Asparagus* species. We also wish to thank Dr. Akira Saito of The Kyushu National Agricultural Experiment Station for experimental help on molecular biobogy.

References

Bui Dang Ha D, Mackenzie IA (1973) The division of protoplasts from *Asparagus officinalis* L. and their growth and differentiation. Protoplasma 78:215–221
Bui Dang Ha D, Norrel B, Masset A (1975) Regeneration of *Asparagus officinalis* L. through callus derived from protoplasts. J Exp Bot 26:263–270
Bytebier B, deBoeck F, De Greve H, Van Montagu M, Hernalsteens JP (1987) T-DNA organization in tumor cultures and transgenic plants of the monocotyledon *Asparagus officinalis* L. Proc Natl Acad Sci USA 84:5345–5349
Cohen SI, Heald FD (1941) A wild and root rot of asparagus caused by *Fusarium oxysporum* (Schlecht.) Plant Dis Rep 25:503–509
Dan Y, Stephens CT (1991) Studies on protoplast culture types and plant regeneration from callus-derived protoplasts of *Asparagus officinalis* L. cv. Lucullus 234. Plant Cell Tissue Organ Cult 27:321–331
Delbreil B, Jullien M (1993) *Agrobacterium*-mediated transformation of *Asparagus officinalis* L. long-term embryogenic calli and regeneration of transgenic plants. Plant Cell Rep 13:372–376
Delbreil B, Jullien M (1996) Genetic transformation in *Asparagus officinalis* L. In: Bajaj YPS (ed) Biotechnology in agriculture and forestry, vol 38. Plant protoplasts and genetic engineering VII. Springer, Berlin Heidelberg New York, pp 164–177
Elmer WH, Ball T, Volokita M, Stephens CT, Sink KC (1989) Plant regeneration from callus-derived protoplasts of asparagus. J Am Soc Hortic Sci 1147:1019–1024
Finkelstein RR, Tenbarge KM, Shumway JE, Crouch ML (1985) Role of ABA in maturation of rapeseed embryos. Plant Physiol 78:630–636
Gleba YY, Momot UP, Okolot AN, Cherep NN, Skarzhynskaya MV, Kotov V (1983) Genetic processes in intergeneric cell hybrid *Atropa* + *Nicotiana*. 1. Genetic constitution of cells of different clonal origin grown in vitro. Theor Appl Genet 65:269–276
Grogan RG, Kimble KA (1959) The association of *Fusarium* wilt with the asparagus decline and replant problem in California. Phytopathology 49:122–125
Grosser JW, Gmitter FG Jr, Chandler JL (1988) Intergeneric somatic hybrid plants from sexually incompatible woody species: *Citrus sinensis* and *Severina disticha*. Theor Appl Genet 75:397–401
Johnston SA, Springer JK, Lewis GD (1979) *Fusarium moniliforme* as a cause of stem and crown rot of asparagus and its association with asparagus decline. Phytopathology 69:778–780
Jose LC-P, Liliana L, Nacyra A-G, Consuelo M-A, Ana MB, Luis H-E (1997) An efficient particle bombardment system for the genetic transformation of asparagus (*Asparagus officinalis* L.). Plant Cell Rep 16:255–260
Kisaka H, Kisaka M, Kanno A, Kameya T (1997) Production and analysis of plants that are somatic hybrids of barley (*Hordeum vulgare* L.) and carrot (*Daucus carota* L.). Theor Appl Genet 94:221–226
Kong Y, Chin CK (1988) Culture of asparugus protoplasts on porous polypropylene membrane. Plant Cell Rep 7:67–69
Kott LS, Beversdorf WD (1990) Enhanced plant regeneration from microspore-derived embryos of *Brassica nupus* by chilling, partial desiccation and age selection. Plant Cell Tissue Organ Cult 23:187–192
Krumbiegel G, Schieder O (1981) Comparison of somatic and sexual incompatibility between *Datura innoxia* and *Atropa belladonna*. Planta 153:466–470
Kunitake H, Mii M (1990) Somatic embryogenesis and plant regeneration from protoplasts of asparagus (*Asparagus officinalis* L.). Plant Cell Rep 7:67–69
Kunitake H, Nakashima T, Mori K, Tanaka M (1995) Plant regeneration from protoplasts of several cultivars of asparagus (*Asparagus officinalis* L.). Bull Saga Pref Agric Res Ctr 29:1–19
Mace ME, Belland AA, Beckman CH (1981) Fungul wilt disease of plants. Academic Press, New York
May RA, Sink KC (1995) Genotype and auxin infleence direct somatic embryogenesis from protoplasts derived from embryogenic cell suspensions. Plant Sci 108:71–84

Miranda M, Motomura T, Ikeda F, Ohgawara T, Saito W, Endo T, Omura M, Moriguchi T (1997) Somatic hybrids obtained by fusion between *Poncirus trifoliata* (2x) and *Fortunella hindsii* (4x) protoplasts. Plant Cell Rep 16:401–405

Mukhopadhyay S, Desjardin Y (1994) Plant regeneration from protoplast-derived somatic embryos of *Asparagus officinalis* L. J Plant Physiol 144:94–99

Murashige T, Skoog F (1962) A revised medium for rapid growth and bioassay with tobacco tissue culture. Physiol Plant 15:473–497

Nakano M, Mii M (1993) Interspecific somatic hybridization in *Dianthus*: selection of hybrids by the use of iodoacetamide inactivation and regeneration ability. Plant Sci 88:203–208

Nakashima T, Kunitake H, Mori K, Tanaka M (1993) The development of utilization method of embryoid and shoot primordia. 5. In vitro mass propagation of asparagus (*Asparagus officinalis* L.) Bull Saga Pref Agric Exp Stn 28:12–29

Ohgawara T, Kobayashi S, Ohgawara E, Uchimiya H, Ishii S (1985) Somatic hybrid plants obtained by protoplast fusion between *Citrus sinensis* and *Poncirus trifoliata*. Theor Appl Genet 71:1–4

Ohgoshi S, Yoshikawa K (1987) Production of asparagus new cultivars using tissue culture. Summary of vegetable crop investigation in Fukushima Prefecture, pp 38–39 (in Japanese)

Rogers SO, Bendich AJ (1985) Extraction of DNA from milligram amounts of fresh, herbarium and mummified plant tissue. Plant Mol Biol 15:77–84

Senaratna T, Kott LS, Beversdorf WD, Mckersie BD (1991) Desiccation of microspore derived embryos of oilseed rape (*Brassica napus* L.). Plant Cell Rep 10:342–344

Shigemoto N, Kohmura H, Imoto M, Irifune K, Morikawa H (1995) Production of asparagus transgenic plants using a particle gun. Proc Annu Meet Bot Sci Jpn Kyoto, 303 pp, (in Japanese)

Terada R, Kyozuka J, Nishibayashi S, Shimamoto K (1987) Plantlet regeneration from somatic hybrids of rice (*Oryza sativa* L.) and barnyard grass (*Echinochloa oryzicola* Vasing). Mol Gen Genet 210:39–43

Uchimiya H, Ohgawara T, Kato H, Akiyama T, Harada H, Sugiura M (1983) Detection of two different nuclear genomes in parasexual hybrids by ribosomal RNA gene analysis. Theor Appl Genet 64:117–118

Welsh J, McCletland M (1990) Fingerprinting genomes using PCR with arbitrary primers. Nucleic Acids Res 18:7213–7218

Wetter L, Dyck J (1983) Isoenzyme analysis of cultured cells and somatic hybrid. In: Sharp WR, Evans DA, Ammirato PV, Yamada Y (eds) Handbook of plant cell culture vol 1. Macmillan, New York, pp 607–628

Williams JGK, Kubelik AE, Levak KJ, Rafalski JA, Tingey SC (1990) DNA polymorphisms amplified by arbitrary primers are useful as genetic markers. Nucleic Acids Res 18:6531–6535

II.3 Somatic Hybridization in *Cichorium intybus* L. (Chicory)

C. RAMBAUD and J. VASSEUR

1 Introduction

1.1 The Plant

Cichorium intybus L., belonging to the Asteraceae family, is composed of three varieties (Longly and Louant 1987; Leteinturier et al. 1991). (1) Industrial chicory is cultivated for the production of roots, which are used as a coffee additive or for inuline production. It comprises Magdebourg and Brunswick chicories belonging to the var. *sativum* Biskoff. (2) Witloof or Bruxelles chicory, whose roots are able to produce etiolated buds under artificial conditions (forcing). Resulting from the improvement of the Magdebourg-type chicory, it corresponds to the var. *foliosum* Biskoff. (3) Wild chicories with large red, green, or variegated leaves which are currently consumed as a salad (Trévise, Vérone, Pain de sucre ou Chioggia). This is the var. *silvestre* Biskoff.

Cichorium intybus was selected for a biennal cycle; during the first year, plants develop rosette leaves and after the passage of a cold period a floral stem is produced. Flowers form capitula, which are grouped together in ears. Each capitule contains 15 or 20 ligulated blue flowers. Fruits are white, black, or variegated achenes. Chicory is an allogamous and entogamous species in which self-incompatibility is not strict; 10 or 20% of produced seeds come from self-fertilization, which can even be up to 33% in some witloof chicory seed samples (Bellamy et al. 1996).

1.2 Industrial Chicory Breeding

Industrial chicory (*Cichorium intybus* L., var. *sativum*) is cultivated for the production of roots, since root yield has increased greatly during the past few years (45 t roots ha^{-1}, in 1992), chicory has become a potential crop for the production of sweets (inulin or fructose). France is actually the first industrial chicory producer with about 200000 t of roots and 30000 ha (Desprez 1993), representing 35% of the world production. Breeder's objectives are focused

---

Laboratoire de Physiologie Cellulaire et Morphogenèse végétales, Université des Sciences et Technologies de Lille 1, Bâtiment SN2, 59655 Villeneuve d'Ascq Cedex, France

on the improvement of this plant for resistance to fructification, a root form convenient for good mechanized harvesting, inuline or soluted dry matter yields, and root preservation.

To obtain homozygous lines, breeders have checked for a high yield of self-compatible genotypes. Brother-sister crosses gave rise to families that had been mixed for heterosis and that thus constitued actual population varieties (e.g., cv. Orchies). This progress was made possible by the easy in vitro cloning of chicory and also by the existence of male sterile plants. Nuclear male sterility was used to create commercial diploid or triploid hybrids. With cytoplasmic male sterility, triploid hybrid varieties could be produced without the cost and financial risk of using a cloned male sterile parent.

In chicory, cytoplasmic male sterility did not exist, by means of either spontaneous, or interspecific hybridization. For this reason, somatic hybridization has been considered in this species because by protoplast fusion it is possible to transfer or induce cytoplasmic male sterility.

1.3 Review of *Cichorium* Protoplast Culture Work

First division of mesophyll protoplasts of *C. intybus* L. and *C. endivia* L. was reported by Binding et al. (1981), who observed that although plant development from protoplasts was feasible in the genus *Cichorium*, necrosis and mitotic arrest of small colonies prevented efficient plant regeneration. Crepy et al. (1982) showed that an apparent toxicity of nitrate was responsible for protoplast-derived colony necrosis and that when nitrate was removed from the culture medium, plantlet regeneration improved. In our laboratory, fertile plants were obtained by Saksi et al. (1986), who showed that each chicory line or variety presents special problems, which have to be overcome by empirically adjusting the isolation and culture procedures. In witloof chicory, a decrease in the sucrose concentration in the early culture and a rapid transfer to a medium with reduced mannitol concentration allows plant regeneration from the cultivar Zoom. In these studies, plant regeneration from protoplasts was always achieved through shoot formation. Embryogenic *Cichorium* hybrid protoplasts were able to undergo somatic embryogenesis, and subsequently developed into plants. However, embryos occurred indirectly via callus proliferation and subsequent differentiation or directly without an intermediate callus stage (Sidikou-Seyni et al. 1992).

2 Somatic Hybridization

In order to develop techniques of somatic hybridization or cybridization from the Magdebourg type of *Cichorium intybus* L., regeneration of plants from cv. Pévèle was carried out. As for other species, in chicory, genetic factors interfere with the ability to regenerate plants. For this reason, culture conditions and media have been studied for this cultivar, and we could show that the preconditioning of the donor plants for protoplast isolation was essential to improving the plating efficiency (Rambaud et al. 1990).

2.1 Donor Plant Culture Conditions

Achenes of *Cichorium intybus* L. cv. Pévèle were surface-sterilized by a solution of 0.1% $HgCl_2$, washed three times in sterile distilled water and placed in Petri dishes on a Heller (1953) culture medium (major and minor salts without $FeCl_3$) supplemented with $19.5\,mg\,l^{-1}$ Fe-EDTA, $20\,g\,l^{-1}$ sucrose, and $6\,g\,l^{-1}$ agar (Biokar type E). Aseptic seedlings were then transferred to culture tubes in the same medium.

Culture tubes were placed in a growth chamber ($24 \pm 1\,°C$ day / $20 \pm 1\,°C$ night). Photoperiod, light intensity, and light quality were analyzed, and it is recommended to culture the plants with a 16-h daylight photoperiod, and to place the culture tubes in the vicinity of Deluxe cool light fluorescent tubes ($50\,\mu mol\,m^{-2}\,s^{-1}$).

The age of the donor plants, clearly defined by the number of days after sowing, was crucial. The highest yields of viable protoplasts with the highest plating efficiency were obtained with Magdebourg chicory plants, which had one or two leaves (11–13 days old).

2.2 Isolation and Culture of Protoplasts

Leaves of chicory were used for protoplast isolation. Ten million protoplasts were obtained from 1 g of fresh leaves. They contained many chloroplasts and the majority of them were 30–40 µm in diameter (Fig. 1A). First division was observed after 24 to 48 h of culture in the MC1 liquid medium (Table 1) and four to ten cell microcalli could be observed after 7 days of culture (Fig. 1D). After this first period, microcalli were transferred to a second medium MC2 (Table 1) to prevent browning and necrosis. The old medium was removed by centrifugation and replaced by fresh medium of the same composition. The cell suspension was then embedded in agarose medium. This transfer was a critical step, as the number of microcalli which survived at that stage was very low. For this second stage, the purity of the gelling agent and also the incorporation of the microcalli in the agar were critical for good plating efficiency. Low melting-point agarose with a static gelation temperature of 37°C (Biorad) was used for this transfer, which was made by dripping 1 ml of the suspension culture resulting from the first phase onto the surface of the agarose just after cooling. Callus proliferation also necessitated a medium with reduced osmoticum, low auxin and cytokinine concentrations, and a suppression of nitrate, which became the inhibitor in the second phase. This second phase led to the proliferation of minicalli (Fig. 1E) with a plating efficiency of about 6%. When minicalli were 2 mm in size, they were transferred to a budding medium in which auxin was omitted, 6-benzylaminopurine (BAP)

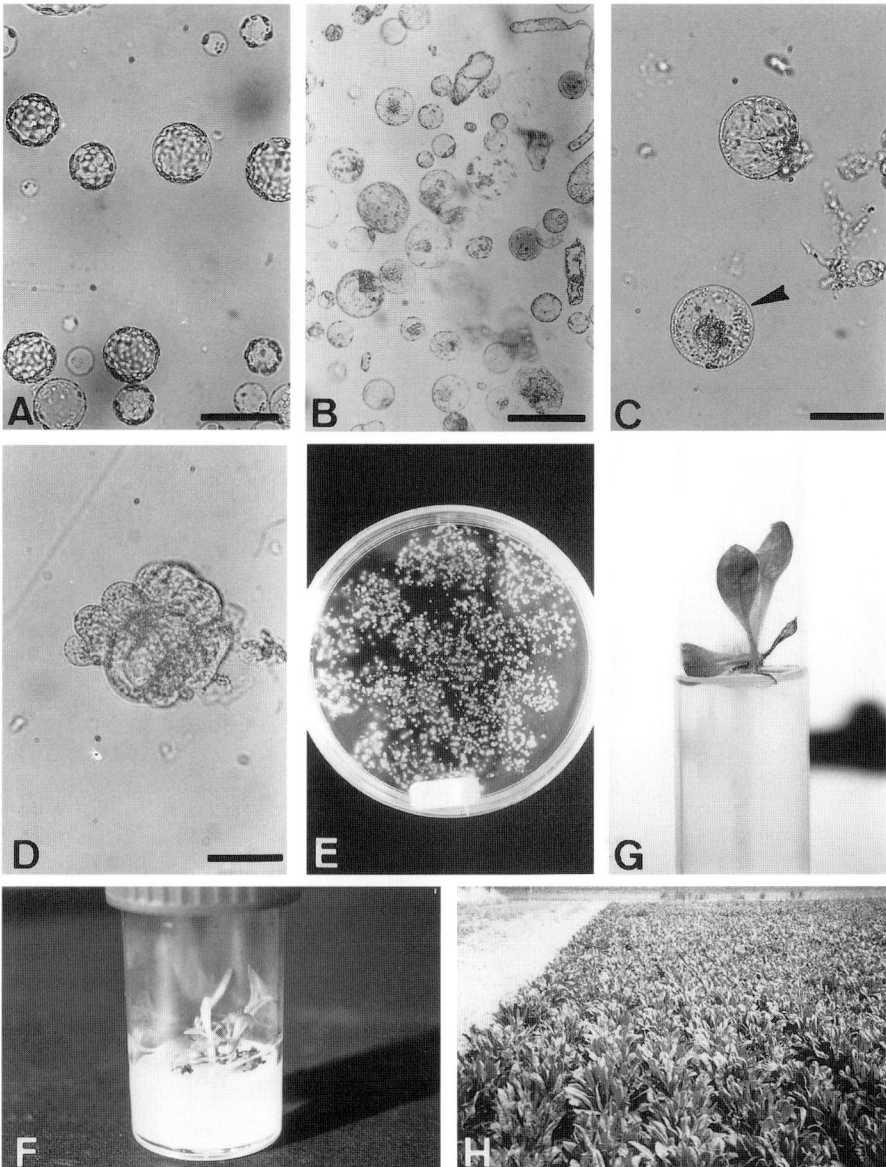

Fig. 1A–H. Different stages in the regeneration of plants from somatic cybridization in chicory. **A** Mesophyll protoplasts of *Cichorium intybus* L. cv. Pévèle. *Bar* 60 µm. **B** Hypocotyl protoplasts of *Helianthus annuus* cv. Mirasol. *Bar* 120 µm. **C** Chicory-sunflower heterokaryocyte. *Bar* 60 µm. **D** Microcallus derived from protoplast after 7 days' culture, in the MC1* medium. *Bar* 60 µm. **E** Population of developing cell colonies derived from agarose-plated chicory protoplasts (90-mm petri dish). **F** Shoots regenerated from a callus which had been transferred from MC3 to MC4 medium. **G** Rooted plantlet of chicory. **H** Regenerated chicory plants grown in the field

Table 1. Composition (mg l^{-1}) of the culture media for protoplasts (MC1), heterokaryocytes (MC1*), microcalli (MC2), callus proliferation (MC3), bud induction (MC4), and rooting of plantlets (H5) of *Cichorium intybus*

Components	Washing solution	MC1	MC1*	MC2	MC3	MC4	H10
CaCl$_2$·2H$_2$O	220	440	440	440	220	220	37.5
KCl	–	–	–	375	–	–	750
KH$_2$PO$_4$	85	85	85	85	85	85	–
KNO$_3$	950	950	950	–	950	950	–
MgSO$_4$·7H$_2$O	185	185	185	185	185	185	250
NaH$_2$PO$_4$·H$_2$O	–	–	–	–	–	–	62.5
NaNO$_3$	–	–	–	–	–	–	600
NH$_4$NO$_3$	825	–	–	–	–	–	–
FeSO$_4$·7H$_2$O	13.9	13.9	13.9	13.9	1.39	13.9	13.9
EDTA·Na$_2$	18.65	18.65	18.65	18.65	18.65	18.65	18.65
Oligoelements	Heller (1953)	Heller (1953)	Heller (1953)	Heller (1953)	Heller (1953)	Heller (1953)	Heller (1953)
Vitamins	Morel and Wetmore (1951)	Morel and Wetmore (1951)	Morel and Wetmore (1951)	Morel and Wetmore (1951)	Morel and Wetmore (1951)	Morel and Wetmore (1951)	–
Myo-inositol	100	250	250	100	100	100	–
Glutamine	–	375	375	750	–	–	–
Sucrose	10000	10000	10000	10000	10000	5000	10000
Mannitol	90000	90000	90000	60000	–	–	–
NAA	–	2	2	0.5	0.5	–	–
BAP	–	1	1	0.5	0.5	0.1	–
Coconut water	–	–	2%	–	–	–	–
Casein hydrolysate	–	–	150	–	–	–	–
MES	–	–	2560	–	–	–	–
Agarose low melt	–	–	–	5000	–	–	–
Agar	–	–	–	–	6000	6000	6000
pH	5.5	5.5	5.5	5.5	5.5	5.5	5.5

was lowered to $0.1\,mg\,l^{-1}$, and sucrose lowered to $5\,g\,l^{-1}$. At this stage, to prevent vitrification of the developing buds, calli were cultured in large vessels (Fig. 1F). This yielded better rooting of the developing buds on a medium with low concentration of mineral salts and without growth regulators (Table 1, H10 medium). About 75% of the calli could give one or many buds that formed roots (Fig. 1G) with a frequency of about 90%.

2.3 Fusion of Protoplasts

Somatic hybridization has been performed in chicory to induce male sterility. First, we had to choose a fusion partner which would fulfil the following requirements: (1) it could give a CMS determinant, (2) it would be phylogenetically distant enough to prevent nuclear fusion, (3) it would be close enough to allow mitochondrial rearrangements between the genomes of the two fused species. With this aim, sunflower was chosen, since it presented the advantage of being recalcitrant to protoplast culture and so facilitating the selection of somatic hybrids. Sunflower protoplasts (Fig. 1B) were prepared from hypocotyls of aseptic seedlings, 6 to 10 days after sowing. For the fusion process, a ratio of 1:3 (sunflower / chicory) was necessary to give the best results.

A PEG method with a high concentration of Ca^{2+} was used for the protoplast fusion (Rambaud et al. 1992). The addition of DMSO had induced a heterokaryocyte rate of about 25%. Heterokarocytes (Fig. 1C) were easily identified by the fact that the sunflower hypocotyl protoplasts were colorless while the mesophyll protoplasts of chicory were green. After 24h of culture, the heterokaryocytes were isolated under a microscope, using a thin-mouthed Pasteur pipet, and then placed on a 96-microwell Nunc plate (at the rate of 12 heterokaryocytes well^{-1}), each well containing $100\,\mu l$ of culture medium. In the MC1 medium optimal for chicory protoplasts cultured at high density (2×10^4 protoplasts ml^{-1}), the heterokaryocytes became dark and died. The addition of casein hydrolysate ($150\,mg\,l^{-1}$), coconut milk (2%) and MES (5mM) allowed the division of heterokaryocytes at rates of 25%. This condition was better than the high density culture, but limited to the number of cultured protoplasts, by manual handling. After development, minicalli with a size of 1mm were transferred to an MC3 propagation medium (Table 1) which allowed further development, reaching a size of 2–3mm, sufficient for budding the calli on a MC4 medium (Table 1). Compared to the 90% regenerants generally found for calli from chicory protoplasts which have not been treated with PEG, the yield of regenerants observed from heterokaryocytes was about 25%. Rooting of the buds on H10 medium (Table 1) was not a critical step (except if buds were vitrified). The plantlets obtained were transferred first to a greenhouse and then to the field (Fig. 1H) for the selection of male sterile plants.

3 Analysis of the Regenerated Plants

All the 600 regenerated plants showed a chicory phenotype, but 16 presented either male sterility or total sterility differently characterized according to the plants (Rambaud et al. 1993). The sterility was often accompanied by a deficiency in vigour. Among the 16 sterile plants, we could distinguish three types of sterility (Fig. 2). In the first one, male sterility was characterized by a lack of anther dehiscence (Fig. 2B) and the absence of pollen or viable pollen revealed by the Alexander test (1969); another type of male sterility was characterized by the complete absence of anthers (Fig. 2C). The last type of male sterility was characterized either by the absence of both the anthers and the style or by the presence of a reduced style (Fig. 2D); when a reduced style occurred, it was possible to obtain descendants as for the male sterile types, but the yields of the seeds differed greatly, according to the plants.

To determine if the observed male sterilities really came from the protoplast fusion and if their origins were cytoplasmic, two types of investigations were made in parallel: (1) molecular analysis of the mitochondrial genome to determine if mitochondrial rearrangements have occurred in the male sterile cybrid, and (2) genetic analysis of the heritability of the male sterile character. Three plants, that were named 411, 523, and 524, gave good yields of seeds (particularly 411), presented more vigor, and were chosen for the genetic and molecular analyses.

3.1 Genetic Analysis

This analysis was made with the 411 plant coming from protoplast fusion. For crosses, only plants presenting flowers without anthers were used. Progenies were analyzed in four generations and showed that a wide range of floral morphologies; fertile, totally sterile, and two types of male sterilities (with brown anthers or without anthers) could be observed. These four types of floral morphologies were observed among the studied generations, but with different ratios (Rambaud et al. 1997). Results showed a floral morphology segregation in the descendants that was not due to a nuclear determinism of the male sterility, because the presence of totally sterile and fertile plants remained at a very low level in the fourth progeny. This was the result of an instability of the character that is becoming stabilized among the generations. This instability could be produced by mitochondrial rearrangements or by an alloplasmic situation due to the heterogeneous genetic composition of the industrial chicory varieties used as pollinators. Moreover, more rapid stabilization of the floral morphology could be observed when homozygous lines of witloof chicory are used as pollinators.

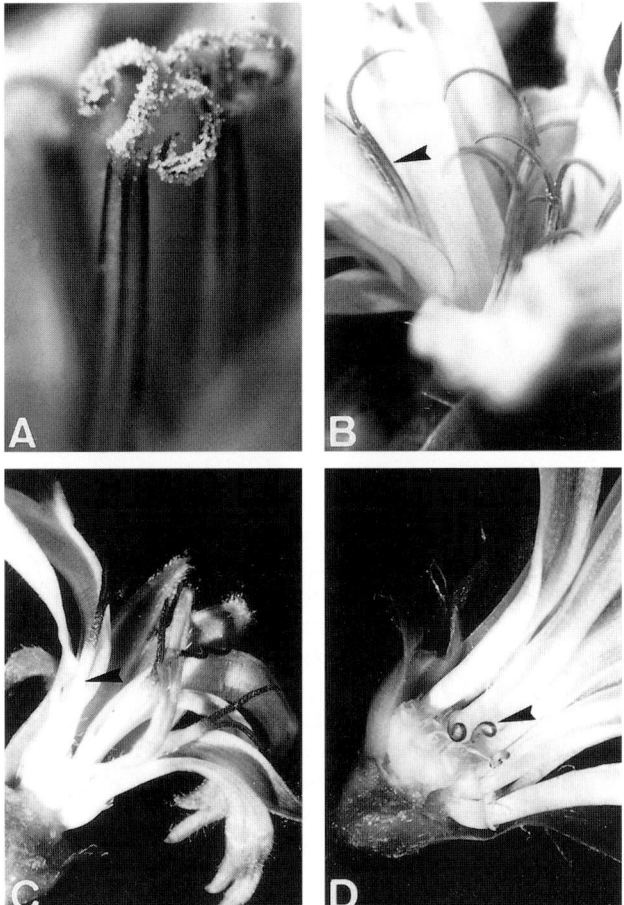

Fig. 2 A–D. Different floral morphology types of chicory plants coming from cybridization. **A** Fertile flower. **B** Male sterile flowers with brown anthers. **C** Male sterile flowers without anther. **D** Sterile flowers with aborted styles and without anther

3.2 Mitochondrial Genome Analysis

Analysis of the mitochondrial DNA with Southern hybridizations showed that all the male sterile or totally sterile plants have a rearranged mitochondrial genome. Rearranged mitochondrial DNA patterns of the cybrids predominantly contained the mitochondrial DNA fragments of *Cichorium intybus*. However, fragments of *Helianthus annuus*, as well as new fragments, were found, reinforcing the evidence that the regenerated plants were the result of the protoplast fusion between chicory and sunflower (Rambaud et al. 1993). All the cybrids tested showed different mitochondrial restriction profiles and were, therefore, different from each other. Molecular analyses were made on the progeny of a particular CMS chicory with the cytotype 411, backcrossed three times with an industrial chicory, and the results suggested an important instability of the mitochondrial genome; among 60 brother-sister plants analyzed, four mitotypes and four floral morphologies could be found without any correlation between the mitotypes and the different floral morphologies (Rambaud et al. 1997). Further analyses made on the fifth generation of backcrosses with witloof chicories showed a general stabilization of the floral morphologies for three cytotypes studied, even if a light instability of the mitochondrial genome persists (Dubreucq et al. 1999).

3.3 CMS Origin

Determinant of the CMS in these cybrids was also investigated (Dubreucq et al. 1999). As a male sterile sunflower was used for the protoplast fusion, we wanted to know if the CMS PET1 sunflower determinant, *orf522*, had been transferred in the male sterile chicory cybrids and so could be responsible for chicory male sterility.

Analyses of three male sterile cytoplasms coming from three different fusion events showed that if *orf522* was transferred in two cytoplasms, 411 and 523, it was absent from the third cytoplasm (524); therefore, *orf522* could not be responsible for CMS in the 524 cytoplasm. In sunflower, the *orf522* gene is cotranscribed with another mitochondrial gene, the *atpA* gene, coding for the subunit α of the ATPase complex; in the male sterile cytoplasms 411 and 523, although *atpA* and *orf522* are not linked together, a weak signal could be observed by RT-PCR using *orf522*-specific oligonucleotides, but the same signals could be observed in reverted or restored 411 plants.

Moreover, when a fertile chicory line, Jupiter, was used as a pollinator, the fertility of the three cytoplasms seemed to be a restorer (Dubreucq et al. 1999). Thus, even if the 411, 523, and 524 cybrids displayed different structures, the mitochondrial CMS determinant was maintained across generations and could be of the same nature because of the restoration results. So we think it unlikely that *orf522* was responsible for CMS in the 411 and 523 cybrids.

Therefore, investigations concerning the mitochondrial determinant of CMS in chicory should continue. The appearance of spontaneous revertant plants coming from the 411 and 523 cytotypes was a good initiation to this study.

4 Summary and Conclusion

The aim of somatic hybridization in chicory was to induce or to transfer CMS in this plant. To do this, conditions were established for efficient plant regeneration from protoplasts of the cultivar Pévèle. We observed that preconditioning and age of the donor plants, nitrogen source, sucrose and auxin in the first step, and rapid transfer in an agarose medium for the second step were determinant in obtaining good regeneration yields.

Male sterile chicory plants were obtained by fusion between chicory mesophyll protoplasts and hypocotyl protoplasts derived from male sterile sunflower plants. The protoplasts were fused by the PEG method. The products were selected manually and cultivated at very low density in a liquid medium. Three to 20% of the heterokaryocytes divided and evolved first into microcalli, then into calli from which budding could be induced. Genetic and molecular analyses revealed a maternal inheritance of the male-sterile character and mitochondrial rearrangements in the cybrids. Three cytoplasms in which cytoplasmic male sterility has been maintained through generations in spite of a light instability of the mitochondrial genomes were particularly studied.

The 411 cytoplasm has been introduced in some commercial varieties of industrial chicory and witloof chicory.

5 Protocol for the Regeneration of Somatic Hybrids

5.1 Isolation of Protoplasts

Leaves of chicory were sliced to 1 mm in thickness and incubated for 5 h 30 min at 30 °C in an enzyme solution containing 3 mg ml^{-1} (w/v) Cellulase Caylase-345 (Cayla Ldt., Toulouse France), 0.5 mg ml^{-1} Pectinase Caylase-M2, and 90 g l^{-1} mannitol. Protoplasts were filtered through a 50-µm pore size stainless filter and then pelleted by centrifugation at 100 g for 15 min. The supernatant was eliminated with a Pasteur pipet until the protoplast concentration was about 10^7 ml^{-1}. For sunflower seedlings, hypocotyls were removed 6–10 days after germination on 10 g l^{-1} sucrose supplemented with 6 g l^{-1} agar, cut into pieces, and incubated in the same maceration solution.

5.2 Protoplast Fusion

Protoplasts were fused according to the method of Kao (1982) with modifications as notified by Rambaud et al. (1992). The solution of PEG was prepared by dissolving 30 g PEG 4000 Serva, 150 mg CaCl$_2$·2H$_2$O, and 10 mg KH$_2$PO$_4$ in H$_2$O; then 10 ml DMSO were added and the total volume was adjusted to 100 ml with H$_2$O and the pH adjusted to 5.5 with KOH. High Ca^{2+} ion solution was made up by mixing equal amounts of reagent A and B before use. Reagent A contains 8 g mannitol, 0.75 g glycine in 100 ml H$_2$O (adjust the pH to 10.5 with NaOH and store at 4 °C in the dark). Reagent B contains 8 g mannitol, 1.47 g CaCl$_2$·2H$_2$O in 100 ml H$_2$O. The fusion process was as follows: 1 vol of the protoplast suspension adjusted to a concentration of 10^7 ml^{-1} was placed in a 55-mm petri dish and 3 vol of the PEG solution were added dropwise, the PEG-protoplast mixture was gently homogenized for 1 min and 2 × 3.5 vol of the Ca^{2+} ion solution were added at 1-min intervals; 3 min later, 3 × 7 vol of the washing solution of protoplasts (Table 1) were added. After incubation for 10–15 min at room temperature, the fusion mixture was centrifuged (8 min at 100 g) and the protoplasts were purified by three new centrifugations in the washing medium (Table 1).

5.3 Postfusion Culture

Fused protoplasts were cultured at 30 °C onto MC1 liquid medium (Table 1). After 1 or 2 days of culture in this medium at a density of 2×10^4 protoplasts ml^{-1}, the isolated heterokaryocytes were cultured at low density ($12\mu l^{-1}$) at 30 °C in a modified MC1 medium, MC1* (Table 1), in which 1-naphtylacetic acid (NAA) was lowered to $0.5\,mg\,l^{-1}$, 2-(N-morpholino)ethanesulphonic acid (MES) (5 mM), casein hydrolysate ($150\,mg\,l^{-1}$), and coconut milk (2%; v/v) were added. One month later, colonies derived from heteroplasmic protoplast fusions were transferred onto a proliferation medium MC3 (Table 1) and then onto a medium for regeneration, MC4 (Table 1). After rooting on Heller (1953) medium H10 (Table 1), plants were transferred to a greenhouse for a few weeks and then transplanted to fields.

Acknowledgments. This research was supported by the Ets Florimond Desprez, INRA Versailles, and the Région Nord-Pas de Calais.

References

Alexander MP (1969) Differential staining of aborted or non-aborted pollen. Stain Tech 41:117–122
Bellamy A, Vedel F, Bannerot H (1996) Varietal identification in *Cichorium intybus* L. and determination of genetic purity of F_1 hybrid seed samples, based on RAPD markers. Plant Breed 115:128–132
Binding H, Nehls R, Kock R, Finger J, Mordhorst G (1981) Comparative studies on protoplast regeneration in herbaceous species of the Dicotyledoneae class. Z Pflanzenphysiol 101: 119–130
Crepy L, Chupeau MC, Chupeau Y (1982) The isolation and culture of leaf protoplasts of *Cichorium intybus* and their regeneration into plants. Z Pflanzenphysiol 107:123–131
Desprez B (1993) Recherches de méthodes d'obtention de plantes haploïdes chez la chicorée (*Cichorium intybus* L.). Thesis, University of South-Paris, Paris, 175 pp
Dubreucq A, Berthe B, Asset JF, Boulidard L, Budar F, Vasseur J, Rambaud C (1999) Analyses of mitochondrial DNA structure and expression in three cytoplasmic male sterile cicories originating from somatic hybridization between fertile chicory and CMS sunflower protoplasts. Theor Appl Genet 99:1094–1105
Heller R (1953) Recherches sur la nutrition minérale des tissus végétaux cultivés in vitro. Ann Sci Nat Bot Biol Veg 14:1–223
Kao KN (1982) Plant protoplast fusion and isolation of heterokayocytes. In: Wetter LRL, Constabel F (eds) Plant tissue culture methods. The National Research Council of Canada Prairie Regional Laboratory Sakatoon Saskatchewa, pp 49–56
Leteinturier J, Cochet JP, Marle M, Benigni M (1991) L'endive: guide pratique. CTIFL, Paris, 271 pp
Longly B, Louant BP (1987) Mécanismes de la reproduction chez la chicorée de Bruxelles: fondements et applications à la sélection. IRSIA, Bruxelles, 108 pp
Morel G, Wetmore RH (1951) Fern callus tissue cultures. Am J Bot 56:619–630
Murashige T, Skoog F (1962) A revised medium for rapid growth and bioassays with tobacco tissue cultures. Physiol Plant 15:473–497
Rambaud C, Dubois J, Vasseur J (1990) Some factors related to protoplast culture and plant regeneration from leaf mesophyll protoplasts of Magdebourg chicory (*Cichorium intybus* L. var. Magdebourg). Agronomie 10:767–772
Rambaud C, Dubois J, Vasseur J (1992) The induction of tetraploidy in chicory (*Cichorium intybus* L. var. Magdebourg) by protoplast fusion. Euphytica 62:63–67
Rambaud C, Dubois J, Vasseur J (1993) Male-sterile chicory cybrids obtained by intergeneric protoplast fusion. Theor Appl Genet 87:347–352

Rambaud C, Bellamy A, Dubreucq A, Bourquin J-C, Vasseur J (1997) Molecular analysis of the fourth progeny of plants derived from a cytoplasmic male sterile chicory cybrid. Plant Breed 116:481–486

Saksi N, Dubois J, Millecamps JL, Vasseur J (1986) Régénération de plantes de Chicorée Witloof cv. Zoom à partir de protoplastes: influence de la nutrition glucidique et azotée. C R Acad Sci Paris 302:165–170

Sidikou-Seyni R, Rambaud C, Dubois J, Vasseur J (1992) Somatic embryogenesis and plant regeneration from protoplasts of *Cichorium intybus* L. x *Cichorium endivia* L. Plant Cell Tissue Organ Cult 29:83–91

II.4 Cybridization in *Citrus unshiu* Marc. (Satsuma Mandarin) and *C. sinensis* (L.) Osbeck (Sweet Orange)

M. YAMAMOTO[1,2], S. KOBAYASHI[3], T. YOSHIOKA[1], and R. MATSUMOTO[1]

1 Introduction

1.1 Morphology, Importance, and Distribution of *Citrus unshiu* and *C. sinensis*

In citrus, male sterility is one of the most important characteristics because seedless fruits can be produced in a cultivar having both male sterility and parthenocarpy (Yamamoto et al. 1995). In many higher plants, male sterility is caused by nuclear-cytoplasmic interaction (Kaul 1988). In citrus also, aborted anthers, which are the strongest male-sterility factor (Iwamasa 1966), are caused by nuclear-cytoplasmic interaction, and only limited cultivars such as *Citrus unshiu* (satsuma mandarin) have sterile cytoplasm (Yamamoto et al. 1997). *C. unshiu*, originating in Japan, is thus an important species for producing seedless cultivars. Furthermore, the species is superior in tree habit, productivity, disease resistance, and fruit quality (early ripening, easy peeling, seedlessness) and adapts to the Japanese climate. Hence, production of *C. unshiu* accounts for approximately 70% of the Japanese citrus production and is spreading all over the world. On the other hand, *C. sinensis* (sweet orange) is the most economically important citrus in the world because of its excellent fruit quality. Although there are four types of *C. sinensis* (common orange, navel orange, blood orange, and acidless orange), all cultivars of *C. sinensis* have been developed from mutations.

1.2 Need for Cybridization

In seedless breeding, it is important to control the aborted anthers of *Citrus* by both nuclear and cytoplasmic genes. Since the cytoplasmic gene is inherited maternally, seed parental cultivars which can produce progenies with aborted anthers are restricted. Sterile cytoplasm has so far been found only in

---

[1] Department of Citriculture, National Institute of Fruit Tree Science, Kuchinotsu, Nagasaki 859-2501, Japan
[2] Present address: Faculty of Agriculture, Kagoshima University, Kagoshima 890-0065, Japan
[3] Persimmon and Grape Research Center, National Institute of Fruit Tree Science, Akitsu, Hiroshima 729-2494, Japan

C. unshiu and Encore (*C. nobilis* × *C. deliciosa*) (Yamamoto et al. 1992, 1997). It is essential to develop many cybrids having sterile cytoplasm to produce various types of seedless cultivars with aborted anthers.

In many higher plants, such as rice (Kadowaki et al. 1986) and maize (Kemble et al. 1980), a strong relationship has been reported between cytoplasmic male sterility (CMS) and mitochondrial DNA (mtDNA). In citrus also, CMS is most likely related to mtDNA. Recently, Saito et al. (1993) demonstrated that *Citrus* cybrids having a nuclear genome of mesophyll parent and mtDNA of a callus parent could be produced by symmetric protoplast fusion. We also performed electrofusion between protoplasts derived from *C. unshiu* callus and *C. sinensis* mesophyll to develop a cybrid having sterile cytoplasm of *C. unshiu* and nuclear genome of *C. sinensis*.

1.3 Brief Review of Past and Present Work

Citrus cybrids were first produced using a donor-recipient system by Vardi et al. (1987). Vardi et al. (1989) also produced cybrids having a *Citrus* nuclear genome and a *Microcitrus* cytoplasmic genome by the same method. They analyzed the cytoplasmic genome of the cybrids and discovered the recombination of mtDNA and segregation of chloroplast DNA (cpDNA) (Table 1).

Ohgawara et al. (1989) produced diploid ($2n = 18$) *C. paradisi*-like plants by symmetric protoplast fusion between nucellar callus of *C. sinensis* and mesophyll of *C. paradisi*. Saito et al. (1993) found that these plants were cybrids having mtDNA of *C. sinensis*. Moreover, Saito et al. (1993) demonstrated that cybrids having the nuclear genome of a mesophyll parent and the mitochondrial genome of a callus parent could be produced by symmetric fusion using *C. sudachi* callus and *C. aurantifolia* or *C. limon* mesophyll. Yamamoto and Kobayashi (1995), Grosser et al. (1996), and Moriguchi et al. (1996) also reported the *Citrus* cybrids produced by symmetric fusion, and that all cybrids had the nuclear genome of mesophyll parents and the mitocondrial genome

Table 1. Citrus cybrids produced by the donor-recipient system

Donor (γ-irradiated)	Recipient (IA-treated)	Composition of genomes			Reference
		Nuclear	mt	cp	
PPT (Poorman X *Poncirus trifoliata*)	*Citrus limon* (L)[a] cv. Villafranca	L	L+[b]	–	Vardi et al. (1987)
PPT (P)	*C. aurantium* (A)	A	P	–	Vardi et al. (1987)
C. aurantium (A)	*C. limon* (L) cv. Villafranca	L	A	–	Vardi et al. (1987)
Microcitrus (M)	*C. jambhili* (J)	J	Re[c]	M or J	Vardi et al. (1989)
Microcitrus (M)	*C. aurantium* (A)	A	Re[c]	–	Vardi et al. (1989)

IA = Iodoacetate, mt = mitochondria, cp = chloroplast.
[a] Abbreviation of species or cultivars.
[b] Additional fragments were observed.
[c] Recombination event was detected.

of callus parents. Recombination/rearrangement of mtDNA in *Citrus* interspecific cybrids produced by symmetric fusion were reported by Moriguchi et al. (1997) and Tokunaga et al. (1999) (Table 2).

Saito et al. (1994) demonstrated that somatic hybrids could be produced using cybrid callus. In citrus protoplast fusion, embryogenic callus from nucellar tissue or embryo of polyembryonic cultivars is necessary. However, genetically true-to-type embryogenic callus cannot be obtained from embryo culture of monoembryonic cultivars because of high heterozygosity. Thus, embryogenic callus of cybrids having a nuclear genome of monoembryonic cultivars will expand the range of combinations in protoplast fusion.

2 Somatic Hybridization

2.1 Isolation and Fusion of Protoplasts

Culture of embryogenic callus and protoplast isolation were conducted according to Kobayashi et al. (1988), who produced *Citrus* somatic hybrids. The density of protoplasts from suspension-cultured cells of *C. unshiu* cv. Juman unshiu and leaves of *C. sinensis* cv. F. N. Washington navel orange was adjusted to $5 \times 10^5 \,\text{ml}^{-1}$ and $1 \times 10^6 \,\text{ml}^{-1}$, respectively (Fig. 1A,B).

Citrus cybrids can be obtained by a donor-recipient system as well as the standard symmetric fusion (Vardi et al. 1987; Saito et al. 1993). In this study, we planned to produce cybrids using the latter because it is quite simple: irradiation and iodoacetate treatment are not necessary, and irradiation might cause damage or mutation of cells.

Although many somatic hybrids were produced by the polyethylene glycol (PEG) method in citrus (Ohgawara et al. 1994), the efficiency of this method is not so high. Saito et al. (1991) and Hidaka and Omura (1992) demonstrated that efficiency of somatic hybrid production was higher in electrofusion than in the PEG method. Therefore, electrofusion was performed in this study. The electrical parameters used in this study were as follows: AC fields, 1 MHz, $125 \,\text{V cm}^{-1}$, 60 s; DC field, $1250 \,\text{V cm}^{-1}$ square-pulse 100 µs, three times at 0.1-s intervals.

2.2 Regeneration of Cybrids

The selection system of somatic hybrids in citrus, in which fused protoplasts were cultured in hormone-free MT medium containing 0.6 M sucrose, was established (Ohgawara et al. 1985). Saito et al. (1993) found that cybrids as well as somatic hybrids could be produced using this selection method.

The fusion products were cultured in hormone-free MT medium (BM) containing 0.6 M sucrose and 0.6% Sea Plaque agarose at a cell density of 1×10^5 cells ml^{-1} (Fig. 1C). Five pale green globular embryoids developed. The

Table 2. Citrus cybrids produced by symmetric fusion

Callus parent	Mesophyll parent	Composition of genomes			Reference
		Nuclear	mt	cp	
Citrus sudachi (S)	C. aurantifolia (Af)[a]	Af	S	–	Saito et al. (1993)
C. sudachi (S)	C. limon (L)	L	S	–	Saito et al. (1993)
C. unshiu (U) cv. Juman	C. sinensis (Si) cv. F. N. Washington	Si	U	U	Yamamoto and Kobayashi (1995)
C. microcarpa (M)	C. aurantium (A) cv. Keen	A	M	M	Grosser et al. (1996)
C. reticulata (R) cv. Cleopatra	C. aurantium (A)	A	R	R or A	Grosser et al. (1996)
C. sinensis (Si) cv. Valencia	C. Limon (L) cv. Femminello	L	Si	–	Grosser et al. (1996)
Seminole (Se) (C. paradisi × C. tangerina)	C. limon (L) cv. Lisbon	L	Se	–	Moriguchi et al. (1996)
C. reticulata (R) cv. Hazzara (Abohar)	C. jambhiri (J)	J	R	–	Moriguchi et al. (1996)
Seminole (Se) (C. paradisi × C. tangerina)	C. jambhiri (J)	J	Se(Re)[b]	–	Moriguchi et al. (1997)
C. unshiu (U) cv. Juman	C. junos (Ju)	Ju	U(Re)[b]	U	Tokunaga et al. (1999)
C. unshiu (U) cv. Juman	C. limon (L) cv. Eureka	L	U(Re)[b]	U	Tokunaga et al. (1999)

mt = mitochondria, cp = chloroplast.
[a] Abbreviation of species or cultivars.
[b] Cybrid has all fragments derived from the callus parent, and additional fragments were detected (recombination/rearrangement event occurred).

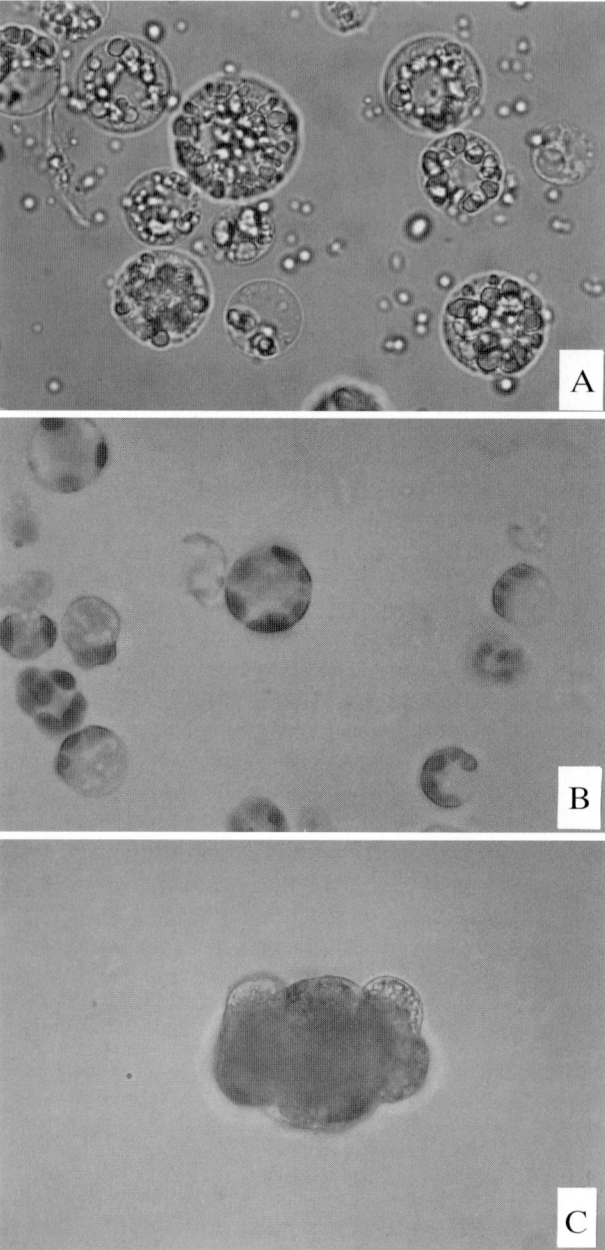

Fig. 1. A Nucellar protoplasts of *Citrus unshiu*. **B** Mesophyll protoplasts of *C. sinensis*. **C** Division of protoplasts

embryoids were transferred to BM containing $500\,mg\,l^{-1}$ malt extract, $40\,mg\,l^{-1}$ adenine, 5% sucrose, and 0.9% agar. One of the five developed into a cotyledonary embryoid after 1 month. The cotyledonary embryoid was transferred to BM containing $10\,mg\,l^{-1}$ GA_3, 2% sucrose, and 0.9% agar.

Fig. 2. Leaf morphology (from *left* to *right*) of *C. unshiu*, the regenerated plant (cybrid), *C. sinensis* (Yamamoto and Kobayashi 1995). Leaf of the plant and *C. sinensis* has a wing, but that of *C. unshiu* has no wing

Within 3 months of culture, the cotyledonary embryoid developed into an entire plant.

Leaf morphology of the regenerated plant was similar to that of the mesophyll parent *C. sinensis*; both the regenerated plant and *C. sinensis* had a wing in the leaves, but *C. unshiu* had no wing (Fig. 2). Chromosome counts showed that the regenerated plant was diploid ($2n = 18$), the same as both parents (Fig. 3).

2.3 Flower and Fruit Characteristics of the Cybrid

The regenerated plant was grafted onto potted *Poncirus trifoliata* rootstock, and grown in an unheated greenhouse. The plant developed flowers and set fruits. The flowering habit of the plant was raceme, the same as *C. sinensis* (Fig. 4). Flower characteristics of the plant were almost identical to those of *C. sinensis* except for the number of stamens (Fig. 5; Table 3). The fruit of the plant also resembled that of *C. sinensis* (Figs. 6, 7). Table 4 shows several fruit characteristics of the plant and its parents. The qualitative characteristics such as fruit shape, presence of navel, fruit peelings, kind of aroma, and flesh color of the plant were identical to those of *C. sinensis*. Some quantitative characteristics of the plant, which were easily affected by environmental conditions, differed slightly from those of *C. sinensis* because the fruit of the plant was the first crop and very few fruit were set.

Top working trees of the plant were also grown in orchards, but these trees did not set fruits. We plan to investigate the various detailed traits of the plant, such as tree and flower characteristics, and disease and cold resistance.

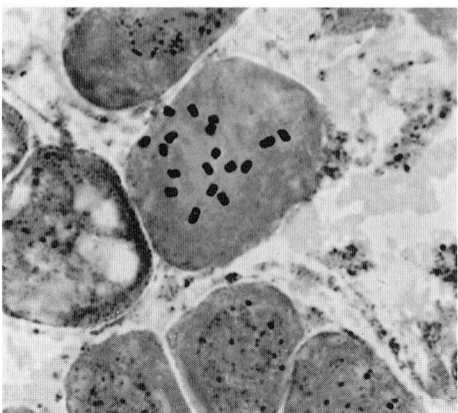

Fig. 3. Metaphase plate from the regenerated plant (cybrid) ($2n = 18$, same as both parents). (Yamamoto and Kobayashi 1995)

Fig. 4. Flowering habit of the regenerated plant (cybrid) which was grafted onto potted *Poncirus trifoliata* in an unheated greenhouse. Flowering habit of the plant is raceme, the same as *C. sinensis*

2.4 DNA Analysis

To clarify the genetic constitution of the regenerated plant, we performed restriction fragment length polymorphisms (RFLP) analysis of its nuclear and cytoplasmic genomes. For nuclear DNA analysis, plasmid pRR217, which contains whole nuclear rDNA sequences of rice (Takaiwa et al. 1984), was used as a probe. The regenerated plant had the same rDNA fragment pattern as *C.*

Fig. 5. Flower morphology of the regenerated plant (cybrid) (*left*) and *C. sinensis* (*right*)

Fig. 6. Fruit of the regenerated plant (cybrid) which was grafted onto potted *Poncirus trifoliata* in unheated greenhouse. First fruit was observed 2 years after grafting

sinensis (Fig. 8). For cpDNA analysis, plasmid pTBa1, which contains the *Nicotiana tabacum* cpDNA fragment (Sugiura et al. 1986), was used as a probe. CpDNA of the regenerated plant was identical to that of *C. unshiu* (Fig. 9). For mitochondrial DNA analysis (mtDNA), we used P9.7 and P12.4 fragments from *Brassica campestris* mtDNA (Palmer and Shields 1984) and *atpA*, *atp6*,

Table 3. Flower characteristics of the regenerated plant (cybrid) and *C. sinensis* cv. F. N. Washington

Plant	Flower weight (g)	Shape of flower	Petal				Stamen		Pollen yield	Shape of ovary	Shape of style
			Length (mm)	Width (mm)	No.	Color	No.	Degree of separation			
Regenerated plant (cybrid)	0.31 ± 0.01	Spindle	14.5 ± 0.3	5.7 ± 0.2	4.2 ± 0.1	White	17.6 ± 0.4	Separate	None	Oblong	Straight
C. sinensis cv. F. N. Washington	0.33 ± 0.01	Spindle	15.4 ± 0.2	6.2 ± 0.2	4.0 ± 0.1	White	20.7 ± 0.6	Separate	None	Oblong	Straight

These plants were grafted onto potted trifoliata orange rootstocks in unheated greenhouse. Date of analysis: April 11, 1995.

Table 4. Fruit characteristics of the regenerated plant (cybrid) and parents, *C. sinensis* cv. F. N. Washington and *C. unshiu* cv. Juman

Plant	Fruit weight (g)	Shape of fruit	D/H[a] index	Navel of fruit	Thickness of rind (mm)	Fruit peeling	Kind of aroma	Flesh color	Brix	Citric acid (%)
Regenerated plant (cybrid)	430	Globose	101	Present	8.0	Slightly difficult	*C. sinensis*	Yellowish orange	9.8	0.91
C. sinensis cv. F. N. Washington	286	Globose	105	Present	3.9	Slightly difficult	*C. sinensis*	Yellowish orange	10.4	1.32
C. unshiu cv. Juman	130	Oblate	149	Absent	4.0	Easy	*C. unshiu*	Orange	12.3	1.45

These plants were grafted onto potted trifoliata orange rootstocks in unheated greenhouse. Date of analysis: January 13, 1997.
[a] Diameter × 100/height.

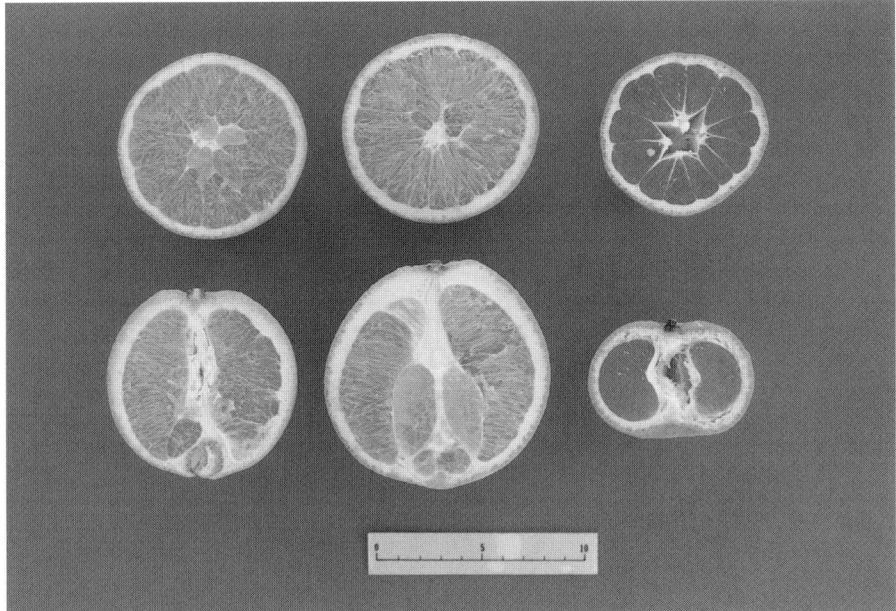

Fig. 7. Cut fruits (from *left* to *right*) of *C. sinensis*, the regenerated plant (cybrid), *C. unshiu*. Fruit shape of the regenerated plant is almost identical to that of *C. sinensis*

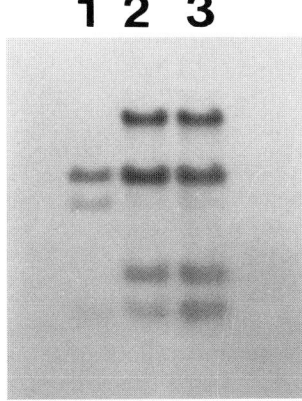

Fig. 8. Southern blot hybridization of SacI digests of total DNA to labeled rDNA fragments. **1** *C. unshiu*; **2** *C. sinensis*; **3** the regenerated plant (cybrid). (Yamamoto and Kobayashi 1995)

atp9, and *cox* I from rice (Kadowaki et al. 1989, 1990a, b; Ishikawa and Kadowaki 1993). In 18 out of 19 probes and restriction enzyme combinations, mtDNA fragment patterns of the regenerated plants were identical to those of *C. unshiu* (Fig. 10). When *atp6* and HindIII were used as the probe-enzyme combination, the regenerated plant had specific fragments derived from both parents (Fig. 10). From these results, we concluded that the regenerated plant

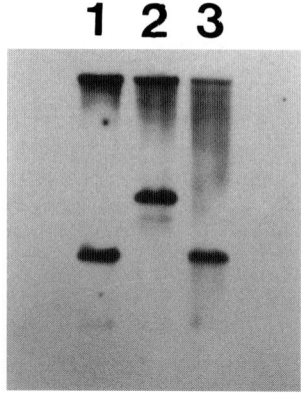

Fig. 9. Southern blot hybridization of PstI digests of total DNA to labeled cpDNA fragments. **1** *C. unshiu*; **2** *C. sinensis*; **3** the regenerated plant (cybrid). (Yamamoto and Kobayashi 1995)

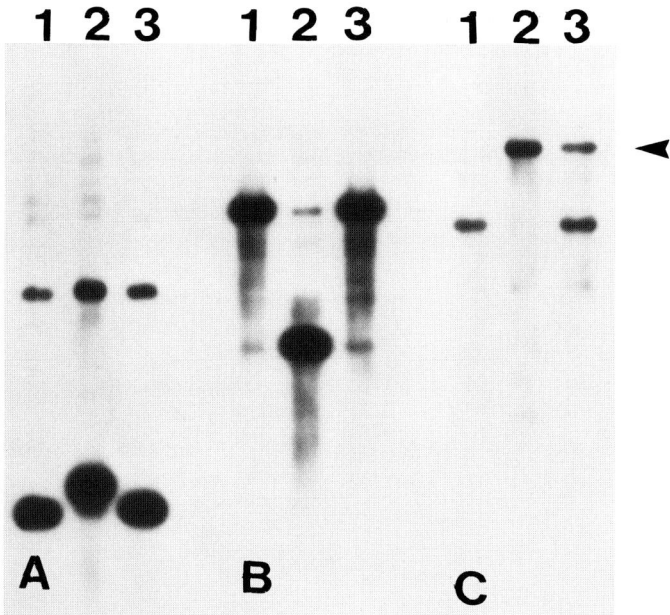

Fig. 10. Southern blot hybridization of restriction endonuclease digests of total DNA to labeled mtDNA fragments (*atp6*). **1** *C. unshiu*; **2** *C. sinensis*; **3** the regenerated plant (cybrid). *A* BamHI digested; *B* EcoRI digested; *C* HindIII digested. *Arrow* indicates the specific fragment of *C. sinensis*

was a cybrid having *C. sinensis* nuclear and *C. unshiu* cytoplasmic genomes. In addition, mtDNA recombination or rearrangement probably occurred in the regenerated plant.

Recombination and/or rearrangement of the mtDNA through protoplast fusion were reported for various higher plants (Waara and Glimelius 1995). In citrus protoplast fusion, recombination or rearrangement of mtDNA occurred in intergeneric combinations. Vardi et al. (1989) produced cybrids between *Citrus* and *Microcitrus* by a donor-recipient system, and reported that recombination occurred in mtDNA. Motomura et al. (1995) reported mtDNA recombination of somatic hybrids between *Citrus* and *Atalantia* or *Severinia*. In contrast, mtDNA of all somatic hybrids and cybrids had been identical to those of callus parents in interspecific combinations (Kobayashi et al. 1991; Saito et al. 1993; Yamamoto and Kobayashi 1995; Grosser et al. 1996; Moriguchi et al. 1996). However, Moriguchi et al. (1997) and Tokunaga et al. (1999) recently reported mtDNA recombination/rearrangement of *Citrus* cybrids through symmetric protoplast fusion. The mtDNA of the cybrid obtained in this study was almost identical to that of the callus parent (*C. unshiu*), but a recombination or rearrangement event was detected. From these results, recombination and/or rearrangement of mtDNA could occur not only in intergeneric protoplast fusion, but also in interspecific protoplast fusion in citrus. We considered that the presence or absence of recombination and/or rearrangement of mtDNA was affected by parental combination, source of protoplasts, and culture conditions.

In contrast to mtDNA, recombined cpDNA has rarely been developed in protoplast fusion in many higher plants (Waara and Glimelius 1995). In cybrids of *Citrus* with a plastome from *Microcitrus* produced by asymmetric fusion, Vardi et al. (1989) showed that cpDNA was completely segregated in the cybrid plants. Kobayashi et al. (1991) also reported that each of 16 somatic hybrids (*C. sinensis* + Murcott tangor) contained either one parental cpDNA or the other. The cpDNA of the cybrid obtained in this study was the same as that of *C. unshiu*. Grosser et al. (1996) also reported cpDNA segregation in *Citrus* cybrids. However, Motomura et al. (1996) detected the rearrangement or recombination of cpDNA in *Citrus* + *Microcitrus* somatic hybrids. We considered that segregation of cpDNA also most likely occurs in protoplast fusion of citrus, though there is an exception. This point should be carefully investigated, because there have been few studies on detailed cpDNA analysis of citrus cybrids and somatic hybrids.

3 Summary and Conclusion

We performed electrofusion to combine *Citrus unshiu* cv. Juman unshiu protoplasts isolated from embryogenic callus with *C. sinensis* cv. F. N. Washington navel orange mesophyll protoplasts. One plant was regenerated from the fusion products. The plant had 18 chromosomes ($2n = 18$ in each parent), and its leaf, flower, and fruit characteristics were almost identical to those of *C. sinensis*. The regenerated plant showed the same nuclear rDNA fragment pattern as that of *C. sinensis*, whereas the plant showed chloroplast and mitochondrial DNA (mtDNA) banding patterns identical to those of *C. unshiu*. From these results, we confirmed the regenerated plant to be a cybrid having the *C. sinensis* nuclear genome and the *C. unshiu* cytoplasmic genome. In addition, the cybrid had specific mtDNA fragments derived from both parents when rice *atp6* and HindIII were used as the probe and restriction enzyme, respectively. This result suggests that mtDNA recombination or rearrangement occurred in this cybrid.

Cytoplasm regulates cytoplasmic male sterility, vital plant functions such as photosynthesis, sugar and fatty acid metabolism, ATP production, and disease resistance. Since there is little cytoplasmic information on citrus, this cybrid offers new information about the cytoplasmic function.

This study demonstrates that cybrids having the nuclear genome derived from the mesophyll cell and the cytoplasmic genome derived from the callus cell can be produced from electrofusion using embryogenic callus of *C. unshiu*. *C. unshiu* has a sterile cytoplasm and is moderately resistant to major diseases such as citrus canker and citrus tristeza virus. Thus, the cybrid produced in this study may be a useful material for breeding and studying the function of cytoplasm.

4 Protocol

Protoplasts were isolated from callus and leaves. Callus derived from nucellar embryo of *C. unshiu* cv. Juman unshiu was used as a source of protoplasts. This culture was maintained for 1 year on agar medium consisting of Murashige and Tucker (MT) (Murashige and Tucker 1969) medium, in which $20\,mg\,l^{-1}$ kinetin only was used as phytohormone. Callus was suspended in a liquid MT medium supplemented with $20\,mg\,l^{-1}$ kinetin. Serial transfer of the callus was carried out every 2 weeks. *C. sinensis* cv. F. N. Washington navel orange was grafted onto a young trifoliate orange by the method of Takahara et al. (1986). These plants were grown in a growth chamber kept at 25°C under $16\,h\,day^{-1}$ illumination with cool fluorescent light (3000 lx). About three fully expanded leaves were harvested from plants after 2 months of grafting.

Protoplasts from both suspension-cultured cells and leaves were isolated by the method of Kobayashi et al. (1988), and suspended in 0.6 M mannitol. Their density was adjusted at $5 \times 10^5\,ml^{-1}$ and $1 \times 10^6\,ml^{-1}$, respectively.

Electrofusion was carried out using a model BE-800 electrofusion apparatus (Kansai denshi, Osaka, Japan) connected to an electrode FTC-33D5 (Shimadzu Co. Ltd., Kyoto, Japan). Callus and mesophyll protoplasts were mixed in equal volumes, and 3 ml of the mixture was transferred to a 60-mm diameter plastic petri dish and induced fusion. The electrical parameters used in this study were as follows: AC fields, 1 MHz, $125\,V\,cm^{-1}$, 60s; DC field, $1250\,V\,cm^{-1}$ square-pulse 100 µs, three times at 0.1-s intervals.

The treated protoplasts were transferred to 10-ml tubes and pelleted by centrifugation at 80 g for 3 min. The supernatant was discarded and the treated protoplasts were cultured in 4 ml medium, which consisted of hormone-free MT medium (BM) containing 0.6 M sucrose and 0.6% Sea Plaque agarose (FMC, Rockland, ME, USA) at a cell density of 1×10^5 cells ml^{-1}, in Falcon dishes (60 × 15 mm).

Embryoids derived from protoplasts were transferred to BM containing $500\,mg\,l^{-1}$ malt extract, $40\,mg\,l^{-1}$ adenine, 5% sucrose, and 0.9% agar. A cotyledonary embryoid which developed was transferred to BM containing $10\,mg\,l^{-1}$ GA_3, 2% sucrose, and 0.9% agar.

Five root tips of regenerated plants pretreated with 8-hydroxyquinoline (2 mM) for 20 h at 10°C were fixed in a mixed solution of ethanol and acetic acid (3:1) for 24 h, and then stained with lacto-propionic orcein for 3 h, according to Oiyama (Oiyama 1981).

Total DNA was extracted from leaves according to Rogers and Bendich (1985). One µg of DNA, digested with restriction endonucleases for 5 h at 37°C, was separated on 0.8% agarose gels, transferred to nitrocellulose filter according to Southern (1975), and baked in vacuo for 2 h at 80°C.

DNA fragments prepared from the following recombinant plasmids were used as probes: plasmid pRR217 contains whole nuclear rDNA sequences of rice (Takaiwa et al. 1984), plasmid pTBa1 contains *Nicotiana tabacum* cpDNA fragment (Sugiura et al. 1986), mtDNA clones contain PstI fragments of *Brassica campestris* mtDNA (Palmer and Shields 1984) and mtDNA oligonucleotides of rice (Kadowaki et al. 1989, 1990a, b; Ishikawa and Kadowaki 1993). Plasmid pRR217, pTBa1, mtDNA clones, and mtDNA oligonucleotides were kindly provided by Drs. K. Oono, M. Sugiura, J. D. Palmer and K. Kadowaki, respectively.

Labeling of probe DNA and visualization of probe-target DNA hybrid were carried out by the ECL method (Amersham).

References

Grosser JW, Gmitter FG, Tusa N, Reforgianto Recupero G, Cucinotta P (1996) Further evidence of a cybridization requirement for plant regeneration from citrus leaf protoplasts following somatic fusion. Plant Cell Rep 15:672–676

Hidaka T, Omura M (1992) Regeneration of somatic hybrid plants obtained by electrical fusion between satsuma mandarin (*Citrus unshiu*) and rough lemon (*C. jambhiri*) or yuzu (*C. junos*). Jpn J Breed 42:79–89

Ishikawa M, Kadowaki K (1993) Excess RNA editing in rice mitochondrial *atp9* transcripts. Plant Cell Physiol 34:959–963

Iwamasa M (1966) Studies on the sterility in genus *Citrus* with special reference to the seedlessness. Bull Hortic Res Stn Jpn B6:1–81

Kadowaki K, Ishige T, Suzuki T, Harada K, Shinjo C (1986) Difference in the characteristics of mitochondrial DNA between normal and male sterile cytoplasms of japonica rice. Jpn J Breed 36:333–339

Kadowaki K, Suzuki T, Kazama S, Oh-fuchi T, Sakamoto W (1989) Nucleotide sequence of the cytochrome oxidase subunit I gene from rice mitochondria. Nucleic Acids Res 17:7519

Kadowaki K, Kazama S, Suzuki T (1990a) Nucleotide sequence of the F1-ATPase α-subunit gene from rice mitochondria. Nucleic Acids Res 18:1302

Kadowaki K, Suzuki T, Kazama S (1990b) A chimeric gene containing the 5′ portion of *atp6* is associated with cytoplasmic male sterility of rice. Mol Gen Genet 224:10–16

Kaul MLH (1988) Gene-cytoplasmic male sterility. In: Male sterility in higher plants. Springer, Berlin Heidelberg New York, pp 97–192

Kemble RJ, Gunn RE, Flavell RB (1980) Classification of normal and male-sterile cytoplasms in maize. II. Electrophoretic analysis of DNA species in mitochondria. Genetics 95:451–458

Kobayashi S, Ohgawara T, Ohgawara E, Oiyama I, Ishii S (1988) A somatic hybrid plant obtained by protoplast fusion between navel orange (*Citrus sinensis*) and satsuma mandarin (*C. unshiu*) Plant Cell Tissue Organ Cult 14:63–69

Kobayashi S, Ohgawara T, Fujiwara K, Oiyama I (1991) Analysis of cytoplasmic genomes in somatic hybrids between navel orange (*Citrus sinensis* Osb.) and 'Murcott' tangor. Theor Appl Genet 82:6–10

Moriguchi T, Hidaka T. Omura M, Motomura M, Akihama T (1996) Genotype and parental combination influence efficiency of cybrid induction in *Citrus* by electrofusion. Hortscience 31:275–278

Moriguchi T, Motomura T, Hidaka T, Akihama T, Omura M (1997) Analysis of mitochondrial genomes among *Citrus* plants produced by the interspecific somatic fusion of 'Seminole' tangelo with rough lemon. Plant Cell Rep 16:397–400

Motomura T, Hidaka T, Moriguchi T, Akihama T, Omura M (1995) Intergeneric somatic hybrids between *Citrus* and *Atalantia* or *Severinia* by electrofusion, and recombination of mitochondrial genomes. Breed Sci 45:309–314

Motomura T, Moriguchi T, Akihama T, Hidaka T, Omura M (1996) Analysis of cytoplasmic genomes in somatic hybrids between 'Hazzara (Abohar)' (*Citrus reticulata* Blanco) and *Microcitrus australis* (Planch.) Swingle. J Jpn Soc Hortic Sci 65:497–503

Murashige T, Tucker DPH (1969) Growth factor requirements of *Citrus* tissue culture. In: Chapman HD (ed) Proc 1st Int Citrus Symp. Univ California, Riverside, vol 3, pp 1155–1161

Ohgawara T, Kobayashi S, Ohgawara E, Uchimiya H, Ishii S (1985) Somatic hybrid plants obtained by protoplast fusion between *Citrus sinensis* and *Poncirus trifoliata*. Theor Appl Genet 71:1–4

Ohgawara T, Kobayashi S, Ishii S, Yoshinaga K, Oiyama I (1989) Somatic hybridization in *Citrus*: navel orange (*C. sinensis* Osb.) and grapefruit (*C. paradisi* macf.). Theor Appl Genet 78:609–612

Ohgawara T, Uchimiya H, Ishii S, Kobayashi S (1994) Somatic hybridization between *Citrus sinensis* and *Poncirus trifoliata*. In: Bajaj YPS (ed) Biotechnology in agriculture and forestry, vol 27. Somatic hybridization in crop improvement I. Springer, Berlin Heidelberg New York, pp 439–454

Oiyama I (1981) A technique for chromosome observation in root tip cells of citrus. Bull Fruit Tree Res Stn D3:1–7

Palmer JD, Shields CR (1984) Tripartite of the *Brassica campestris* mitochondrial genome. Nature 307:437–440

Rogers SO, Bendich AJ (1985) Extraction DNA from milligram amounts of fresh, herbarium, and mummified plant tissues. Plant Mol Biol 5:69–76

Saito W, Ohgawara T, Shimizu J, Ishii S (1991) Acid citrus somatic hybrids between sudachi (*Citrus sudachi* Hort. ex Shirai) and lime (*C. aurantifolia* Swing.) produced by electrofusion. Plant Sci 77:125–130

Saito W, Ohgawara T, Shimizu J, Ishii S, Kobayashi S (1993) *Citrus* cybrid regeneration following cell fusion between nucellar cells and mesophyll cells. Plant Sci 88:195–201

Saito W, Ohgawara T, Shimizu J, Kobayashi S (1994) Somatic hybridization in *Citrus* using embryogenic cybrid callus. Plant Sci 99:89–95

Southern EM (1975) Detection of specific sequences among DNA fragments separated by gel electrophoresis. J Mol Biol 98:503–517

Sugiura M, Shinozaki K, Zaita N, Kusuda M, Kumano M (1986) Clone bank of the tobacco (*Nicotiana tabacum*) chloroplast genome as a set of overlapping restriction endonuclease fragments: mapping of 11 ribosomal protein genes. Plant Sci 44:211–216

Takahara T, Okudai N, Kuhara S (1986) Elimination of citrus viruses by semi-micrografting. Bull Fruit Tree Res Stn D8:13–24

Takaiwa F, Oono K, Sugiura M (1984) The complete nucleotide sequence of a rice 17S rRNA gene. Nucleic Acids Res 12:5441–5448

Tokunaga T, Yamao M, Takenaka M, Akai T, Hasebe H, Kobayashi S (1999) Cybrid plants produced by electrofusion between satsuma mandarin (*Citrus unshiu*) and yuzu (*C. junos*) or lemon (*C. limon*), and recombination of mitochondrial genomes. Plant Biotechnol 16:297–301

Vardi A, Breiman A, Galun E (1987) *Citrus* cybrids: production by donor-recipient protoplast-fusion and verification by mitochondrial-DNA restriction profiles. Theor Appl Genet 75:51–58

Vardi A, Arzee-Gonen P, Frydman-Shani A, Bleichman S, Galun E (1989) Protoplast-fusion-mediated transfer of organellas from *Microcitrus* into *Citrus* and regeneration of novel alloplasmic trees. Theor Appl Genet 78:741–747

Waara S, Glimelius K (1995) The potential of somatic hybridization in crop breeding. Euphytica 85:217–233

Yamamoto M, Kobayashi S (1995) A cybrid plant produced by electrofusion between *Citrus unshiu* (satsuma mandarin) and *C. sinensis* (sweet orange). Plant Tissue Cult Lett 12:131–137

Yamamoto M, Okudai N, Matsumoto R (1992) Segregation for aborted anthers in hybrid seedlings using *Citrus nobilis* × *C. deliciosa* cv. Encore as the seed parent. J Jpn Soc Hortic Sci 60:785–789

Yamamoto M, Matsumoto R, Yamada Y (1995) Relationship between sterility and seedlessness in citrus. J Jpn Soc Hortic Sci 63:335–339

Yamamoto M, Matsumoto R, Okudai N, Yamada Y (1997) Aborted anthers of *Citrus* result from gene-cytoplasmic male sterility. Sci Hortic 70:9–14

II.5 Somatic Hybridization in *Cucumis*

C.I. JARL

1 Introduction

1.1 The Crops

Almost one fifth of the world's vegetable production consists of the harvests of different members of the family Cucurbitaceae. The most important species for human use in this group are different types of pumpkins, squash and gourds (e.g., *Luffa* sp. *Lagenaria* sp. *Cucurbita pepo*, *C. mixta*, *C. moschata*, *C. maxima*), cucumbers and gherkins (*Cucumis sativus* and *C. anguria*), cantaloupes and other types of melon (*Cucumis melo*), and watermelon (*Citrullus vulgaris*). Among others, the watermelon is responsible for almost half of both the production and the acreage. Table 1 shows that various Cucurbitaceae species are produced in most parts of the world, although the distribution varies (FAOSTAT). Although the cucurbits, being vegetable crops, do not belong to the major staple crops, they are still of great importance as a complement to the diet in many countries.

During the 25 years from 1962 to 1997, the production of the different crops more than doubled. This was accomplished to some degree by an increase in acreage cultivated by cucumber and gherkins (25% increase) and melons (50%), while the area grown with pumpkins, gourd, and squash has actually diminished (Table 2; FAOSTAT). The most important factor in increasing the production is the dramatic improvement in yield ha^{-1} in the different crops, accomplished by intensive breeding programs.

The yield is, however, still substantially diminished by different diseases to which the cucurbits are prone. The economic losses caused by virus diseases are especially damaging, as no efficient treatment is available. In cucumber and melon, virus diseases such as watermelon mosaic virus, zucchini yellow mosaic virus, and papaya ringspot virus are responsible for a significant loss each year. Great efforts have been invested in breeding programs to improve the disease resistance of the different crops. In the genus *Cucumis*, the available genetic variation in the different cultivated crops is not sufficient (e.g., Knerr et al. 1989; den Nijs and Custers 1990). It is therefore of interest to extend the gene pool to wild relatives in which valuable sources of disease

Plant Biology, Lund University, P.O. Box 117, 221 00 Lund, Sweden

Table 1. Data obtained from FAO database (http://apps.fao.org) on area harvested, yield, and production of different cucurbits on the different continents in 1997

	Pumpkins, squash, gourds	Cucumber and gherkins	Watermelons	Cantaloupes and other melons
1997: Area harvest (HA)				
North America	2320	67320	74706	58015
South America	60230	4650	131749	40450
Oceania	14196	1420	4989	3692
Africa	86820	28275	145250	57980
Europe	160350	241659	278121	142367
Asia	726136	1180247	1781279	655188
World	1127623	1557576	2466503	1038978
1997: Yield (HG/HA)				
North America	163793	170915	247330	184475
South America	138812	164467	87471	105566
Oceania	126243	156056	170976	212013
Africa	158535	141485	169251	180250
Europe	119355	144946	128983	171407
Asia	122049	171696	203842	175064
World	121225	165706	186831	167699
1997: Production (MT)				
North America	38000	1150600	1847700	1070230
South America	836062	76477	1152416	427013
Oceania	179215	22160	85300	78275
Africa	1376400	400050	2458364	1045090
Europe	1913864	3502741	3587300	2440273
Asia	8862391	20264430	36309950	11469860
World	13669610	25809980	46081920	17420480

Table 2. Data obtained from FAO database (http://apps.fao.org) on area harvested, yield, and production of different cucurbits total in the world in 1962 and 1997, respectively

	Pumpkins, squash, gourds	Cucumber and gherkins	Watermelons	Cantaloupes and other melons
Area harvest (HA)				
1962	1328661	1022017	1992766	575022
1997	1127623	1557576	2466503	1038978
Yield (HG/HA)				
1962	42523	91701	95685	116351
1997	121225	165706	186831	167669
Production (MT)				
1962	5649815	9371964	19067730	6690427
1997	13669610	25809980	46081920	17420480

resistance have been found (e.g., Kroon et al. 1979). Resistance to several diseases affecting both cucumber and melon can be found in wild relatives (Provvidenti 1987; Provvidenti et al. 1978, 1984; Herrington et al. 1991; Fuchs and Gonsalves 1995; Pan and More 1996).

1.2 Need for Somatic Hybridization

Gene transfer among different species in *Cucumis* is limited due to strong sexual incompatibility barriers (Deakin et al. 1971). These sexual barriers may probably be related to the origin of the species to some extent. Two geographical and cytogenetically different groups have been identified within the genus *Cucumis* (Whitaker and Bermis 1976). Most evidence suggests that the group of cucumber and gherkins originated from India. This is the only species within the genus with a chromosome number of $2n = 2x = 14$. The closely related member of this group, *C. hardwickii*, can be sexually crossed, producing viable offspring with *C. sativus*. *C. hardwickii* is sometimes not considered as a separate species, but named *C. sativus*. var. hardwickii. The other group, *Melo*, defined as a subgenus by Jeffrey (1980), originated in the southern parts of Africa and has also another chromosome number, $2n = 2x = 24$. The different species within this group can be divided further into four cross-sterile groups, Anguria, Hirsutus, Metaliferous, and Melo (den Nijs and Custers 1990). Within the group Anguria, uni- and/or bidirectional crosses have been accomplished. Some of the members of this group are resistant to diseases which severely affect the commercially most interesting *Cucumis* species, cucumber and melon.

Crosses between cucumber and melon or between these two and the other *Cucumis* groups will in most cases only give rise to seeds or embryos unable to germinate, despite several different approaches, including embryo rescue, irradiation treatments, and in vitro fertilization. Also crossings between cucumber or melon with other genera of Cucurbitaceae are usually prevented by the incompatibility barriers (see e.g. Deakin et al. 1971; Chatterjee and More 1991a,b; Lebeda et al. 1996).

In conclusion, there is a great need for the transfer of the available sources of disease resistance, particularly to cucumber and melon. In sexual crossings, this has been proven extremely difficult due to the incompatibility barriers. Also different types of manipulation such as embryo rescue or pollen irradiation have not overcome the problem. Hence, the need for the alternative method of hybridization by protoplast fusion is great.

1.3 State of the Art

In several species of Cucurbitaceae, plants can be regenerated from different organs and in some cases by organogenesis as well as by embryogenesis (for review see Jelaska 1986; Wehner et al. 1990). There is some evidence that the capacity to regenerate in melon is partially genetically inherited (Molina and Nuez 1995).

The availability of a successful regeneration has been exploited also in practical applications. Due to high seed prices in cucumber, it is of great interest to develop methods for efficient production of somatic embryos, which has also been successful (e.g., Pellinen et al. 1997). One of the commercially interesting metabolites, the cucurbitacines, has also been produced by in vitro cultures of cucurbits (Halaweish and Tallamy 1998).

Fertile plants can be regenerated from protoplasts readily from several different cultivars of cucumber (Orczyk and Malepszy 1985; Trulson and Shahin 1986; Colijn-Hooymans et al. 1988). Protoplast regeneration has been described in melon, but with rather low frequencies and often using Cantaloupe Charentais, a cultivar of diminishing commercial value (Moreno et al. 1985, 1986; Li et al. 1990; Debeaujon and Branchard 1992; Tabei et al. 1992). A few publications report successful plant regeneration from several commercially interesting cultivars (Bokelmann et al. 1990, 1991). Quite a lot of work has been invested in several labs to accomplish interspecific protoplast fusions among different *Cucumis* species. However, the rate of success has so far been quite limited. In fusion products of melon and a pumpkin hybrid, the transferred pumpkin character was lost during the first divisions (Yamaguchi and Shiga 1993). In Jarl et al. (1995), calli could be regenerated from fusion products. When fused protoplasts were isolated from Polish accessions of melon and cucumber, division usually stopped prior to the microcalli stage (Fellner et al. 1996). In the two latter publications, limitation of the incompatibility reaction by irradiation of one partner was attempted. Irradiation has been used successfully in several publications to fragment the genetic material and to enhance recombination and fertility in transformation experiments (e.g., Jarl and Rietveld 1996) as well as in asymmetric hybridizations in different plant genera, e.g., *Brassica* (e.g., Sjödin and Glimelius 1989; Forsberg et al. 1998), *Lycopersicon* (Melzer and O'Connell 1992), *Nicotiana* (e.g., Bates et al. 1987), and *Solanum* (Jarl et al. 1999).

2 Somatic Hybridization

In the following, our attempt to introduce virus resistance into melon by protoplast fusions will be described (Jarl et al. 1995). The source of resistance was a Chinese accession in which resistance to virus diseases as well as to powdery mildew had been found. To limit the amount of genetic material transferred from cucumber, this partner were irradiated prior to fusion. For details of methods, see Protocol (Sect. 4).

2.1 Isolation of Protoplasts

In vitro-grown material of cucumber as well as of melon was used. Protoplasts could be isolated in high yields from several different cultivars and from different plant organs. However, in our hands, protoplasts isolated from young cotyledons showed the highest frequency of plant regeneration. Hence, this material was used subsequently. By a 2–3 day preculture of the plant material, yields as well as viability of the isolated protoplasts were markedly improved. To enable selection of heterokaryons after fusion, the cucumber parent was bleached by germination on a norfluorazon-containing medium,

Somatic Hybridization in *Cucumis* 143

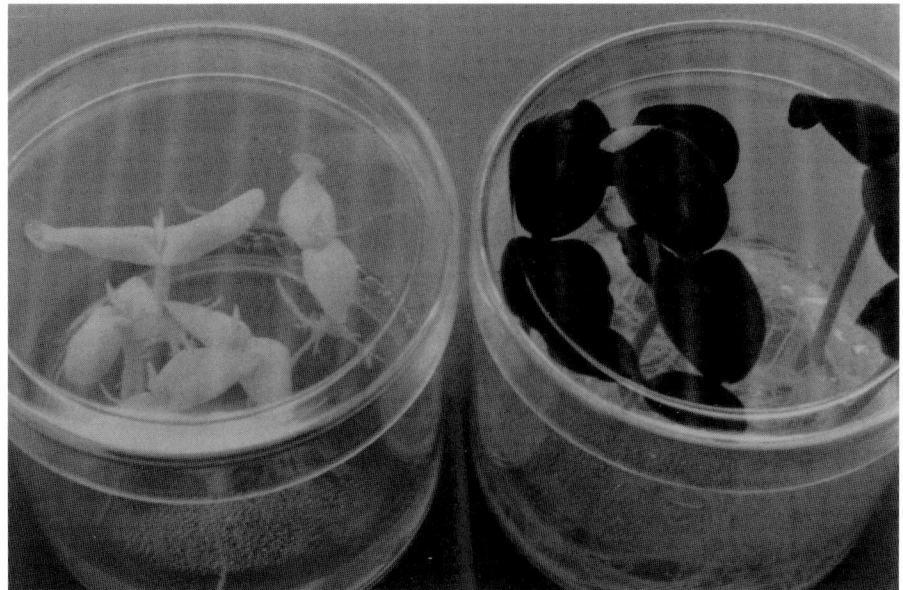

Fig. 1. One-week-old green seedlings of melon (*right*) and white seedlings of cucumber, bleached by SAN (*left*)

resulting in chlorophyll-less protoplasts (Fig. 1). After isolation, the white cucumber protoplasts were stained by FDA (fluoroscein diacetate).

As an attempt to limit the incompatibility reactions in the heterokaryons, experiments were carried out in which the genetic material of the donor was fragmented by X-ray irradiation. As there is no clear relationship between irradiation doses and extent of fragmentation, two different doses were tested: the lethal dose (1000 Gy) and half the lethal dose (500 Gy).

2.2 Protoplast Fusion

Initially, polyethylene glycol (PEG)-mediated fusion was compared with electrofusion. Electrofusion performed as described below gave higher fusion frequencies (more than 10%), as well as better survival of the cells after fusion. Also, after chemical fusion, there was considerable cell aggregation obstructing cell sorting after fusion. After fusion, the protoplasts were cultured for 2–3 days in culture medium before the selection was carried out. This was to allow the protoplasts to recover from the fusion treatment and also to avoid selecting unviable heterokaryons, as these would often be degraded within the first few days.

The fusion was performed between green melon protoplasts and white cucumber protoplasts, stained with FDA. This enabled selection of the heterokaryons, using the red autofluorescence of chlorophyll and the yellow FDA fluorescence combined in the fusion products. Selection was done visually in a fluorescence microscope with a micromanipulator.

2.3 Isolation, Culture, and Regeneration of Protoplasts

All development of culture regimes was done with unfused melon protoplasts. Five different genotypes were compared: three breeding lines: Z1, Z2, Z3 (Zaadunie BV, The Netherlands) and two commercial cultivars: Galia and Charentais types (Zaadunie BV, The Netherlands).

It was important to keep the protoplast density at about $1–5 \times 10^4$ protoplasts ml^{-1}. As 100–200 heterokaryons could be isolated from a fusion experiment, a system had to be developed to culture small volumes of 20–30 µl. Using microtiter wells (Greiner, 96 × 6) proved to be superior to other systems, e.g., using agarose microbeads. It was crucial to add a small volume, 25 µl, each week, to prevent drying out. For the initial protoplast culture, B5 medium (Gamborg et al. 1968) was the best combination of nutrient salts. The vitamin composition and hormone combination in medium KM8p (Kao and Michayluk 1975) proved beneficial. Osmolarity was varied between 350–610 mOsm kg^{-1} by different combinations of mannitol, sucrose, and glucose. Highest PE (plating efficiency) was obtained using 250 mM mannitol, 170 mM glucose, and 34 mM sucrose with 75% division after 1 week (Fig. 2). During the first 3 weeks after isolation this medium was added to the cultures. After this period the cultures were diluted 1:2 with the same medium in which mannitol was omitted, thus decreasing the osmolarity to 300 mOsm kg^{-1}. At this stage, the cultures were solidified by adding agarose and placed under dim light to induce greening. As shoot-initiation medium and the shoot-development media, B5 medium with different hormone combinations was tested. As auxins, IAA (3-indoleacetic acid) was used in concentrations varying between 0 and 2.5 mg l^{-1}, or NAA (naphtalene acetic acid) at 0–0.1 mg l^{-1}. Four cytokinins were compared, single or in combination, 0–10.0 mg l^{-1} kin (kinetin), 0–5.0 mg l^{-1} benzylaminopurine (BA), 0–0.5 mg l^{-1} 2-iP (isopentenyl adenine) and/or 0–5.0 mg l^{-1} zeatin. Additions of 0.1 mg l^{-1} abscisic acid (ABA) or 0.1 mg l^{-1} giberellic acid (GA$_3$) were also tested. All different combinations of shoot-initiation medium and shoot-development medium were investigated. Best results were obtained with 0.5 mg l^{-1} zeatin in the shoot-initiation medium, followed by a shoot-development medium with 0.1 mg l^{-1} IAA and 0.2 mg l^{-1} BA. With this combination, shoot regeneration frequency with the best responding genotypes exceeded 75%. The genotypes tested varied substantially in shoot regeneration and in response to the tested media (Fig. 3). Different concentrations of sucrose were compared, with 3% sucrose resulting in the best shoot regeneration. Also different gelling agents were compared and could be seen to have a significant influence (Fig. 4). Using 0.8% agarose in the shoot-induction medium, followed by 0.5% Gelrite in the shoot-

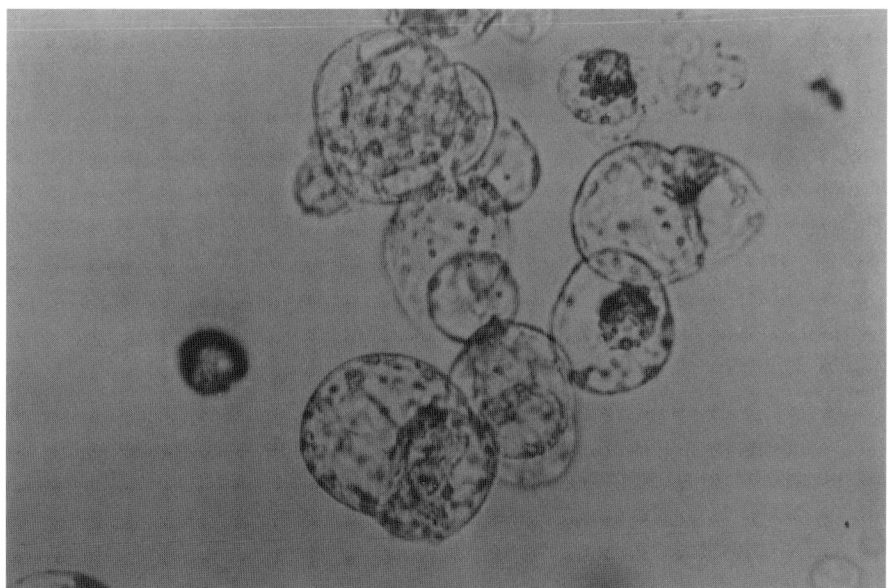

Fig. 2. Dividing melon protoplasts after 4 days of culture

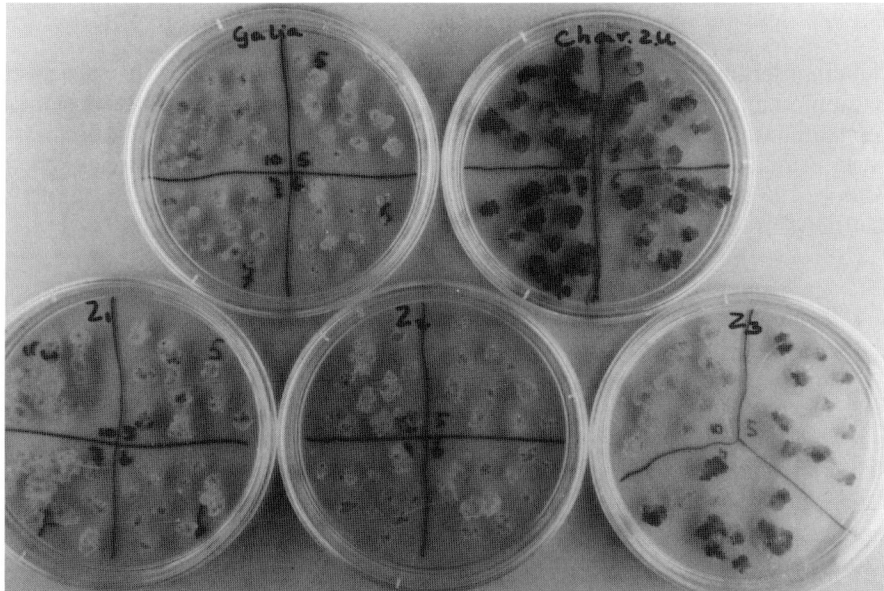

Fig. 3. Comarison of shoot regeneration between the five tested melon cultivars, two commercial cultivars of Galia and Charentais type *top*, and three breeding lines (Z1, Z2, and Z3) *below*

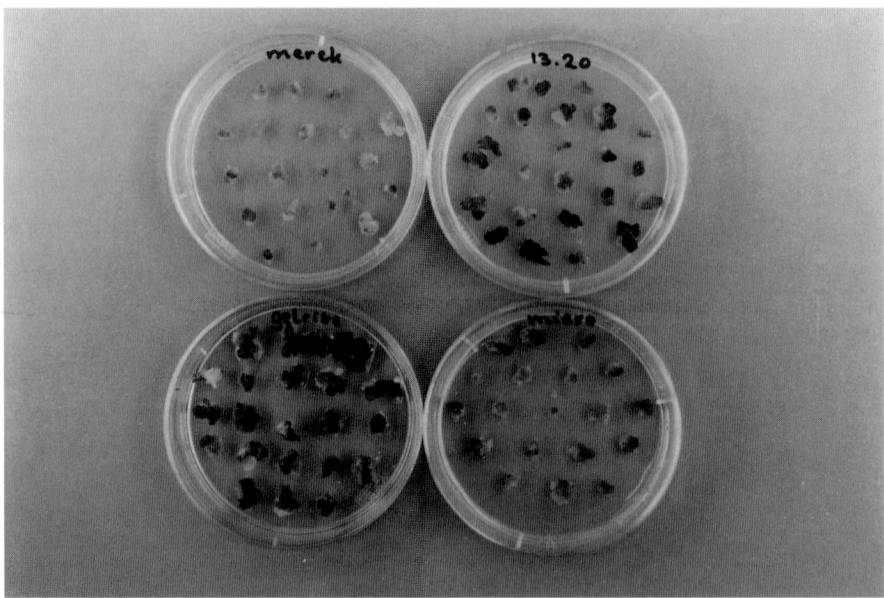

Fig. 4. Comparison of the influence of different gelling agents (Merck Agar, Agarose 13.20, Gelrite, and Micro agar) on shoot regeneration in melon, Charentais type

development medium, ensured a good development of initiated shoots as well as minimizing vitrification. Shoots regenerated from protoplasts were generally difficult to root and transfer to the greenhouse; 40–90% of the shoots regenerated from protoplasts could be rooted in MS medium (Murashige and Skoog 1962) supplemented with $0.03\,mg\,l^{-1}$ NAA, 3% glucose, and 0.5% Gelrite (Fig. 5).

2.4 Regeneration of Somatic Hybrids

A total of 18000 heterokaryons from 17 separate fusion experiments were selected by micromanipulation. Only 1.5% (248) of the heterokaryons showed one cell division. In one fusion experiment, calli were obtained from the selected heterokaryons. Those calli grew well and shoot primordia developed (see Fig. 6). However, no further development was observed, and plants were never obtained. DNA analysis using RAPD primers of these calli showed the same pattern as melon, with the occurrence of new bands corresponding to equivalent bands in cucumber. This might indicate the transfer of certain genetic material from cucumber to melon in these calli. In asymmetric fusions, the goal is to transfer only limited amounts of donor DNA to the receptor. Detection of small amounts of unknown DNA sequences is very difficult by molecular analyses.

Fig. 5. Rooted melon plant in soil (Jarl et al. 1995, with kind permission of Kluwer Academic)

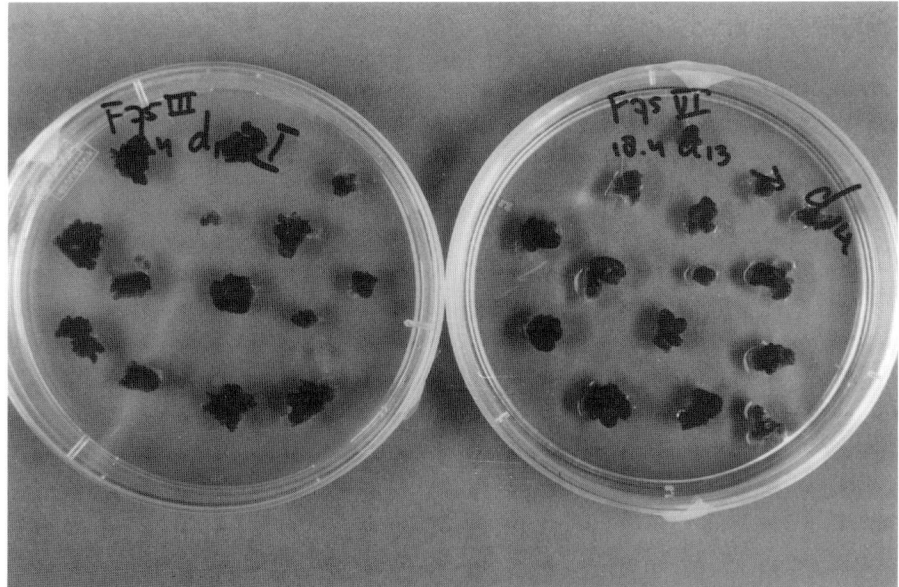

Fig. 6. Shoot primordia on calli obtained from two fusion experiments

To our knowledge, no other publications report successful interspecific fusions in Cucurbitaceae by protoplast fusions, in spite of attempts by several authors.

3 Summary and Conclusion

Although we used a highly efficient system for protoplast regeneration as well as protoplast fusion and hybrid selection, no plants could be obtained from the selected heterokaryons. One explanation for the failure to develop could be that some kind of somatic incompatibility reaction might have taken place in the hybrid calli, preventing the regeneration of shoots. Sexual incompatibility can be divided in pre-and postzygotic incompatibility. Prezygotic incompatibility involves reactions preventing the formation of a zygote and is thus circumvented by using protoplast fusions. However, if the failure to develop hybrid plants is due to mechanisms taking place in the zygote, this reaction can also be expected to take place in heterokaryons obtained by cell fusion. Such postzygotic incompatibility might involve structural interactions of the mitotic apparatus as well as the interphase cytoskeleton, cell-cycle regulation, or interactions of nuclei and cytoplasms (Harms 1983). Such incompatibility reactions may involve also cytoplasmic characters and traits, which may not overcome the irradiation treatment used to limit the amount of nuclear material transferred.

Possibly, in order to transfer characters efficiently within Cucurbitaceae, one solution is to isolate the genes of useful characteristics in the donor species. Subsequently, such isolated genes might be used to transform the acceptor species. As plant regeneration from in vitro material can be obtained readily in Cucurbitaceae, this should offer a workable approach.

4 Protocol

Sterilized seeds of melon and cucumber are germinated on a half-strength MS medium. If heterokaryons are selected visually, $1.5\,\text{mg}\,\text{l}^{-1}$ norfluorazon (SAN 9789) is added to the germination media of cucumber. Seven-day-old cotyledons of both species are put on preculture medium (MS medium with $1\,\text{mg}\,\text{l}^{-1}$ NAA and $1\,\text{mg}\,\text{l}^{-1}$ BA) 2 days prior to isolation. Protoplasts are isolated overnight in isolation medium (1/4 MS salts, 0.4M mannitol, 0.1 M glycine, 0.8% Cellulase RS Onozuka, and 0.4% Macerozyme R10 Onozuka, both from Yakult-Honsha, Japan). Protoplasts are washed three times with fusion solution (0.5M mannitol and 1mM $CaCl_2$) and the density is adjusted to 8×10^5 protoplasts ml^{-1}. Prior to fusion, the cucumber protoplasts are irradiated with 500 or 1000 Gy and stained for 30 min with FDA ($1\,\mu\text{g}\,\text{ml}^{-1}$). Parameters for electrofusion with our equipment (a square-wave generator and a flat stainless-steel fusion chamber with electrodes 3 mm apart) were as follows. Alignment was obtained by 1 Mhz, $60\text{--}100\,\text{V}\,\text{cm}^{-1}$ for 30s. Fusion was induced by three to six pulses (500μs) of $750\,\text{V}\,\text{cm}^{-1}$ with intervals of 1s. Fused protoplasts are cultured for 2–3 days in culture medium (B5 salts, KM8p vitamins, 35mM sucrose, 0.25M mannitol, 0.17M glucose, $100\,\text{mg}\,\text{l}^{-1}$ myoinositol, $1\,\text{g}\,\text{l}^{-1}$ glycine, $1\,\text{mg}\,\text{l}^{-1}$ NAA, $0.2\,\text{mg}\,\text{l}^{-1}$ 2,4-D, $0.5\,\text{mg}\,\text{l}^{-1}$ BA) prior to selection of heterokaryons. Selection is carried out with micromanipulator or by a cell sorter. Selected heterokaryons are cultured in microtiter wells (Greiner, 96 × 6-mm wells) in small volumes of culture medium at 5×10^3 protoplasts ml^{-1}; 25 µl culture medium is added each week. After 1 month, microcalli are transferred to larger wells (Greiner, 24 × 16-mm wells); 0.5 ml culture medium is added in which mannitol is omitted, and the cultures are solidified by adding agarose to a final concentration of 0.2% Cultures are placed under dim light (22 µmol m^{-2} s^{-1}) for another month, after which minicalli are transferred to a shoot-initiation medium (MS medium with 0.5 mg l^{-1} zeatin). Light intensity is now increased to 64 µmol m^{-2} s^{-1}. After 3 weeks, calli with developing shoot primordia are put on shoot-development medium (MS medium with 0.1 mg l^{-1} IAA, 0.2 mg l^{-1} BA, and 0.5% Gelrite). Subculturing to the same media is made every 3 weeks until shoots are sufficiently developed (10–25 mm) to be transferred to rooting medium (0.03 mg l^{-1} NAA, 0.5% Gelrite, and 3% glucose). Again, subculturing is done until root development is sufficient to allow transfer to soil and greenhouse.

References

Bates GW, Hasenkampf CA, Contolini CL, Piastuch WC (1987) Asymmetric hybridization in *Nicotiana* by fusion of irradiated protoplasts. Theor Appl Genet 74:718–726

Bokelmann GS, Jarl CI, Kool AJ (1990) An efficient method for plant regeneration of protoplasts of commercial varieties of melon. In: Nijkamp HJJ, van der Plas LHW, van Artrijk J (eds) Abstr VIIth Int Congr on plant tissue and organ culture. Kluwer Dordrecht, p 8

Bokelmann GS, Jarl CI, Kool AJ (1991) Plant regeneration of protoplasts of different cultivars of melon (*Cucumis melo* L.). Cuc Gen Coop Rep 14:78–80

Chatterjee M, More TA (1991a) Techniques to overcome barrier of interspecific hybridization in *Cucumis*. Cuc Gen Coop Rep 14:66–68

Chatterjee M, More TA (1991b) Interspecific hybridization in *Cucumis* spp. Cuc Gen Coop Rep 14:69–70

Colijn-Hooymans CM, Bouwer R, Orczyk W, Dons JMM (1988) Plant regeneration from cucumber (*Cucumis sativus* L.) protoplasts. Plant Sci 57:63–71

Deakin JR, Bohn GW, Whitaker TW (1971) Interspecific hybridization in *Cucumis*. Econ Bot 25:195–211

Debeaujon I, Branchard M (1992) Induction of somatic embryogenesis and caulogenesis from cotyledon and leaf protoplast-derived colonies of melon (*Cucumis melo* L.) Plant Cell Rep 12:37–40

den Nijs APM, Custers JBM (1990) Introducing resistances into cucumbers by interspecific hybridization. In: Bates DM, Robinson RW, Jeffrey C (eds) Biology and utilization of the Cucurbitaceae. Cornell University Press, New York, pp 382–396

FAOSTAT, FAO Statistical Databases, http://apps.fao.org/page/collections

Fellner M, Binarová P, Lebeda A (1996) Isolation and fusion of *Cucumis sativus* × *Cucumis melo* protoplasts. Cucurbits towards 2000. Proc Vth Eucarpia meeting on cucurbit genetics and breeding, Malaga, Spain

Forsberg J, Dixelius C, Lagercrantz U, Glimelius K (1998) UV dose-dependent DNA elimination in asymmetric somatic hybrids between *Brassica napus* and *Arabidopsis thaliana*. Plant Sci 131:65–76

Fuchs M, Gonsalves D (1995) Resistance of transgenic hybrid squash ZW-20 expressing the coat protein genes of zucchini yellow mosaic virus and watermelon mosaic virus 2 to mixed infection by both potyviruses. Bio-Technology 13:1446–1473

Gamborg OL, Miller RA, Ojima K (1968) Nutrient requirements of suspension cultures of soybean root cells. Exp Cell Res 50:151–158

Halaweish FT, Tallamy DW (1998) Production of cucurbitacins by cucurbit cell cultures. Plant Sci 131:209–218

Harms CT (1983) Somatic incompatibility in the development of higher plant somatic hybrids. Q Rev Biol 58:325–353

Herrington ME, Prytz S, Brown P (1991) Resistance to papaya ringspot virus-W, zucchini yellow mosaic virus, and watermelon mosaic virus-2 in *C. maxima*. Cuc Gen Coop Rep 14:123–124

Jarl CI, Rietveld EM (1996) Transformation efficiencies and progeny analysis after varying different parameters of direct gene transfer of *Nicotiana tabacum* protoplasts. Physiol Plant 98:550–556

Jarl CI, Bokelmann GS, de Haas JM (1995) Protoplast regeneration and fusion in *Cucumis*: melon × cucumber. Plant Cell Tissue Organ Cult 43:259–265

Jarl CI, Rietveld EM, de Haas JM (1999) Transfer of fungal resistance through interspecific somatic hybridisation between eggplant and *Solanum torvum*. Plant Cell Rep 18:791–796

Jeffrey C (1980) A review of the Cucurbitaceae. J Linn Soc Bot 81:233–247

Jelaska S (1986) Cucurbits. In: Bajaj YPS (ed) Biotechnology in agriculture and forestry, vol 2. Crops I. Springer, Berlin Heidelberg New York, pp 371–386

Kap KN, Michayluk MR (1975) Nutritional requirements for growth of *Vicia hajastana* cells and protoplasts at very low population density in liquid media. Planta 126:105–110

Knerr LD, Staub LD, Holder DJ, May BP (1989) Genetic diversity in *Cucumis sativus* L. assessed by variation at 18 allozyme coding loci. Theor Appl Genet 78:119–128

Kroon GH, Custers JBM, Kho YO, den Nijs APM, Varekamp HQ (1979) Interspecific hybridization in *Cucumis* (L.). Need for genetic variation, biosystematic relations and possibilities to overcome crossability barriers. Euphytica 28:723–728

Lebeda A, Kristkova E, Kubalakova (1996) Interspecific hybridization of *Cucumis sativus* × *Cucumis melo* as a potential way to transfer resistance to *Pseudoperonospora cubensis*. Cucurbits towards 2000. Proc Vth Eucarpia Meeting on Cucurbit genetics and breeding, Malaga, Spain

Li R, Sun Y, Zhang L, Li X (1990) Plant regeneration from cotyledon protoplasts of Xinjiang muskmelon. Plant Cell Rep 9:199–203

Melzer JM, O'Connell MA (1992) Effect of radiation dose on the production of and the extent of asymmetry in tomato asymmetric somatic hybrids. Theor Appl Genet 83:337–344

Molina RV, Nuez F (1995) Characterization and classification of different genotypes in a population of *Cucumis melo* based on their ability to regenerate shoots from leaf explants. Plant Cell Tissue Organ Cult 43:249–257

Moreno V, García-Sogo M, Granell I, García-Sogo B, Roig LA (1985) Plant regeneration from calluses of melon (*Cucumis melo* L., cv. Amarillo Oro) Plant Cell Tissue Organ Cult 5:139–146

Moreno V, Zubeldia L, Garcá-Sogo B, Nuez F, Roig LA (1986) Somatic embryogenesis in protoplast-derived cells of melon (*Cucumis melo* L. cv. Amarillo Oro). In: Jensen CJ, Odenbach W, Schieder OJ (eds) Genetic manipulation in plant breeding. Proc Int Symp Berlin 1985. de Gruyter, Berlin, pp 491–493

Murashige T, Skoog F (1962) A revised medium for rapid growth and bio-assays with tobacco tissue culture. Physiol Plant 15:473–497

Orczyk W, Malepszy S (1985) In vitro culture of *Cucumis sativus* L. V. Stabilizing effect of glycine on leaf protoplasts. Plant Cell Rep 4:269–273

Pan RS, More TA (1996) Screening of melon (*Cucumis melo* L.) germplasm for multiple disease resistance. Euphytica 88:125–128

Pellinen TP, Sorvari S, Tahvonen R, Sewon P (1997) Somatic embryogenesis in cucumber (*Cucumis sativus* L.) callus and suspension cultures. J Appl Bot 71:116–118

Provvidenti R (1987) Inheritance of resistance to a strain of zucchini yellow mosaic virus in cucumber. HortScience 22(1):102–103

Provvidenti R, Robinson RW, Munger HM (1978) Resistance in feral species to six viruses infecting Cucurbita. Plant Dis Rep 62:329–326

Provvidenti R, Gonsalves D, Humaydan HS (1984) Occurrence of zucchini yellow mosaic virus in cucurbits from Connecticut, New York, Florida and California. Plant Dis 68:443–446

Sjödin C, Glimelius K (1989) Transfer of resistance against *Phoma lingam*. to *Brassica napus* by asymmetric somatic hybridization combined with toxin selection. Theor Appl Genet 78:513–520

Tabei Y, Nishio T, Kanno T (1992) Shoot regeneration from cotyledonary protoplasts of melon (*Cucumis melo* L. cv. Charentais). J Jpn Soc Hortic Sci 61(2):317–322

Trulson AJ, Shahin EA (1986) In vitro plant regeneration in the genus *Cucumis*. Plant Sci 47:35–43

Wehner TC, Cade RM, Locy RD (1990) Cell, tissue, and organ culture techniques for genetic improvement of cucurbits. In: Bates DM, Robinson, RW, Jeffrey C (eds) Biology and utilization of the Cucurbitaceae. Cornell University Press, Ithaca, pp 367–381

Whitaker TW, Bermis WP (1976) Cucurbits. In: Simmonds NW (ed) Evolution of crop plants. Longman, New York, pp 64–69

Yamaguchi J, Shiga T (1993) Characteristics of regenerated plants via protoplast electrofusion between melon *Cucumis melo* and pumpkin interspecific hybrid *Cucumis maxima* × *Cucurbita moschata*. Jpn J Breed 43:173–182

II.6 Somatic Hybridization in *Diospyros* (Persimmon)

M. TAMURA[1] and R. TAO[2]

1 Introduction

Persimmon (*Diospyros kaki*) is distributed in the temperate regions of East Asia, and has been cultured for centuries in China, Korea, and Japan (Tao and Sugiura 1992; Tamura et al. 1995a). It is an important species for fruit production and has gained popularity in the temperate parts of Asia. Recently, persimmon has been gaining popularity throughout the world and its culture is spreading outside Asia, including Australia, New Zealand, the USA, Brazil, Israel, and Italy. Production yield and area are increasing in these countries, and fruit exports from Oceania to Asian countries are expanding. With this recent worldwide interest, development of rapid breeding systems to improve fruit quality is highly desired.

Progress in persimmon breeding has been slow, because crossing has been hampered by the scarcity of commercially important cultivars carrying perfect and/or male flowers. In addition, there is a strong crossing block between *D. kaki* and other *Diospyros* spp., because *D. kaki* cultivars are polyploid, hexaploid ($2n = 90$, $x = 15$), or nonaploid ($2n = 135$, $x = 15$) (Zhuang et al. 1990), while most other *Diospyros* species are diploid ($2n = 30$, $x = 15$) (Tamura et al. 1998a). Somatic hybridization could break these barriers and produce new cultivars with improved qualities. Furthermore, increased ploidy level, not existing naturally in *Diospyros*, can be produced. Polyploidization often offers some beneficial traits such as large fruits, polyploid vigor, thicker stems, or short internodes.

Although it is necessary to establish the regeneration system from protoplasts for somatic hybridization, an efficient system has been established for persimmon. The first study of regeneration from persimmon protoplasts was reported by Tao et al. (1991). Tamura et al. (1993) modified the method and established an efficient regeneration. Chromosome-doubling plants were also obtained by colchicine treatment of persimmon protoplasts (Tamura et al. 1996). By using the regeneration system, intraspecific somatic hybrids were produced by Tamura et al. (1995b) using these regeneration techniques, and interspecific somatic hybrids were also obtained (Tamura et al. 1998b). In this chapter, these results of somatic hybridization of persimmon are summarized.

---

[1] Plant Biotechnology Laboratory, Institute for Fundamental Research, Suntory Ltd., Shimamoto-cho, Osaka, 618-8503, Japan
[2] Graduate School of Agriculture, Kyoto University, Kyoto 606-8502, Japan

2 Somatic Hybridization

2.1 Isolation of Protoplasts

Protoplasts were isolated from callus culture. The calli were induced from leaf primordia in dormant winter buds on MS agar medium (Murashige and Skoog 1962) with a half-strength nitrate (1/2N), containing 10µM zeatin and 1µM IAA. The calli were subcultured every 6 weeks to the same fresh medium and used for protoplast isolation.

Conditions for protoplast isolation were tested for *D. glandulosa*, *D. lotus*, and seven cultivars of *D. kaki*. Optimal conditions are summarized in Table 1. Optimal osmotic potential for all cultivars and species was 0.7M mannitol, except for *D. kaki* cv. Miyazakitanenashi and *D. lotus*, for which the optimal mannitol was 0.7–0.8M and 0.9M, respectively. Optimal concentrations of Cellulase Onozuka RS (Yakult Co. Ltd. Tokyo, Japan) were 0.5–1% for all cultivars and species tested (Tamura 1997). Optimal Macerozyme R10 (Yakult Co. Ltd. Tokyo, Japan) concentration for protoplast isolation was 0.02–0.2% and varied with cultivars and species. Calli subcultured for 1 to 2 weeks gave the highest protoplast yields for all cultivars and species tested. The calli of this stage contained the highest number of actively dividing cells (G_2 phase), which probably led to high protoplast yield.

Protoplast yields varied with cultivars or species. *D. kaki* cvs. Jiro and Suruga gave a higher yield of $2-4 \times 10^6$ protoplasts g^{-1} than the other cultivars ($1-2 \times 10^6$ protoplasts g^{-1}). The yields of *D. glandulosa* and *D. lotus* were $4-6 \times 10^5$ protoplasts g^{-1}, lower than those of *D. kaki* cultivars.

2.2 Fusion of Protoplasts

D. kaki cvs. Jiro ($2n = 90$, $x = 15$) and Suruga ($2n = 90$, $x = 15$) were used for intraspecific somatic hybridization because they gave higher protoplast yields.

Table 1. Optimal protoplast isolation conditions for *D. glandulosa*, *D. lotus*, and seven cultivars of *D. kaki*

Species or cultivar	Concentrations in maceration media			Days after subculture
	Mannitol (M)	Cellulase RS (%)	Macerozyme R10 (%)	
D. kaki				
Jiro	0.7	0.5–1.0	0.05–0.1	14
Suruga	0.7	0.5	0.1	7–14
Mushirodagosho	0.7	0.5–1.0	0.05	7
Fuyu	0.7	0.5	0.2	7
Hiratanenashi	0.7	1.0	0.02	7
Saijo	0.7	0.5	0.2	7–14
Miyazakitanenashi	0.7–0.8	1.0	0.2	7
D. glandulosa	0.7	0.5	0.2	7–14
D. lotus	0.9	1.0	0.2	7–14

In addition, although they both bear the most desirable fruits type of pollination constant and nonastringent (PCNA), both only bear female flowers and crossing is impossible between the two cultivars. Since PCNA is known as a recessive trait (Yamada 1993), resultant somatic hybrids would be expected to be PCNA. Dodecaploid PCNA plants produced could be used as new cultivars by themselves or could be used as mother plants to produce nonaploid seedless PCNA-type cultivars by crossing with other PCNA hexaploid cultivars.

D. kaki cv. Jiro and *D. glandulosa* were selected for interspecific somatic hybrids. *D. glandulosa* was diploid and considered as a candidate for rootstock for persimmon in Thailand. Agriculturally important traits such as drought resistance could be introduced into *D. kaki* by somatic hybridization.

Protoplasts were electrically fused to produce intra- and interspecific somatic hybrids. The isolated callus protoplasts were suspended in 1–3 mM $CaCl_2$ supplemented with 0.6 M mannitol. A higher concentration of $CaCl_2$ in the suspension inhibited the formation of single pairs or pearl chain when an alternating current field was applied. It appeared that a single direct current pulse was enough to induce cell fusion and subsequent pulses were not effective.

2.3 Culture of Fused Protoplasts

The fused protoplasts were cultured in a modified KM8p agarose medium at a density of 5×10^5 protoplasts ml^{-1} (Kao and Michayluk 1975; Tamura et al. 1993, 1995b). Interestingly, cell division of protoplasts was enhanced by the fusion treatment. With electrically treated protoplasts, the first division occurred after 3–4 days, while it took more than week for nontreated control Jiro and Suruga protoplasts, and *D. glandulosa* protoplasts did not divide. It has also been demonstrated that exposure to a short electric pulse stimulates cell division and regeneration of several plant species (Rathore and Goldsworthy 1985a,b; Rech et al. 1987; Ochatt et al. 1988; Keller et al. 1997). After 2 weeks, the cells were transferred to agarose-bead culture (Shillito et al. 1983; Tao et al. 1991). Microcalli, which were released from agarose blocks after 3 months, were transferred to a modified KM8 agar medium. After 6 weeks, the calli were subcultured onto MS (1/2 N) agar medium supplemented with 1 µM IAA and 10 µM zeatin.

2.4 Selection of Somatic Hybrid Calli

Nuclear DNA contents of somatic hybrid calli should be the sum of both parents' DNA contents, if the fusion products are obtained by a one-to-one combination. Thus, in our studies, the relative nuclear DNA content of the calli recovered from fusion-treated protoplasts was determined by flow cytometry to select somatic hybrids.

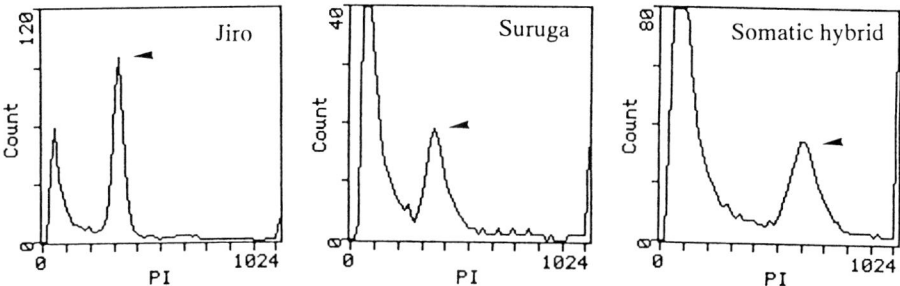

Fig. 1. Flow cytometric histograms of relative nuclear DNA content of nuclei isolated from Jiro, Suruga, and somatic hybrid. *Arrows* indicate the peak of relative nuclear DNA content (2C)

2.4.1 Intraspecific Somatic Hybrids

The relative nuclear DNA contents were twice as much as those of parental calli among about 17% callus lines (22 of 127) (Fig. 1). These callus lines seemed to be somatic hybrids between either of three combinations; Jiro and Suruga, Jiro and Jiro, or Suruga and Suruga.

2.4.2 Interspecific Somatic Hybrids

The relative nuclear DNA contents in 149 of the 166 callus lines (about 90%) were the sum of those of Jiro ($6x$) and *D. glandulosa* ($2x$) (Fig. 2A–C), and seemed to be interspecific hybrids ($8x$) between *D. kaki* cv. Jiro and *D. glandulosa*. The nuclear DNA contents of 26 of these $8x$ callus lines doubled after four subcultures ($16x$) probably because of spontaneous chromosome doubling (Fig. 2G). Ten of the 166 lines had a nuclear DNA content two times higher ($12x$) than Jiro (Fig. 2D). Six chimeric callus lines, one was chimera with $6x$ and $8x$ cells and five were chimera with $2x$ and $8x$ cells (Fig. 2E,F), were also obtained. However, $8x$ cells appeared to have a higher growth rate than $6x$ or $2x$ cells, and they became nonchimeric calli with only $8x$ cells.

RAPD analysis confirmed that the $8x$ and $16x$ callus lines were interspecific somatic hybrids, while the $12x$ lines were somatic hybrids between Jiro protoplasts or the products from spontaneous chromosome doubling. All callus lines including 123 $8x$ and 26 $16x$ lines yielded bands specific to both Jiro and *D. glandulosa* (Fig. 3). The callus lines of $12x$ showed only the band specific to Jiro.

2.5 Regeneration of Somatic Hybrids

2.5.1 Intraspecific Somatic Hybrid

Nine of the 22 somatic hybrid callus lines were regenerated to form shoots. RAPD analysis was performed to determine the lineage. All shoot lines regen-

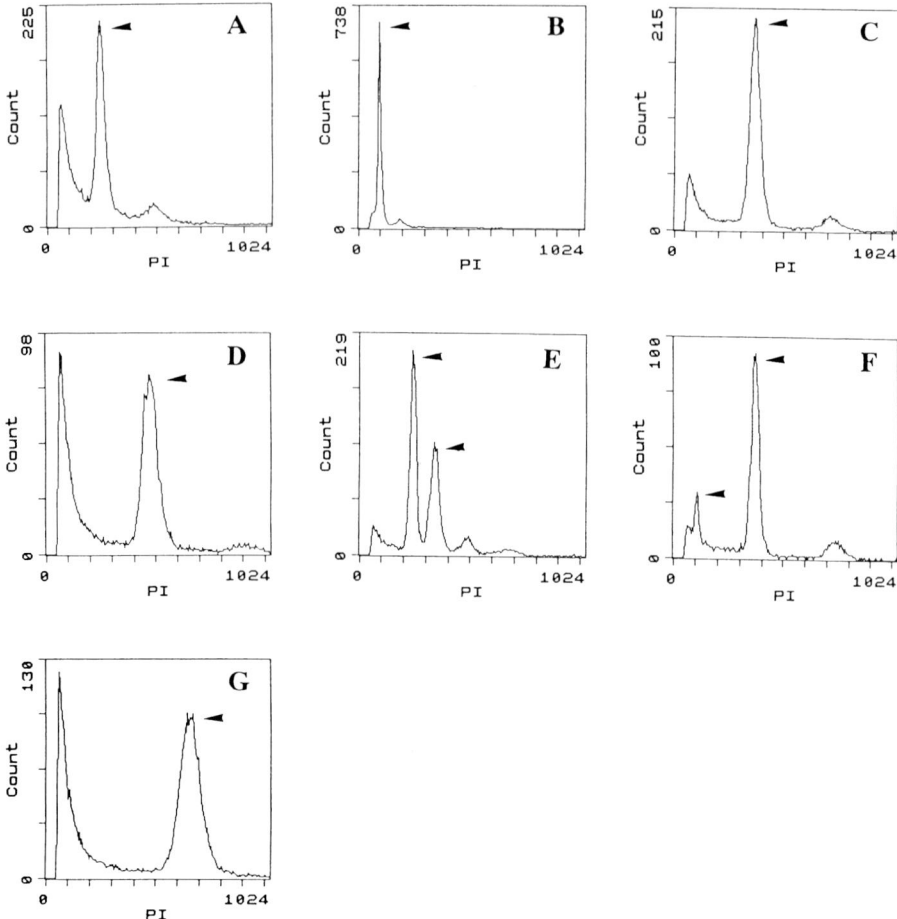

Fig. 2A–G. Flow cytometric histograms of various calluses derived from electrofused protoplasts. **A** *D. kaki* cv. Jiro. **B** *D. glandulosa*. **C** Interspecific hybrid (8*x*). **D** Hybrid between Jiro and Jiro. **E** Chimera with Jiro (6*x*) and interspecific hybrid (8*x*) cells. **F** Chimera with *D. glandulosa* (2*x*) and interspecific hybrid (2*x*) cells. **G** Interspecific hybrid (16*x*). *Arrows* indicate the peak of relative nuclear DNA content (2C)

erated from somatic hybrid calli had a combination of both Jiro and Suruga banding profiles. indicating that they are somatic hybrids between Jiro and Suruga (Fig. 4). After rooting treatment, six of the nine shoot lines rooted, and rooting percentage was between those of parents (Table 2). The chromosome number of root-tip cells of intraspecific somatic hybrids was $2n = 180$ (Fig. 5), which was twice as many as those of parental cultivars ($2n = 90, x = 15$). Some morphological differences between hybrids and parental plants were observed. Somatic hybrids had round, big, leaves and thick stems (Fig. 6).

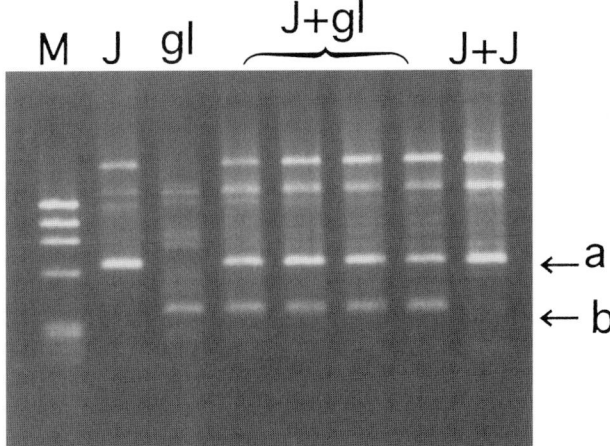

Fig. 3. Nuclear DNA analysis of interspecific hybrid callus by RAPD (*J* D. kaki cv. Jiro; *gl* D. glandulosa; *J + gl* interspecific hybrids between Jiro and D. glandulosa; *J + J* somatic hybrids between Jiro and Jiro; *M* DNA size marker (λ/HindIII digest). *Arrows* indicate bands specific to 'Jiro' (*a*) and D. glandulosa (*b*)

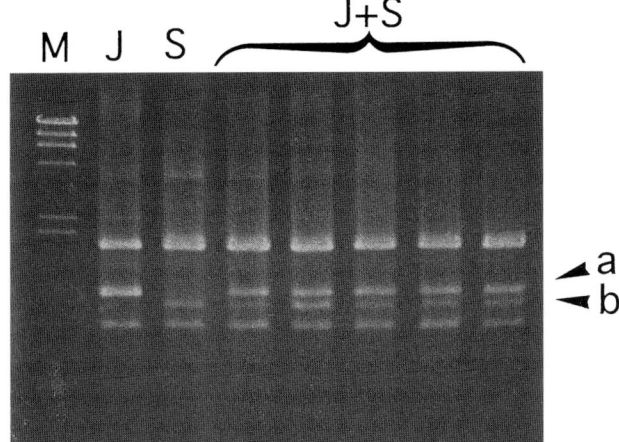

Fig. 4. RAPD analysis of somatic hybrids and their parents. *J* Jiro; *S* Suruga; *J + S* somatic hybrids; *M* DNA size marker (λ/HindIII digest). *Arrows* indicate bands specific to Jiro (*a*) and Suruga (*b*)

Table 2. Rooting rate of intraspecific somatic hybrids and their parents

	Parents		Shoot line								
	Jiro	Suruga	A	B	C	D	E	F	G	H	I
No. of explants	29	20	25	25	16	31	32	8	13	5	1
No. of rooted plants	16	0	10	9	2	6	4	2	0	0	0
Rooting rate (%)	55.0	0	40.0	36.0	12.5	19.4	12.5	25.0	0	0	0

Data taken after 50 days in culture.

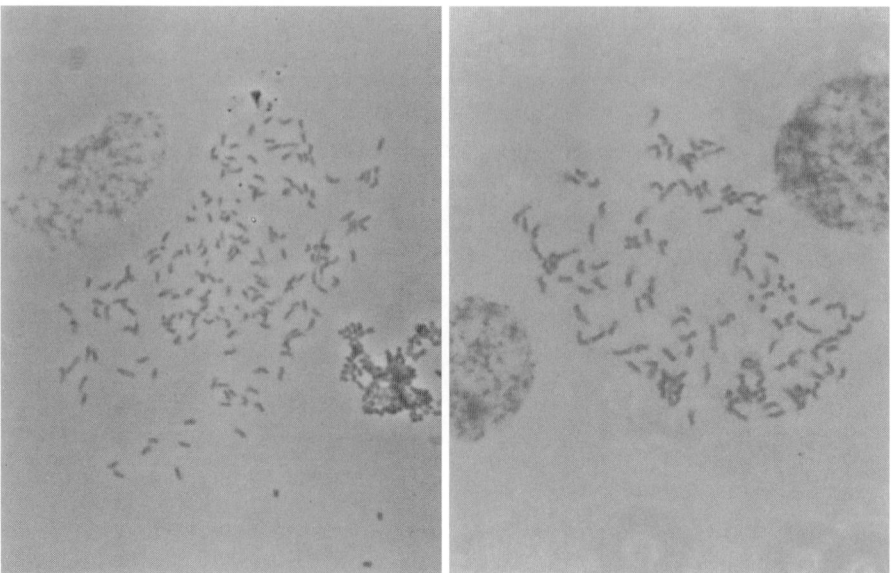

Fig. 5. Chromosomes of intra- (*left*) and interspecific (*right*) somatic hybrids

Fig. 6. Intraspecific somatic hybrid plant between Jiro and Suruga (*left*) and Jiro plant (*right*)

2.5.2 Interspecific Somatic Hybrid

The percentage of regeneration from interspecific hybrid callus was higher for $8x$ callus lines (50.4%, 62 of 123) than for $16x$ lines (3.8%, 1 of 26), probably because of the adverse effects of too many chromosomes in the $16x$ lines.

Fig. 7. cpDNA analysis by PCR-RFLP *J D. kaki* cv. Jiro; *gl D. glandulosa; J + gl* interspecific hybrid; *M* DNA size marker (φX174/*Hae*III digest)

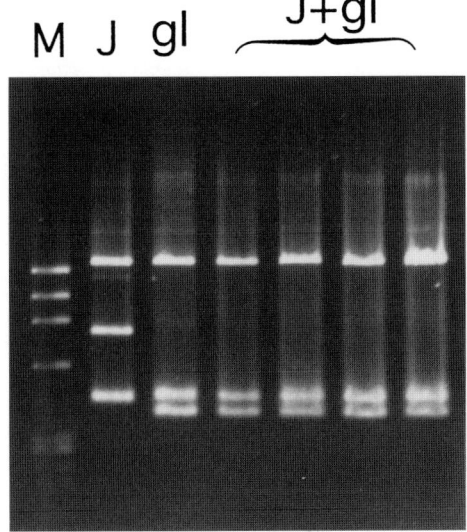

Selective elimination of cytoplasmic genes after symmetric cell fusion was often observed in *Citrus* (Saito et al. 1993; Yamamoto and Kobayashi 1995; Grosser et al. 1996). In persimmon, elimination of one parental chloroplast genome was confirmed by PCR-RFLP analysis of interspecific shoots. The PCR-RFLP banding pattern of chloroplast DNA indicated that all interspecific somatic shoot lines had only the chloroplast genome of *D. glandulosa* (Fig. 7). Kobayashi et al. (1991) and Ohgawara et al. (1994) also reported that chloroplast genomes in somatic hybrid plants obtained from electrofusion were from only one of their parents. In our study, because only somatic hybrids with *D. glandulosa* chloroplast genome were obtained, chloroplasts of *D. glandulosa* might be more stably retained in the cells of somatic hybrids.

Chromosomes of root-tip cells of 8*x* interspecific somatic hybrids was $2n = 120$, the sum of *D. kaki* cv. Jiro ($2n = 90$) and *D. glandulosa* ($2n = 30$) (Fig. 5). No elimination of the nuclear chromosomes appeared to occur in these 8*x* lines. No rooting was observed in 16*x* hybrids. The interspecific hybrids obtained were morphologically similar to Jiro (Fig. 8).

3 Summary and Conclusion

Conditions of callus protoplast isolation were examined with seven cultivars of *D. kaki* and two other *Diospyros* species. The optimal isolation condition for each cultivar or species did not differ greatly, except for Macerozyme concentration. For intraspecific somatic hybridization, hexaploid commercially

Fig. 8. Interspecific somatic hybrid plant between *D. kaki* cv. Jiro and *D. glandulosa*

important cultivars of *D. kaki*, cvs. Jiro and Suruga, were selected. Diploid species of *D. glandulosa* and *D. kaki* cv. Jiro were selected for interspecific somatic hybridization. Protoplasts were fused electrically and cultured as already described (Tamura et al. 1995a,b). Intra- and interspecific hybrid calli could be selected by determining the relative nuclear DNA content using a flow cytometer. Somatic hybridization was also confirmed by nuclear DNA analysis using RAPD. Chromosome numbers of root tip cells of intra- and interspecific hybrids were $2n = 12x = 180$, and $2n = 8x = 120$, respectively. Hybrid vigor was observed in intraspecific somatic hybrid plants.

4 Protocol

4.1 Protoplast Isolation, Fusion and Culture

Induce calli from primordial leaves of *D. kaki* cvs. Jiro and Suruga, and *D. glandulosa*. Plasmolyze calli subcultured for 1–2 weeks in CPW solution (pH 5.6) (Draper et al. 1988) with 0.7 M mannitol for 1 h. Digest calli in CPW solution (pH 5.6) containing 0.5% Cellulase Onozuka RS, Macerozyme R10 (Jiro and Suruga: 0.05%, *D. glandulosa*: 0.2%), 5 mM MES, 1% PVP-10, and 0.7 M mannitol for 17 h at 27 °C. Enzyme digestion was facilitated by a rotary shaking of 100 rpm. After washing protoplasts twice with CPW solution with 0.7 M mannitol, adjust protoplast density to 1×10^6 cells ml^{-1} with solution containing 0.6 M mannitol and 1 mM CaCl$_2$. Mix Jiro and Suruga, or Jiro and *D. glandulosa* protoplast suspension and apply an alternate current of 35 V cm^{-1} at 2 MHz for 10–20 s and subsequent DC pulse of 0.4 kV cm^{-1} for 500 µs in a fusion chamber SSH-04 using a Somatic Hybridizer SSH-10 (Shimadzu, Kyoto, Japan). Embed protoplasts at 5×10^5 protoplasts ml^{-1} in KM8p agarose medium with modified concentrations of NH$_4$NO$_3$ (150 mg ml^{-1}), NAA (10 µM), zeatin (1 µM), glucose (0.5 M), and 2 mM glutamine. Transfer cells to agarose-bead culture after 2 weeks. After 3 months of culture, transfer the protoplast-derived microcalli onto KM8 agar medium with a modified concentration of NH_4NO_3 (150 mg l^{-1}), NAA (1 µM), zeatin (1 µM) and 2 mM glutamine. After 6 weeks, subculture the calli on MS (1/2 N) agar medium containing 1 µM IAA and 10 µM zeatin.

4.2 Flow Cytometric Analysis

Cut the calli derived from protoplasts into pieces in 10 mM Tris-HCl (pH 7) containing 0.1% Triton X-100, 100 mM NaCl, and 10 mM Na$_2$EDTA. After filtering the solution through a 20-µM nylon mesh to remove debris, stain nuclei with propidium iodide (PI) (100 µg ml^{-1}) and analyze the relative nuclear DNA content by flow cytometry (Coulter Epics Elite or XL, FL, USA). Select the callus lines with the sum of the relative nuclear DNA content of both parents, and transfer them to MS medium containing 0.1 µM IAA and 10 µM zeatin under a 16-h photoperiod to induce adventitious buds.

4.3 RAPD and PCR-RFLP Analysis

Transfer adventitious buds to MS medium with 5 µM zeatin to enhance the elongation of shoots. Extract total DNA and use it for PCR template DNA. PCR reaction mixture is consists of 10 mM Tris-HCl (pH 8.9), 1.5 mM MgCl$_2$, 80 mM KCl, 500 µg ml^{-1} BSA, 0.1% sodium cholate, 0.1% Triton X-100, 200 µM of NTP mix, 0.25 µM primer (Jiro + Suruga: OPH-11, Jiro + *D. glandulosa*: OPI-15, Operon Technologies, CA, USA), 20 ng of template DNA, and 1 unit of Tth DNA polymerase (Toyobo, Osaka, Japan) in 20 µl. Amplify with a PCR machine (Astec PC700, Kyoto, Japan) programmed for 40 cycles of 1 min at 92 °C, 1 min at 40 °C and 2 min at 70 °C. Confirm whether the bands specific to both parents are detected in somatic hybrids.

For analyzing the chloroplast DNA of interspecific somatic hybrids, amplify a variable chloroplast DNA region of 3.2 kb (Ogihara et al. 1991) by PCR. PCR reaction mixture is consisted of 10 mM Tris-HCl (pH 8.3), 50 mM KCl, 1.5 mM MgCl$_2$, 200 µM dNTP mix (Pharmacia Biotech, Tokyo, Japan), 10 pmol of each primer (5′-ATGTCACCACAAACAGAAACTAAAGCAAGT-3′ and 5′-ACTACAGATCTCATACTACCCC-3′), 200 ng of template DNA, and 1 unit of Taq DNA polymerase (Takara, Shiga, Japan) in 100 µl. Amplify by PCR with an initial denaturation of 2 min at 94 °C, and 35 cycles of 1 min at 94 °C, 1 min at 55 °C and 4 min at 70 °C (Yonemori et al. 1996). Digest amplified products with *sty*I (Gibco BRL, MD, USA) and detect banding pattern by electrophoresis.

4.4 Rooting and Chromosome Counting

For rooting, dip the basal ends of shoots in a 50% aqueous IBA ethanol solution (1.5 mM) for a few seconds, and plant them on 1/2 MS (1/2 N) medium. Culture in darkness for the first 10 days of culture and then under the illumination of 60 µmol m^{-2} s^{-1} of 16-h photoperiod. Immerse root tips in distilled water at 4 °C for 24 h and fix them in a solution of acetic acid and methanol (1:1) for 1 h. After hydrolyzing in 5 N HCl at 20 °C for 40 min, stain in Feulgen solution at 4 °C for 30 min and digest cell wall at 37 °C for 70 min with an enzyme solution containing 4% Cellulase Onozuka RS, 1% Pectolyase Y23, 0.07 M KCl, and 7.5 mM Na$_2$EDTA (pH 4). Crush in acetic acid: methanol (1:3) on the slides. Count chromosome numbers under a microscope.

References

Draper J, Scott R, Amitage P, Walden R (1988) Plant genetic transformation and gene expression. Blackwell Scientific, Oxford, pp 148–149

Grosser LW, Gmitter FG Jr, Tusa N, Reforgiato RG, Cucinotta P (1996) Further evidence of a cybridization requirement for plant regeneration from citrus leaf protoplasts following somatic fusion. Plant Cell Rep 15:672–676

Kao KN, Michayluk MR (1975) Nutritional requirements for growth of *Vicia hajastana* cells and protoplasts at a very low population density in liquid media. Planta 126:105–110

Keller A, Coster HGL, Schnabl H, Mahaworasilpa TL (1997) Influence of electrical treatment and cell fusion on cell proliferation capacity of sunflower protoplasts in very low density culture. Plant Sci 126:79–86

Kobayashi S, Ohgawara T, Fujiwara K, Oiyama I (1991) Analysis of cytoplasmic genomes in somatic hybrids between navel orange (*Citrus sinensis* Osb.) and Murcott tangor. Theor Appl Genet 82:6–10

Murashige T, Skoog F (1962) A revised medium for rapid growth and bioassay with bobacco tissue culture. Physiol Plant 15:473–497

Ochatt SJ, Chand PK, Rech EL, Davey MR, Power JB (1988) Electroporation-mediated improvement of plant regeneration from colt cherry (*Prunus avium* × *Pseudocerasus*) protoplasts. Plant Sci 54:165–169

Ogihara YT, Terachi T, Sasakuma T (1991) Molecular analysis of the hot spot region related to length mutations in wheat chloroplast DNAs. 1. Nucleotide divergene of genes and intergenic spacer regions located in the hot spot region. Genetics 129:873–884

Ohgawara T, Uchimiya H, Isii S, Kobayasi S (1994) Somatic hybridization between *Citrus sinensis* and *Poncirus trifoliata*. In: Bajaj YSP (ed) Biotechnology in agriculture and forestry, vol 27. Somatic hybridization in crop improvement 1. Springer, Berlin Heidelberg New York, pp 439–454

Rathore KS, Goldsworthy A (1985a) Electrical control of growth in plant tissue cultures. Bio-Technology 3:253–254

Rathore KS, Goldsworthy A (1985b) Electrical control of shoot regeneration in plant tissue cultures. BioTechnology 3:1107–1109

Rech EL, Ochatt SJ, Chand PK, Power JB, Davey MR (1987) Electro-enhancement of division of plant protoplast-derived cells. Protoplasma 141:169–176

Saito W, Ohgawara T, Shimizu S, Ishii S (1993) Citrus cybrid regeneration following cell fusion between nucellar cells and mesophyll cells. Plant Sci 88:195–201

Shillito RD, Paszkowski J, Potrykus I (1983) Agarose plating and a bead type culture technique enable and stimulate development of protoplast derived colonies in a number of plant species. Plant Cell Rep 2:244–247

Tamura M (1997) Ploidy manipulation through protoplast culture of persimmon. Acta Hortic 436:135–142

Tamura M, Tao A, Sugiura A (1995a) Regeneration of plants from protoplasts of *Diospyros kaki* L. (Japanese persimmon). In: Bajaj YPS (ed) Biotechnology in agriculture and forestry, vol 34. Plant protoplasts and genetic engineering 6. Springer, Berlin Heidelberg New York, pp 43–54

Tamura M, Tao R, Sugiura A (1995b) Regeneration of somatic hybrids from electrofused protoplasts of Japanese persimmon (*Diospyros kaki* L.). Plant Sci 108:101–107

Tamura M, Tao R, Sugiura A (1996) Production of dodecaploid plants of Japanese persimmon (*Diospyros kaki* L.) by colchicine treatment of protoplasts. Plant Cell Rep 15:470–473

Tamura M, Tao R, Sugiura A (1993) Improved protoplast culture and plant regeneration of Japanese persimmon (*Diospyros kaki* L.). Jpn J Breed 43:239–245

Tamura M, Tao R, Yonemori K, Utsunomiya N, Sugiura A (1998a) Ploidy level and genome size of several *Diospyros* species. J Jpn Soc Hortic Sci 67:306–312

Tamura M, Tao R, Sugiura A (1998b) Production of somatic hybrids between *Diospyros glandulosa* and *D. kaki* by protoplast fusion. Plant Cell Tissue Organ Cult 54:85–91

Tao R, Sugiura A (1992) Micropropagation of Japanese persimmon (*Diospyros kaki* L.). In: Bajaj YPS (ed) Biotechnology in agriculture and forestry, vol 18. High-tech and micropropagation 2. Springer, Berlin Heigelberg New York, pp 426–440

Tao R, Tamura M, Yonemori K, Sugiura A (1991) Plant regeneration from callus protoplasts of adult Japanese persimmon (*Diospyros kaki* L.). Plant Sci 79:119–125

Yamada M (1993) Persimmon breeding in Japan. JARQ 27:33–37

Yamamoto M, Kobayashi S (1995) A cybrid plant produced by electrofusion between *Citrus unshiu* (satsuma mandarin) and *C. sinensis* (sweet orange). Plant Tissue Cult Lett 12:131–137

Yonemori K, Parfitt DE, Kanzaki S, Sugiura A, Utsunomiya N, Subhadrabandhu S (1996) RFLP analysis of an amplified region of cpDNA for phylogeny of the genus *Diospyros*. J Jpn Soc Hortic Sci 64:771–777

Zhuang DH, Kitajima A, Ishida M, Sobajima Y (1990) Chromosome numbers of *Diospyros kaki* cultivars. J Jpn Soc Hortic Sci 59:289–297 (in Japanese with English summary)

II.7 Somatic Hybridization in *Ipomoea* (Sweet Potato) Species

M.M. Belarmino[1] and T. Sasahara[2]

1 Introduction

1.1 Importance and Distribution

Sweet potato (*Ipomoea batatas*) is a trailing vine of the morning glory or bineweed family, Convulvulaceae. *Ipomoea* is a pantropical genus with more than 500 species recognized, and displays a polyploid series of $2n = 30, 60$, and 90 (Austin 1987). The domesticated sweet potato is the only hexaploid in the section that produces edible storage roots. The sweet potato grows in a wide range of climates and soils. It can tolerate a low fertilizer input and irregular water supply; hence, the sweet potato is a popular crop for small farmers in many parts of the tropics and subtropics.

Among the important food crops, sweet potato ranks fifth in both economic importance and contribution to the caloric and protein intake of diets in developing countries. The storage roots can produce 25–28% carbohydrates (Martin 1984) and 2–10% protein on a dry weight basis (Yang et al. 1975; Hattori et al. 1985). While their protein content is relatively low, the protein quality is extraordinarily high (Horton et al. 1989). Sweet potato is also an important source of starch for distilleries and other industrial uses.

1.2 Need for Somatic Hybridization

The major goals of sweet potato breeding programs are improvements of yield, storage ability, disease and insect resistance, and improved culinary properties (Lin et al. 1983). Many commercial sweet potato varieties have been obtained by sexual hybridization using lines or cultivars within the same species. The improvement of sweet potato through conventional breeding methods, however, has been very slow compared with other crops such as corn or soybean (Henderson et al. 1983). The slow progress is attributed to sexual incompatibilities and sterilities between cultivars, closely related wild species,

---

[1] Tissue Culture Laboratory, Department of Horticulture, Visayas State College of Agriculture, Baybay, Leyte 6521-A, Philippines
[2] Laboratory of Plant Breeding, Faculty of Agriculture, Yamagata University, Tsuruoka 997-8555, Japan

and distant relatives. Hybrids cannot even be obtained from sexual crosses between cultivars belonging to the same incompatible group. This limits the genetic resources that most breeders have utilized for such traits as resistance to diseases, insects, and nematodes. In addition, very limited explorations have been made with other species or genera as a means of transferring desirable germplasm into the cultivated sweet potato (Iwanaga 1988; Freyre et al. 1991; Iwanaga et al. 1991). Incompatibility has caused numerous problems in hybridizing desirable parents within the batatas group and, there are probably even greater problems in interspecific or intergeneric hybridization (Kowyama et al. 1980; Shiotani et al. 1990). Some attempts to overcome incompatibility such as bud pollination, old flower pollination, and various physical as well as chemical treatments of the stigma before pollination have failed (Fujise 1964).

Somatic hybridization through protoplast fusion has been an effective technique in overcoming constraints of sexual crosses (Glimelius et al. 1991). This technique offers great possibilities for achieving interspecific and wide crosses in *Ipomoea*, in the hope of transferring desirable genes from wild species into the cultivated sweet potato. Thus, protoplast fusion technique serves as an important tool for the production of sweet potato hybrids that are difficult or impossible to obtain by conventional breeding methods.

1.3 Brief Review of Previous Work

The development of methods for the isolation and culture of protoplasts and plant regeneration from protoplast-derived callus of *Ipomoea* species (Murata and Miyagi 1986; Murata et al. 1987, 1994; Sihachackr and Ducreux 1987; Liu et al. 1991, 1992; Belarmino et al. 1994) led to the commencement of protopast fusion studies. Some workers suggested that protoplast fusion could be applied between cross-incompatible sweet potato cultivars (Murata et al. 1993), and between sweet potato and a wild relative, *I. triloba* L. (Liu et al. 1992); however, they failed to validate the hybridity of the regenerated plants. Recently, Wang et al. (1997) have demonstrated the production of somatic hybrids between sweet potato cultivars that belong to the same cross-incompatible group. Fusion of protoplasts was carried out using polyethylene glycol (PEG) solution, and two hybrid plants were obtained from 393 fusion-derived calli. The hybridity of the regenerants was confirmed by their leaf morphology, chromosome number, and RAPD assay. The hybrids grew slowly, and showed flower abnormalities and low pollen fertility – phenomena that have been reported in somatic hybrids of other crops (Gleddie et al. 1986; Handley et al. 1986; Sihachakr et al. 1988; Guri et al. 1991) and that are attributed to meiotic irregularities (Guri et al. 1991). The absence of suitable selection methods in these studies has made it difficult to establish a reliable hybridization system.

The success in somatic hybridization of sweet potato depends mainly on the presence of an efficient hybrid selection method, now that protoplast-to-plant regeneration systems have been established in sweet potato and some related wild species. An efficient and universal selection system needs to be exploited for the identification of the somatic hybrids at an early stage. Inactivation of the protoplasts of one parental species with metabolic inhibitors (e.g., iodoacetamide,) or identification by a characteristic morphology such as pigmentation or callus quality, can be a useful selection method when no genetic marker exists. The procedure for somatic hybridization between the cultivated sweet potato and related wild species, *I. trifida* Don. and *I. lacunosa* L. (Belarmino et al. 1996), and a wide cross between sweet potato and *Tagetes erecta* L. (Belarmino and Sasahara 1996) are described below. These studies do not claim to be comprehensive, but rather present a simple protoplast fusion system that utilizes X-ray irradiation of wild species (donor parent) and iodoacetamide (IOA) treatment of *I. batatas* (recipient parent). In the wide cross between *I. batatas* and *I. erecta* L., a strategy for the selection of hybrid cells is also presented. Overall, these advances are expected to contribute to efforts at broadening the genetic diversity of sweet potato – of prime importance in sweet potato breeding.

2 Protoplast Isolation

2.1 *Ipomoea* Protoplasts

Three-week-old in vitro plants of *I. batatas* and wild relatives, *I. trifida* Don and *I. lacunosa* L., were used as source of protoplasts. Since the leaf lamina resisted enzyme digestion (Sihachakr and Ducreux 1987), petioles and stems (Bidney and Shepard 1980; Murata and Miyaji 1986; Murata et al. 1987; Kokubu and Sato 1988; Liu et al. 1992; Sihachakr and Ducreux 1993) and cell suspension or callus (Wu and Ma 1979; Otani and Shimada 1988; Murata et al. 1994) have been most frequently used for protoplast isolation in *Ipomoea* species. A few cases of successful isolation of mesophyll protoplasts were attributed to mechanical pretreatment of leaf lamina (Murata et al. 1987) and a long digestion period using strong enzymatic composition (Otani et al. 1987).

Isolation of protoplasts from stems and petioles was facilitated by plasmolyzing tissues for 1h in 20ml of MS basal medium (Murashige and Skoog 1962) supplemented with 0.38M mannitol, prior to tissue digestion in an enzyme solution consisting of $10\,gl^{-1}$ (w/v) Cellulase Onozuka RS (Yakult Honsha Co. Ltd., Tokyo), $2\,gl^{-1}$ Macerozyme R-10 (Seishin Co. Ltd., Tokyo), $1.0\,gl^{-1}$ Pectolyase Y-23 (Seishin Co. Ltd., Tokyo), 6.0mM 2-N-morpholinoethane sulfonic acid (MES), 0.5mM $CaCl_2 \cdot 2H_2O$, 0.38M mannitol, and 0.03M sucrose (pH 5.8). Preplasmolysis aids in the recovery of viable protoplasts by decreasing leakage of electrolytes (Cocking 1972), preventing the uptake of exogenous enzymes into the cytoplasm and lessening osmotic shock during isolation (Ruesink 1980). The optimal condition for enzyme digestion was obtained when approximately 2g (FW) of stem and petiole tissue were incu-

Table 1. Isolation and culture of stem and petiole protoplasts from *Ipomoea batatas* and related wild species

	I. batatas	*I. trifida*	*I. lacunosa*
Protoplast yield g^{-1} FW	$1-2 \times 10^5$	5.0×10^5	$4-6 \times 10^5$
Planting density (ml^{-1})	1.0×10^5	1.0×10^5	2.0×10^5
Cell division (%) at day 15	16.5	19.2	7.1
Colony formation (%) at day 40	2.2	4.6	0.8
Type of embryogenic callus	Yellow, opaque firm, nodular surface	Dark yellow firm or compact nodular surface	Yellow white friable irregular surface
Type of regenerated plant	Normal	Normal	Norma

bated overnight in 15 ml of the enzyme solution at 27 °C in the dark without agitation. The digested mixtures were filtered through 100-µm nylon mesh and to the filtrate was added an equal volume of 0.06 M sucrose. The free protoplasts were harvested by centrifugation at 600 rpm for 4 min (Liu et al. 1991). Using this protocol, *I. trifida* and *I. lacunosa* released more protoplasts than *I. batatas* (Table 1). The protoplast yield can be increased by maintaining in vitro plants under low light (8 W m^{-1} 24 h) (Shepard 1981; Wallin and Johansson 1989). The species and genotype have been known to influence yield and culture of protoplasts. In *Ipomoea*, Evans blue tests revealed 85.2 to 89.5% viable protoplasts from sweet potato and 75.4–81.2% from wild relatives. The protoplasts varied in sizes and contained few chloroplasts (Fig. 1A). Since the protoplasts that settled on the bottom of the petri dishes were denser and rich in cytoplasm and showed high viability and capacity for cell division (Sihachakr and Ducreux 1993), they were used for fusion studies.

2.2 *Tagetes erecta* Protoplasts

Leaves from 7-day-old in vitro-grown seedlings were used as source of protoplasts. About 1 g leaf strips was plasmolyzed for 30 min in 15 ml of CPW7W solution containing 0.38 M mannitol. Agitation of leaf tissue-enzyme mixture was inefficient and reduced the viability of leaf protoplasts. Thus, leaf pieces were incubated stationarily in 5 ml of an enzyme solution which consisted of 10 g l^{-1} (w/v) Cellulase Onozuka RS (Yakult Honsha Co. Ltd., Tokyo), 2.0 g l^{-1} Macerozyme R-10, 1.0 g l^{-1} Pectolyze Y-23, 6.0 mM MES dissolved in CPW7M solution, 5.0 mM CaCl$_2$·2H$_2$O, 0.11 M mannitol, and 0.03 M sucrose, pH 5.8. Efficient collection of protoplast pellet was done by centrifugation at 100 g for 5 min. Thorough washing of isolated protoplasts with CPW7M prior to suspension in culture medium is important to remove traces of enzyme solution that may hinder or slow down the regeneration of cell wall. Mesophyll protoplasts of *T. erecta* showed high yield (4.2×10^5 g^{-1} FW of leaf tissue) and viability (83.3%). The chloroplast-rich protoplasts of *T. erecta* were distinctly

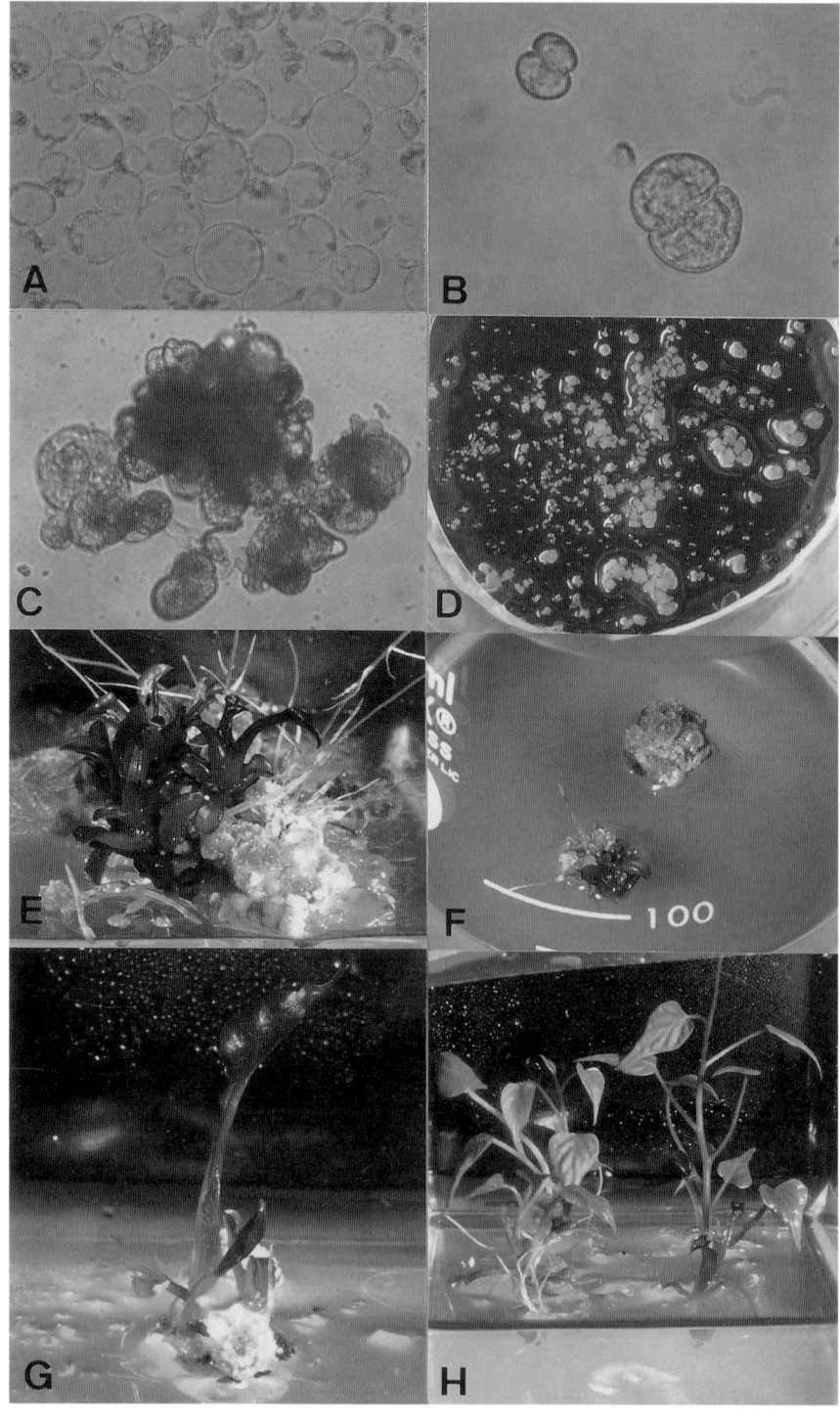

discriminated from the stem and petiole protoplasts of *I. batatas*; hence, these features provided a useful marker for the selection of fused protoplasts.

3 Protoplast Culture and Plant Regeneration

3.1 *Ipomoea* Protoplasts

There are several reports on the culture of *Ipomoea* protoplasts. Several workers suggested the importance of supplements in the culture medium for initial cell division, callus formation, and plant regeneration (for review see Sihachakr and Ducreux 1993). Some vitamins, such as folic acid (Von Arnold and Erickson 1978) and biotin, appeared to promote cell division, while casamino acids enhanced the formation of compact colonies and microcalluses (Kao and Michayluk 1974; Ozaki 1993). Likewise, moderate stimulation of somatic embryogenesis by casamino acids and glutamine was observed (Shekwat and Galston 1983; Grimes and Hodges 1990; Shetty and Asano 1991). Positive effects of GA_3 on plant formation were also reported (Pattat-Ochatt et al. 1988; Takayagi et al. 1991; Ochatt et al. 1992; Shimonishi et al. 1992).

Liquid culture medium appeared to be suitable for the initial culture of *I. batatas* protoplasts (Sihachakr and Ducreux 1987; Murata et al. 1994). Thus, *Ipomoea* protoplasts were resuspended in 2 ml of liquid medium containing modified MS salts (Murata et al. 1987), 500 µM myo-inisitol, 5 µM thiamine-HCl, 10 µM nicotinic acid, 5 µM pyridoxine-HCl, 50 mg l^{-1} folic acid, 0.49 M mannitol, and 0.03 M sucrose, pH 5.8. The ammonium ions needed to be reduced to prevent deleterious effects on protoplasts as reported in *Solanum tuberosum* (Shepard and Totten 1977). The density of protoplasts and the combination of auxin and cytokinin were equally important during the initial culture phase. *I. batatas* and *I. trifida* protoplasts can be cultured at low density (1×10^5 ml^{-1} of medium) on modified MS medium supplemented with a combination of 0.5 mg l^{-1} 2,4-D and 1.0 mg l^{-1} kinetin (Liu et al. 1992; Belarmino et al. 1994). On the contrary, *I. lacunosa* required higher protoplast density (2×10^5 ml^{-1}) for initial culture in the same medium. Protoplasts of *I. batatas* and *I. trifida* needed 5 days' incubation to initiate first mitotic cell division (Fig. 1B) and 7 days for *I. lacunosa*. Gradual reduction of the osmotic concentration was critical for sustained cell division. The mannitol supplement in the medium

Fig. 1A–H. Plants from stem and petiole protoplasts of *Ipomoea* species. **A** Protoplasts isolated from stems and petioles (×192). **B** First division of protoplast-derived cell, 5 days in culture (×288). **C** A protoplast-derived cell colony, 21 days in culture. **D** Compact minicalli plated on agar-solidified MS medium, 30 days in culture (×0.79). **E** Plant regeneration from protoplast-derived calli of *Ipomoea lacunosa* (×0.76), *I. trifida* (×0.92), and *I. batatas* (×0.76). **F** Protoplast-derived plants of *I. batatas* (×1.38) **G, H** see text page 170

was gradually decreased from 0.49 M to 0.3, 0.1, and 0 M at 1-week intervals by dilution with fresh medium lacking mannitol. Sucrose, on the other hand, was increased from 0.03 to 0.09 M after 3 weeks or after colonies were formed (Fig. 1C). *I. batatas* and *I. trifida* formed clusters of 8–12 cells and compact colonies (1–2 mm in diameter) after 2 weeks of culture on fresh medium. *I. lacunosa* cells, however, needed 3 weeks to form 6–10 cell clusters on the same culture medium. Growth of *I. lacunosa* colonies was expedited when cultures were shaken at 60 rpm in darkness. Moreover, the *Ipomoea* species varied somewhat in their auxin and cytokinin requirement for callus production. The combination of $0.2\,mg\,l^{-1}$ 2,4-D and $0.5\,mg\,l^{-1}$ kinetin induced the formation of firm and yellow minicalli (1–2 mm in diameter) from *I. batatas* (Fig. 1D) and *I. trifida*, whereas the combination of $0.2\,mg\,l^{-1}$ IAA and $0.2\,mg\,l^{-1}$ BA enhanced the production of slightly compact and yellow minicalli from *I. lacunosa*. These minicalli were cultured further on MS medium supplemented with 0.09 M sucrose, $50\,mg\,l^{-1}$ casamino acids, 0.2–$0.5\,mg\,l^{-1}$ 2,4-D, $1.0\,mg\,l^{-1}$ kinetin, and $1.0\,mg\,l^{-1}$ ABA, for embryogenesis. Following 3 weeks of incubation at 27°C in darkness, opaque and yellow calli with firm and irregular or nodular surface – typical of an embryogenic sweet potato callus (Chee and Cantliffe 1988) – were produced. ABA has been reported to enhance shoot initiation from protoplast-derived callus of *I. batatas* (Murata et al. 1987), potato (Shepard 1980), and *Hevea brasiliensis* (Etienne et al. 1993). Our results on *I. batatas* and *I. trifida* were similar, however, to those with *Lolium perenne* L. (Faiz and Torello 1992) ABA has no beneficial effect on *I. lacunosa* L. Shoot differentiation from embryogenic calli of *I. batatas* and *I. trifida* was observed 2 weeks after culture on MS medium containing $800\,mg\,l^{-1}$ glutamine, $1.0\,mg\,l^{-1}$ kinetin or $2.0\,mg\,l^{-1}$ BA, and $1.0\,mg\,l^{-1}$ GA_3. *I. lacunos*, on the other hand, regenerated plants on medium lacking kinetin or BA. To induce shoot formation, the calli were initially incubated in the dark for 2 weeks followed by 16 h of daylight ($8\,W\,m^{-2}$) at 27°C. The *Ipomoea* species also differed in regeneration (Table 2), which further confirms the strong variability in the genus *Ipomoea* (Belarmino et al. 1992a). Contrary to the well-rooted *I. lacunosa* plants (Fig. 1E), regenerated *I.batatas* (Fig. 1F) and *I. trifida* (Fig. 1G) plants developed few roots. Rooting was improved when plants were detached from the callus and transferred on agar-solidified MS medium lacking plant growth regulators. Plants arising from embryogenic callus can be distinguised from organogenic shoots by their ability to produce secondary and tertiary proliferations typical of somatic embryos shown in *Brassica* (Gupta et al. 1990; Kirti and Chopra 1990). The protoplast-derived plants morphologically resembled their parents (Fig. 1H) and easily acclimatized.

3.2 *Tagetes erecta* Protoplasts

Mesophyll protoplasts of *T. erecta* started to divide in culture medium used for sweet potato, but after 2 weeks, the dividing cells turned brown and necrosed. Browning of in vitro cultures of certain marigold species, i.e., *T. erecta* and *T. patula*, have been reported (Ketel et al. 1985; Ketel 1986, 1987),

Table 2. Callus formation and plant regeneration from untreated stem and petiole protoplasts of *Ipomoea batatas* and related wild species

Ipomoea species	Callus medium (mg l^{-1})	Regeneration medium (mg l^{-1})	Total calli	No. of calli with plants	No. of regenerated plants
I. batatas and. *I. trifida*	0.2 2,4-D+ 1.0 kinetin+ 1.0 ABA	2.0 BA+ 1.0 GA$_3$	28	4 ± 0.03[a]	8 ± 0.06
	0.2 2,4-D+ 1.0 kinetin+ 1.0 ABA	1.0 Kinetin+ 1.0 GA$_3$	30	2 ± 0.02	9 ± 0.09
	0.5 2,4-D+ 1.0 kinetin+ 1.0 ABA	1.0 Kinetin+ 1.0 GA$_3$	25	1 ± 0.02	4 ± 0.08
I. lacunosa	0.2 IAA+ 1.0 BA	0.5 GA$_3$	28	6 ± 0.04	22 ± 0.12
	0.2 2,4-D+ 1.0 BA	1.0 GA$_3$	25	3 ± 0.02	16 ± 0.14
	0.2 IAA+ 2.0 BA	1.0 GA$_3$	25	2 ± 0.02	11 ± 0.14
	1.0 IAA+	1.0 GA$_3$	25	1 ± 0.02	9 ± 0.18

[a] Mean and standard error.

and suspected to be due to the oxidation of phenol-like substances (Wiermann 1981) and other growth-inhibiting substances that are secreted into the culture medium (Johansson et al. 1982). For sustained cell divisions, the mesophyll protoplasts were embedded in agarose and nursed with either leaf callus-derived cell suspension of *T. erecta* or shoot tip-derived cell suspension of sweet potato (M.M. Belarmino, unpubl). The sweet potato nurse cells were more effective in sustaining cell division and colony formation (M.M. Belarmino, unpubl). Browning of the developing colonies was a constant problem in *T. erecta* cultures (Belarmino et al. 1992b). Browning was reduced by inclusion of 1.0 mg l^{-1} cysteine in the liquid medium and by frequent replenishment of the medium at 5-day intervals. Protoplast-derived callus of *T. erecta* was characteristically fine, watery and brown – typical of a nonembryogenic callus. Microscope examination of the callus revealed highly vacuolated cells and large intercellular spaces. The calli continued to proliferate on MS medium supplemented with different combinations of auxin and cytokinin, but failed to produce plants upon transfer on hormone-free MS medium. Thus, further optimization of suitable culture conditions for protoplast culture is needed to establish a protoplast-to-plant regeneration system in *T. erecta*.

4 Interspecific Hybridization Between Sweet Potato and Wild Relatives

4.1 Fusion Treatment

Selection of fused hybrid protoplasts was performed by means of the metabolic and physical complementation of IOA-treated protoplasts of *I. batatas* cultivars, Shirosatsuma and Kanto 101 (cytoplasmic recipient), and X-ray-irradiated protoplasts of the wild species, *I. trifida* and *I. lacunosa* (cytoplasmic donor). Division of *I. batatas* cells was completely inhibited by the treatment of 10 mM IOA solution cotaining 0.38 M mannitol, 0.03 M sucrose, and 0.5 mM $CaCl_2 \cdot 2H_2O$ (pH 5.8) for 15 min at 4 °C in darkness (Table 3). Inactivation of protoplasts by IOA treatments was also reported in *Solanum officinarum* (Kemble et al. 1986), *Medicago sativa* and *Onobrychis vicifolia* (Li et al. 1993). Meanwhile, X-ray irradiation of *I. trifida* and *I. lacunosa* protoplasts at 10 krad inhibited further growth of colonies (Table 4).

Table 3. Effect of iodoacetamide (IOA) tretment on cell division and colony formation of *Ipomoea batatas* cultivar

IOA conc. (mM)	Cell division[a] (%)		Colony formation[b]	
	Kanto 101	Shirosatsuma	Kanto 101	Shirosatsuma
	Mean standard error			
0.0	13.1 ± 1.1	16.0 ± 0.9	170.0 ± 3.6	217.0 ± 4.2
2.5	4.5 ± 0.4	2.0 ± 0.0	65.0 ± 2.3	78.0 ± 1.5
5.0	1.5 ± 0.2	0.5 ± 0.3	34.0 ± 2.1	25.0 ± 2.1
10.0	0.0	0.0	0.0	0.0
15.0	0.0	0.0	0.0	0.0
20.0	0.0	0.0	0.0	0.0

[a] Percent cell division 15 days after incubation.
[b] Frequency of colony formation as shown by the number of visible protocalli per petri dish.

Table 4. Effect of X-ray irradiation on cell division and colony formation of wild *Ipomoea* species[a]

X-ray dose (krad)	Cell division[b]		Colony formation[c]	
	I. trifida	*I. lacunosa*	*I. trifida*	*I. lacunosa*
	Mean standard error			
0.0	16.0 ± 1.0[d]	7.0 ± 0.4	220.0 ± 2.1	120.0 ± 1.5
2.5	19.0 ± 0.8	9.0 ± 0.4	83.0 ± 1.2	77.0 ± 3.6
5.0	8.0 ± 1.0	4.0 ± 0.3	37.0 ± 1.0	15.0 ± 2.1
10.0	1.5 ± 0.0	0.5 ± 0.0	5.0 ± 0.0	8.0 ± 0.0
15.0	0.0	0.0	0.0	0.0
20.0	0.0	0.0	0.0	0.0

[a] Protoplasts were cultured at a density of $2 \times 10^5 \, ml^{-1}$.
[b] Percent cell division 15 days after incubation.
[c] Frequency of colony formation as shown by the number of visible protocalli per petri dish.
[d] Data are the mean and standard error of three observations.

The fusion of *I. batatas* and wild species was performed by using a commercial electrofuser (Shimazu Model SSH-2, Tokyo), and using PEG 6000 according to the small-scale fusion method of Kao et al. (1974) with minor modifications. The fusion combinations consisted of *I. batatas* cv. Shrisatsuma × *I. trifida*, Shirosatsuma × *I. lacunosa*, *I. batatas* cv. Kanto 101 × *I. trifida*, and Kanto 101 × *I. lacunosa*. Electrofusion was conducted by mixing X-ray-irradiated and IOA-treated protoplasts in a 1:1 ratio at a final concentration of $2 \times 10^5 \, ml^{-1}$ in a fusing solution of 2.5 mM $CaCl_2 \cdot 2H_2O$ and 0.49 M mannitol at pH 5.8 (Belarmino et al. 1996). An aliquot of this mixture (1 ml) was transferred to a mass fusion chamber connected to the electrofuser. An alternating current (AC) source of 1 MHz and $100 \, V \, cm^{-1}$ output voltage was used to line up the protoplasts, and 15 µs direct current (DC) pulses of $3.5 \, kV \, cm^{-1}$ were applied three times to induce protoplast fusion. PEG-mediated fusion, on the other hand, was conducted by mixing *I. batatas* protoplasts ($2 \times 10^5 \, ml^{-1}$) and wild *Ipomoea* protoplasts ($3 \times 10^5 \, ml^{-1}$), in a final volume of 0.3 ml W_5 solution (Negrutiu et al. 1986). Fusion was facilitated by adding a drop (0.2 ml) of $300 \, g \, l^{-1}$ PEG 6000 solution the protoplast mixture and incubating for 15 min at 22 °C in the dark. After incubation, the protoplasts were washed by centrifugation two times with W_5 solution.

4.2 Culture of Fused Protoplasts

The fused protoplasts were cultured on MS medium with modified NH_4NO_3 (0.25 mM) and KNO_3 (9.9 mM), and containing 0.11 mM folic acid, 0.02 mM biotin, 29.2 mM sucrose, 0.38 M mannitol, $1.0 \, mg \, l^{-1}$ 2,4-D and $0.5 \, mg \, l^{-1}$ kinetin at pH 5.8 (Belarmino et al. 1994). The frequency of heterofusion (Fig. 2A) and homofusion was 2% after application of an electric pulse and 4% after PEG treatment, regardless of the donor-recipient combinations. Protoplasts divided (Fig. 2B) and formed six to ten cell clusters (Fig. 2C) 15 days after fusion treatment. PEG-treated protoplasts produced a high number of colonies (protocalli) compared with the electrically fused protoplasts (Table 5). Generally, the induction of embryogenesis from fusion-derived calluses (Fig. 2D) was difficult resulting in low regeneration frequency (Table 5). Only three plants (Fig. 2E) were regenerated out of 32 calli from the electrically fused combination of *I. batatas* cv. Shirosatsuma × *I. trifida*, whereas no plants were produced from other fusion combinations of *I. batatas* and *I. lacunosa*. The putative interspecific hybrids of Shirosatsuma × *I. trifida* easily acclimatized and were grown to maturity. For unknown reasons, the interspecific somatic hybrids did not flower in the greenhouse or field.

4.3 Chromosome and Isozyme Analysis

The chromsome number of parental species, i.e., *I. batatas* ($2n = 6x = 90$), *I. trifida* ($2n = 2x = 90$) and *I. lacunosa* ($2n = 2x = 30$) were compared with the fusion products using callus samples. Ten random callus samples from each of

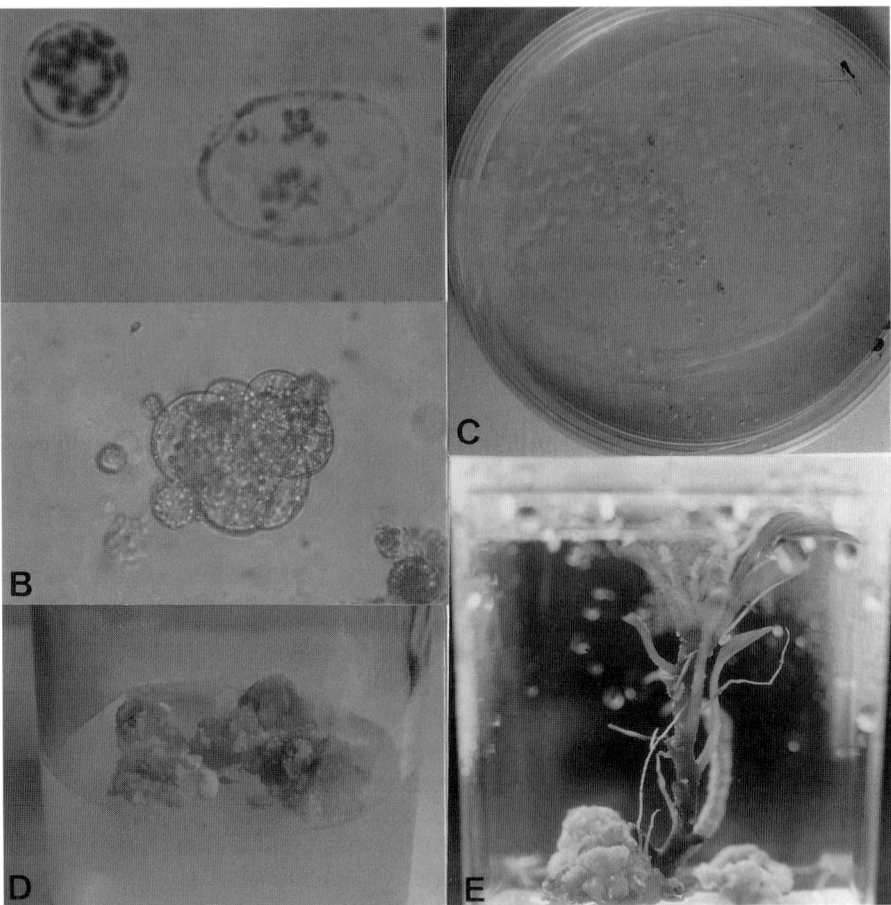

Fig. 2A–E. Plant regeneration from interspecific hybrid calli between IOA-treated *Ipomoea batatas* (recipient) and X-ray-irradiated *I. trifida* (donor). **A** Heterokaryon, 30 min after incubation (×510). **B** Cell clusters, 15 days after incubation (×390). **C** Protocalli on agar-solidified MS medium, 25 days after incubation. **D** Protoplast-derived calli (×0.8). **E** Interspecific hybrid plant (×1.0)

the fusion combinations revealed chromosome numbers varying from $2n = 120–160$ for Shirosatsuma × *I. trifida*, $2n = 98–120$ for Shirosatsuma × *I. lacunosa*, $2n = 108–172$ for Kanto 101 × *I. trifida*, and $2n = 101–111$ for Kanto 101 × *I. lacunosa* combinations. Generally, the hybrid cells possess chromosome numbers that were lower than the sum of the two parents. We suppose that chromosome elimination from wild species occurred after X-ray irradiation (Dudits et al. 1980) or spontaneous loss of chromosomes from both parents during the in vitro passage (Negrutiu et al. 1989).

Peroxidase banding patterns of three plant-generating calli that were produced by electrofusion indicated that these calli were hybrids between

Table 5. Callus formation and plant regeneration from fused[a] and untreated protoplasts of *Ipomoea batatas* cultivars and related wild species

Fusion combination and parental species	Total protoplasts cultured	No. of protocalli	Total calli transferred	Calli with plantlets	Total regenerated plants
Electrofusion					
Shirosatsuma × *I. trifida*	4×10^5	81 ± 2[b]	32	3	3[c]
Shirosatsuma × *I. lacunosa*	4×10^5	63 ± 1	42	0	0
Kanto 101 × *I. trifida*	4×10^5	72 ± 4	50	0	0
Kanto 101 × *I. lacunosa*	4×10^5	54 ± 2	40	0	0
PEG-mediated fusion					
Shirosatsuma × *I. trifida*	4×10^5	120 ± 2	80	0	0
Shirosatsuma × *I. lacunosa*	4×10^5	101 ± 2	60	0	0
Kanto 101 × *I. trifida*	6×10^5	152 ± 3	120	0	0
Kanto 101 × *I. lacunosa*	6×10^5	90 ± 2	50	0	0
Parents (untreated)					
Shirosatsuma	2×10^5	250 ± 3	83	7	21
Kanto 101	2×10^5	200 ± 6	100	9	26
I. trifida	2×10^5	210 ± 2	55	5	10
I. lacunosa	2×10^5	30 ± 3	103	10	42

[a] Fusion between 10mM IOA-treated protoplasts of sweet potato cultivars and 10krad X-irradiated protoplasts of wild species.
[b] Mean standard error.
[c] These three did not flower, therefore produced no seeds.

Shirosatsuma × *I. trifida* due to the presence of one specific band from the *I. trifida* parent (Fig. 3, lane 4). Furthermore, peroxidase-band patterns of leaves from putative hybrids showed that two of them contained two specific bands that corresponded to those of *I. trifida* (Fig. 4). Although calli in the fusion combination of Shirosatsuma × *I. lacunosa* did not regenerate plants, peroxidase banding patterns showed incorporation of isozymes specific for *I. lacunosa* into the hybrid calli (Fig. 5). The peroxidase isozymes are related to many physiological processes of plant development (Cordewener et al. 1991; McDougall et al. 1993), and expression of isozymes may differ among different organs and with the aging of the plant. However, the banding patterns of peroxidase isozymes in leaves at three physiological ages, i.e., 1 month after regeneration in vitro, 2 and 6 months after acclimatization, were stable, indicating that the band patterns were not a result of an alteration of peroxidase activity during culture but rather a reliable indicator of the hybridity of *I. batatas* and *I. trifida*.

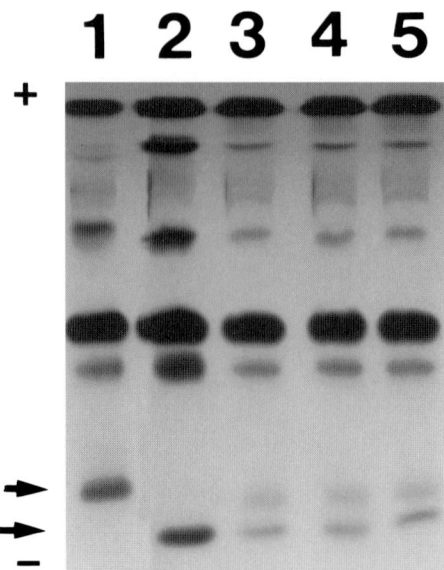

Fig. 3. Peroxidase isozyme band patterns from hybrid and parental calli. *Arrows* indicate bands specific to hybrid or parental calli. Lanes: *1 I. batatas*; *2 I. trifida*; *3–5 I. batatas* × *I. trifida* hybrid calli

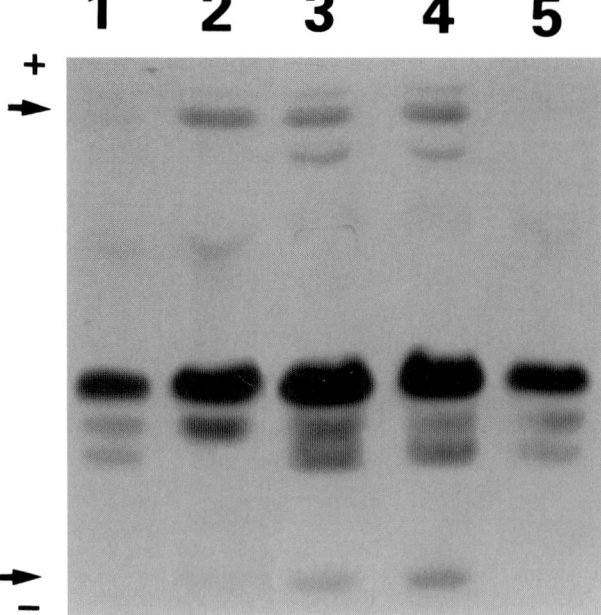

Fig. 4. Peroxidase isozyme band patterns from leaves of hybrid and parental plants. *Arrows* indicate bands specific to hybrid or parental plants. Lanes: *1 I. batatas*; *2 I. trifida*; *3–5 I. batatas* × *I. trifida* hybrid plants

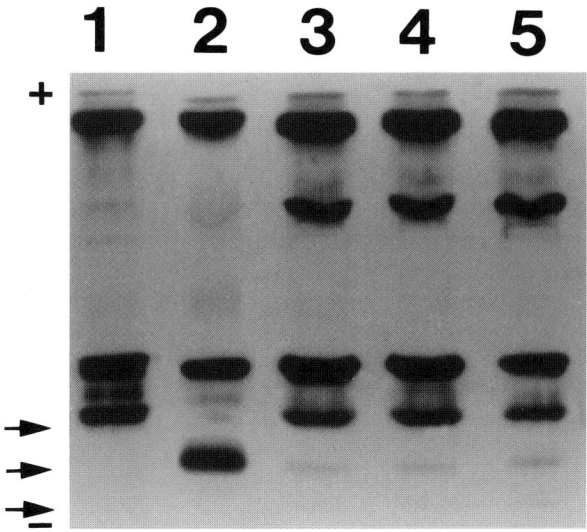

Fig. 5. Peroxidase isozyme band patterns from calli of hybrid and parental plants. *Arrows* indicate bands specific to hybrid or parental plants. Lanes: *1 I. batatas; 2 I. lacunosa; 3–5 I. batatas × I. lacunosa* hybrid calli

5 Asymmetric Protoplast Fusion Between Sweet Potato (*Ipomoea batatas* Lam.) and African Marigold (*Tagetes erecta* L.)

5.1 Fusion Treatment

Somatic hybridization between *I. batatas* Lam and *T. erecta* L. was conducted by electrofusion (Shimazu SSH-2, Tokyo) and the PEG method. To facilitate hybrid selection, stem and petiole protoplasts of *I. batatas* (recipient) were treated with IOA prior to fusion. The division of *I. batatas* cells was completely inhibited by treatment with 10mM IOA for 15min (Belarmino et al. 1996). Cell division of IOA-treated *I. batatas* protoplasts was also never observed when they were subsequently cocultured (using culture plate insert; Millicell-HA 12mm diameter, 0.45µM pore size, Millipore USA) with nontreated *T. erecta* protoplasts after fusion. Thus, for fusion, *T. erecta* mesophyll protoplasts (Fig. 6A) were mixed with 10mM IOA-treated *I. batatas* protoplasts (Fig. 6B) at 1:2 (v/v) ratio.

Optimization of the suitable electrical conditions was critical for efficient electrofusion. Protoplast fusion was induced by a 20-µs DC square pulse of $1.5\,kV\,cm^{-1}$ connected to a mass fusion chamber with parallel stainless steel electrodes. An AC field lower than 1.5kV resulted in low protoplast adhesion and a voltage higher than 1.5kV caused protoplast destruction when DC pulses were added. On the other hand, PEG-mediated protoplast fusion was carried out using PEG 6000 (Belarmino et al. 1996). The protoplasts from both species were resuspended in W_5 solution (Negrutiu et al. 1986) at a density of

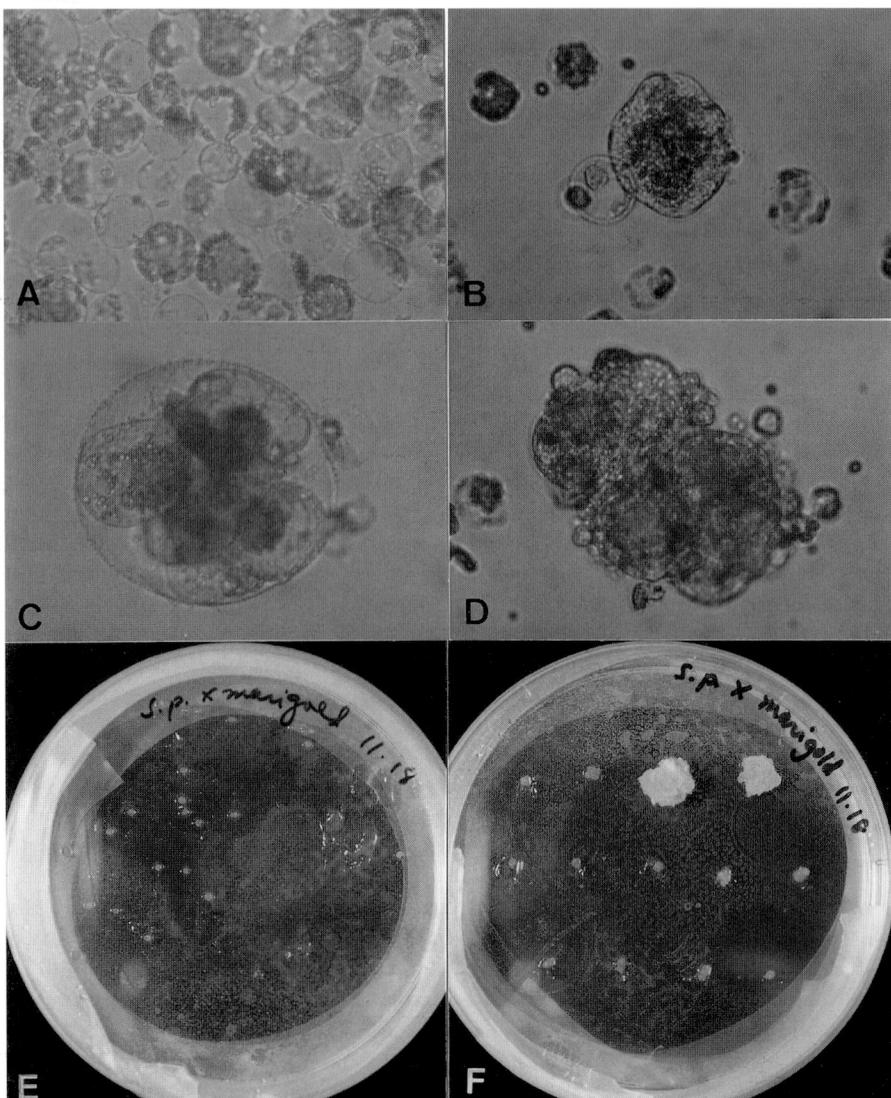

Fig. 6A–F. Production of somatic hybrid calli from PEG-mediated fusion of IOA-treated stem and petiole protoplasts of *Ipomoea batatas* and leaf protoplasts of *Tagetes erecta*. **A** Leaf from protoplasts of *T. erecta* (×260). **B** Heterokaryon, 30 min after PEG treatment (×260). **C** Cell cluster (×390). **D** Hybrid cell microcolony, 21 days after incubation (×1.5). **E** Protocalli, 30 days after incubation (×1.1). **F** Hybrid calli proliferating vigorously on agar-solidified MS medium

$1 \times 10^5 \text{ml}^{-1}$. Protoplast suspensions of both species were mixed in equal volumes and one ml of the mixture was further mixed with an equal amount of PEG solution [30% (v/v) PEG and 0.1 M $CaCl_2 \cdot H_2O$ in 50 mM HEPES buffer, pH 6.5]. After incubation at 25 °C for 30 min, 10 ml of washing solution

(0.1 M $CaCl_2 \cdot H_2O$ in 0.6 M mannitol solution, pH 9) was added gradually. The PEG-treated protoplasts were collected by centrifugation (100 g for 4 min) and washed once with 0.6 M mannitol solution. Maintenance of the osmotic concentration by using the same osmotic agent is important for obtaining viable protoplasts after PEG treatment.

5.2 Culture of Fused Protoplasts

Fusion-treated protoplasts were cultured at a density of $2 \times 10^5 \text{ml}^{-1}$ in 56 × 14-mm Nunclon petri dishes, each containing 2.0 ml of modified MS medium used for *I. batatas* protoplasts (Belarmino et al. 1996). PEG treatment induced 2–4% fusion of the *I. batatas* and *T. erecta* protoplasts (Table 6), whereas ≤0.1% was obtained by electrofusion. The low frequency of electrically fused protoplasts may be due to the sensitivity of protoplasts to electric pulse (Yamanaka et al. 1992) that led to the rupture of *T. erecta* mesophyll protoplasts. Thus, there is a need to further optimize the electrofusion technique to increase fusion frequency of *I. batatas* × *T. erecta* protoplasts. The PEG treatment, on the other hand, effectively induced adhesion and fusion of *I. batatas* and *T. erecta* protoplasts. The *T. erecta* leaf protoplast permits visual identification of heterokaryons and the differences in totipotency provided the basis for partial selection strategy. Fused protoplasts exhibited an enlarged size and cytoplasmic strands (Fig. 6C; Yamanaka et al. 1992). PEG-treated protoplasts were cultured in the same way as *I. batatas* × wild species. After 12 days of incubation, cell division occurred, and subsequently colonies were formed after 21 days (Fig. 6D). These colonies can be plated on agar-solidified MS medium supplemented with 1.0mg l^{-1} 2,4-D and 0.2mg l^{-1} kinetin to produce compact yellow protocalli (Fig. 6E). On subsequent transfers on the same medium, the putative hybrid calli proliferated rapidly compared to other calli and those of control parents (Fig. 6F), suggesting that at least some hybrid vigor was expressed during the callus proliferation stage. Similar observations have been reported in intraspecific combination involving *Solanum tuberosum* (Austin et al. 1985; Waara et al. 1989), in interspecific combination of *Datura* (Schieder 1982) and *Brassica* species (Taguchi and Kameya 1986), as well as in intergeneric combination between *Lycopersicon esculentum* and *Solanum muricatum* (Sakamoto and Taguchi 1991). In the *I. batatas* × *T. erecta* fusion, the hybrid vigor of calli was not an indication of regenerability. Sustained growth of hybrid calli was observed, but plants were not produced upon transfer to a range of regeneration media. This reflects the difficulty of combining useful genes from two phylogenetically unrelated plants.

5.3 Characterization of Hybrid Calli

Somatic chromosome analyses of calli obtained from *I. batatas* × *T. erecta* fusion were carried out for the identification of hybrids. The chromosome number of fusion-derived calli ranged from $2n$ = 98–120 (Table 6), which was

Table 6. Fusion frequency, callus formation, and chromosome number of somatic hybrid calli from PEG fusion of iodoacetamide-treated stem and petiole protoplasts of *Ipomoea batatas* and untreated leaf protoplasts of *Tagetes erecta* L.

Fusion experiments	Frequency[a] of fusion (%)	First mitotic cell division	No. of microcolonies[b]	No. of calli[c]	Chrom. no.[d] ($2n = \ldots x$)
1	2.0 ± 0.4	–	0	0	*
2	4.5 ± 0.2	+	5	0	*
3	2.0 ± 0.2	+	3	0	*
4	3.5 ± 0.3	+	13	2	98; 110
5	4.0 ± 0.4	+	8	1	120
6	3.0 ± 0.2	+	11	2	104; 106
7	2.0 ± 0.3	–	0	0	*
8	4.5 ± 0.5	+	12	2	112; 118
9	3.5 ± 0.3	+	10	1	98
10	4.0 ± 0.4	+	5	0	*
11	*	+	8	0	*
12	*	+	10	1	120
13	*	+	7	0	*
14	*	+	10	0	*
Total		+	102	9	

[a] Checked 1–2 h after PEG treatment; * not checked.
[b] After 21 days of incubation.
[c] After 7 days of culture on the callus proliferation medium.
[d] Average of four observations.

less than the sum of *I. batatas* ($2n = 90$) and *T. erecta* ($2n = 48$). Thus, the somatic hybrid calli obtained by PEG-mediated protoplast fusion were aneuploid. It would appear that a certain number of chromosomes were eliminated during protoplast culture (Negrutiu et al. 1989). Due to the difficulty of identifying the individual chromosomes from each species, it is still unclear whether specific elimination of parental chromosomes occurred.

Isozyme analysis was used to validate the hybridity of *I. batatas* × *T. erecta* calli. Among several isozymes tested, zymograms of peroxidase revealed a distinct difference between *I. batatas* and *T. erecta* parental calli, while fusion-derived calli had common bands of both parents (Fig. 7). Restriction endonuclease analysis of rDNA provided additional evidence for hybridization. *Bam*HI was shown to be the best for discriminating between rDNA fragments of *I. batatas* and *T. erecta*. The hybrid calli contained the specific DNA fragments from both parents (Fig. 8).

6 Summary and Conclusion

Asymmetric hybridization between *I. batatas* and related species, *I. trifida* and *I. lacunosa*, was conducted in an attempt to transfer disease resistance genes of the wild species into the cultivated sweet potato. Of particular interest is *I.*

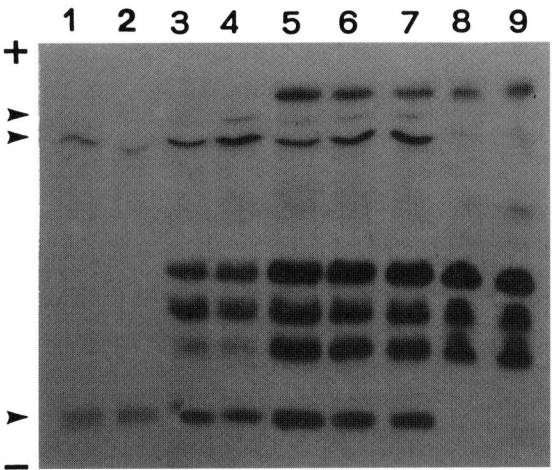

Fig. 7. Zymogram of peroxidase isozyme from hybrid and parental calli. *Arrows* indicate bands that are specific to either hybrid or parental calli. Lanes: *1–2 T. erecta; 3–7 hybrid calli; 8–9 I. batatas*

trifida, since it contains dominant resistant genes against the root-knot nematode (Shiotani et al. 1993; Tokui et al. 1993), and is known to possess commercially useful traits (Iwanaga 1988). Fusion of IOA-treated *I. batatas* protoplasts (cytoplasmic recipient) and X-ray-irradiated *I. trifida* protoplasts (cytoplasmic donor) resulted in *I. batatas* × *I. trifida* hybrid plants, and *I. batatas* × *I. lacunosa* hybrid calli. Chromosome counts and peroxidase analyses of hybrid plants and calli revealed the integration of fragmented chromosomes of wild species into the *I. batatas* nuclear genome. Although *I. trifida* is closely related to *I. batatas* in morphological and agronomic traits, *I. batatas* × *I. trifida* hybrids did not flower. In addition, plants were not regenerated from *I. batatas* × *I. lacunosa* hybrid calli, suggesting an unbalanced genome, resulting in loss of morphogenesis (Xu et al. 1993).

Furthermore, asymmetric fusion between *I. batatas* and a phylogenetically remote species, *T. erecta*, was conducted to explore the possibility of incorporating nematode resistance genes of *T. erecta* (Uhlenbroek and Bijiloo 1958; Obayashi 1989) into the cultivated sweet potato. The chloroplast-containing protoplasts of *T. erecta* facilitate the selection of fused protoplasts. Also, the modified MS medium for the initial culture of *I. batatas* protoplasts can serve as partial selection medium in such a way that only somatic hybrid cells divide and form callus. Thus, the inactivation of *I. batatas* protoplasts with IOA prior to fusion results in cell division and callus formation specific only to heterokaryon-derived cells. The selection medium combined with hybrid vigor of the fusion-derived calluses provides a full selection system for *I. batatas* and *T. erecta* somatic hybrids. Using this system, hybrid calli were obtained from IOA-treated stem and petiole protoplasts of *I. batatas* and nontreated mesophyll protoplasts of *T. erecta*, demonstrating the possibility of applying wide somatic hybridization in the genus *Ipomoea*.

Many problems remain to be solved in the somatic hybridization of *Ipomoea* species. The poor recovery of fertile hybrid plants from the inter-

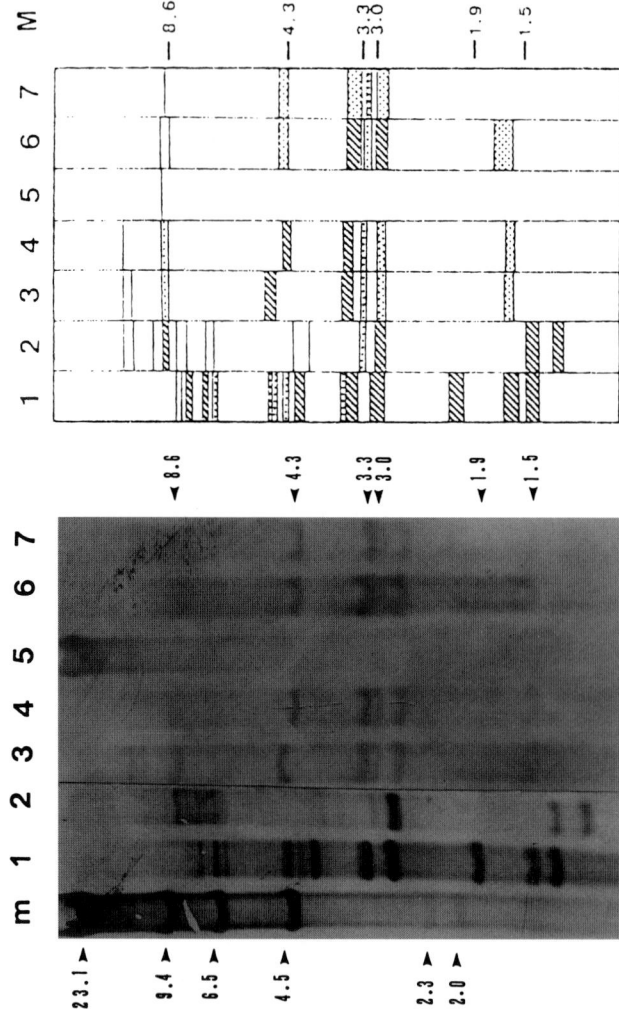

Fig. 8. Southern blot hybridization of BamHI-digested total DNA of *Ipomoea batatas* × *Tagetes erecta* hybrid and parental calli. A nonradioactive labeling method was used for rDNA probe pRR217. Lanes: *1 T. erecta*; *2 I. batatas*; *3–7 I. batatas* × *T. erecta* hybrid calli. Hind-digested lamda DNA was used as size marker *m*

specific fusion of *I. batatas* and wild species still hampers the efficient utilization of this technique in sweet potato breeding programs. For practical utilization of somatic hybrids, it is important to induce flowering or restore fertility of the interspecific somatic hybrids, select useful traits such as disease resistance, and eliminate undesirable characters such as the production of fibrous roots in place of the edible storage roots. Finally, there is a need to optimize suitable conditions for the production of highly regenerable calli in the wide cross between *I. batatas* and *T. erecta*.

7 Protocol for Asymmetric Protoplast Fusion between Sweet Potato (*Ipomoea batatas* Lam.) and Wild Relative, *I. trifida* Don

1. For protoplast isolation, use 2 g (FW) of stems and petioles (2–3 mm segments) from 3-week-old in vitro *Imopoea* plants.
2. Plasmolyze tissues for 1 h in 20 ml of MS medium containing 0.38 M mannitol, then replace with 15 ml of filter-sterilized enzyme solution consisting of $2 g l^{-1}$ Macerozyme R-10, $1 g l^{-1}$ Pectolyase Y-23, 6 mM 2-N-morpholino-ethane sulfonic acid, 0.5 mM $CaCl_2 \cdot 2H_2O$, 0.38 M mannitol, and 0.03 M sucrose (pH 5.8), and incubate without shaking at 27 °C in the dark.
3. Collect protoplasts by floating in 0.03 M sucrose and centrifuging at 600 rpm for 4 min. Wash isolated protoplasts two times with W_5 solution (pH 5.6) by resuspending and centrifuging at 600 rpm for 4 min.
4. To select hybrid cells efficiently, trate *I. batatas* protoplasts with 10 mM IOA solution containing 0.5 M mannitol and 10.0 mM $CaCl_2 \cdot H_2O$ (pH 5.8) for 15 min at 4 °C and expose the *I. trifida* (or related wild species) to more than 10 krad of X-ray irradiation.
5. Mix IOA-treated and X-ray-irradiated protoplasts (1:1 ratio) at a final density of $2 \times 10^5 ml^{-1}$ in a fusing solution of 2.5 mM $CaCl_2 \cdot 2H_2O$ and 0.49 M mannitol at pH 5.8. Transfer 1 ml of an aliquot of this mixture into a mass fusion chamber connected to a commercial electrofuser (Shimazu SSH-1, Kyoto, Japan).
6. Apply AC field at $100 V cm^{-1}$, 1 MHz AC for 5–15 s for protoplast alignment, then DC square pulses at $3.5 kV cm^{-1}$ for 15 s for protoplast fusion.
7. Fusion-treated protoplasts should be kept at room temperature for 1 h, subsequently collected by centrifugation at 600 rpm for 4 min.
8. Protoplasts are resuspended in 2 ml of MS medium containing modified salts (0.25 mM NH_4NO_3; 9 mM KNO_3), 4500 μM myoinositol, 5 μM thiamine-HCl, 10 μM nicotinic acid, 5 μM pyridoxine-HCl, 0.11 mM folic acid, 0.02 mM biotin, 29.9 mM sucrose, 0.384 M mannitol, 4.5 μM 2,4-D, and 2.3 μM kinetin at pH 5.8. Culture dishes are maintained at 27 °C in darkness.
9. After 2 weeks, colonies (clusters of six to ten cells) formed are transferred on agar-solidified MS medium containing 2.3 μM kinetin and 3.8 μM ABA, or in liquid MS medium supplemented with 4.5 μM 2,4-D, for the formation of embryogenic callus. Shoot and root differentiation occurred upon transfer to hormone-free MS medium. Regenerated plants can be micropropagated by single-node stem culture on the same medium.
10. Hybridity of the regenerated plants can be confirmed by chromosome counts, isozyme (e.g., peroxidae), and DNA analysis.

Acknowledgments. The authors are grateful to Prof. Toshinori Abe for his valuable suggestions. This work was partially funded by the Japan Ministry of Education, Culture and Science through a scholarship grant to M.M. Belarmino.

References

Austin S (1987) The taxonomy, evolution and genetic diversity of sweet potatoes and wild relatives. In: Exploration, maintenance and utilization of sweet potato genetic resources. Rep 1st Sweet Potato Plan Conf Int Potato Ctr (CIP) Lima, Peru

Austin S, Bear M, Ehlenfelt M, Kazmierczak PJ, Helgeson JP (1985) Intra-specific fusions in *Solanum tuberosum*. Theor Appl Genet 71:172–175

Belarmino MM, Sasahara T (1996) Asymmetric protoplast fusion between sweet potato (*Ipomoea batatas* L.) and African marigold (*Tagetes erecta* L.). Philipp J Crop Sci 21(1, 2):19–27

Belarmino MM, Abe T, Sasahara T (1992a) Efficient plant regeneration from leaf calluses of *Ipomoea* batatas (L.) Lam. and its related species. Jpn J Breed 42:109–114

Belarmino MM, Abe T, Sasahara T (1992b) Callus induction and plant regeneration in African marigold (*Tagetes erecta* L.). Jpn J Breed 42(4):835–841

Belarmino MM, Abe T, Sasahara T (1994) Plant regeneration from stem and petiole protoplasts of sweet potato (*Ipomoea batatas*) and its wild relative, *I. lacunosa*. Plant Cell Tissue Organ Cult 37:145–150

Belarmino MM, Abe T, Sasahara T (1996) Asymmetric protoplast fusion between potato and its relatives, and plant regeneration. Plant Cell Tissue Organ Cult 46:195–202

Bidney DL, Shepard JF (1980) Colony development from sweet potato petiole protoplasts and mesophyll cells. Plant Sci Lett 18:335–360

Binding H, Nehls R (1978) Somatic hybridization of *Vicia faba* + *Petunia hybrida*. Mol Gen Genet 164:137–143

Chee RP, Cantliffe DJ (1988) Somatic embryony patterns and plant regeneration in *Ipomoea batatas* Poir. in vitro. Dev Biol 24:955–958

Cocking EC (1972) Plant cell protoplasts-isolation and development. Annu Rev Plant Physiol 23:29–50

Cordewener J, Boojij H, van der Zandt F, van der Kammen, de Vries S (1991) Tunimycin-inherited carrot somatic embryogenesis can be rstored by secreted cationic peroxidase isozymes. Planta 184:478–486

Dudits D, Fejer O, hadlaczky G, Koncz, Lazar GB, Horvath G (1980) Intergeneric gene transfer mediated by plant protoplast fusion. Mol Gen Genet 179:283–288

Etienne H, Sota B, Montoro P, Miginiac E, Carron PM (1993) Relations between exogenous growth regulators and endogenous indole-3-acetic acid and abscisic acid in the expression of somatic embryogenesis in *Hevea brasiliensis* (Mull. Arg.) Plant Sci 88:91–96

Faiz ZO, Torello WA (1992) Plant regeneration from callus and protoplasts of perennial ryegrass (*Lolium perenne* L.) J Plant Physiol 140:101–105

Freyre R, Orjeda G, Iwanaga M (1991) Use of *Ipomoea trifida* (H. B. K.) G. Don germplasm for sweet potato improvement. 2. Fertlity of synthetic hexaploids and triploids with $2n$ gametes of *I. trifida*, and their interspecific crossability with sweet potato. Genome 34:209–214

Fujise K (1964) Studies on flowering, seed setting and self- and cross-incompatibility in the varieties of sweet potato. Bull Kyushu Agric Exp Stn 9:123–246

Gleddie S, Keller WA, Setterfield G (1986) Production and characterization of somatic hybrids between *Solanum melongena* L. and *S. sisymbriifolium* Lam. Theor Appl Genet 71:613–621

Glimelius K, Fahlesson J, Langgren M, Sjodin C, Sungerg E (1991) Gene transfer via somatic hybridization in plants. TibTech 9:24–30

Grimes HD, Hodges TK (1990) The inorganic $NO_3:NH_4$ ratio influences plant regeneration and auxin sensitivity in primary callus derived from immature embryos of indica rice (*Oryza sativa* L.) J Plant Physiol 136:362–367

Gupta V, Agnihotri A, Jagannathan V (1990) Plant regeneration from callus and protoplasts of *Brassica nigra* (IC 257) through somatic embryogenesis. Plant Cell Rep 9:427–430

Guri A, Dunbar LJ, Sink KC (1991) Somatic hybridization between selected *Lycopersicon* and *Solanum* species. Plant Cell Rep 10:76–80

Handley LW, Nickels RL, Cameron MW, Moore PP, Sink KC (1986) Somatic hybrid between *Lycopersicon esculentum* and *Solanum lycopersicoides*. Theor Appl Genet 71:691–697

Hattori T, Nakagawa T, Maeshima M, Nakamura K, Asahi T (1985) Molecular cloning and nucleotide sequence of cDNA for sporamin, the major soluble protein of sweet potato tuberous roots. Plant Mol Biol 5:313–320

Henderson JM, Phills BR, Whatley BT (1983) Sweet potato. In: Evans DA, Sharp WR, Ammirato PV, Yamada Y (eds) Handbook of plant cell culture, vol 1. Macmillan, New York, pp 302–326

Horton D, Prain G, Gregory P (1989) high-level investment returns for global sweet potato research and development. Circular 17(3) ISSN 0256-8632, CIP, Lima, Peru

Iwanaga M (1988) Use of wild germplasm for sweet potato breeding. In: Exploitation, maintenance, and utilization of sweet potato genetic resources (pp 199–210). Rep 1st Sweet Potato Conf 1987, In Potato Cent CIP, Lima, Peru

Iwanaga M, Freyre R, Orjeda G (1991) Use of *Ipomoea trifida* germplasm for sweet potato improvement. I. Development of synthetic hexaploids of *I. trifida* by ploidy level manipulations. Genome 34:201–208

Johansson L, Anderson B, Erickson T (1982) Improvement of anther culture technique: Cultivated charcoal bound in agar medium in combination with liquid medium and elevated CO_2 concentration. Physiol Plant 54:24–30

Kao KN, Michayluk MR (1974) A method for high frequency intergeneric fusion of plant protoplasts. Planta 115:355–367

Kao KN, Constabel F, Michayluk MR, Gamborg OL (1974) Plant protoplast fusion and growth of intergeneric hybrid cells. Planta 120:215–227

Kemble RJ, Barsby TL, Wong RSC, Shepard JF (1986) Mitochondrial DNA rearrangements in somatic hybrids of *Solanum tuberosum* and *Solanum brevidens*. Theor Appl Genet 72:787–793

Ketel DH (1986) Morphological differentiation and occurrence of thiophenes in leaf callus cultures from *Tagetes* species: relation to the growth medium of the plants. Physiol Plant 66: 392–396

Ketel DH (1987) Distribution and accumulation of thiophenes in plants calluses of different *Tagetes* species. J Exp Bot 38:322–330

Ketel DH, Breteller H, De Groot B (1985) Effect of explant origin on growth and differentiation of calluses from *Tagetes* species. J Physiol 118:327–333

Kirti PB, Chopra VL (1990) Rapid plant regeneration through organogenesis and somatic embryogenesis from cultured protoplasts. Plant Cell Tissue Organ Cult 20:65–67

Kokubu T, Sato M (1988) Isolation and culture of petiole protoplasts of sweet potato, *Ipomoea batatas* (L.) Lam. and its related species. Mem Fac Agric Kagoshima Univ 24:83–89

Kowyama Y, Shimano N, Kawase T (1980) Genetic analysis of incompatibility in the diploid *Ipomoea* species closely related to the sweet potato. Theor Appl Genet 58:149–155

Li YG, Tanner GJ, Delves AC, Larkin PJ (1993) Asymmetric somatic hybrid plants between *Medicago sativa* L. (alfalfa, lucerne) and *Onobrychis viciflia* scop. (sainfoin). Theor Appl Genet 87:455–463

Lin SSM, Peet CR, Chen DM, Lo HF (1983) Breeding goals for sweet potato in Asia and the Pacific-a survey of sweet potato production and utilization. In: Martin FW (ed) Proc Am Soc Hortic Sci 27 (B):42–60

Liu Q, Kokubu T, Sato M (1991) Plant regeneration from *Ipomoea trifida* protoplasts. Jpn J Breed 41:103–108

Liu Q, Kokubu T, Sato M (1992) Shoot regeneration from protoplast fusion of sweet potato and related species. Jpn J Breed 42 (Suppl 1):88–89

Martin FW (1984) Handbook of tropical crops. CRC, Boca Raton, 296 pp

McDougall G, Millam S, Davidson D (1993) Alterations in surface-sterilized peroxidases during in vitro root development of explants of *Linum ustatissimum*. Plant Cell Tissue Organ Cult 32:101–107

Murashige T, Skoog F (1962) A revised medium for rapid growth and bio-assays with tobacco tissue cultures. Physiol Plant 15:473–497

Murata T, Miyaji Y (1986) Regeneration of plants from stem callus of sweet potato. Jpn J Breed 36 (Suppl 1):24–25

Murata T, Hoshino K, Miyaji Y (1987) Callus formation and plant regeneration from petiole protoplasts of sweet potato (*Ipomoea batatas* Poir). Plant Cell Rep 3:112–115

Murata T, Fukuoka H, Kishimoto M (1993) Plant regeneration from fused cells of sweet potato. Breed Sci 43 (Suppl 1):109

Murata T, Fukuoka H, Kishimoto M (1994) Plant regeneration from mesophyll and cell suspension protoplasts of sweet potato, *Ipomoea* batatas (L.) Lam. Breed Sci 44:35–40

Negrutiu I, DeBrouwer D, Watts JW, Sidirov VI, Dirks R, Jacobs M (1986) Fusion of plant protoplasts: a study using auxotrophic mutants of *Nicotiana plumbaginifolia*, Viviani. Theor Appl Genet 72:279–286

Negrutiu I, Mouras A, Gleba YY, Sidorov V, Hinnisdaels S, Famelaer I, Jacobs M (1989) Symmetric versus asymmetric fusion combinations in higher plants. In: Bajaj YPS (ed) Biotechnology in agriculture and forestry, vol 8. Plant protoplasts and genetic engineering I. Springer, Berlin Heidelberg New York, pp 304–319

Obayashi N (1989) Studies on the methods for controlling the root-lesion nematode, *Pratylenchus penetrans* Cobb infecting Japanese radish. Bull Kanagawa Hortic Exp Stn 39:87–90

Ochatt SJ, Cheevreau F, Gallet M (1992) Organogenesis from Passe Cranssane and Old Home pear (*Pyrus communis* L.) protoplasts and isoenzymatic trueness-to-type of the regenerated plants. Theor Appl Genet 83:1013–1018

Otani M, Shimada T (1988) Plant regeneration from leaf calluses of *Ipomoea trichocarpa* Ell. Jpn J Breed 38:205–211

Otani M, Shimada T, Niizeki H (1987) Mesophyll protoplast culture of sweet potato (*Ipomoea batatas* L.). Plant Sci 53:157–160

Ozaki K (1993) Plantlet formation from the callus of primary leaf in sword bean [*Canavalia gladiata* (JACQ) DC]. Plant Tissue Cult Lett 10(1):45–48

Pattat-Ochatt EM, Ochatt SJ, Power JB (1988) Plant regeneration from protoplasts of apple rootstocks and scion varieties (*Malus* × *domestica* Borkh). J Plant Physiol 133:460–465

Ruesink AW (1980) Protoplasts of plant cells. Methods Enzymol 69:69–84

Sakamoto S, Taguchi T (1991) Regeneration of intergeneric somatic hybrid plants between *Lycopersicum esculentum* and *Solanum muricatum*. Theor Appl Genet 81:509–513

Schieder O (1982) Somatic hybridization, a new method for plant improvement. In: Vasil IV, Scrowfort WR, Kenneth JF (eds) Plant improvement and somatic cell genetics. Academic Press, New York, pp 239–253

Shekwat NS, Galston AW (1983) Mesophyll cell protoplasts of fenugreek (*Trigonella fonumgraecum*): isolation, culture and shoot regeneration. Plant Cell Rep 2:119–121

Shepard JF (1980) Abscissic acid-enhanced shoot initiation in protoplast-derived calluses of potato. Plant Sci Lett 18:327–333

Shepard JF (1981) Protoplasts as source of disease resistance in plants. Annu Rev Plant Physiol 19:145–166

Shepard JF, Totten RE (1977) Mesophyll cell protoplasts of potato: isolation, proliferation, and plant regeneration. Plant Physiol 60:313–316

Shetty K, Asano Y (1991) The influence of organic nitrogen sources on the induction of embryogenic callus in *Agostis alba* L. J Plant Physiol 139:82–85

Shimonoishi K, Karube M, Kukimura H (1992) Rapid embryogenesis by NAA and plant regeneration in sweet potato. Jpn J Breed 42 (Suppl 1):60–61

Shiotani I, Yoshida S, Kawase T (1990) Numerical taxonomic analysis and crossability of diploid *Ipomoea* species related to the sweet potato. Jpn J Breed 40:159–174

Shiotani I, Ooyagi S, Okumura S, Tokui M (1993) Resistance of *Ipomoea trifida* polyploid strains to the root-knot nematode. Jpn J Breed 43 (Suppl 2):37

Sihachakr D, Ducreux G (1987) Plant regeneration from protoplast culture of sweet potato (*Ipomoea batatas* Lam.). Plant Cell Rep 6:326–328

Sihachakr D, Ducreux G (1993) Regeneration of plants from protoplasts of sweet potato (*Ipomoea batatas* L. Lam.). In: Bajaj YPS (ed) Biotechnology in agriculture and forestry, vol 23. Springer, Berlin Heidelberg New York, pp 43–59

Sihachakr D, Haicour R, Serraf I, Barrientos E, Herbreteau F, Ducreux G, Rossignol L, Souvannavong V (1988) Electrofusion for the production of somatic hybrid plants of *Solanum melongena* and *S. khasianum* C. B. Clark. Plant Sci 57:215–223

Tabaeizadeh Z, Ferl RJ, Vasil IK (1986) Somatic hybridization in the *Graminae*: *Saccharaum officinarum* L (sugar-cane) + *Pennisetum americanum* (L.) K. Schum. (pearl millet). Proc Natl Acad Sci USA 83:5616–5619

Taguchi T, Kameya T (1986) Production of virus-free sweet potato plants. In: Int Working Group on sweet potato viruses (IWGSP). Newslett Issue No 1, pp 8–9

Takayagi R, Kitaura T, Miura Y (1991) Effect of giberellin on somatic embryogenesis of sweet potato. Jpn J Breed 41 (Suppl 1):28–29

Tokui M, Nakamura M, Takahashi E, Shiotani (1993) Dominant genes controlling resistance to root-knot nematode in the *Ipomoea trifida* strains. Jpn J Breed 43 (Suppl 2):247

Uhlenbroek F, Bijiloo B (1958) Investigation on nematicides. I, isolation and structure of a nematicidal principle occurring in *Tagetes* root. Rec Trav Chim 77:1004–1009

Von Arnold S, Eriksson T (1978) A revised medium for growth of pea mesophyll protoplasts. Physiol Plant 39:257–260

Waara S, Tegelstrom H, Wallin A, Eriksson T (1989) Somatic hybridization between anther-derived dihaploid clones of potato (*Solanum tuberosum* L.) and the identification of hybrid plants by isozyme analysis. Theor Appl Genet 77:49–56

Wallin A, Johansson L (1989) Plant regeneration from leaf mesophyll protoplasts of in vitro cultured shoots of a columnar apple. J Plant Physiol 135:565–570

Wang J, Sakai T, Taura S, Sato M, Kokubu T (1997) Production of somatic hybrid between cultivars of sweet potato, *Ipomoea batatas* (L.) Lam. in the same cross-incompatible group. Breed Sci 47:135–139

Wiermann R (1981) Secondary plant products and cell and tissue differentiation. In: Stumpf PK, Conn EE (eds) The biochemistry of plants-a comprehensive treatise, vol 7. Academic Press, New York, pp 86–116

Wu YY, Ma TP (1979) Isolation, culture and callus formation of *Ipomoea batatas* protoplasts. Acta Bot Sin 21:334–338

Xu YS, Murto M, Dunckley R, Jones MGM, Pehu E (1993) Production of asymmetric hybrids between *Solanum tuberosum* and irradiated *S. brevidens*. Theor Appl Genet 85:729–734

Yamanaka H, Kuginuki K, Tsuguo K, Nishio T (1992) Efficient production of somatic hybrids between *Raphanus sativus* and *Brassica oleracea*. Jpn J Breed 42:329–339

Yang TH, Tsai YC, Hseu CT, Ko HS, Chen SW, Blackwell RQ (1975) Protein content and its amino acid distribution of locally produced rice and sweet potato in Taiwan. J Chin Agric Chem Soc 13:132–138

II.8 Somatic Hybridization Between *Lycopersicon esculentum* Mill. (Tomato) and Wild Nontuberous *Solanum* Species

T. Gavrilenko

1 Introduction

The wild nontuberous *Solanum* species of the *Etuberosa* series (*S. brevidens* Phil. and *S. etuberosum* Lindl.) are reported to carry resistance to viral and bacterial infections and tolerance to frost (Ross and Rowe 1969; Jones 1979; Austin et al. 1988; Gibson et al. 1990). These species are reproductively isolated (Pandey 1962; Wann and Johnson 1963). However, protoplast fusion technique can overcome barriers of sexual incompatibility (Austin et al. 1985; Helgeson et al. 1986; Fish et al. 1987; Gavrilenko et al. 1992). Although protoplast cultures of the wild species *S. brevidens* and *S. etuberosum* are of great utility, there are only single studies (Haberlach et al. 1985) concerning intraspecific variation with respect to their morphogenetic capacity. At the same time, genotypes with high morphogenetic potential might be successfully used in various biotechnological programs including somatic hybridization.

The production of somatic hybrid plants by protoplast fusion is a useful method for combining genetic materials. In view of the limited genetic resources in cultivated tomato (*Lycopersicon esculentum* Mill.) germplasm, wild *Solanum* species of the *Etuberosa* series are an important source for increasing resistance to biotic and abiotic stresses in tomato. In the present study, genotypic differences in protoplast culture response were detected among eight samples of wild nontuberous *Solanum* species. A sexual hybrid line (*S. brevidens* x *S. etuberosum*) with higher morphogenetic potential was successfully used in intergeneric somatic hybridization to increase genotypic variability in cultivated tomato.

N.I. Vavilov Institute of Plant Industry, B. Morskaya Street, 42, 190000 St. Petersburg, Russia

2 Methods of Protoplast Isolation, Protoplast Culture, and Plant Regeneration

2.1 Plant Material

Seeds of tomato, *Lycopersicon esculentum* cv. Tamina (k-4505), *S. etuberosum* (k-9141, k-21382, k-8808), *S. brevidens* (k-21263, k-21265, k-21265-v), and two lines of sexual hybrids (*S. brevidens* × *S. etuberosum*, k-11671) and (*S. etuberosum* × *S. brevidens*, k-11673), were obtained from the N.I. Vavilov Institute of Plant Industry, Russia. Seeds of these samples were sterilized with 10% hydrogen peroxide for 20 min, washed three times with sterile water, and transferred on hormone-free MS medium (Murashige and Skoog 1962) containing 1% sucrose and 0.8% agar. In vitro plants were grown under 5000 lx (16-h photoperiod, 25 °C).

2.2 Protoplast Isolation

Leaves of 30-day-old plants of *Solanum* species were sliced and incubated in an enzyme solution containing 0.2% Cellulase Onozuka R-10 (Serva), 0.1% Driselase (Kyowa Hakko Kogyo Co. Ltd., Japan), 0.5 M sucrose, and 5 mM $CaCl_2$ (pH 5.6). Mesophyll protoplasts of tomato were isolated in an enzyme solution containing 0.6% Cellulysin (Calbiochem) and 0.1% Macerase (Calbiochem) with the same concentration of sucrose and $CaCl_2$. Incubation was performed in the dark at 23–25 °C for 16 h. Protoplasts were separated from undigested leaf tissues by filtration through 290- and 50-μm nylon sieves. The filtrate was centrifugated at 100 g for 5 min to float the viable protoplasts on the top of the solution. The floating protoplasts were resuspended in a washing medium W_5 (Menczel et al. 1981), and were then precipitated with two subsequent centrifugations at 60 g for 3 min.

2.3 Protoplast Culture and Shoot Regeneration

The purified mesophyll protoplasts were embedded into agar layer CL medium over a reservoir of Res medium according to the method of Haberlach et al. (1985). The petri dishes were incubated at 25 °C in the dark. One month after initial plating, the bilayer agar containing protoplast-derived microcalli was cut into blocks and transplanted to Cul medium (Haberlach et al. 1985); the cultures were transferred to light (1000 lx) and a daylength of 16 h. When calli had become green, they were transferred to Dif regeneration medium (Haberlach et al. 1985). Calli with regenerated shoots were transferred onto hormone-free MS medium.

The morphogenetic potential of the wild *Solanum* species studied was determined from the following indices, which were calculated for each genotype from the results of three or four independent experiments: (1) protoplast plating efficiency, i.e., the percentage of protoplasts that divided by day 14 of culture, and (2) the regeneration capacity of protoclones, i.e., the ratio of protoclones forming at least one shoot to the total number of protoclones transferred to the regeneration medium.

2.4 Selection of the Genotypes of Wild Nontuberous *Solanum* Species with High Morphogenetic Potential

Eight genotypes of wild nontuberous *Solanum* species belonging to the series *Etuberosa* were examined for protoplast culture and shoot regeneration to select genotypes with higher morphogenetic potential for further somatic hybridization experiments. The yield of protoplasts varied from 5×10^5 to 5×10^6 cells g^{-1} FW of leaves. In all genotypes the viable protoplasts were isolated. Cells started to divide within 4-8 days. The results of protoplasts culture and regeneration are shown in Table 1.

A positive correlation between indices of protoplast plating efficiency and regeneration capacity was observed. Genotypes belonging to the *Etuberosa* series showed a significant regeneration capacity when the protoplast plating efficiency was sufficiently high.

Our results demonstrated significant ($p \leq 0.05$) intraspecific variation in morphogenetic potential, that was manifested especially clearly for *S. brevidens* (Table 1). Among eight genotypes studied, three (k-21382, k-11671, k-11673) showed a higher protoplast plating efficiency of more than 50%. Of these three genotypes, two had hybrid nature; they were presented by sexual interspecific hybrid lines. Sexual hybrid line *S. brevidens* × *S. etuberosum*, k-11671 was involved in intergeneric protoplast fusion experiments with cultivated tomato.

Table 1. Protoplast plating efficiency and regeneration capacity in eight samples belonging to the *Etuberosa* series of *Solanum* genus

Sample	Protoplast plating efficiency (%)	Regeneration capacity (%)
S. etuberosum		
k-9141	42.0 ± 2.8	40.6 ± 1.9
k-8808	31.1 ± 4.4	50.2 ± 2.2
k-21382	58.2 ± 2.4	57.7 ± 3.8
S. brevidens		
k-21263	6.0 ± 1.6	3.2 ± 1.4
k-21265-v	30.6 ± 3.1	5.8 ± 2.0
k-21265	22.8 ± 2.3	31.1 ± 6.2
Sexual hybrids		
S. brevidens × *S. etuberosum*		
k-11671	59.4 ± 3.5	67.6 ± 3.2
S. etuberosum × *S. brevidens*		
k-11673	68.1 ± 2.3	76.4 ± 2.1

3 Production and Analysis of Intergeneric Somatic Hybrids

3.1 Protoplast Fusion

To obtain intergeneric somatic hybrids, protoplasts of the sexual hybrid *S. brevidens* × *S. etuberosum*, k-11671, which showed a higher morphogenetic potential, were fused with leaf protoplasts of tomato cv. Tamina using polyethylene glycol-dimethyl sulfoxide solutions at a high pH, as described by Menczel et al. (1981). Fusion solutions were filter-sterilized. Parental protoplasts were mixed at a ratio of 1:1, washed and resuspended ($60g$, 3 min) in a washing medium W_5 at a cell density of 7×10^5 to $2 \times 10^6 \, ml^{-1}$. The fused protoplasts were embedded in the Cl medium and layered onto Res medium and cultured as described above. Nonfused parental protoplasts were cultured, as controls, under the same conditions.

3.2 Selection of Intergeneric Somatic Hybrids

The results of protoplast fusion and control experiments are presented in Table 2. The control experiments with the tomato parent showed that protoplasts of *L. esculentum* did not divide in the cell plating medium Cl, and later died. In the control plates containing protoplasts of the wild parent, *S. brevidens* × *S. etuberosum*, intensive cell divisions were observed.

In the fusion-treatment petri dishes, putative hybrid calli were selected on the basis of their faster growth and their morphological peculiarities. Mainly, the calli that arose in the fusion experiments were small, granular and yellow-green in color; they were morphologically identical to control (*S. brevidens* × *S. etuberosum*) cell colonies. However, calli with new characters appeared rarely in the fused protoplast cultures. These cell colonies had pale green coloring, they grew rapidly, and formed large, dense colonies. No calli with these characters were found in either control plates (Fig. 1). Putative hybrid colonies were transferred individually to the surface of fresh Cul medium and then onto

Table 2. Results of control and intergeneric protoplast fusion experiments

Protoplast culture	No. of selected colonies	No. of calli with regenerants (%)	No. of hybrids obtained	
			Morphologically abnormal shoots in vitro (%)	No. of plants growing in greenhouse
Intergeneric fusion *L. esculentum* (+) (*S. brevidens* × *S. etuberosum*)	26	6 (23%)	50%	12
Control *S. brevidens* × *S. etuberosum*		(58%)	0.6%	

Fig. 1. Selection of hybrid calli; colonies plated into Cl/Res medium and transferred onto Cul medium. Among small microcalli there are four large putative hybrid colonies

Dif regeneration medium. Of 26 selected cell colonies, only six could form regenerants. Half of the regenerants derived from the faster-growing calli had abnormal morphology and distorted shoots with misshapen leaves (Table 2). Morphologically abnormal regenerants were transferred onto MSGK-medium (MS + 0.5 mg l^{-1} GA$_3$ + 0.1 mg l^{-1} NAA) for further development and root formation. Twelve rooted shoots were treated as shoot cultures in vitro and were transferred to sterilized soil. Of 12 hybrid plants only 9 were able to grow in vivo.

3.3 Isoenzyme and RAPD Analyses

Hybridity was confirmed by isoenzyme and RAPD analyses. Leaves from in vitro-growing putative somatic hybrids and parental plants were used for molecular analyses of peroxidases and esterases according to the methods described by Gavrilenko et al. (1992). Analyses by RAPD using total DNA were according to Bulat et al. (1993) and to Gavrilenko et al. (1994), using oligonucleotide primers synthesized by Bulat et al. (1992) and using random decamer primers from Operon Technologies, USA.

All 12 regenerants derived from the faster-growing cell colonies contained the species-specific components from the both parents, which confirms the efficiency of the hybrid selection scheme (Figs. 2, 3). Biochemical and molecular analyses detected genetic heterogeneity in the hybrid material. Some hybrids had changes in their isoenzyme and RAPD profiles; they did not contain the individual parental bands or they had some novel components

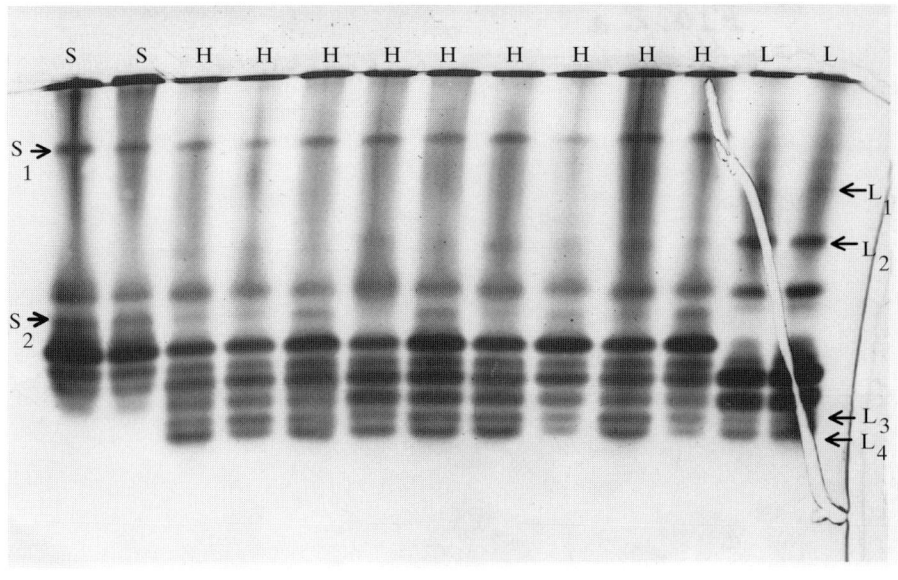
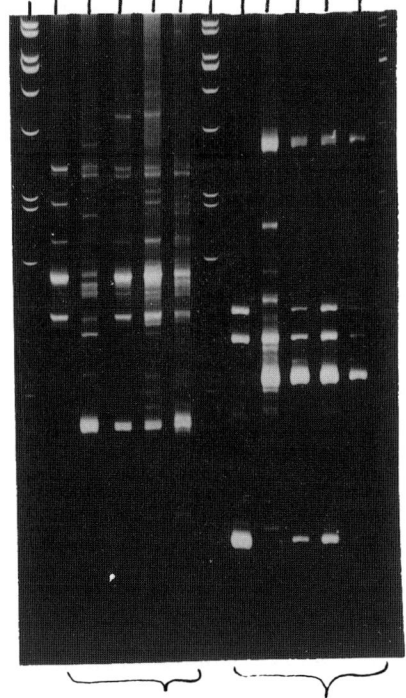

Fig. 2a,b. Identification of the intergeneric somatic hybrids *L. esculentum* (+) (*S. brevidens* × *S. etuberosum*). **a** Acrylamide gel electrophoresis of peroxidases of *L. esculentum* cv. Tamina, wild parent (*S. brevidens* × *S. etuberosum*), mixtures of extracts from leaves of parental samples and intergeneric somatic hybrids. The species-specific components are indicated by *arrows* (*L* tomato; *S S. brevidens* × *S. etuberosum*). **b** RAPD profiles generated by primers 3-2 and 3-3. Parental species: *lane 1* (tomato); *lane 2* (*S. brevidens* × *S. etuberosum*); *lanes 3–5* intergeneric somatic hybrids; *lane 0* molecular marker. (Bulat et al. 1993)

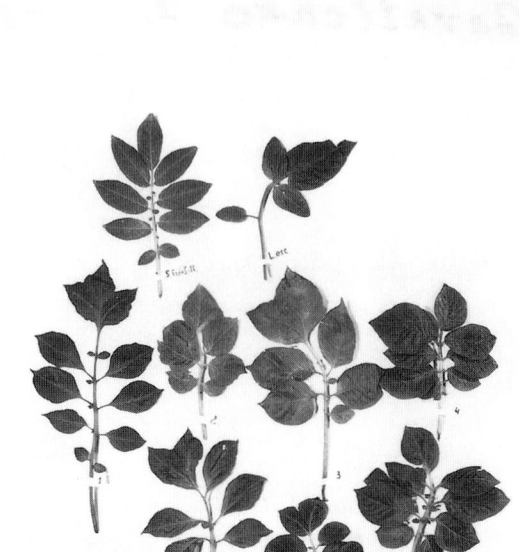

Fig. 3. Leaves of parental lines: *L. esculentum* cv. Tamina (*upper right*), (*S. brevidens* × *S. etuberosum*) (*upper left*), and intergeneric somatic hybrids (*two lower rows*)

(Figs. 2, 3; Table 3). Most probably, the changes at the molecular level are the result of somaclonal variation.

3.4 Cytological Analysis

Chromosome counting was performed in root-tip cells according to Abramova (1988). Both parents, *L. esculentum* and sexual hybrid *S. brevidens* × *S. etuberosum*, were diploids ($2n = 2x = 24$). Therefore, the expected chromosome number in somatic hybrids generated from the fusion of somatic parental cells should be tetraploid, $2n = 4x = 48$. All nine intergeneric somatic hybrids studied by cytological analysis were hexaploids. Chromosome counting revealed a considerable variation in chromosome numbers of the somatic hybrids; most of them were aneuploids (Fig. 4; Table 3).

Hexaploid hybrids could result from the fusions between three protoplasts. Another mechanism is also possible, i.e., the fusions between parental protoplasts are at different phases of the cell cycle. It is possible that cytogenetic instability of the intergeneric somatic hybrids *L. esculentum* (+)

Table 3. Phenotypic and genetic variation of intergeneric somatic hybrids *L. esculentum* (+) (*S. brevidens* × *S. etuberosum*)

Genotype	Chromosome no.	Leaf indexes lateral	Leaf indexes terminal	No. of leaflets	Shape of terminal leaflet	Corolla shape	Corolla color	Pollen viability (%)	Type of peroxidase spectrum[a]
Parental species:									
L. esculentum	24	1.6 ± 0.2	2.2 ± 0.3	4	I	S	Y	92.1	L($L_1 - L_4$)
(*S. brevidens* × *S. etuberosum*)	24	2.7 ± 0.1	2.9 ± 0.1	10	N	R	DV	18.6	S($S_1 - S_2$)
Somatic hybrids									
1d	70–82	1.0 ± 0.2	1.6 ± 0.2	4	E	S	LV	4.5	H
1f	70–82	1.1 ± 0.1	1.4 ± 0.2	6	E	R	BV	10.5	H – L_1
1e	68–72	0.9 ± 0.1	1.2 ± 0.2	4	E	–	–	–	H – L_1
2a	72	1.0 ± 0.1	1.3 ± 0.3	4	E	R	DV	28.9	H – L_1
2b	78	1.0 ± 0.1	1.2 ± 0.5	6	I	S	DV	3.7	H – L_1
2c	62	1.1 ± 0.1	1.7 ± 0.2	4	E	R	DV	8.1	H
4a	63–72	1.2 ± 0.1	1.3 ± 0.3	6	E	P	DV	7.6	H – $L_1 - L_2$
4b	72	1.0 ± 0.1	1.4 ± 0.2	4	I	P	BV	18.4	H – $L_1 - L_2$
4c	72	0.9 ± 0.3	1.5 ± 0.3	6	E	R	DV	–	H – $L_1 - L_2$

Note: Shape of lateral leaflets: I – ivy-leaves; N – narrow-oval; E – egg-shaped. Corolla shape: S – stellate; R – rotate; P – pentagonal. Corolla color – Y – yellow; DV – dark-violet; BV – blue-violet; LV – light-violet. «–» – non-flowering hybrids.
[a] Specific components are given in brackets as indicated in Fig. 2a; H – hybrid type, including the sum of all parental components.

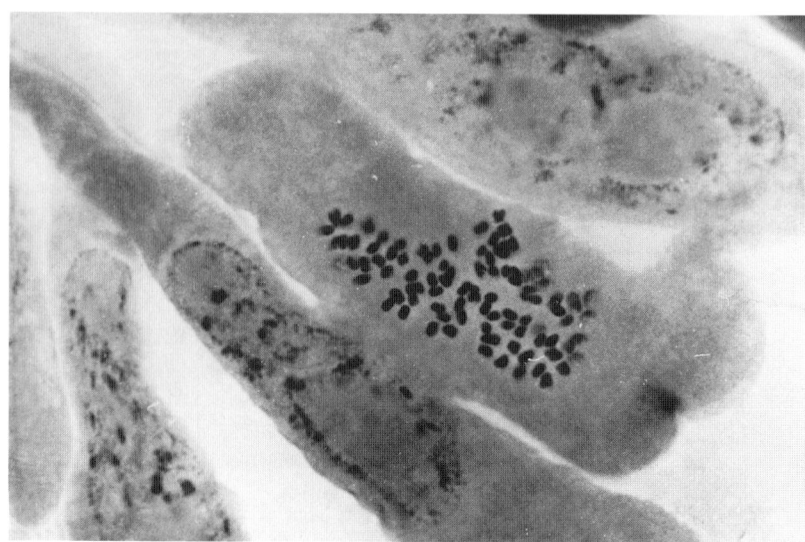

Fig. 4. Root-tip metaphase of intergeneric somatic hybrid 1f

(*S. brevidens* × *S. etuberosum*) is the reason for their limited regeneration capacity and high frequency of morphologically abnormal regenerants in vitro (Table 2).

3.5 Morphological Assessment and Fertility

Parental and hybrid plants were grown under standard greenhouse conditions. Hybrids were intermediate between the parents in general morphology (Fig. 3). At the same time, somatic hybrids showed a high phenotypic variation (Table 3). None of the hybrids was entirely identical in phenotype.

The fertility of parent species and somatic hybrids was studied by the acetocarmine method. Pollen viability varied in a wide range and did not depend on the chromosome numbers of the hybrids (Table 3).

4 Summary and Conclusion

Morphogenetic potential of eight samples of wild nontuberous *Solanum* species of the *Etuberosa* series was evaluated under protoplast culture conditions. Significant intraspecific variation in morphogenetic capacity was shown. The sexual interspecific hybrid line (*S. brevidens* × *S. etuberosum*), which had high protoplast plating efficiency and high regeneration capacity, was selected and involved in protoplast fusion experiments with cultivated tomato.

Intergeneric somatic hybrids between sexually incompatible species were obtained following the chemical fusion of tomato (*L. esculentum*) mesophyll protoplasts with leaf protoplasts of the sexual hybrid (*S. brevidens* × *S. etuberosum*). Putative somatic hybrid cell colonies were selected on the basis of their faster growth. Putative somatic hybrids were identified by isoenzyme and RAPD analyses.

Somatic hybridization does not only make it possible to overcome sexual incompatibility, protoplast fusion is also an effective method to increase genetic variability in hybrid material. The genetic variability of somatic hybrids detected at the molecular, chromosome, and plant levels might be due to somaclonal variation or to interactions between parental genomes.

The potential of intergeneric *L. esculentum* (+) (*S. brevidens* × *S. etuberosum*) hybrids will be assessed in tomato-breeding programs as a basis for the introgression into tomato of genes for resistance to viral diseases and for frost tolerance.

5 Protocol

Leaf protoplasts of wild *Solanum* species and tomato were isolated from 1-month-old plants grown in vitro. Leaves of *Solanum* species were treated for 16 h in the dark at 23–25 °C in enzyme solution contained 0.2% Cellulase Onozuka R-10 (Serva), 0.1% Driselase (Kyowa Hakko Kogyo Co. Ltd., Japan), 0.5 M sucrose, and 5 mM $CaCl_2$ (pH 5.6). Mesophyll protoplasts of tomato were isolated in an enzyme solution containing 0.6% Cellulysin (Calbiochem) and 0.1% Macerase (Calbiochem) with the same concentration of sucrose and $CaCl_2$. The protoplasts were sieved through nylon screens (290 and 50 µm) and centrifugated for 5 min at 100 g. Following flotation, the protoplasts were resuspended in W_5 (Menczel et al. 1981) salt solution and precipitated two times by centrifugation for 3 min at 60 g.

Protoplast suspensions of parental lines were mixed in a 1:1 ratio and pelleted (60 g, 3 min) in washing medium W_5 at a cell density of 7×10^5 to 2×10^6 ml^{-1}. Protoplasts were fused by means of chemical fusion using PEG-DMSO fusion solutions as described by Menzcel et al. (1981). A mixture of parental protoplasts was transferred to 40-mm plastic petri dishes and treated with 50 µl of 30% PEG (MW 4000, Merck), 0.3 M glucose, and 66 mM $CaCl_2$. The petri dishes were incubated at 25 °C in the dark. Control dishes were also prepared for both parental protoplast cultures.

Fusion-treated protoplasts were embedded into agar layer Cl medium over a reservoir of Res medium according to the method of Haberlach et al. (1985). When colonies were formed after 1 month of culture, the bilayer agar was cut into blocks and transplanted onto the Cul medium, (Haberlach et al. 1985); the cultures were transferred to light (1000 lx) and a day length of 16 h.

In the present study, leaf mesophyll protoplasts of the sexual hybrid (*S. brevidens* x *S. etuberosum*) were fused with leaf protoplasts of tomato nondividing in Cl/Res medium. Putative intergeneric somatic hybrids could be selected by heterosis. Putative hybrid colonies were transferred individually to the surface of fresh Cul medium. When calli had become green, they were transferred to the Dif regeneration medium (Haberlach et al. 1985). The shoots emerging from the parental calli were placed on the plant-growing hormone-free medium MS. Morphologically abnormal hybrid regenerants were transferred onto MSGK medium (MS + 0.5 mg l^{-1} GA_3 + 0.1 mg l^{-1} NAA) for further development and root formation. The rooted regenerants were transplanted to sterilized soil for greenhouse testing.

References

Abramova L (1988) Methodical guide. Chromosome counting and chromosome morphology. VIR, Leningrad, 23 pp

Austin S, Baer M, Helgeson J (1985) Transfer of resistance to potato leaf roll virus from *Solanum brevidens* into *Solanum tuberosum* by somatic fusion. Plant Sci 39:75–82

Austin S, Lojkowska E, Ehlenfeldt MK, Keiman A, Helgeson J (1988) Fertile interspecific somatic hybrids of *Solanum*: a novel source of resistance to *Erwinia* soft rot. Phytopathology 78: 1216–1220

Bulat S, Kaboev O, Mironenko N (1992) Universal primer polymerase chain reaction for the study of genomes. Genetika (Russian) 28:19–28

Bulat S, Gavrilenko T, Mironenko N (1993) The genome analysis of intergeneric somatic hybrids by the universally primed polymerase chain reaction technique. The XVth Int Botanical Congr, Tokyo, 549 pp

Fish N, Karp A, Jones MGK (1987) Improved isolation of dihaploid *Solanum tuberosum* protoplasts and the production of somatic hybrids between dihaploid *S. tuberosum* and *S. brevidens*. In Vitro 23:575–580

Gavrilenko T, Barbakar N, Pavlov A (1992) Somatic hybridization between *Lycopersicon esculentum* and non-tuberous *Solanum* species of the *Etuberosa* series. Plant Sci 86:203–204

Gavrilenko T, Dorokhov D, Nikulenkova T (1994) Characteristics of intergeneric somatic hybrids of *Lycopersicon esculentum* Mill. and non-tuberous potato species of the *Etuberosa* series. Russ J Genet 30:1388–1396

Gibson R, Pehu E, Woods R, Jones MJK (1990) Resistance to potato virus Y and potato virus X in *Solanum brevidens*. Ann Appl Biol 116:151–156

Haberlach G, Cohen B, Reichert N, Baer A, Towill L, Helgeson JP (1985) Isolation, culture and regeneration of protoplasts from potato and several related *Solanum* species. Plant Sci 39:67–74

Helgeson JP, Hunt G, Haberlach G, Austin S (1986) Somatic hybrids between *Solanum brevidens* and *Solanum tuberosum*: expression of late blight resistance gene and potato leaf roll resistance. Plant Cell Rep 3:212–214

Jones RAC (1979) Resistance to potato leaf roll virus in *Solanum brevidens*. Potato Res 22:149–152

Medgyesy P, Menczel L, Maliga P (1980) the use of cytoplasmic streptomycin resistance. Mol Gen Genet 179:693–698

Menczel L, Nagy F, Kiss Z, Maliga P (1981) Streptomycin-resistant and -sensitive somatic hybrids of *Nicotiana tabacum* into *Nicotiana knightiana*: correlation of resistance to *N. tabacum* plastids. Theor Appl Genet 59:191–195

Murashige T, Skoog F (1962) A revised medium for rapid growth and bioassays with tobacco tissue cultures. Physiol Plant 15:473–497

Pandey KK (1962) Interspecific incompatibility in *Solanum* species. Am J Bot 49:874–882

Ross R, Rowe P (1969) Utilizing the frost resistance of diploid *Solanum* species. Am Potato J 46:5–13

Wann EV, Johnson KW (1963) Intergeneric hybridization involving species of *Solanum* and *Lycopersicon*. Bot Gaz 124:451–455

II.9 Somatic Hybridization Between *Lycopersicon esculentum* Mill. (Tomato) and *Solanum melongena* L. (Eggplant)

V.M. SAMOYLOV[1] and K.C. SINK

1 Introduction

Biotechnologies for plant improvement are of interest because they circumvent the limitations of traditional crop breeding. For example, somatic hybridization overcomes sexual incompatibility between species, and many somatic hybrids between combinations of a wide range of plant species have been reported (Bajaj 1994), including eggplant (Sihachakr et al. 1994).

The genus *Lycopersicon* is a well-documented resource of valuable agronomic traits (Daunay et al. 1991), and consequently it would be expedient to introduce some of these traits into the gene pool of *Solanum melongena*, eggplant, via somatic hybridization. However, the possibility of combining genomes of sexually incompatible plant species by protoplast fusion often has major limitations to the integration and/or elimination of genetic material in somatic hybrid plants. These changes in genetic component often cause difficulties in plant regeneration and the production of sterile plants in many cases. Eggplant (+) *L. esculentum* × *L. pennellii*, nonregenerable hybrid calli have been previously described by Guri et al. (1991). In the same fusion combination [*S. melongena* (+) *L. esculentum* × *L. pennellii*] Liu et al. (1995) reported the regeneration of somatic hybrid plants with leaf shape similar to tomato and a branching pattern that resembled eggplant despite the irradiation of tomato protoplasts with a lethal dose of γ-irradiation, which confirmed that in intergeneric fusion combinations the irradiation dose seems of minor importance in controlling the elimination of donor chromosomes and the ultimate genome composition of regenerated hybrid plants (Gleba et al. 1988; Wolters et al. 1991; Spangenberg et al. 1994). Irradiation of donor protoplasts provides direction to the elimination of donor chromosomes in fusion combinations of taxonomically remote species, although the mechanisms of such elimination are not yet understood. Therefore, the objective of this study was to monitor donor chromosome flow to recipient eggplant in order to obtain asymmetric somatic hybrids. Such analysis was conducted in the production of highly asymmetric somatic hybrid plants following fusion of γ-irradiated donor protoplasts of *L. esculentum* × *L. pennellii* (EP) with recipient protoplasts of *S. melongena* as reported herein.

Department of Horticulture, Michigan State University, East Lansing, Michigan 48824-1325, USA
[1] *Present address:* Syngenta, 3054 Cornwallis Rd, Research Triangle Park, North Carolina 27709-2257, USA

2 Somatic Hybridization

2.1 Protoplast Culture and Plant Regeneration

Somatic hybrid calli were recovered following PEG/DMSO fusion using all γ-irradiation levels (100, 250, 500, 750, and 1000 Gy) on donor protoplasts of EP (+) eggplant. Our experiments confirmed earlier results from our laboratory where kanamycin at $25\,mg\,l^{-1}$ completely inhibited growth of *S. melongena* explants and calli, and simultaneously had no adverse influence on the growth of EP. Testing variable doses of γ-rays on EP protoplasts revealed that 100 Gy and higher prevented colony formation.

Initially, plating efficiency of eggplant protoplasts, when handled according to Guri and Izhar (1984), was 20–25%. However, only 900–1000 microcalli (0.1–1 mm in diameter) were recovered per control fusion (EP and eggplant protoplasts mixed in a 1:1 ratio at 5×10^6 protoplast ml^{-1}). Consequently, several experiments were carried out to improve the plating efficiency of fusion products using different culture media and embedding protoplasts in 1.4% alginate. Optimum results were achieved with modified KM culture medium (see Protocol; Sect. 6) combined with alginate culture that yielded up to 1.4×10^4 cell colonies per fusion. Subsequently, in fusions with γ-irradiated EP donor protoplasts, 208 total KmR$^+$ calli were selected, among which 123, 9, 6, 53, and 17 were obtained from 100, 250, 500, 750, and 1000 Gy levels, respectively (Table 1). The vast majority of calli produced in all irradiated donor-recipient fusion combinations turned pale white and died within 15 months postfusion on selective medium. Only callus H18, selected from the 100-Gy irradiation, after 12 months in culture regenerated multiple shoots among which three, H18-1, H18-2, and H18-3, taken from independent sites, were removed, rooted, and transferred to the greenhouse. Further attempts to induce shoot regeneration from somatic hybrid calli from all irradiated donor-recipient fusion combinations were unsuccessful over a 2-year period.

Table 1. Recovery of calli and analyses conducted following somatic hybridization between different irradiation levels on donor EP protoplasts fused with eggplant. (Samoylov and Sink 1996)

Irradiation dose (Gy)	Fusion ratio EP:E	Surviving calli (no.)	KmR$^+$ calli (no.)	No. of calli analyzed		
				PCR	RAPD	Flow cytometry
100	1:1	5.2×10^3	123	11	11	5
250	1:2	1.4×10^4	9	5	5	2
500	1:3	3.7×10^3	6	5	3	–
750	1:3	1.3×10^4	53	5	5	4
1000	1:3	1.2×10^4	17	4	4	4

– Not determined

2.2 Greenhouse Performance

Morphologically, the hybrid plants resembled eggplant, but growth was slower than in the fusion parents. Leaf shape and morphology of the hybrid plants closely resembled that of eggplant (Fig. 1A). After 8 months' growth in a greenhouse the three somatic hybrids flowered. Eggplant has dark purple flowers composed of five fused petals with five anthers per flower. Flowers of EP have five yellow separate petals and five anthers. The somatic hybrids produced dark purple flowers, but each was usually composed of six separate petals and six anthers (Fig. 1B). Most hybrid flowers produced an abnormal style, stigma, and anthers void of pollen.

Fig. 1A,B. Leaves (**A**) and flowers (**B**) of *S. melongena* (*left*), somatic hybrid H18-1 (*center*) and *L. esculentum* × *L. pennellii* (EP). (Samoylov et al. 1996)

3 Analysis of Somatic Hybrids

3.1 PCR, RAPD, and Southern Analysis

Out of the 208 kanamycin-resistant calli recovered from the fusion of irradiated EP protoplasts with eggplant protoplasts, 30 were analyzed (Table 1). Each of these calli had the 255-bp *Npt*II fragment. For RAPD analysis, eight primers (OPH-01, OPH-20, OPI-01, OPI-02, OPI-07, OPI-10, OPK-01, OPK-20), which produced the most distinct differences in banding profiles between the parents were used. Among 11 calli analyzed from the 100-Gy donor EP (+) eggplant protoplast fusions, five had RAPD patterns similar to tomato, although two of them, H4 and H16, contained additional polymorphic bands which were not present in either fusion parent. The other six calli (H11, H14, H17, H18, H20, and H21) were confirmed as hybrids although they had banding patterns mostly from eggplant. Banding profiles of kanamycin-resistant calli selected and analyzed from higher doses of irradiation, 5, 3, 5, and 4 from 250, 500, 750, and 1000 Gy, respectively, although predominately eggplant, also had EP bands. Two calli, H11 and H18 (100 Gy), confirmed as hybrids by RAPD analysis were further analyzed for the presence of tomato DNA by Southern hybridization. Such hybridizations with the tomato species-specific repeat pTHG2 and the potato rDNA probe revealed the presence of limited amounts of EP DNA. Also, five other calli obtained from 100 Gy, mentioned previously as having only tomato RAPD pattern(s), had tomato hybridization profiles (Samoylov and Sink 1996).

3.2 Flow Cytometric Analysis

Nuclear DNA content of eggplant and tomato (Arumuganathan and Earle 1991) is close to that of chicken red blood cells (CRBC; Galbraith et al. 1983). Thus, CRBC were used as a standard for mouse thymocytes (Thy) (Fig. 2A) and somatic hybrid calli (Fig. 2D). Conversely, Thy were used as the internal standard to determine the nuclear DNA content of the parental lines involved in somatic hybridization (Fig. 2B,C). The nuclear DNA content of ethanol-fixed Thy was found to be 6.137 ± 0.262 pg/2C. Therefore, the nuclear DNA of *S. melongena* Nuclei (2.479 ± 0.058 pg/2C) and EP (2.017 ± 0.189 pg/2C) was calculated. The range in nuclear DNA content of hybrid calli from different doses of γ-irradiation of donor EP was from 5.498 pg/2C (H18) to 20.163 pg/2C (H11). All calli tested (Table 2) were polyploids (Fig. 2E), for instance H18 – $4.4n$, H11 – $16.3n$ (100 Gy); H251 – $9.2n$, H252 – $9.6n$ (250 Gy); H754 – $6.7n$, H751 – $9.2n$ (750 Gy); H1003 – $8.7n$, and H1002 – $9.2n$ (1000 Gy). Hybrid plants regenerated only from callus H18 with 5.49 ng/2C, a DNA content close to the $4n$ of eggplant and such calli only occurred when the donor was irradiated with 100 Gy. Analysis of hybrid plant H18-1 revealed that it had a 2C value of 5.47 pg.

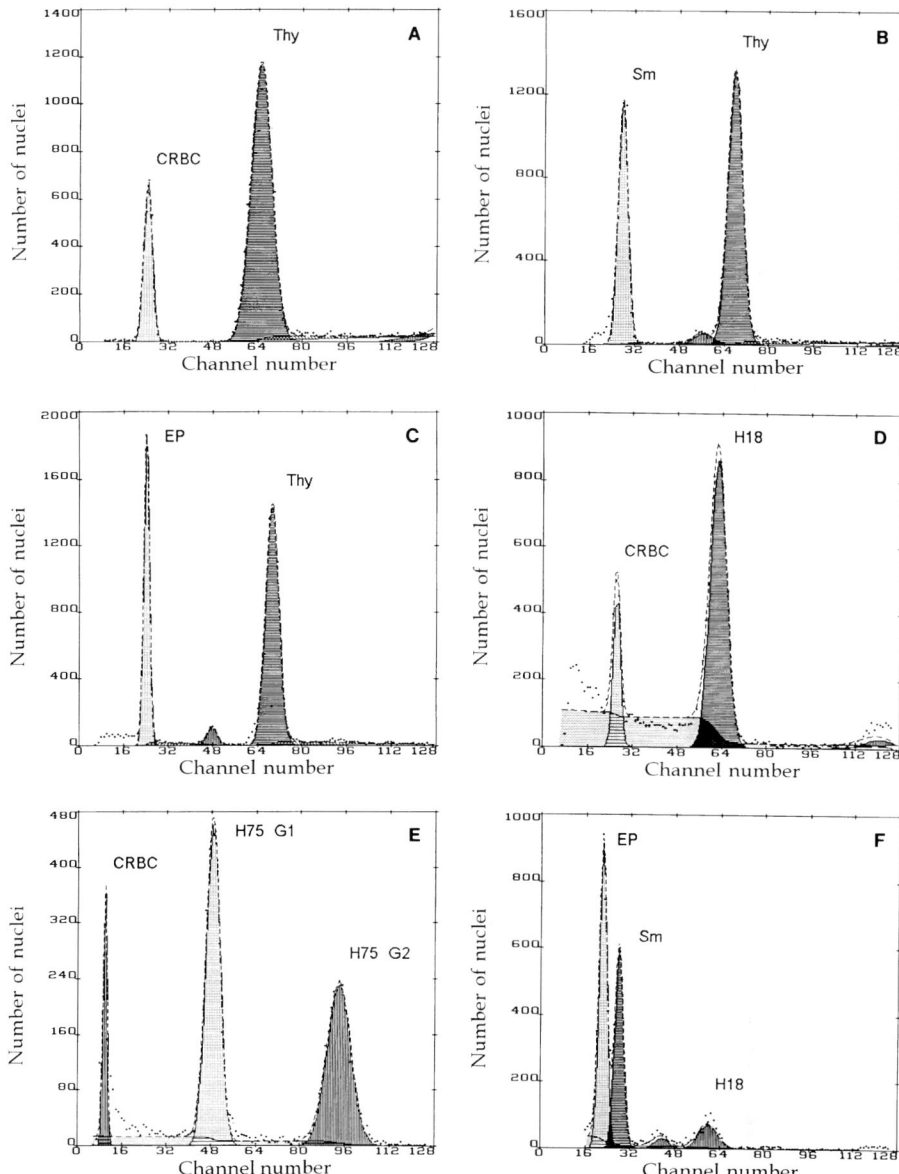

Fig. 2A–F. Flow cytometry histograms of nuclei. **A** Chicken red blood cells (*CRBC*) and mouse thymocytes (*Thy*). **B** *S. melongena* ($2n = 2x = 24$) (*Sm*) and *Thy*. **C** *L. esculentum* × *L. pennellii* ($2n = 2x = 24$) (*EP*) and *Thy*. **D** CRBC and a somatic hybrid callus *H18* (100 Gy). **E** *CRBC* and somatic hybrid callus *H75* (750 Gy). **F** Parental EP, Sm and somatic hybrid callus H18. (Samoylov and Sink 1996)

Table 2. Nuclear DNA content of somatic hybrid calli selected after somatic hybridization between different irradiation levels on donor EP protoplasts fused with eggplant. (Samoylov and Sink 1996)

Irradiation dose (Gy)	Genotype	Nuclear DNA content (2 C/pg)
Plants	S. melongena	2.479 ± 0.058
	EP	2.017 ± 0.189
Calli		
100	H11	20.163
	H14	12.719
	H17	5.688
	H18[a]	5.498
250	H251	11.371
	H252	11.903
750	H751	11.415
	H752	10.805
	H753	11.225
	H754	8.244
1000	H1001	11.447
	H1002	11.346
	H1003	10.814
	H1004	11.143

[a] H18 regenerated hybrid plants H18-1 to -3.

3.3 Dot Blot Analysis

The amount of EP DNA in seven calli and two hybrid plants, all obtained from the 100-Gy exposure, was determined by dot blot analysis (Fig. 3B). The relative amount of L. esculentum repetitive sequences in genomes of L. pennellii, L. esculentum × L. pennellii (EP), and eggplant was first calculated to score the amount of EP genome in each hybrid calli. Tomato species-specific 452-bp repeat pTHG2 is randomly located in all chromosomes of L. esculentum (Zabel et al. 1985); thus, the value of hybridization of pTHG2 to L. esculentum DNA was used as the standard at 100%. The value of hybridization of probe pTHG2 to L. pennellii and EP DNAs relative to L. esculentum DNA was found to be 100%. The hybridization signal of the pTHG2 probe to S. melongena DNA was negligible (Fig. 3A). The mean of the background was subtracted from the values obtained. Thereafter, the percent of tomato repeats in somatic hybrid calli was determined by a comparison of the degree of hybridization of pTHG2 to DNA from the hybrid calli with that hybridizing to EP (Fig. 3B). Calculations of EP chromosomes in somatic hybrid calli were based on the assumption that a symmetrical hybrid with all chromosomes of eggplant (E) and EP should have a DNA content of 4.496 pg and corresponding dot-blot value of about 44.86% of that obtained for an equal amount of EP genome. Thus, one EP chromosome should produce an average dot blot value of 1.869%. Hence, the percentage of DNA and, consequently, the number of EP chromosomes in hybrid calli was calculated and found to be for H11 8, for H14 6.2; for H17 3.8, for H18 6.29, and for H20 13.2. Somatic hybrid

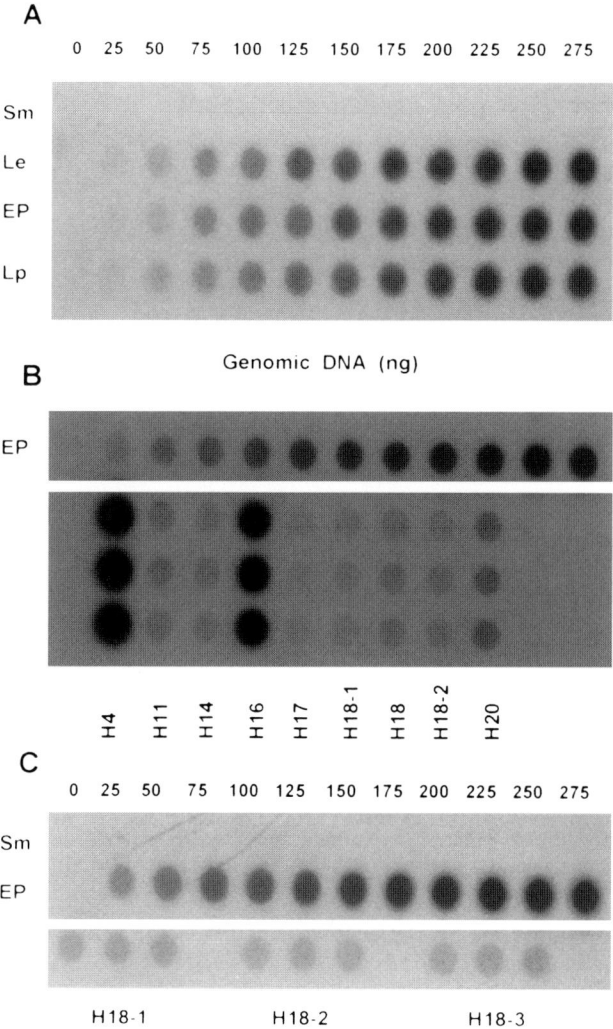

Fig. 3A–C. Dot blot hybridization of genomic DNAs with tomato species-specific probe pTHG2. **A** *Sm S. melongena; Le L. esculentum; Lp L. pennellii; EP L. esculentum* × *L. pennellii.* **B** *EP L. esculentum* × *L. pennellii* and hybrid calli and plants obtained from the 100 Gy experiment. **C** *Sm S. melongena; EP L. esculentum* × *L. pennellii* and somatic hybrid plants H18-1 to H18-3. DNA from somatic hybrids was loaded in triplicate. (Samoylov and Sink 1996; Samoylov et al. 1996)

plants H18-1 and H18-2 contained 4.77 and 5.14 EP chromosomes, respectively, and were regenerated from calli H18 with 6.29 EP chromosomes. Calli H4 and H16, from 100 Gy, were confirmed as tomato by the presence of multiple copies of tomato DNA that corresponded to 118.9 and 107.2 EP chromosomes, respectively.

The percent of EP DNA in plants H18-1, -2, and -3 was 6.23, 5.41, and 5.95%, respectively. Consequently, the amount of EP DNA in hybrid plants H18-1, -2, and -3 was calculated and found to be equivalent to 3.59, 2.90, and 3.19 average size EP chromosomes, respectively.

3.4 RFLP Analysis

3.4.1 Nuclear Genome Constitution

A set of 31 tomato genomic clones (Tanksley et al. 1992) were used to analyze the composition of *L. esculentum*- and *L. pennellii*-specific chromosomes in the asymmetric somatic hybrid plants. The mapped positions of the clones are shown in Fig. 4. Each probe was hybridized to DNA from *L. esculentum*, *L. pennellii*, EP, *S. melongena*, and plants H18-1, -2, and -3. To determine useful polymorphisms, the DNAs from the parental genotypes were digested with restriction enzymes *Dra*I, *Eco*RI, *Eco*RV, or *Hind*III. With these four enzymes, polymorphisms were observed between genotypes with all test probes except TG95 and TG96. Only distinct polymorphic bands between *L. esculentum* and *L. pennellii*, which were also present in EP, were used to score the presence or absence of the corresponding chromosomes in the asymmetric somatic hybrid plants. Although polymorphisms between *L. esculentum* and *L. pennellii* were not found with probes TG95 and TG95, only one of the parents had a unique band, so they were scored as not present due to the absence of the corresponding band in the somatic hybrid plants. RFLP analysis revealed that the somatic hybrid plants possessed only eight to ten fragments out of 12 *L. esculentum* and 12 *L. pennellii* specific chromosomes (Fig. 4). Loci of both species of EP were evenly represented: four to five from *L. esculentum* and four to five from *L. pennellii*. All somatic hybrid plants retained locus TG22, chromosome four, from both species of EP. Loci TG31 and TG79 of *L. esculentum* chromosome two and *L. pennellii* chromosome nine, respectively, were missing in hybrid plant H18-1.

3.4.2 Determination of Organelles

Hybridization of the P-2 probe with blots containing *Eco*RI digested total DNA revealed that all three hybrids contained only cpDNA fragments specific for *S. melongena* (Fig. 5). The mtDNA genome in the asymmetric somatic hybrids was predominantly from eggplant; however, some eggplant-specific polymorphic bands were missing in plants H18-1, -2, and -3 (Fig. 6).

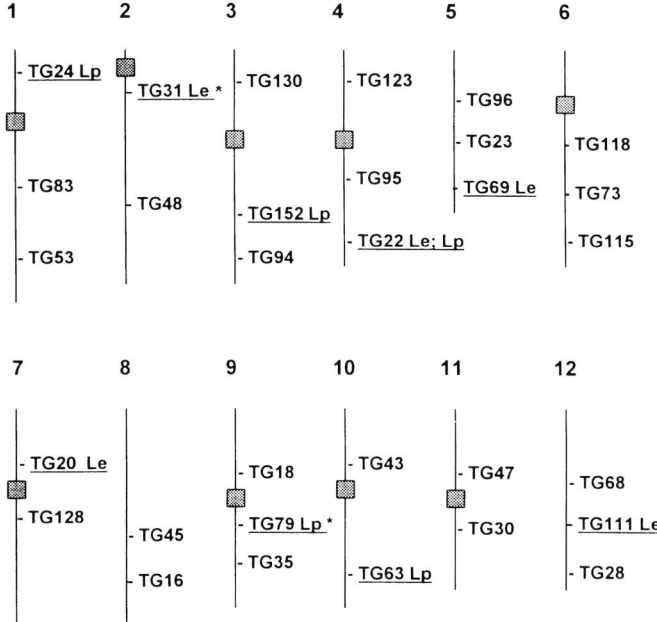

Fig. 4. Tomato map indicating approximate location of selected RFLP markers used for determination of *L. esculentum* (*Le*)- and *L. pennellii* (*Lp*)-specific chromosomes in asymmetric somatic hybrid plants H18-1, H18-2, and H18-3. The *shaded areas* indicate putative centromere location. *Underlined markers* are present. * not present in plant H18-1. (Samoylov et al. 1996)

4 Elimination of the Donor Genome

The nuclear DNA content of hybrid calli resulting from all doses of γ-irradiation on donor EP was similar and was not correlated with the dose applied. The majority of somatic hybrid calli analyzed were 5–9n polyploids. However, somatic hybrid plants with eggplant morphology regenerated only from one callus with a ploidy level close to 4n, and such calli occurred only in the 100-Gy donor EP (+) eggplant protoplast fusions. This morphogenic callus (H18) had 5.49ng/2C, which, in fact, is the 4n of eggplant and the amount of DNA equivalent to about 6.29 EP chromosomes.

It has been shown that species-specific repetitive DNA probes used as dot blots would measure the amount of donor DNA in asymmetric hybrids (Imamura et al. 1987). Analysis of *Nicotiana* asymmetric somatic hybrids by dot blots and chromosome counts (Piastuch and Bates 1990; Kovtun et al. 1993) revealed that the dot blot value was correlated with the number of chromosomes in the hybrids. Furthermore, Daunay et al. (1993), in a study of somatic hybrids of eggplant, and Schoenmakers et al. (1993), describing hybrids between tomato and potato, both reported a correlation between

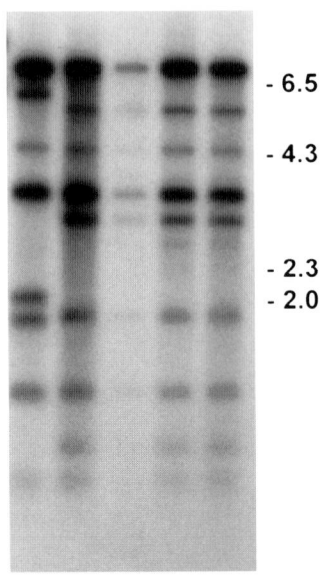

Fig. 5. Southern hybridization of cpDNA from *L. esculentum* × *L. pennelii* (*EP*), *S. melongena* (*Sm*) and somatic hybrid plants H18-1 (*1*), H18-2 (*2*), and H18-3 (*3*). Total genomic DNA was digested with *Eco*RI and hybridized with the 21.9-kb tomato chloroplast *Pst*I fragment (P-2). (Samoylov et al. 1996)

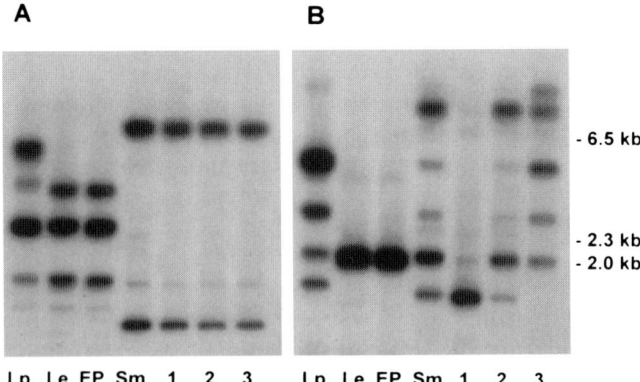

Fig. 6A,B. Southern hybridization of mtDNA from *L. esculentum* × *L. pennelii* (*EP*), *S. melongena* (*Sm*), and somatic hybrid plants H18-1 (*1*), H18-2 (*2*), and H18-3 (*3*). Total DNA was digested with *Hind*III (**A**) or *Eco*RI (**B**) and hybridized with the pZmE1 probe. (Samoylov et al. 1996)

chromosome counts and flow cytometry data. Such analysis used in this study permitted an estimation of the genome composition of selected EP (+) eggplant calli. Flow cytometry data for the callus from which plants were regenerated, line H18, revealed that it had a DNA content of 5.49 pg, which is 1.22-fold larger than expected for the symmetric hybrid (4.496 pg), and approximately two-fold larger than eggplant DNA content (E 4C = 4.958 pg). Thus, in regard to the morphology of regenerated plants and the phenomenon of polyploidization of the recipient genome which frequently occurs in "gamma" hybrids (Famelaer et al. 1989), hybrid H18 was an asymmetric callus with 4C of eggplant and a low amount of EP genome. By calculation, H18 had 4.958 pg of eggplant DNA and 0.532 pg of tomato DNA, which corresponds to the amount of DNA in 48 eggplant and 6.33 EP chromosomes, respectively. Thus, the genome of H18 contains ~9.69% EP genome.

The theoretical dot blot value for this progenitor callus would be 11.83% tomato, and, in fact, the dot blot analysis value was 11.768%. The 11.768% corresponds to 6.29 EP chromosomes, which is close to the theoretical number of 6.33. These results indicate that flow cytometry analysis coupled with dot blot hybridization allowed determination of the ploidy level of hybrids and the amounts of donor and recipient DNA present. Thus, the number of donor and recipient chromosomes in asymmetric hybrids may be calculated. This procedure is particularly valuable when chromosome counting by cytology is difficult and/or cannot be applied when the chromosome morphology is similar, as in the case of tomato and eggplant. Discrepancies, in our experiments, between the flow cytometry and dot blot data are in the range of those found in other studies. They might also be related either to experimental errors of dot blot and/or flow cytometry, or to an undetected nonrandom distribution of the repetitive sequences in the plant genome and/or the average DNA content of the chromosomes used for calculations.

Hybrid plants maintained a ploidy level close to $4n$ but, according to dot blot analysis, the relative amount of donor EP DNA in somatic hybrid plants H18-1, -2 decreased from 4.77 and 5.14 to 3.59 and 2.90 average-sized EP chromosomes, respectively, over a 6-month period. These results correlate with those obtained by Derks et al. (1992) that the nuclear DNA content could vary, even among individual shoots from the same hybrid callus. These authors suggested that during callus growth the amount of nuclear DNA content per cell can change, leading to a nonhomogeneous DNA distribution in a particular callus. Decrease of the donor DNA by 23–34% in somatic hybrid calli within 6 months, culture was also observed by Trick et al. (1994).

RFLP analysis is an effective means of characterizing qualitatively the genome composition of asymmetric somatic hybrids (Wijbrandi et al. 1990a). Although RFLPs allow determination of the specific composition of donor and recipient chromosomes in asymmetric somatic hybrid plants (Melzer and O'Connell 1992), the presence of an RFLP marker does not necessarily imply the presence of that particular intact chromosome (Puite and Schaart 1993). Moreover, study of tomato (+) potato fusion hybrids by genomic in situ hybridization revealed that, when tomato chromosomes are present in duplicate, they are difficult to detect by RFLP (Jacobsen et al. 1995). To study the nuclear composition of somatic hybrid plants H18-1, -2, -3 and determine which of the specific chromosomes of *L. esculentum* and *L. pennellii* were eliminated/retained, we selected two to three RFLP markers located on opposite arms of each species chromosome set. Such analysis revealed that no intact chromosomes of *L. esculentum* or *L. pennellii* were present. Although the regeneration of plants H18-1, -2, and -3 was from one callus, loci TG31 and TG79 of *L. esculentum* chromosome two and *L. pennellii* chromosome nine, respectively, were missing in hybrid plant H18-1.

Comparison of the dot blot and RFLP data permitted an estimate of the average size of EP fragments retained in somatic hybrid plants. According to dot blot analysis the amounts of EP DNA retained in somatic hybrid plants H18-1, -2, -3 were equivalent to 3.59, 2.90, and 3.19 average-sized EP chromosomes, respectively. The above amounts of EP DNA are represented by eight (H18-1) to ten (H18-2 and -3) fragments of *L. esculentum*- and *L. pennellii*-specific chromosomes. Thus, the average size of chromosome fragments retained in plants H18-1, -2, and -3 is equivalent to 0.449, 0.290, and 0.319 average-sized EP chromosome, respectively. Moreover, the established map position of the *Npt*II selectable marker gene, in the T-DNA insert located on chromosome 12 of *L. esculentum* (Chyi et al. 1986), combined with a saturated molecular map of tomato (Tanksley et al. 1992), allowed an approximate determination by RFLP of the tagged chromosome fragment size in EP. TG markers 68 and 28 distal and proximal to the *Npt*II locus, respectively, were not present in plants H18-1, -2, or -3. Thus, the size of the tagged fragment retained in the three asymmetric somatic hybrid plants does not exceed 53.0 cM.

In fusion experiments with taxonomically remote species, irradiation treatment of donor protoplasts is now commonly employed to direct chromosome elimination for the creation of asymmetric hybrids. Although a correlation between level of irradiation and extent of asymmetry was not observed in asymmetric hybrids between γ-irradiated, range of 100–1000 Gy, donor *Atropa belladonna* (+) *Nicotiana tabacum* (Gleba et al. 1988). Similarly, Wolters et al. (1991) found no difference in the elimination of potato DNA from *S. tuberosum* (+) *L. esculentum* asymmetric hybrids when γ-irradiation doses of 50 and 500 Gy were compared. Spangenberg et al. (1994) also found that the degree of elimination of donor chromosomes from X-ray-irradiated Italian ryegrass protoplasts was not dose-dependent for asymmetric somatic hybrids in the range of 25–250 Gy. However, a correlation was found between increase in irradiation dose and further fragmentation of double-stranded plant DNA (Schoenmakers et al. 1994). Derks et al. (1992) reported the absence of intact donor chromosomes in the genome composition of asymmetric somatic hybrids obtained with 300 Gy given to the donor. Our results suggest that even a relatively low dose of 100 Gy was effective in directing chromosome elimination and breakage of the donor genome to the extent that only eight to ten fragments of 0.290 to 0.449 average-sized EP chromosome were retained in somatic hybrid plants. These results support the idea that in somatic fusion experiments with phylogenetically remote species, irradiation directs the elimination of the donor (Gleba et al. 1988), but the degree of elimination may be effected by other factors. Indeed, such factor(s) infrequently allowed regeneration of fertile somatic hybrid plants possessing only one donor chromosome (Dudits et al. 1987). Moreover, highly asymmetric plants have even been obtained without any pretreatment of the donor genome (Babiychuk et al. 1992).

Flow cytometric analysis of the parental species and asymmetric somatic hybrids, on the other hand, permitted monitoring of chromosome elimination in asymmetric hybrid calli between phylogenetically remote species. Based on that analysis, we suggest that in fusion experiments with phylogenetically remote species irradiation of the donor protoplasts, in fact, directs chromosome elimination, but the degree of elimination may depend on the size of the donor and recipient genomes, e.g., ratio (donor:recipient) between DNA content of the species involved in fusion. For instance, Wolters et al. (1991) reported a limited degree of donor DNA elimination in somatic hybrid calli obtained after γ-irradiated protoplasts of *Solanum tuberosum* were fused with *L. esculentum*. These two species are characterized by a donor:recipient DNA ratio of 1.8:1 [DNA contents of *S. tuberosum* and *L. esculentum* are 3.31–3.86 ($2n = 4x$) and 1.88–2.07 pg, respectively (Arumuganathan and Earle 1991)]. In our fusion experiments between *L. esculentum* × *L. pennellii* (+) *S. melongena* with donor:recipient DNA ratio 1:1.22 (DNA contents are 2.017 and 2.479 pg, respectively), only one callus with 6.29 donor chromosomes resulted in the regeneration of asymmetric hybrid plants. Conversely, for fusion experiments with species characterized with a higher DNA ratio for the recipient, for instance, *Daucus carota* (donor) and *Nicotiana tabacum* (recipient), have a DNA ratio 1:9.4 (DNA contents of *D. carota* and *N. tabacum* are 0.98 and 8.75–9.63 pg, respectively: Arumuganathan and Earle 1991) and Dudits et al. (1987) reported regeneration of fertile somatic hybrid plants, possessing only one chromosome of the donor. Also, probably, spontaneous extensive chromosome elimination in intergeneric somatic hybrid plants between *Atropa belladonna* (donor) and *N. tabacum* (recipient), obtained without any pretreatment of the donor genome (Babiychuk et al. 1992) may be explained by this theory. In fact, when irradiation is used, an increased phylogenetic distance between the fusion partners does not necessarily promote the loss of donor chromosomes (McCabe et al. 1993). This relationship is in good agreement with our hypothesis, because the species-specific size of the nuclear genome does not depend on the phylogenetic distance between species.

5 Polyploidization and Elimination of the Donor Genome

In asymmetric somatic fusion experiments, the phenomenon of polyploidization of the recipient genome has been frequently observed, and several explanations have been proposed. For instance, Wijbrandi et al. (1990a) suggested that hybrid cells with a high proportion of the donor genome were better-balanced (around the tri-, penta- or heptaploid levels) and therefore more viable than cells with a low proportion of donor genome. Puite and Schaart (1993) found that the percentage of donor DNA in hybrid calli was dependent on the amount of recipient DNA: the level of donor DNA was the highest when there was a relatively large amount of recipient-genome DNA present. These authors suggested that polyploidization of the recipient genome may be a necessary mechanism by which the proliferation at high numbers of the irradiation-damaged donor chromosomes occurs. In our study (Samoylov and Sink 1996), flow cytometric analysis revealed that the vast majority of somatic hybrid calli that did not regenerate shoots were 5–9n polyploids. Correlation of an increase in the amount of donor EP DNA with ploidy level of asymmetric somatic hybrid calli/plants also has been observed. The three asymmetric somatic hybrid plants obtained were regenerated after 12 months in culture only from callus with a ploidy level close to 4n, and such calli occurred only when donor EP had received 100 Gy. Plants H18-1, -2, and -3, after 8 months of growth in a greenhouse, maintained the near tetraploid level and after 11 months, still appeared to be polyploid. The phenomenon of polyploidization of the recipient genome in asymmetric fusion experiments is in good agreement with our hypothesis (Samoylov and Sink 1996) that in fusion experiments with phylogenetically remote species irradiation of the donor protoplasts, in fact, directs chromosome elimination, but the degree of elimination may depend on the size of the donor and recipient genomes, e.g., ratio (donor:recipient) between DNA content of the species pair involved in fusion. Irradiation treatment of one of the genomes prevents normal proliferation of its chromosomes in somatic hybrids; thus, irradiation determines the genome of which species, involved in fusion, should be eliminated. Therefore, the irradiated genome becomes a "donor" and the nonirradiated a "recipient". Proliferation of the irradiated donor genome/chromosomes introduced via somatic hybridization is a perturbation event for the recipient genome, which may hamper normal functions. Thus, to restore normal functioning of the recipient genome, the elimination of the donor genome should take place in somatic hybrid cells. However, according to our hypothesis, elimination depends on the size of the donor and recipient genomes, e.g., ratio (donor:recipient) between DNA content of species involved in fusion. Thus, in fusion experiments between species having similar nuclear DNA contents, on the one hand, the recipient genome does not have a quantitative DNA advantage over the donor genome. Conversely, the elimination of the donor genome already has been determined by irradiation. Consequently, the recipient genome may synthesize a quantitative DNA advantage of the genome to eliminate the donor. The recipient genome may create this advantage by duplication of its genome, and thus polyploidization of the recipient genome in asymmetric somatic hybrids occurs. Moreover, if the recipient is not able to eliminate irradiated DNA of the donor, the polyploid level of the asymmetric somatic hybrids remains relatively stable. Thus, polyploidization in many cases may hinder shoot-regeneration frequency of somatic hybrids, especially when genomes of taxonomically remote species are mixed. Conversely, for fusion experiments with species characterized with a higher DNA ratio for the recipient, for example, *Daucus carota* (donor) and *Nicotiana tabacum* (recipient), we may assume that *N. tabacum* is already ~10n polyploid relative to *D. carota* (nuclear DNA ratio 9.4:1, DNA contents of *N. tabacum* and *D. carota* are 8.75–9.63 and 0.98 pg, respectively: Arumuganathan and Earle 1991) and fertile somatic hybrid plants with only one chromosome of the donor were obtained (Dudits et al. 1987). Hinnisdaels et al. (1991) also reported the production of highly asymmetric fertile somatic hybrids between *N. plumbaginifolia* and *Petunia hybrida* (DNA contents of *N. plumbaginifolia* and *P. hybrida* are 4.74 and 2.64 pg, respeciitvely: Arumuganathan and Earle 1991); however, somatic hybrids containing *Petunia* chromosomes or fragments maintained a tetraploid *Nicotiana* level. More recent results on partial genome transfer through microprotoplast fusion (Rutgers et al. 1997) are also in good correlation with our hypothesis.

Molecular analyses of somatic hybrids obtained in this study allowed the suggestion of a hypothesis to explain a mechanism of chromosome elimination in taxonomically remote species; however, such a hypothesis has to be further tested by somatic fusion experiments involving species with similar and different-sized genomes. Revealing the elimination mechanisms in somatic hybrid plants will lead to predetermination of donor chromosome elimination and with support from other methods such as sexual crosses and embryo rescue may result in fertile somatic hybrid plants with desired traits. Thereby, somatic hybrid plants between remote species may became a useful source of genetic variability for crop improvement.

6 Protocol

6.1 Plant Materials

Seeds of *S. melongena* L. (line 410, $2n = 2x = 24$) were provided by Dr. S. Izhar, ARO, Israel, and those of cv. Imperial Black Beauty were obtained from Stokes Seeds Inc., Buffalo, NY. Seeds of *L. esculentum* cv. VF36 and *L. pennellii* LA716 were provided by Dr. C.M. Rick, Tomato Genetics Resource Center, University of California. Kanamycin-resistant plant A54 of the sexual hybrid between *Lycopersicon esculentum* Mill. (cv. VF36, $2n = 2x = 24$) × *L. pennellii* L. (cv. LA716, $2n = 2x = 24$), herein EP, was provided by Dr. R. Jorgensen, DNAP, Oakland, CA. EP has a T-DNA insert mapped to *L. esculentum* chromosome 12 (Chyi et al. 1986).

6.2 Protoplast Isolation, Irradiation, and Fusion

Protoplasts were isolated from the leaves of 3–4 week-old seedlings of *S. melongena* according to Guri and Izhar (1984). Protoplasts of EP, from plants grown in vitro under standard conditions, were isolated as described by Tan et al. (1987) with some modifications. Plants 3–4 weeks old were transferred to the dark for 24–48 h, and then pretreated at 4 °C for 4–12 h in the dark. The concentration of the enzymes Cellulysin (Calbiochem) and Macerozyme R10 (Serva) was increased to 0.7 and 0.2%, respectively.

Prior to fusion, protoplasts of EP were irradiated with 100, 250, 500, 750, or 1000 Gy of ^{60}Co-γ-rays (9.6 Gy min^{-1}) at 4–10 °C. After irradiation, protoplasts were washed once with W_5 solution and mixed with those of *S. melongena* in a 1:1 or 1:3 ratio to give a final density of 1×10^6 protoplasts ml^{-1}. Protoplasts were fused using the PEG-DMSO procedure (30% PEG, MW 4000).

6.3 Protoplast Culture and Plant Regeneration

After fusion, protoplasts were cultured in the dark at 28 °C for 2–3 days in Kao medium (1977) modified with MS Fe-EDTA and 0.5 g l^{-1} MES added. On the 3rd day of culture cells were embedded in 1.4% alginate and cultured in 3 ml of medium (60 × 15-mm petri dishes) in the dark at 28 °C. Two weeks later the alginate sections were transferred to 100 × 15-mm Petri dishes and 10 ml of fresh medium plus kanamycin at a rate of 25 mg l^{-1} were added. Four weeks after fusion the alginate sections were dissolved in 0.04 M sodium citrate and the released microcalli were cultured with 10–15 ml of C medium (Shepard 1982), containing 25 mg l^{-1} kanamycin and exposed to light (200 µE m^{-2} s^{-1}). Two to 3 weeks later green calli (1–2 mm in diam) were retrieved individually and placed on MS (Murashige and Skoog 1962) medium supplemented with 2% sucrose, 2 mg l^{-1} zeatin, 0.1 mg l^{-1} IAA, 25 mg l^{-1} kanamycin, and 0.6% Noble agar (Difco) (Guri and Sink 1988). When leaf-like primordia appeared, the calli were subcultured on MS, as above, but supplemented with 100 mg l^{-1} adenine sulfate, 20 ml l^{-1} coconut milk, 0.01 mg l^{-1} biotin, 0.5 mg l^{-1} folic acid, and 0.05 mg l^{-1} GA$_3$. Regenerated shoots were transferred to solid MS with 0.2 mg l^{-1} IAA for rooting. After 2–3 weeks rooted plants were transferred to soil and placed in a greenhouse.

References

Arumuganathan K, Earle ED (1991) Nuclear DNA content of some important plant species. Plant Mol Biol Rep 9:208–218

Babiychuk E, Kushnir S, Gleba YY (1992) Spontaneous extensive chromosome elimination in somatic hybrids between somatically congruent species *Nicotiana tabacum* L. and *Atropa belladonna* L. Theor Appl Genet 84:87–91

Bajaj YPS (1994) Somatic hybridization – a rich source of genetic variability. In: Bajaj YPS (ed) Biotechnology in agriculture and forestry, vol 27. Somatic hybridization in crop improvement I. Spinger, Berlin Heidelberg New York, pp 3–32

Chyi Y, Jorgensen RA, Goldstein D, Tanksley SD, Loaiza-Figueroa F (1986) Locations and stability of *Agrobacterium*-mediated T-DNA insertions in the *Lycopersicon* genome. Mol Gen Genet 204:64–69

Daunay MC, Lester RN, Laterrot H (1991) The use of wild species for the genetic improvement of eggplant (*Solanum melongena* L.) and tomato (*Lycopersicon esculentum* Mill.) In: Hawkes L, Lester RN, Nee M, Estrada N (eds) *Solanaceae* III: taxonomy, chemistry, evolution. Royal Botanical Gardens of Kew and Linnean Society of London, London, pp 389–412

Daunay MC, Chaput MH, Sihachakr D, Allot M, Vedel F, Ducreux G (1993) Production and characterization of fertile somatic hybrids of eggplant (*Solanum melongena* L.) with *Solanum aethiopicum* L. Theor Appl Genet 85:841–850

Derks FHM, Hakkert JC, Verbeek WHJ, Colijn-Hooymans CM (1992) Genome composition of asymmetric hybrids in relation to the phylogenetic distance between the parents. Nucleus-chloroplast interaction. Theor Appl Genet 84:930–940

Dudits D, Maroy E, Praznovszky T, Olah Z, Gyorgyey J, Cella R (1987) Transfer of resistance traits from carrot into tobacco by asymmetric somatic hybridization: regeneration of fertile plants. Proc Natl Acad Sci USA 84:8434–8438

Famelaer I, Gleba YY, Sidorov VA, Kaleda VA, Parokonny AS, Boryshuk NV, Cherep NN, Negrutiu I, Jacobs M (1989) Intrageneric asymmetric hybrids between *Nicotiana plunbaginifolia* and *Nicotiana sylvestris* obtained by gamma-fusion. Plant Sci 61:105–117

Galbraith DW, Harkins KR, Maddox JM, Ayres NM, Sharma DP, Firoozabady E (1983) Rapid-flow cytometric analysis of the cell cycle in intact plant tissues. Science 220:1049–1051

Gleba YY, Hinnisdaels S, Sidorov VA, Kaleda VA, Parokonny AS, Boryshuk NV, Cherep NN, Negrutiu I, Jacobs M (1988) Intergeneric asymmetric hybrids between *Nicotiana plumbaginifolia* and *Atropa belladonna* obtained by gamma-fusion. Theor Appl Genet 76:760–766

Guri A, Izhar S (1984) Improved efficiency of plant regeneration from protoplasts of eggplant (*Solanum melongena* L.). Plant Cell Rep 3:247–249

Guri A, Sink KC (1988) Interspecific somatic hybrid plants between eggplant (*Solanum melongena*) and *Solanum torvum*. Theor Appl Genet 76:490–496

Guri A, Dunbar LJ, Sink KC (1991) Somatic hybridization between selected *Lycopersicon* and *Solanum* species. Plant Cell Rep 10:76–80

Hinnisdaels S, Bariller L, Mouras A, Sidorov V, Del-Favero J, Veuskens J, Negrutiu I, Jacobs M (1991) Highly asymmetric intergeneric nuclear hybrids between *Nicotiana* and *Petunia*: evidence for recombinogenic and translocation events in somatic hybrid plants after gamma-fusion. Theor Appl Genet 82:609–614

Hinnisdaels S, Jacobs M, Negrutiu I (1994) Asymmetric somatic hybrids. In: Bajaj YPS (ed) Biotechnology in agriculture and forestry, vol 27. Somatic hybridization in crop improvement I. Spinger, Berlin Heidelberg New York, pp 57–71

Imamura J, Saul MW, Potrykus I (1987) X-ray irradiation promoted asymmetric somatic hybridization and molecular analysis of the products. Theor Appl Genet 74:445–450

Jacobsen E, de Jong JH, Kamstra SA, van den Berg PMMM, Ramanna MS (1995) Genomic in situ hybridization (GISH) and RFLP analysis for the identification of alien chromosomes in the backcross progeny of potato (+) tomato fusion hybrids. Heredity 74:250–257

Kao KN (1977) Chromosomal behavior in somatic hybrids of soybean-*Nicotiana glauca*. Mol Gen Genet 150:225–230

Kovtun YV, Korostash MA, Butsko YV, Gleba YY (1993) Amplification of repetitive DNA from *Nicotiana plumbaginifolia* in asymmetric somatic hybrids between *Nicotiana sylvestris* and *Nicotiana plumbaginifolia*. Theor Appl Genet 86:221–228

Liu KB, Li YM, Sink KC (1995) Asymmetric somatic hybrid plants between an interspecific *Lycopersicon* hybrid and *Solanum melongena*. Plant Cell Rep 14:652–656

McCabe PF, Dunbar LJ, Guri A, Sink KC (1993) T-DNA-tagged chromosome 12 in donor *Lycopersicon esculentum* × *L. pennellii* is retained in asymmetric somatic hybrids with recipient *Solanum lycopersicoides*. Theor Appl Genet 86:377–382

Melzer JM, O'Connell MA (1992) Effect of radiation dose on the production of and the extent of asymmetry in tomato asymmetric somatic hybrids. Theor Appl Genet 83:337–344

Murashige T, Skoog F (1962) A revised medium for rapid growth and bioassays with tobacco cultures. Physiol Plant 15:473–497

Piastuch WC, Bates GW (1990) Chromosomal analysis of *Nicotiana* asymmetric somatic hybrids by dot blotting and in situ hybridization. Mol Gen Genet 222:97–103

Puite KJ, Schaart JG (1993) Nuclear genomic composition of asymmetric fusion products between irradiated transgenic *Solanum brevidens* and *S. tuberosum*: limited elimination of donor chromosomes and polyploidization of the recipient genome. Theor Appl Genet 86:237–244

Rutgers E, Ramulu KS, Dijkhuis P, Blaas J, Krens FA, Verhoeven HA (1997) Identification and molecular analysis of transgenic potato chromosomes transferred to tomato through micriprotoplast fusion. Theor Appl Genet 94:1053–1059

Samoylov VM, Sink KC (1996) The role of irradiation dose and DNA content of somatic hybrid calli in producing asymmetric plants between an interspecific tomato hybrid and eggplant. Theor Appl Genet 92:850–857

Samoylov VM, Izhar S, Sink KC (1996) Donor chromosome elimination and organelle composition of asymmetric somatic hybrid plants between an interspecific tomato hybrid and eggplant. Theor Appl Genet 93:268–274

Schoenmakers HCH, Wolters AMA, Nobel EM, de Klein CMJ, Koornneef M (1993) Allotriploid somatic hybrids of diploid tomato (*Lycopersicon esculentum* Mill.) and monoploid potato (*Solanum tuberosum* L.). Theor Appl Genet 87:328–336

Schoenmakers HCH, Van der Meulen-Muisers JJM, Koornneef M (1994) Asymmetric fusion between protoplasts of tomato (*Lycopersicon esculentum* Mill.) and gamma-irradiated protoplasts of potato (*Solanum tuberosum* L.): the effects of gamma irradiation. Mol Gen Genet 242:313–320

Shepard JF (1982) Cultivar-dependent cultural refinements in potato protoplast regeneration. Plant Sci Lett 26:127–132

Sihachakr D, Duanay MC, Serraf I, Chaput MH, Mussio I, Haicour R, Rossignol L, Ducreux G (1994) Somatic hybridization of eggplant (*Solanum melongena* L.) with its close and wild relatives. Biotechnol Agric For 27:255–278

Spangenberg G, Valles MP, Wang ZY, Montavon P, Nagel J, Potrykus I (1994) Asymmetric somatic hybridization between tall fescue (*Festuca arundinaceae* Schreb.) and irradiated Italian ryegrass (*Lolium multiflorum* Lam.) protoplasts. Theor Appl Genet 88:509–519

Tan M-LC, Rietveld EM, van Marrewijk GAM, Kool AJ (1987) Regeneration of leaf protoplasts of tomato cultivars (*L. esculentum*): factors important for efficient protoplast culture and plant regeneration. Plant Cell Rep 6:172–175

Tanksley SD, Ganal MW, Prince JP, de Vicente MC, Bonierbale MW, Broun P, Fulton TM, Giovannoni JJ, Grandillo S, Martin GB, Messeguer R, Miller JC, Miller L, Paterson AH, Pineda O, Roder MS, Wing RA, Wu W, Young ND (1992) High-density molecular linkage maps of the tomato and potato genomes. Genetics 132:1141–1160

Trick H, Zelcer A, Bates GW (1994) Chromosome elimination in asymmetric somatic hybrids: effect of gamma dose and time in culture. Theor Appl Genet 88:965–972

Wijbrandi J, Zabel P, Koornneef M (1990a) Restriction fragment length polymorphism analysis of somatic hybrids between *Lycopersicon esculentum* and irradiated *L. peruvianum*: evidence for limited donor genome elimination and extensive chromosome rearrangements. Mol Gen Genet 222:270–277

Wijbrandi J, Posthuma A, Kok JM, Rijken R, Vos JGM, Koornneef M (1990b) Asymmetric somatic hybrids between *Lycopersicon esculentum* and irradiated *Lycopersicon peruvianum*. Theor Appl Genet 80:305–312

Wolters AMA, Schoenmakers HCH, van der Meulen-Muisers JJM, van der Knaap E, Derks FHM, Koornneef M, Zelcer A (1991) Limited DNA elimination from the irradiated potato parent in fusion products of albino *Lycopersicon esculentum* and *Solanum tuberosum*. Theor Appl Genet 83:225–232

Zabel P, Meyer D, van de Stolpe O, van der Zaal B, Ramanna MS, Koornneef M, Krens F, Hille J (1985) Towards the construction of artificial chromosomes for tomato. In: van Vloten-Doting L, Groot GSP, Hall TC (eds) Molecular form and function of the plant genome. NATO ASI. Ser A Life Sci 83:609–624

II.10 Somatic Hybridization Between *Lycopersicon esculentum* Mill. (Tomato) and *Solanum ochranthum* Dun.

J.R. Stommel, R.S. Kobayashi, and S.L. Sinden

1 Introduction

Solanum ochranthum is a recently discovered woody vine-like tropical nightshade found mainly in wet swampy habitats of the northern Andes. *S. ochranthum* is a member of the section Petota, subsection Potatoe, series Juglandifolia. Species in the series Juglandifolia have a combination of *Lycopersicon* characters that differentiate them from other *Solanum* species. *S. ochranthum* and other members of the series Juglandifolia do not form tubers, have pinnately divided leaves, and have flowers that are yellow in color, similar to *Lycopersicon* spp. Flowers of most other species in section *Petota*, subsection *Potatoe*, are pink to blue or white. *Solanum ochranthum* is considered to be a relative of the cultivated tomato *Lycopersicon esculentum* (Rick 1988; Spooner et al. 1993).

Many improvements in disease resistance and fruit quality of tomato can be attributed to the use of wild germplasm in breeding programs (Rick 1988). Additionally, other useful attributes such as tolerance to environmental stresses and resistance to insect and other pests have been identified in close relatives of the cultivated tomato (Rick 1986). Field observations of *S. ochranthum* in both its native habitat and in test plots suggest that this species is a valuable new source of insect, bacterial, fungal, and viral resistance for tomato improvement (Rick 1986; Rick et al. 1990). Tingey et al. (1981) discovered that *S. ochranthum* possesses type-B glandular trichomes which confer resistance to small insects and mites in some other *Solanum* species (Tingey and Gibson 1978; Tingey and Sinden 1982). Moretti et al. (1990) screened tomato relatives for resistance to greenhouse insect pests and detected resistance to leaf miner, *Liriomyza trifolii*, in *S. ochranthum*. Resistance to the tomato late blight pathogen, *Phytophthora infestans*, was detected in *S. ochranthum*, suggesting that *S. ochranthum* could be a promising source of late blight resistance genes for resistance breeding in tomato (Kobayashi et al. 1994b).

S. ochranthum has not yet been utilized in tomato-improvement programs because *S. ochranthum* is sexually incompatible with *L. esculentum* and

---

US Department of Agriculture, Agricultural Research Service, Plant Sciences Institute, Vegetable Laboratory, Beltsville, MD 20705, USA

appears to be genetically isolated from all other *Lycopersicon* and *Solanum* species. Attempts to cross *S. ochranthum* with tomato and nine other *Lycopersicon* or *Solanum* species have all been unsuccessful (Rick 1979; Rick et al. 1990). Protoplast fusion has been used to bridge crossing barriers between other wild *Solanum* species and their solanaceous crop relatives to transfer resistance genes into the cultivated crops (Helgeson et al. 1993). There are reports of somatic hybrids from more than ten combinations of *Lycopersicon* and other *Solanum* species (see Hanson et al. 1989; Lefrançois et al. 1993; Watanabe et al. 1995 for review), indicating the possibility of successful gene transfer from *S. ochranthum* to various *Lycopersicon* and *Solanum* species through protoplast fusion.

To utilize somatic hybridization as a means of transferring genes between incompatible species, suitable methods to obtain and culture durable protoplasts from both parental plants are needed. Regeneration capacity of the protoplasts from both parents is not, however, a prerequisite for obtaining somatic hybrids (Möllers and Wenzel 1992; Lefrançois et al. 1993). Even if *S. ochranthum* protoplasts did not readily regenerate, it seemed likely that somatic hybrids between *S. ochranthum* and *L. esculentum* could be obtained if good-quality protoplasts were isolated from *S. ochranthum* plants and fused with protoplasts from tomato genotypes which were readily regenerated (Shahin 1984; Neidz et al. 1985; Tan et al. 1987; Kobayashi et al. 1996; see Koblitz 1991 for a review).

We describe here a procedure for the isolation and culture of *S. ochranthum* protoplasts and fusion conditions for obtaining somatic hybrids between *S. ochranthum* and selected *L. esculentum* genotypes.

2 Isolation of Protoplasts

Axenic shoot-tip cuttings of *S. ochranthum*, LA 2117, obtained from C.M. Rick, University of California, Davis, California, were propagated on a medium containing MS salts, Staba vitamins (Staba 1969), $100\,mg\,l^{-1}$ casein, 3% sucrose (OM), or OM with 0.6% activated charcoal (OM + AC) to provide the protoplast source plants for isolation, fusion, and regeneration experiments. Axenic plant cultures were grown at 27°C under cool white fluorescent light ($60\,\mu mol\,m^{-2}\,s^{-1}$) with 16-h day length. *S. ochranthum* was difficult to micropropagate. Axenic nodal cuttings frequently failed to grow on any of four propagation media tested: OM, OM + AC, tomato propagation medium (TPM) as described by Tan et al. (1987), or potato propagation medium (PPM) as described by Haberlach et al. (1985). Axenic shoot-tip cuttings grew better and more consistently than nodal cuttings on all media tested.

Root growth of *S. ochranthum* axenic plants was poor on all media. OM and OM + AC media were superior to either TPM or PPM for micropropagating *S. ochranthum* protoplast source plants. Addition of charcoal to the OM medium improved root growth and resulted in taller, larger *S. ochranthum*

Table 1. Mean plant height, root dry weight and protoplast yield ($\times 10^4$ protoplasts g^{-1} FW) of *S. ochranthum* cultured on OM or OM + AC

Propagation medium	Plant height (cm)	Root dry wt (g)	Protoplast yield ($\times 10^4$)
OM	5.2	0.026	392.7
OM + AC	11.6	0.044	532.2
Significance[a]	**	*	NS

NS, *, ** [a] Significance determined by pairwise comparison t-test in four replicated experiments. Nonsignificant or significant at $P = 0.05$ or 0.01, respectively.

plantlets, compared to growth on OM medium without charcoal (Table 1). The addition of charcoal did not, however, significantly affect protoplast yield.

Protoplasts from 4–5 week old in vitro cultures were isolated using methods modified from Bellini et al. (1990). Leaf tissue was digested in I_{10} medium (pH 6.5) containing 0.2% Macerozyme R-10, 1% Cellulase Onozuka R-10 (Yakult Pharmaceutical Ind., Tokyo, Japan), and 0.5% Driselase (Sigma Chemical Co., St. Louis, MO). The digest was filtered through cheesecloth and protoplasts collected by centrifugation at $100g$. Protoplasts were washed once in washing medium (14 mM $CaCl_2 \cdot 2H_2O$, 260 mM KCl, 3.5 mM MES, pH 5.6) and finally collected by flotation in 10% sucrose rinse medium as described by Shepard (1980). On average, 6×10^5 protoplast g^{-1} FW were obtained from leaves of plants cultured on all three propagation media tested (Fig. 1).

3 Culture of Protoplasts

Protoplast culture conditions for *S. ochranthum* were optimized for callus formation as a prerequisite to performing fusion experiments with *L. esculentum*. Protoplasts were cultured initially in 0.2-ml agarose beads (solidified with 0.4% Sea Plaque agarose, FMC Corp, Rockland, ME) of LCM medium (Tan et al. 1987), C_1 medium (Bellini et al. 1990), or modified CL (Shepard 1980) containing 1/4 salts with $1 mg l^{-1}$ NAA, $0.5 mg l^{-1}$ BAP and $0.5 mg l^{-1}$ 2,4-D in the dark at 24 °C. Protoplasts were cultured at an initial plating density of $0.5–2 \times 10^5$ protoplasts ml^{-1}. When protoplasts were cultured in liquid medium as compared to agarose beads, many protoplasts turned brown and fewer divided. After continuous cell division began, approximately 4–7 days, 0.1 ml of liquid culture medium was added around each agarose bead. Fresh liquid culture medium was added every 2–7 days. When the protoplasts had developed into 8–16 cell microcalli, the cultures were illuminated for 12-h day lengths with cool white fluorescent light ($60 \mu mol m^{-2} s^{-1}$) at 24 °C. Developing microcalli (0.5–1 mm) were transferred to either greening medium (Tan et al.

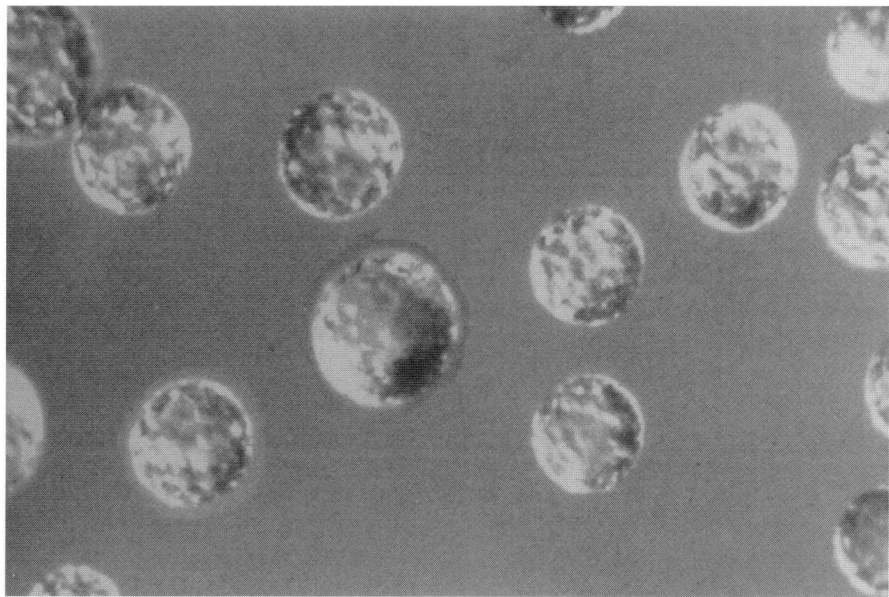

Fig. 1. Freshly isolated mesophyll protoplasts

1987) or C medium (Shepard 1980) and cultured under continuous light at 24 °C. Generally, calli turned from light yellow-green to bright green on either C or greening medium. Calli cultured on greening medium were soft and friable, whereas calli cultured on C medium exhibited more compact growth habit. After 2–4 weeks on C or greening medium, the calli were transferred to shoot-induction medium (Tan et al. 1987) or to D medium (Shepard 1980) and cultured under 16-h day length at 24 °C.

Percent plating efficiency (number of dividing protoplasts/total number of protoplasts at the time of plating × 100) was determined 14 days after isolation. Plating efficiency and callus formation were influenced by both source plant propagation medium and the protoplast plating medium. Protoplasts isolated from plantlets grown on TPM had the highest plating efficiency, but protoplasts isolated from plantlets grown on OM + AC had the highest percentage of callus formation (Table 2). Among the three protoplast culture media examined, only small differences in mean protoplast plating efficiencies were observed, but there were large differences among the culture media in the number of developing calli 1 mm or larger in diameter (Table 2). More protoplasts developed to the callus stage when cultured in CL than when cultured in LCM or C_1 medium. A number of calli did form when protoplasts were cultured in C_1; however, they were slower in forming than when protoplasts were cultured in CL medium. Although healthy green calli that became or remained bright green on the appropriate regeneration medium were obtained using all three protocols, none of more than 500 calli in the *S. ochranthum* protoplast

Table 2. Effect of source plant culture medium and protoplast culture medium on percent plating efficiency and callus development of *S. ochranthum* protoplasts

Protoplast culture medium	Source plant culture medium					
	OM		OM + AC		TPM	
	Percent plating efficiency	Degree of callus formation[a]	Percent plating efficiency	Degree of callus formation[a]	Percent plating efficiency	Degree of callus formation[a]
LCM	3.0 ± 1.2	0	5.1 ± 1.6	+	12.2 ± 7.3	–
C_1	1.1 ± 0.2	+	5.2 ± 2.5	++	6.6 ± 4.8	++
CL	3.6 ± 1.7	++	7.0 ± 1.7	++++	6.3 ± –	++

[a] Callus development rating (from 1×10^5 protoplast cultured): –, no calli (>1 mm); +, 1–50 calli; ++, 51–100 calli; +++, 101–200 calli; ++++, more than 200 calli.

culture experiments regenerated. Of the three culture protocols tested and the four source plant propagation media, the OM + AC propagation medium used to grow source plants, coupled with the protoplast culture protocol modified from Shepard (1980), gave the highest percentage of healthy green calli on regeneration medium.

4 Fusion of Protoplasts and Regeneration of Somatic Hybrids

Using the described protoplast isolation methods, leaf protoplasts of *S. ochranthum* and *L. esculentum* PI 367942 (North Central Regional Plant Introduction Station, Ames, Iowa) or the *L. esculentum* hybrid, 8611 [(*L. esculentum* VFNT cherry × *L. peruvianum*) BC to VFNT cherry] (D. Pratt, University of California, Davis, Calif.), were obtained for fusion experiments. *L. esculentum* PI 367942, a red cherry-type tomato, was axenically propagated from seed as described by Neidz and Sink (1988). *L. esculentum* 8611, was propagated by axenic shoot-tip cuttings on TPM medium. Protoplasts from both *L. esculentum* genotypes were isolated from 3–4 week old plantlets.

The protoplasts from *S. ochranthum* were chemically fused with either *L. esculentum* PI 367942 or *L. esculentum* 8611 protoplasts with polyethylene glycol (PEG MW 8000, Sigma Chemical Co., St. Louis), using methods established by Austin et al. (1985) and described below (Fig. 2). Protoplasts (8–10 $\times 10^5$ of each species) were mixed with enough fusion medium (0.1 M glucose 3.5 mM $CaCl_2$ and 0.7 mM KH_2PO_4, pH 5.7) to adjust the density to 4–5 $\times 10^5$ protoplasts ml^{-1}. Throughout the fusion and dilution procedure, the fusion mixture was continually mixed by gentle swirling. Fusion medium containing 25% PEG was added slowly (dropwise), over 10 min, to the protoplasts. The protoplasts were incubated in PEG (final concentration of 20%) for 10–12 additional min. The PEG mixture was then diluted with an equal volume of 80 mM $CaCl_2$ solution (pH 10 with NaOH) by dropwise addition of the dilu-

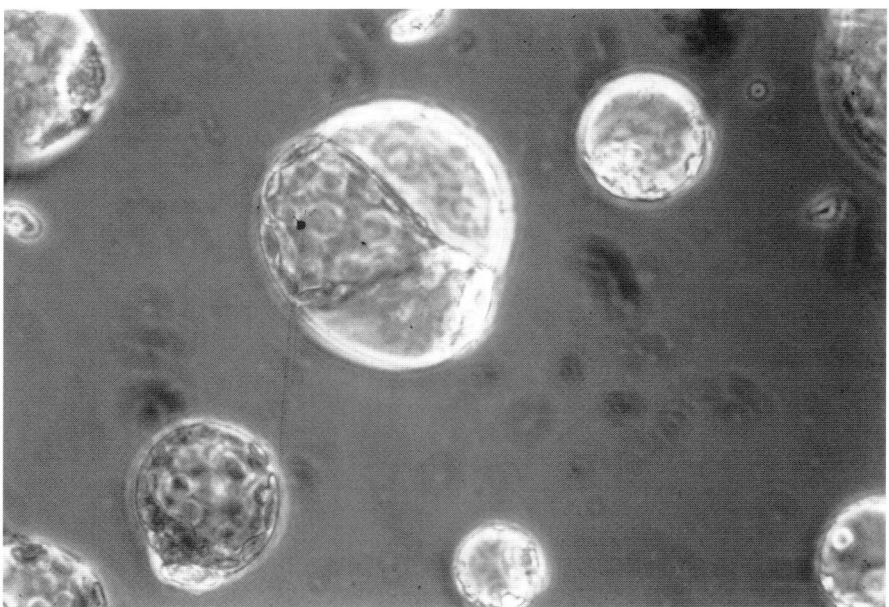

Fig. 2. Fused protoplasts of *L. esculentum* 8611 and *S. ochranthum*

tion solution, and the fusion mixture was incubated for an additional 8–10 min. Two volumes of washing medium containing 0.6 M mannitol and 50 mM $CaCl_2$ (pH 5.7) was then slowly added and the mixture incubated for 1 h. The protoplasts were collected by centrifugation at 50 g for 10 min and washed two additional times. PEG-treated protoplasts were cultured at initial densities ranging from $0.5-4 \times 10^5 \, ml^{-1}$ in agarose beads with modified Shephard's CL and developing calli (Fig. 3) transferred to C medium as described here for *S. ochranthum* protoplast culture prior to culture on shoot-induction medium.

5 Regeneration and Characterization of Somatic Hybrids

Regenerants from the fusion experiments were obtained by culture of callus on D (regeneration) medium as described for *S. ochranthum*. Regenerants were observed beginning about 3–4 weeks after transfer of calli to D medium. Tetraploid and hexaploid somatic hybrids as well as diploid and tetraploid tomato regenerants were obtained (Fig. 4). No *S. ochranthum* regenerants were obtained. The regeneration efficiency was dependent on the genotype of the *L. esculentum* parent used in the fusion. When *S. ochranthum* was fused with the *L. esculentum* hybrid, 8611, 43% of the calli regenerated. Thirty-five percent (65 of 185) of these regenerating calli were somatic hybrids. When *S.*

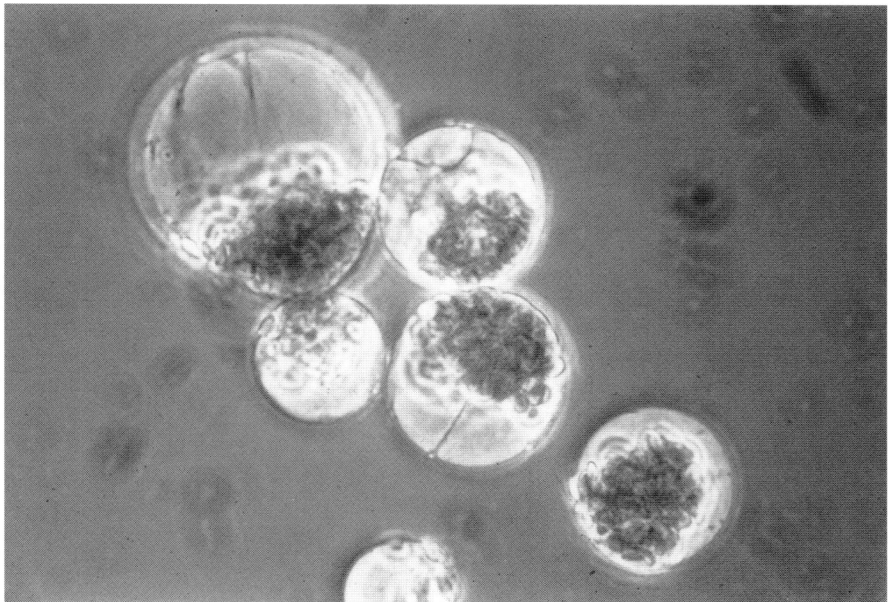

Fig. 3. Dividing protoplasts of *S. ochranthum* + *L. esculentum* 8611 somatic hybrid

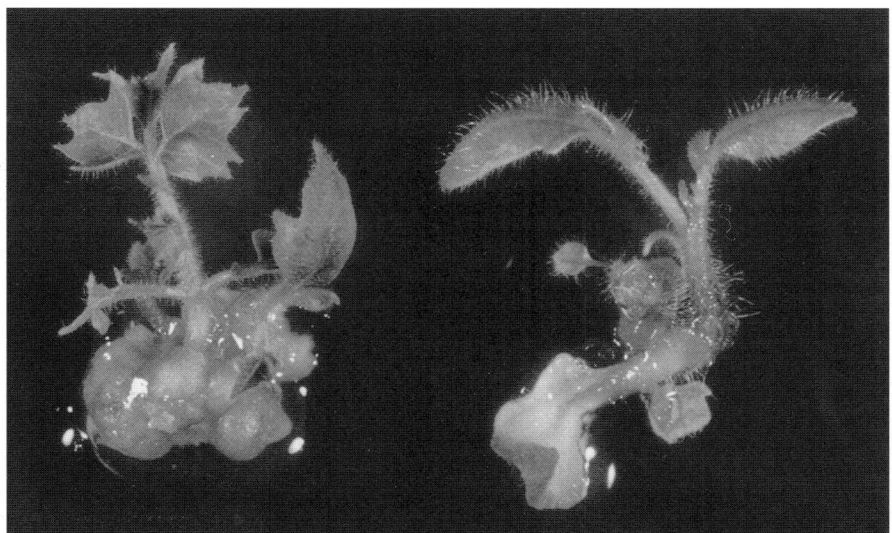

Fig. 4. Protoplast-derived regenerants of *L. esculentum* 8611 (*left*) and *S. ochranthum* + *L. esculentum* 8611 somatic hybrid (*right*)

ochranthum was fused with PI 367942, only 5% of the calli regenerated but all ten regenerants were somatic hybrids. The surprisingly high percentages of somatic hybrids obtained from the fusion experiments are most likely the result of rigorous selection for hybrid cells by the severe chemical fusion conditions, as noted by Austin et al. (1985), and the inability of nonfused *S. ochranthum* protoplasts to regenerate. Austin et al. (1985) obtained only somatic hybrids and no regenerated parent plants from their fusion experiments.

Morphologically, the somatic hybrids appeared intermediate to the two parents. *S. ochranthum* plants are characterized by pinnately compound leaves with leaflets entire and purple pigment in stems, while *L. esculentum* has notched leaflets and green stems. The hybrids have lobed leaflets and stems with purple pigmentation (Fig. 5). The hybridity of the fusion regenerants was confirmed by peroxidase isozyme (Fig. 6) and RAPD markers (Figs. 7 and 8).

The peroxidase banding patterns of the somatic hybrids contained bands present in the patterns of both parents (Fig. 6). Band A, present in the tomato parent and the somatic hybrids, is not present in the *S. ochranthum* parent. Bands B and C are present in the *S. ochranthum* parent and the somatic hybrids, but not in the tomato parent. Similar evidence was obtained using RAPD markers. Two random primers (OPB-08, OPB-09; Operon

Fig. 5A–C. Leaf and flowers of **A** *L. esculentum* 8611, **B** *S. ochranthum*, **C** *S. ochranthum* + *L. esculentum* 8611 somatic hybrid

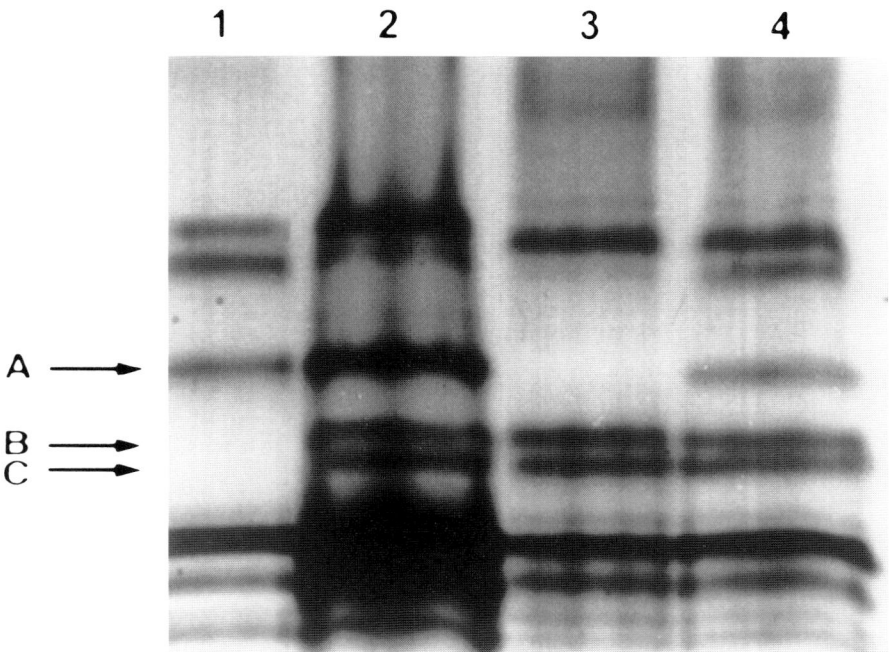

Fig. 6. Peroxidase zymogram of *L. esculentum* 8611, *S. ochranthum* + *L. esculentum* 8611 somatic hybrid, *S. ochranthum*, and a mixture of *L. esculentum* 8611 and *S. ochranthum* extracts (*lanes 1–4*, respectively)

Technologies, Alameda, CA) were selected because they showed distinctly different banding patterns for the two parents. The somatic hybrids' banding patterns resembled the combined patterns of the two parents (Figs. 7 and 8), whereas the *L. esculentum* protoplast regenerants had the same banding patterns as that of the *L. esculentum* parent.

For peroxidase isozyme analysis, sample extracts were made from leaves and stems of 4-week-old in vitro-grown plants. Tissue extraction and electrophoresis were carried out using previously described methods (Kobayashi et al. 1994a). Briefly, the tissue was ground with buffer ($1\,g\,ml^{-1}$) containing 0.1M Tris (pH 8), 32.5mM glutathione, 7.5% (w/v) ascorbic acid, 20% (w/v) sucrose, and 1% bromophenol blue as tracking dye. Isozymes were separated on 7.5% (w/v) polyacrylamide Tris-HCl gels by electrophoresis using Tris/glycine, pH 8.5, running buffer in a Mini-PROTEAN II vertical slab cell (Biorad, Richmond, CA).

For RAPD analysis, DNA extraction was performed according to Edwards et al. (1991) with the following modifications. Lyophilized leaf tissue was powdered without buffer, then 400µl of extraction buffer (Dellaporta et al. 1983) was added, the sample vortexed and incubated for 10min at 65°C. Potassium acetate (130µl; 5M) was added, the sample gently mixed and incu-

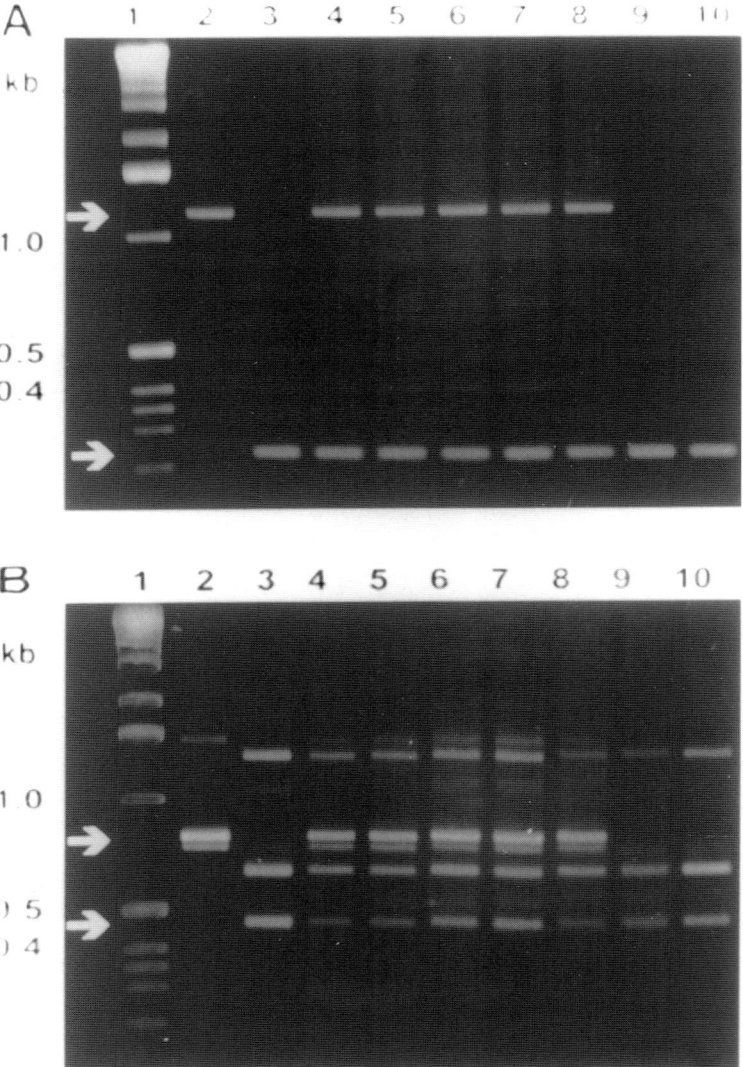

Fig. 7A,B. RAPD amplification products generated by primer **A** OPB-08 and **B** OPB-09 for *S. ochranthum*, *L. esculentum* 8611, and fusion regenerants. *Lane 1* contains BRL 1-kb molecular weight markers; *lane 2* contains *S. ochranthum*; *lane 3* contains *L. esculentum* 8611; *lanes 4–8* contain *S. ochranthum* + *L. esculentum* 8611 somatic hybrids; *lanes 9–10* contain *L. esculentum* 8611 regenerants. Polymorphic fragments are indicated by *arrows*

bated on ice for 20 min. The extracts were then centrifuged at 13 800 g for 10 min. The supernatant was transferred to a fresh tube and an equal volume of ice-cold isopropanol + 13 μl of ammonium acetate (10 M) was added. The tube was incubated at −20 °C for 20 min to allow precipitation of the nucleic acids. The precipitate was collected by centrifugation at 13 800 g. The pellet was

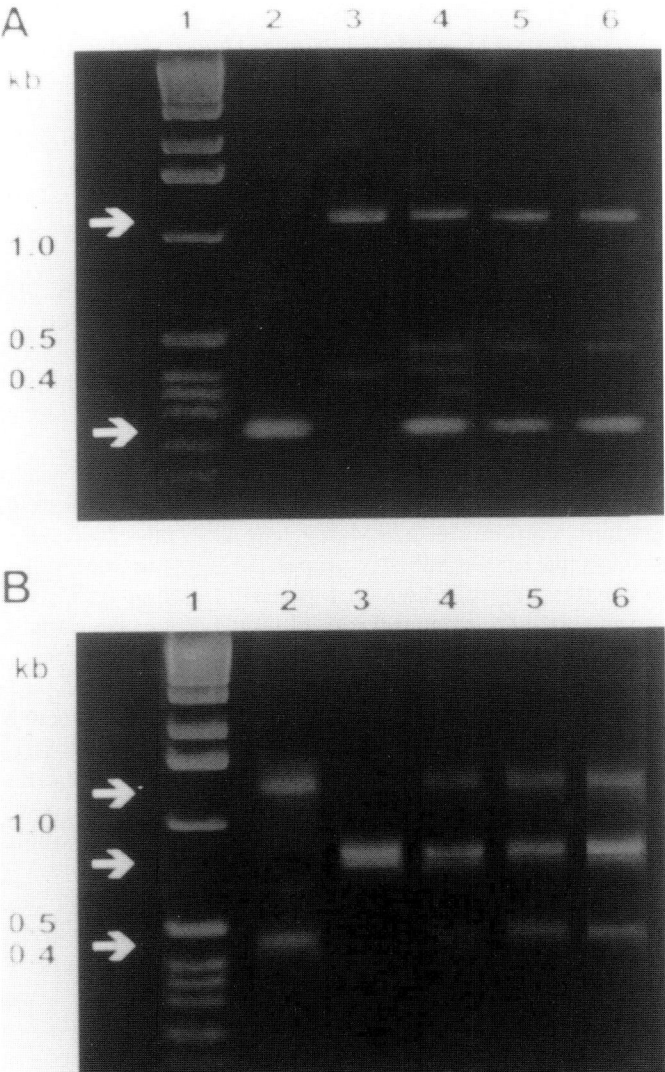

Fig. 8A,B. RAPD amplification products generated by primer **A** OPB-08 and **B** OPB-09 for *S. ochranthum*, *L. esculentum* PI 367942, and their somatic hybrids. *Lane 1* contains BRL 1-kb molecular weight markers; *lane 2* contains *L. esculentum* PI 367942; *lane 3* contains *S. ochranthum*; *lanes 4–6* contains *S. ochranthum* + *L. esculentum* PI 367942 somatic hybrids

washed with 70% ethanol, air dried, and redissolved in 400 µl of TE (50 mM Tris, 10 mM EDTA, pH 8). RNase A (final concentration of 50 µg ml^{-1}) was added and the solution was incubated at 37 °C for 30 min. Then, 400 µl of chloroform:isoamyl alcohol (24:1) was added and the solution mixed by inverting the tube continuously for 5 min. The aqueous phase was transferred to a fresh tube containing 90μl sodium acetate (1M) and 360μl chilled isopropanol, and incubated at −20°C for 20min. The precipitated DNA was collected by centrifugation and the pellet was resuspended in 50μl of sterile water. Random 10-mer oligonucleotides (Operon Technologies, Alameda, CA) were used as primers for PCR amplification. The PCR amplification conditions reported by Levi et al. (1993) were followed, except that amplification was conducted for 45 cycles.

Twenty-three percent (7 of 30) of *S. ochranthum* + *L. esculentum* somatic hybrids examined here were male fertile, as evidenced by the stainability and germination of the pollen. Tetraploid somatic hybrids exhibited a higher degree of stainable pollen (15–73%) in comparison with pollen of hexaploid regenerants (<4%). Attempts to backcross *S. ochranthum* + *L. esculentum* somatic hybrids to tomato have not yet been successful. Additional crosses between male fertile *S. ochranthum* + *L. esculentum* tetraploid somatic hybrids and diploid or tetraploid tomato must be examined.

Pollen viability determinations were carried out using a modification of the method described by Abdul-Baki (1992). Pollen samples were germinated in liquid medium containing 0.29M fructose, 1.27mM $Ca(NO_3)_2$, 0.16mM H_3BO_3, and 0.99mM KNO_3, adjusted to pH 5.2, and the percentage of pollen exhibiting normal pollen tube growth was scored. Pollen viability ascertained via staining was carried out by addition of 0.001% (w/v) fluorescein diacetate (FDA) to germination medium and scoring of fluorescing pollen grains under an ultraviolet source.

Fertile somatic hybrids obtained via protoplast fusion can be used in breeding and genetic programs to further introgress genes from sexually incompatible species into adapted cultivars. For nuclear inherited traits, a backcross or similar program and male or female fertility of the somatic hybrid is required to fully exploit genes from the donor species. Although limited in number, self-set or backcross progeny of *Solanum* + *Lycopersicon* somatic hybrids have been reported. Somatic hybrids of tomato + *S. lycopersicoides* were reported to be fertile and develop viable seeds upon self-pollination (Hossain et al. 1994). Jacobsen et al. (1994) demonstrated that choice of male and female parents influenced the success of backcrossing tetraploid and hexaploid potato + tomato somatic hybrids to potato and that embryo rescue facilitated recovery of backcross progeny. Somatic hybrids of tomato + *S. etuberosum* reportedly produced fruit upon self-pollination, but fruit contained abortive seeds (Gavrilenko et al. 1992). Introgression of the genetic variation present in *S. ochranthum* into the tomato genome is necessary to exploit this sexually isolated species in tomato improvement programs. Only when backcross progeny of *S. ochranthum* + *L. esculentum* can be obtained will recombination between these homoeologous chromosomes be of practical use.

6 Summary and Conclusion

Plants grown in vitro were a suitable source for the isolation of *S. ochranthum* protoplasts. Plants grown on an MS-based propagation medium containing activated charcoal were significantly taller, had higher root dry weight, and protoplasts obtained from these plants had the highest frequency of callus formation compared to plants grown on other media tested. Protoplasts of *S. ochranthum* and protoplasts recovered from fusion experiments developed best when cultured in agarose beads using our modifications of Shepard's protocol (1980) for potato protoplast culture and regeneration. Somatic hybrids can be readily obtained from the fusion of *S. ochranthum* protoplasts with the protoplasts of *L. esculentum* using the methods we have described. Hybridization of *S. ochranthum* with other tomato genotypes or other *Lycopersicon* or *Solanum* species by protoplast fusion may also be possible using the described methods. The prospects for the utilization of *S. ochranthum* genes in the improvement of tomato now appear quite promising.

7 Protocol

7.1 Protoplast Isolation

1. Harvest leaves from axenic shoot tip cultures grown 5 weeks postsubculturing on OM + AC medium at 27 °C under cool white fluorescent light (60 μmol m^{-2} s^{-1}) with 16-h daylength. OM + AC consists of MS salts, Staba vitamins (Staba 1969), 100 mg l^{-1} casein, 3% sucrose, and 0.6% activated charcoal)
2. Feather the leaf tissue by cutting the leaf blade into 2-mm-wide strips. Place the 1–2 g leaf tissue in 10 ml of enzyme solution in a 100 × 25-mm petri plate. The enzyme solution consists of I$_{10}$ medium (pH 6.5) (Bellini et al. 1990) containing 0.2% Macerozyme R-10, 1% Cellulase Onozuka R-10 (Yakult Pharmaceutical Ind., Tokyo, Japan) and 0.5% Driselase (Sigma Chemical Co., St. Louis). Incubate tissues overnight ($\approx$16–20 h) in the dark at 22 °C.
3. Separate the protoplasts from undigested plant debris by filtering the digest through two layers of cheesecloth. Centrifuge the filtrates at 100 g for 10 min and discard the supernatant. Resuspend the protoplasts in 10 ml wash medium (14 mM CaCl$_2 \cdot$ 2H$_2$O, 260 mM KCl, 3.5 mM MES, pH 5.6) and centrifuge as above. Collect protoplasts by flotation in 10% sucrose rinse medium as described by Shepard (1980).

7.2 Protoplast Fusion (Using the Procedures of Austin et al. 1985 with Modifications)

1. Adjust density of protoplasts to 8.5 × 10^5 ml^{-1} of each species with 10% sucrose rinse medium. Add 1 ml of each species to 2 ml of 2× concentration fusion medium containing 0.2 M glucose 7 mM CaCl$_2$ and 1.4 mM KH$_2$PO$_4$, pH 5.7 in a 250-ml beaker.
2. Add slowly (dropwise) 14 ml of 1× fusion medium containing 25% PEG 8000 (Sigma Chemical Co., St. Louis) over 10 min while gently swirling the beaker. Incubate the protoplasts in PEG for 10–12 additional min, with gentle swirling. The beaker is continuously swirled throughout the fusion and dilution steps.

3. Dilute the protoplast/PEG fusion mixture with an equal volume of 80 mM $CaCl_2$ dilution medium, adjusted to pH 10 with NaOH, by dropwise addition of the dilution medium over 10 min. Incubate with gentle swirling for an additional 8 min.
4. Add slowly over 5 min, 27 ml of wash medium containing 0.6 M mannitol and 50 mM $CaCl_2$ (pH 5.7). Transfer the mixture to sterile 50-ml Oak Ridge centrifuge tubes and incubate for 1 h.
5. Centrifuge the protoplasts at 50 g for 10 min and discard supernatant. Rinse the pelleted protoplasts with washing medium two additional times.

7.3 Protoplast Culture and Plant Regeneration

1. Resuspend in a small volume of liquid protoplast culture medium, Shepard's (1980) CL containing 1/4 of the salts concentration.
2. Count the protoplasts using a hemacytometer to determine the number of intact protoplasts recovered.
3. Culture the protoplasts at an initial density of 2×10^5 intact protoplasts ml^{-1} in 0.2 ml agarose beads (0.4% Sea Plaque agarose, FMC Corp, Rockland, ME) of modified CL (Shepard 1980) containing 1/4 salts concentration with 1 mg l^{-1} NAA, 0.5 mg l^{-1} BAP and 0.5 mg l^{-1} 2,4-D in the dark at 24 °C.
4. After approximately 4–7 days, add 0.1 ml of liquid culture medium around each agarose bead. Add an additional 0.1 ml of fresh liquid culture medium every 4 days. After 2 weeks, illuminate the cultures with cool white fluorescent light ($60 \mu mol \, m^{-2} s^{-1}$) at 24 °C for 12-h day length.
5. Transfer developing microcalli (0.5–1 mm) onto C medium (Shepard 1980) and culture under continuous light at 24 °C.
6. Transfer the calli to D medium (Shepard 1980) for regeneration, when the calli become bright green (2–4 weeks on C medium), and culture under 16-h day length at 24 °C.
7. Transfer regenerated shoots to propagation medium (Haberlach et al. 1985) for rooting. Transfer rooted plantlets to the greenhouse.

References

Abdul-Baki AA (1992) Determination of pollen viability in tomatoes. J Am Soc Hortic Sci 117:473–476
Austin S, Baer M, Helgeson JP (1985) Transfer of resistance to potato leaf roll virus from *Solanum brevidens* into *Solanum tuberosum* by somatic fusion. Plant Sci 39:75–82
Bellini C, Chupeau M, Gervais M, Vastra G, Chupeau Y (1990) Importance of myo-inositol, calcium, and ammonium for the viability and division of tomato (*Lycopersicon esculentum*) protoplasts. Plant Cell Tissue Organ Cult 23:27–37
Dellaporta SL, Wood J, Hicks JB (1983) A plant DNA minipreparation: version II. Plant Mol Biol Rep 1:19
Edwards K, Johnstone C, Thompson C (1991) A simple and rapid method for the preparation of plant genomic DNA for PCR analysis. Nucleic Acids Res 19:1349
Gavrilenko TA, Barbakar NI, Pavlov AV (1992) Somatic hybridization between *Lycopersicon esculentum* and non-tuberous *Solanum* species of the *Etuberosa* series. Plant Sci 86:203–214
Haberlach GT, Cohen BA, Reichert NA, Baer MA, Towill LE, Helgeson P (1985) Isolation, culture and regeneration of protoplasts from potato and several related *Solanum* species. Plant Sci 39:67–74
Hanson MR, O'Connell MA, Vidair C (1989) Somatic hybridization in tomato. In: Bajaj YPS (ed) Biotechnology in agriculture and forestry, vol 8. Plant protoplasts and genetic engineering I. Springer, Berlin Heidelberg New York, pp 320–335

Helgeson JP, Haberlach GT, Ehlenfeldt MK, Hunt G, Pohlman JD, Austin S (1993) Sexual progeny of somatic hybrids between potato and *Solanum brevidens*: potential for use in breeding programs. Am Potato J 70:437–452

Hossain M, Imanishi S, Matsumoto A (1994) Production of somatic hybrids between tomato (*Lycopersicon esculentum*) and night shade (*Solanum lycopersicoides*) by electrofusion. Breed Sci 44:405–412

Jacobsen E, Daniel MK, Bergervoet-van Deelen JEM, Huigen DJ, Ramanna MS (1994) The first and second backcross progeny of the intergeneric fusion hybrids of potato and tomato after crossing with potato. Theor Appl Genet 88:181–186

Kobayashi RS, Bouwkamp JC, Sinden SL (1994a) Interspecific hybrids from cross incompatible relatives of sweet-potato. Euphytica 80:159–164

Kobayashi RS, Deahl KD, Stommel JR, Sinden SL (1994b) Evaluation of *Solanum ochranthum* as a potential source of late blight resistance. Tomato Genet Coop Rep 44:15–16

Kobayashi RS, Stommel JR, Sinden SL (1996) Somatic hybridization between *Solanum ochranthum* and *Lycopersicon esculentum*. Plant Cell Tissue Organ Cult 45:73–78

Koblitz H (1991) Protoplast culture and somatic hybridization in *Lycopersicon*. In: Kalloo G (ed) Genetic improvement of tomato. Springer, Berlin Heidelberg New York, pp 247–258

Lefrançois C, Chupeau Y, Bourgin JP (1993) Sexual and somatic hybridization in the genus *Lycopersicon*. Theor Appl Genet 86:533–546

Levi A, Rowland LJ, Hartung JS (1993) Production of reliable randomly amplified polymorphic DNA (RAPD) markers from DNA of woody plants. HortScience 28:1188–1190

Möllers C, Wenzel G (1992) Somatic hybridization of dihaploid potato protoplasts as a tool for potato breeding. Bot Acta 105:133–139

Moretti A, Laterrot H, Bordat D (1990) Observations of *Solanum ochranthum* and *S. juglandifolium*. Tomato Gen Coop Rep 40:25

Neidz RP, Sink KC (1988) Multifactor analysis of environmental preconditioning of tomato seedlings on protoplast culture and development. J Plant Physiol 133:385–391

Neidz RP, Rutter SM, Handley LW, Sink KC (1985) Plant regeneration from leaf protoplasts of six tomato cultivars. Plant Sci 39:199–204

Rick CM (1979) Biosystematic studies in *Lycopersicon* and closely related species of *Solanum*. In: Hawkes JG, Lester RN, Skelding AD (eds) The biology and taxonomy of the Solanaceae. Academic Press, New York, pp 667–678

Rick CM (1986) Germplasm resources in the wild tomato species. Acta Hortic 190:39–47

Rick CM (1988) Tomato-like nightshades: affinities, autoecology, and breeders' opportunities. Econ Bot 42:145–154

Rick CM, DeVerna JW, Chetelat RT (1990) Experimental introgression to the cultivated tomato from related wild nightshades. In: Bennett AB, O'Neill SD (eds) Horticultural biotechnology. Wiley-Liss, New York, pp 19–30

Shahin EA (1984) Isolation and culture of protoplasts: tomato. In: Vasil IK (ed) Cell culture and somatic cell genetics of plants, vol 1. Academic Press, Orlando, pp 370–380

Shepard JF (1980) Mutant selection and plant regeneration from potato mesophyll protoplasts. In: Rubenstein I, Gengenbach RL, Green CE (eds) Genetic improvement of crops/emergent techniques. Univ of Minnesota Press, Minneapolis, pp 185–219

Spooner DM, Anderson GJ, Jansen RK (1993) Chloroplast DNA evidence for the interrelationships of tomatoes, potatoes and pepinos (Solanaceae). Am J Bot 80:676–688

Staba JE (1969) Plant tissue culture as a technique for the phytochemist. Recent Adv Phytochem 2:80

Tan MMC, Rietveld EM, van Marrewijk GAM, Kool AJ (1987) Regeneration of leaf mesophyll protoplasts of tomato cultivars (*L. esculentum*): factors important for efficient protoplast culture and plant regeneration. Plant Cell Rep 6:172–175

Tingey WM, Gibson RW (1978) Feeding and mobility of the potato leafhopper impaired by glandular trichomes of *Solanum berthaultii* and *S. polyadenium*. J Econ Entomol 71:856–858

Tingey WM, Sinden SL (1982) Glandular pubescence, glycoalkaloid composition and resistance to the green peach aphid, potato leafhopper and potato flea beetle in *Solanum berthaultii*. Am Potato J 59:95–106

Tingey WM, Mehlenbacher SA, Laubengayer JE (1981) Occurrence of glandular trichomes in wild *Solanum* species. Am Potato J 58:81–83

Watanabe KN, Orrilo M, Vega S, Valkonen JPT, Pehu E, Hurtado A, Tanksley SD (1995) Overcoming crossing barriers between nontuber-bearing and tuber-bearing *Solanum* species: towards potato germplasm enhancement with a broad spectrum of solanaceous genetic resources. Genome 38:27–35

II.11 Somatic Hybridization Between *Solanum melongena* L. (Eggplant) and *Solanum sanitwongsei* Craib.

H. Asao[1], S. Arai[1], and M. Hirai[2]

1 Introduction

Eggplant (*Solanum melongena* L.) is an economically important nontuberous solanaceous crop in tropical and warm temperate regions. The origin of the eggplant is supposed to be in India. It is cultivated on about 1.31×10^6 ha worldwide, yielding 21.24 million t in 1999, with China (6.27×10^5 ha, 11.03 million t), India (4.25×10^5 ha, 6.10 million t), Turkey (0.33×10^5 ha, 0.85 million t), Egypt (4.25×10^5 ha, 0.57 million t) and Japan (0.15×10^5 ha, 0.49 million t) being the main producers (FAO 1999).

Eggplant production is sometimes severely curtailed due to infection by *Pseudomonas solanacearum*. To overcome this problem, several rootstocks have been used. *S. sanitwongsei* shows a high level of resistance to *P. solanacearum*. However, this stock has not been used widely because of the slow growth of the young plant on the rootstock. Thus, the breeding of rootstocks on which grafted *S. melongena* grows vigorously is highly requested. Somatic hybridization by protoplast fusion is expected to provide a new possibility to provide such desirable genetic characters to rootstock plants. Prerequisite protocols for regenerating plants from protoplast cultures in both eggplant and *S. sanitwongsei* have been developed in our group (Asao et al. 1989). Furthermore, the method of cell selection using a wilt-inducing product secreted by a virulent strain of *P. solanacearum* has also been described (Asao et al. 1992). The combination of these two techniques with protoplast fusion would allow us to breed desirable rootstocks for *S. melongena* graftings. Thus far, somatic hybrids of *S. melongena* with other related species have been produced by protoplast fusion. These were *S. sisymbriifolium* (Gleddie et al. 1986), *S. khasianum* (Sihachakr et al. 1988), *S. torvum* (Guri and Sink 1988a; Sihachakr et al. 1989), *S. nigrum* (Guri and Sink 1988b), *S. integrifolium* (Kameya et al. 1990), and *S. aethiopicum* (Daunay et al. 1993). Most of the somatic hybrids proved to be sterile and could not be used for breeding as rootstocks for eggplant production.

In this chapter our attempts to produce somatic hybrids between *S. melongena* and *S. sanitwongsei* through electrofusion are described, in which

---

[1] Nara Prefectural Agricultural Experiment Station, 88 Shijyo-cho, Kashihara-shi, Nara 634-0813, Japan
[2] National Research Institute of Vegetables, Ornamental Plants and Tea, 360 Ano-cho, Mie, 514-2392, Japan

we obtained a fertile somatic hybrid. Further, in order to evaluate the agricultural potential of the somatic hybrid and its offspring, the plants were tested for their field resistance to *P. solanacearum*, and the feasibility of using them as rootstocks (Asao et al. 1994) after their morphological characters and pollen viability were examined.

2 Somatic Hybridization

2.1 Protoplast Isolation

Seeds of *S. melongena* L.(cv. Senryou II) and *S. sanitwongsei* Crib (cv. Karehen) were aseptically sown on agar-solidified ($8gl^{-1}$), 1/2 MS medium (Murashige and Skoog 1962) containing $15gl^{-1}$ sucrose, at pH 5.8. The protoplasts were prepared from cotyledons which were taken from 2- to 3-week-old plants. The lamina strips were incubated overnight at 25 °C, in the dark, in an enzyme solution containing salt CPW (Xu et al. 1981), 0.3% Cellulase Onozuka R-10 (Yakult Co., Ltd., Japan), 0.06% Macerozyme R-10 (Yakult Co., Ltd., Japan), 0.5 M mannitol, and 10 mM 2-(N-morpholino) ethanesulfonic acid (MES), pH 5.8. Released protoplasts were filtrated through a 100-μm nylon mesh and then mixed with W_5 solution (Menczel et al. 1981) and centrifuged ($100g$, 3 min). The pellet was further washed twice with W_5 solution by centrifugation ($100g$, 3 min). Prior to fusion, the protoplasts were washed once with a 0.45-M glucose solution supplemented with 0.2 M $CaCl_2 \cdot 2H_2O$, then suspended in the same solution at a density of 5×10^4 protoplasts ml^{-1}.

2.2 Protoplast Fusion and Culture

Protoplasts from cotyledons of *S. melongena* and *S. sanitwongsei* were mixed in a ratio of 5:1, and then a 1-ml aliquot of the mixture was pipetted into a 15 × 60-mm petri dish. Thereafter, an immersible electrode (Shimadzu FTC33D5) was placed in the protoplast suspension. In order to align the protoplasts, an AC field at $200 V cm^{-1}$ and 1 MHz was applied for 30 s. Subsequently, two DC square pulses of field strength $0.75 KV cm^{-1}$ for 20 μs (Shimadzu SSH-2) were applied to achieve protoplast fusion.

Mesophyll protoplasts from *S. melongena* and *S. sanitwongsei* were electrofused (Fig. 1A). Immediately after fusion, 1 ml culture medium was slowly added to the protoplast mixture. The culture medium consisted of KM (Kao and Michayluk 1975) supplemented with $0.5 mgl^{-1}$ 2,4-dichlorophenoxyacetic acid (2,4-D), $1 mgl^{-1}$ kinetin, $1 mgl^{-1}$ α-naphthaleneacetic acid (NAA), $1000 mgl^{-1}$ wilt-inducing product (WIP) secreted by a virulent strain of *P. solanacearum* (Tanigawa et al. 1991), 0.45 M glucose as an osmoticum, and 10 mM MES at pH 5.8. The protoplasts were initially cultured in the dark at 25 °C for 7 days. Afterwards, they were exposed to illumination by fluorescent

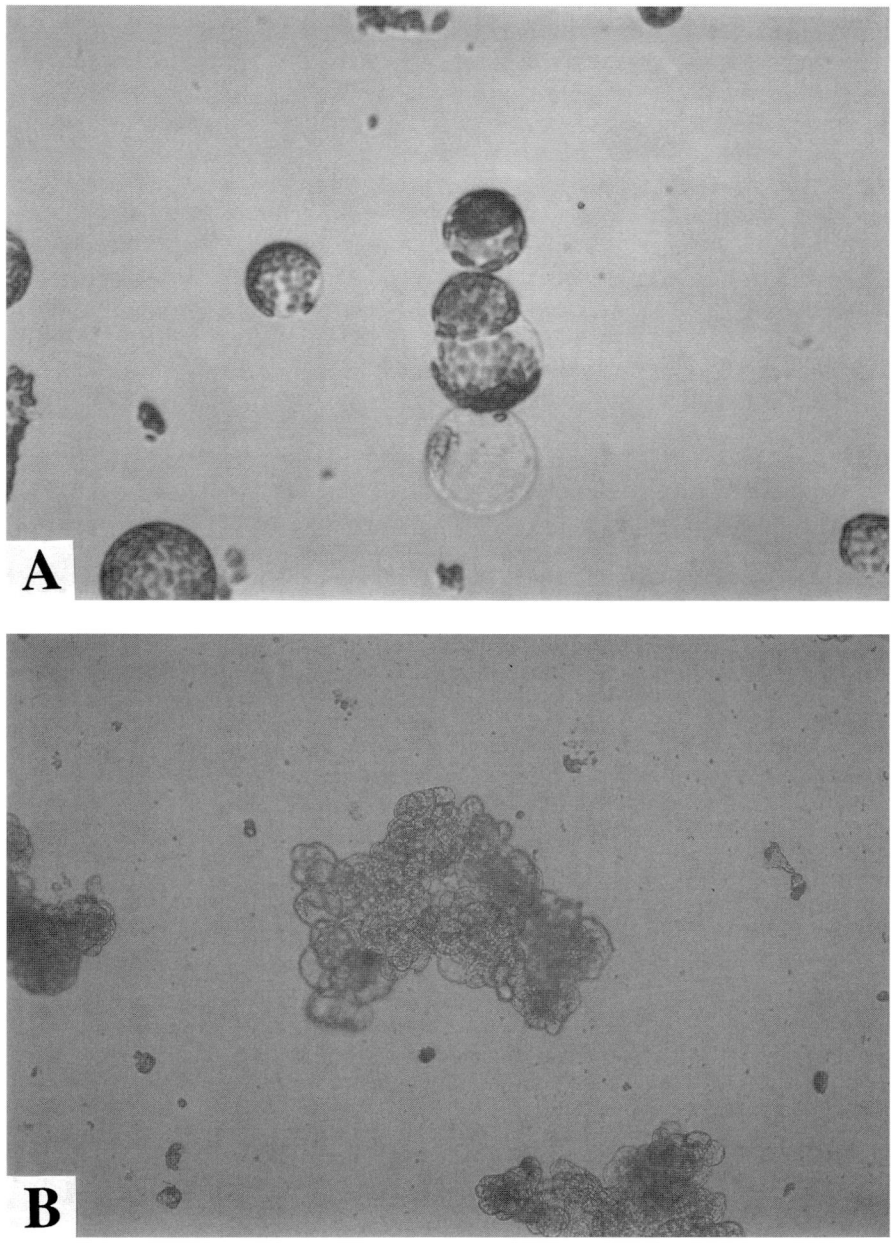

Fig. 1A–C. Electrofusion and culture of protoplasts. **A** Electrofusion of protoplasts. **B** Colony formation after 2 weeks of culture. **C** Shoot regeneration after 8 weeks of culture (see p. 236)

Fig. 1A–C. *Continued*

light for 16 h day^{-1} at 500 lx. Every week, the cultures were diluted twice with the same medium without the osmoticum and the auxin. About 4 weeks after the fusion, when the protoplast-derived colonies reached approximately 1 mm, the culture medium was replaced by C medium (Shepard and Totten 1977). The green calli on the C medium were transferred to the shoot regeneration medium containing MS salts and vitamins (Murashige and Skoog 1962), 3 mg l^{-1} zeatin, 0.2 mg l^{-1} 3-indoleacetic acid (IAA), 30 g l^{-1} sucrose, and 2 g l^{-1} gellan gum. Shoots were excised from the calli and rooted on hormone-free MS medium under illumination of 3000 lx (16 h day^{-1}) at 25 °C.

First cell division occurred after about 4 days, and some of the cells divided and developed further into colonies (Fig. 1B). After about 6 weeks of culture, the small calli formed were transferred to the regeneration medium. After another 2 to 3 weeks, green shoots were regenerated from the callus (Fig. 1C). Based on the plant morphology, we selected a total of eight putative somatic hybrid plants, i.e., 6.7% of the 120 regenerated plants obtained in the fusion treatment. These plants were then cultivated in a field where the soil was contaminated with *P. solanacearum* to assess field resistance to *P. solanacearum*. After this stage only one of the eight putative somatic hybrid plants survived.

2.3 Characteristics of the Somatic Hybrid and Offspring (S1–S5 Plants) from Self-Polination of the Hybrid

Morphological observation of both parents and the somatic hybrid is summarized in Table 1. The somatic hybrid showed intermediate characteristics between both parents in terms of gross morphology (Fig. 2A), leaf shape (Fig. 2B), flower size and color (Fig. 2C), and fruit shape and color (Fig. 2D). In particular, an immature fruit of the somatic hybrid was black as in eggplant, while that of *S. sanitwongsei* was green. The mature fruit turned orange, as in the case of *S. sanitwongsei*, while that of eggplant turned yellow. The pollen viability of the parents exceeded 90%, while that of the hybrid was 82.3% and this plant set fruits and seeds. Cytological analysis revealed that the chromosome number of the hybrid was $2n = 4x = 48$, i.e., the sum of both parents. The same chromosome number was also found in S_1 plants.

DNA was extracted from the leaves according to the method of Edwards et al. (1991) to examine genetic composition of the hybrids. The reaction mixture (10μl) for the polymerase chain reaction (PCR) was composed of 10mM Tris-HCl (pH 8.9), 80mM KCl, 1.5mM $MgCl_2$, 0.1% sodium cholate, 0.1% Triton X-100, 0.2mM dATP, 0.2mM dGTP, 0.2mM dCTP, 0.2mM dTTP, 0.2μM primer, 10 to 30ng template DNA, and 0.5 unit Tth DNA Polymerase (Toyobo). The primer RA 12–17 (5'-CGTCGGGGAGAA-3') was used. Amplification was carried out in a Program Temp Control System PC-700 (Astec) with preheating at 94°C for 30s, 45 cycles at 94°C for 30s, at 60°C for 2min, at 72°C for 3min, and postheating at 72°C for 7min. After all the cycles were completed, 10μl of the products were analyzed by agarose gel electrophoresis. The band patterns of the hybrid revealed by the RAPD assay were different from those of both parents, but bands from both parents were included, thus providing evidence of hybridity (Fig. 3).

Table 1. Morphological characteristics and chromosome numbers of the parental species and the somatic hybrid

	S. melomgena	Somatic hybrid	*S. sanitwongsei*
Leaf shape	Lobed	Slightly divided	Deeply divided
Anthocyanin on stem	Present	Minimal amount	Absent
Flowering	Solitary flower	Cyme (4–7 flowers)	Cyme (4–7 flowers)
Flower diameter (cm)	5.0	4.2	2.6
Fruit size (cm)			
Height	13.0	2.9	1.0
Width	5.0	2.7	1.1
Fruit color			
Immaturity	Black	Black	Green
Maturity	Yellow	Orange	Orange
Seed number fruit^{-1}	300<	133	53
Pollen viability (%)	98.0	82.3	93.7
Chromosome no.	24	48	24

Fig. 2A–D. Characteristics of plant morphology of *S. melongena* (*left*), the somatic hybrid (*center*), and *S. sanitwongsei* (*right*). **A** Plant morphology. **B** Leaf shape. **C** Flower shape. **D** Shape of immature fruit (*upper*) and mature fruit (*lower*)

Somatic Hybridization Between *S. melongena* L. (Eggplant) and *S. sanitwongsei* Craib. 239

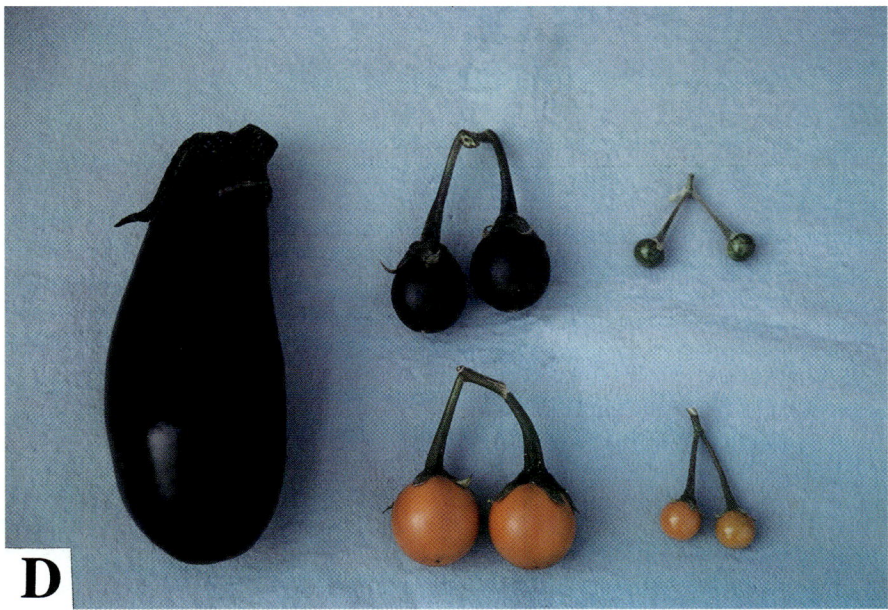

Fig. 2A–D. *Continued*

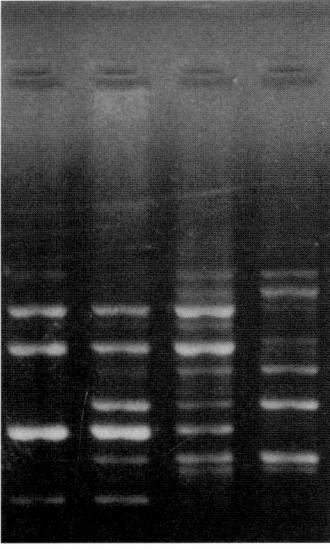

Fig. 3. Screening of somatic hybrids by RAPD markers. *1* S. sanitwongsei; *2* the somatic hybrid; *3* S_1 plant; *4* S. melongena

Table 2. Effect of temperature on the germination and the diameter of the hypocotyl

Plant	Germination (%)[a]		Diameter of the hypocotyl (mm)[b]	
	20°C	25°C	20°C	25°C
S_1 plants[c]	71.3	76.6	0.94	1.09
S. sanitwongsei	12.5	71.9	0.58	0.79
S. integrifolium	98.4	100.0	0.97	1.14
S. melongena	82.8	96.9	0.95	1.18

[a] 13 days after seeding.
[b] 18 days after seeding.
[c] Offspring from the self-pollinated fertile hybrid plant.

The seeds obtained from selfing of the hybrid were immersed in 100 mg l^{-1} gibberellin A_3 for 1 day, and then they were sown in a growth chamber (5000 lx) at 20 and 25°C. The germination rate, the diameter of the hypocotyl, and plant height were scored. The germination rate of *S. sanitwongsei* was lower than that of the other species, especially at 20°C, while the germination rate of the S_1 seeds was higher than that of *S. sanitwongsei* regardless of temperature (Table 2). The hypocotyl of the S_1 plants was about 1.5 times thicker than that of *S. sanitwongsei* and as thick as that of *S. melongena* and *S. integrifolium* (Table 2). The S_1 plants were also taller than *S. sanitwongsei*.

The fruit production of *S. melongena* grafted on S_1 hybrid plants was compared with that of the plants grafted on *S. sanitwongsei* and on *S. integrifolium*

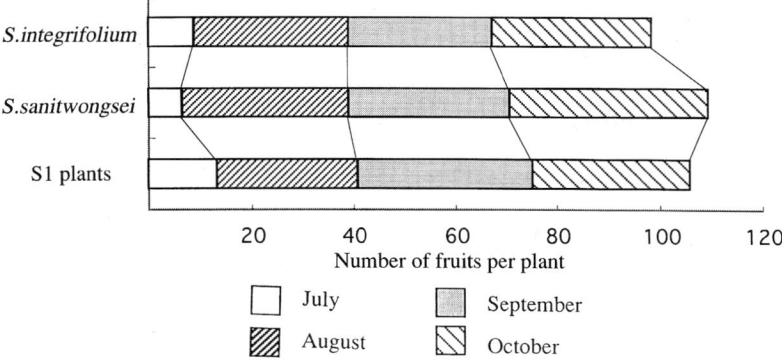

Fig. 4. Fruit yield of *S. melongena* grafted on S_1 plants, *S. sanitwongsei*, and *S. integrifolium*

Poir. Fruit yields were investigated from July to October in 1993. The yield of eggplants grafted on S_1 plants in July was higher than that of eggplants grafted on *S. integrifolium* and *S. sanitwongsei*, while the total fruit yield was similar to that of the other species (Fig. 4).

The somatic hybrid was propagated by cutting, and used as rootstock of *S. melongena*. Likewise, offspring (S_1–S_5 plants) from self-pollination of the hybrid were used as rootstocks of *S. melongena*. The grafted plants were cultivated in a field infested with *P. solanacearum*. *S. melongena* plants grafted on *S. sanitwongsei* and *S. integrifolium* were also grown in the same field as a control. In the field test (1992–1997) on contaminated soil, about 100% of the eggplants grafted on *S. integrifolium* died within 45 days after transplanting. However, the appearance of the symptoms on eggplants grafted on the somatic hybrids and their progeny (S_1–S_5 plants) was delayed and only 12.5~43.8% of the eggplants died by infection by *P. solanacearum*. The resistance to *P. solanacearum* of the somatic hybrids and S_1 plants was comparable to that of *S. sanitwongsei*, and S_2–S_5 plants were more resistant to *P. solanacearum* than *S. sanitwongsei* (Table 3).

3 Summary and Conclusion

In this study, $1000\,\text{mg}\,\text{l}^{-1}$ of wilt-inducing product (WIP) was added to the protoplast culture medium in order to obtain hybrids resistant to *P. solanacearum*. As the protoplasts of *S. melongena* seldom divided in the medium supplemented with $1000\,\text{mg}\,\text{l}^{-1}$ WIP (Asao et al. 1992), it was likely that the regenerated plants were somatic hybrids or *S. sanitwongsei* plants. Since only eight plants were regenerated in the present study, WIP applied to the fused protoplasts may have killed most of the unfused eggplant protoplasts. However, we

Table 3. Resistance of eggplants on soil contaminated with *Pseudomonas solanacearum*

Plant	Mortality (%)[d]					
	1992	1993	1994	1995	1996	1997
Rootstock[a]						
Somatic hybrid[b] and its progeny	22.2	12.5	29.2	21.4	43.8	37.5
S. sanitwongsei	22.2	12.5	58.3	57.1	77.8	81.2
S. integrifolium	100.0	81.3	100.0	100.0	100.0	100.0
Control plant[c]						
S. melongena	100.0	100.0	100.0	100.0	100.0	100.0

[a] *S. melongena* was grafted on each rootstock.
[b] 1992: somatic hybrid, 1993: S_1 plants (offspring from the self-pollinated fertile hybrid plant), 1994: S_2 plants, 1995: S_3 plants, 1996: S_4 plants, 1997: S_5 plants.
[c] Seedlings without grafting.
[d] Mortality was scored at 70 days after transplanting.

analyzed only one of them. The effectiveness of the present strategy requires further studies. As the plating efficiency of *S. sanitwongsei* was higher than that of *S. melongena*, the protoplasts of *S. sanitwongsei* and *S. melongena* were mixed in the ratio of 1:5, respectively, prior to fusion. A similar experiment was reported by Guri and Sink (1988a,b).

Several papers on protoplast fusion have been published in eggplants, including electrofusion (Sihachakr et al. 1988, 1989, Daunay et al. 1993), polyethylene glycol treatment (Gleddie et al. 1985, 1986, Guri and Sink 1988a,b), and dextran treatment (Kameya et al. 1990). Although it is generally recognized that somatic hybridization would overcome the sexual barrier between eggplant and other wild species, the utilization of the hybrid plants in breeding programs of eggplant has been limited because of their sterility. Up to the present, Kameya et al. (1990) and Daunay et al. (1993) have described fertile somatic hybrids between *S. melongena* and *S. integrifolium*, and between *S. melongena* and *S. aethiopicum*, respectively.

Since few seeds were obtained from the crosses between *S. melongena* and *S. sanitwongsei*, and all of them were empty seeds (unpubl. data), the hybrid cannot be used in the breeding program of a rootstock for eggplant. On the contrary, this study showed that a fertile hybrid of *S. melongena* and *S. sanitwongsei* could be produced by protoplast electrofusion. The somatic hybrid had 48 chromosomes, corresponding to the sum of the chromosomes of the parents. The same chromosome number was also observed in the S_1 plants. The fertility of the present hybrid may be attributed to two sets of genome in a single cell. These findings suggest that fertility can be maintained in the progeny of this hybrid. Therefore, the present somatic hybrid could be used in breeding programs for the development of resistance to *P. solanacearum*.

We previously released a line of *S. sanitwongsei*, designated Karehen, as a rootstock resistant to bacterial wilt of eggplant. However, germination and seedling growth of this species are slow, and the resulting seedlings display slender hypocotyls. Therefore, the time from seedling to grafting is prolonged.

Moreover, the scion showed slow growth even after grafting, which resulted in a low productivity of the rootstock in the early stage. Somatic hybridization with *S. melongena* considerably modified the growth habit of the rootstock. The rapid germination and growth of the present hybrid and S_1 plants may be inherited from *S. melongena*. The current study also showed the superiority of the S_1 plants as rootstocks for eggplant production.

4 Protocol

1. Protoplasts of *S. melongena* and *S. sanitwongsei* are prepared from cotyledons which are taken from 2- to 3-week-old plants.
2. Cut leaves into 1-mm strips with a scalpel and soak in an enzyme solution overnight in the dark at 25 °C. Incubate protoplasts with an enzyme solution consisting of CPW salt, 0.3% Cellulase R-10, 0.06% Macerozyme R-10, 0.5 M mannitol, and 10 mM MES, pH 5.8.
3. Filter the enzyme-protoplasts suspension through a 100-μm nylon mesh. Add the W_5 solution and centrifuge at 100 g for 3 min. Wash isolated protoplasts with the W_5 solution and centrifuge at 100 g for 3 min.
4. Prior to fusion, wash once with a 0.45 M glucose solution supplemented with 0.2 M $CaCl_2 \cdot 2H_2O$, and suspend in the same solution at a density of 5×10^4 protoplasts ml^{-1}.
5. Mix protoplasts of *S. melongena* and *S. sanitwongsei* (5:1 ratio), pipette 1-ml aliquots of the mixture into a petri dish of the size of 15×60 mm. Place an immersible electrode in the protoplast suspension.
6. Apply AC field at 200 V cm^{-1} and 1 MHz for 30 s for protoplast alignment, then two DC square pulses at 0.75 KV cm^{-1} for 20 μs for protoplast fusion.
7. Add 1 ml culture medium of KM supplemented with 0.5 mg l^{-1} 2,4-D, 1 mg l^{-1} kinetin, 1 mg l^{-1} NAA, 1000 mg l^{-1} wilt-inducing product (WIP), 0.45 M glucose, and 10 mM MES at pH 5.8.
8. Keep the cultures in darkness at 25 °C for 7 days. Subsequently, the cultures are exposed to a 16 h day^{-1} illumination at 500 lx.
9. About 4 weeks after fusion, transfer calli that reach appoximately 1 mm in diameter to C medium. Then, transfer the resultant green calli to regeneration medium consisted of MS salts and vitamins, 3 mg l^{-1} zeatin, 0.2 mg l^{-1} IAA, 30 g l^{-1} sucrose, and 2 g l^{-1} gellan gum.
10. Excise shoots from regenerating calli and transfer to hormone-free MS medium under illumination of 3000 lx (16 h day^{-1}) 25 °C.

References

Asao H, Tanigawa M, Arai S, Kobatake H (1989) Plant regeneration from mesophyll protoplasts of the eggplant and its wild species. Bull Nara Agric Exp Stn 20:73–78

Asao H, Tanigawa M, Okayama K, Arai S (1992) Breeding of resistant *Solanum* spp. for bacterial wilt by cell selection using a wilt-inducing product. Bull Nara Agric Exp Stn 23:7–12

Asao H, Arai S, Sato T, Hirai M (1994) Characteristics of a somatic hybrid between *Solanum melongena* L. and *Solanum saniywongsei* Craib. Breed Sci 44:301–305

Daunay MC, Chaput MH, Sihachakr D, Allot M, Vedel F, Ducreux G (1993) Production and characterization of fertile somatic hybrids of eggplant (*Solanum melongena* L.) with *Solanum aethiopicum* L. Theor Appl Genet 85:841–850

Edwards K, Johnstone C, Thompson C (1991) A simple and rapid method for the preparation of plant genomic DNA for PCR analysis. Nucleic Acids Res 19:1349

FAO (1999) FAO FAOSTAT

Gleddie S, Fassuliotis G, Keller WA, Setterfield G (1985) Somatic hybridization as a potential method of transferring nematode and mite resistance into eggplant. Z Pflanzenzücht 94: 352–355

Gleddie S, Keller WA, Setterfield G (1986) Production and characterization of somatic hybrids between *Solanum melongena* L. and *S. sisymbriifolium* Lam. Theor Appl Genet 71:613–621

Guri A, Sink KC (1988a) Organelle composition in somatic hybrids between an atrazine resistant biotype of *Solanum nigrum* and *Solanum melongena*. Plant Sci 58:51–58

Guri A, Sink KC (1988b) Interspecific somatic hybrid plants between eggplant (*Solanum melongena*) and *Solanum torvum*. Theor Appl Genet 76:490–496

Kameya T, Miyazawa N, Toki S (1990) Production of somatic hybrids between *Solanum melongena* L. and *S. integrifolium* Poir. Jpn J Breed 40:429–434

Kao KN, Michayluk MR (1975) Nutritional requirements for growth of *Vicia hajastana* cell and protoplasts at a very low population density in liquid media. Planta 126:105–110

Menczel L, Nagy I, Kizz ZR, Maliga P (1981) Streptomycin-resistant and -sensitive somatic hybrids of *Nicotiana tabacum* + *Nicotiana knightiana*: correlation of resistance to *N. tabacum* plastid. Theor Appl Genet 59:191–195

Murashige T, Skoog F (1962) A revised medium for rapid growth and bioassays with tobacco cultures. Physiol Plant 15:473–497

Shepard JF, Totten RE (1977) Mesophyll cell protoplasts of potato. Plant Physiol 60:313–316

Sihachakr D, Haicour R, Serraf I, Barrientos E, Herbreteau C, Ducreux G, Rossignol L, Souvannavong V (1988) Electrofusion for the production of somatic hybrid plants of *Solanum melongena* L. and *Solanum khasianum* C.B. Clark. Plant Sci 57:215–223

Sihachakr D, Haicour R, Chaput MH, Barrientos E, Ducreux G, Rossignol L (1989) Somatic hybrid plants produced by electrofusion between *Solanum melongena* L. and *Solanum torvum* Sw. Theor Appl Genet 77:1–6

Tanigawa M, Asao H, Okayama K (1991) Wilt-inducing activity and components of exopolysaccharide produced by a virulent *Pseudomonas solanacearum*. Bull Nara Agric Exp Stn 22:23–27

Xu ZH, Davey MR, Cooking EC (1981) Isolation and sustained division of *Phaseolous aureus* (mung bean) root protoplasts. Z Pflanzenphysiol 104:289–298

II.12 Somatic Hybridization Between *Solanum commersonii* Dun. and *S. tuberosum* L. (Potato)

T. CARDI

1 Introduction

The large and very diversified *Solanum* genus includes seven cultivated species and more than 200 wild tuber-bearing species with basic chromosome number $x = 12$ and ploidy level ranging from $2x$ ($2n = 24$) to $6x$ ($2n = 72$). *Solanum tuberosum* subsp. *tuberosum* ($2n = 4x = 48$), derived from crossing, chromosome doubling, and diversification of some South-American species, includes the common potato crop, and is adapted to long-day conditions (Matsubayashi 1991; Hawkes 1994). Nowadays, the common potato is grown worldwide for various food and industrial purposes, and it is the most important noncereal crop (Bajaj 1987; Zuba and Binding 1989). Traditional breeding of *S. tuberosum* (*tbr*), however, is a difficult task, due to high heterozygosity, tetrasomic inheritance, self-incompatibility, and sterility in cultivated genotypes. Also because it is vegetatively propagated, its genetic variability is rather limited. Fortunately, a number of innovative breeding and biotechnological approaches proved successful in genetic manipulation of *S. tuberosum* to induce and preserve novel genetic variability, produce disease-free plants, manipulate ploidy, and introduce genes from wild species (Bajaj 1987; Zuba and Binding 1989; Bradshaw and Mackay 1994).

The wild *Solanum* species have long been recognized as a valuable source of genes for the improvement of the common potato, but only very few have been used for such a purpose (Hawkes 1994). *Solanum commersonii* (*cmm*), with diploid ($2n = 2x = 24$) and triploid ($2n = 3x = 36$) forms, diversified in Argentina, Uruguay, Paraguay, and South Brazil, belongs to the series Commersoniana. Typically, it has a bushy growth habit and star-shaped (stellate) flowers. Some differentiation has been evidenced in both nuclear and cytoplasmic (chloroplast and mitochondrial) genomes of *cmm* and *tbr* (Hosaka et al. 1984; Matsubayashi 1991; Perl et al. 1991; Hawkes 1994; Scotti et al. 1998; Bastia et al. 1999b, 2000a; Cardi et al. 1999a).

A number of stress-resistance genes and other useful agronomic characteristics are present in *S. commersonii*, as recently summarized by Cardi (1999), but they are not readily accessible due to the fact that *cmm* and *tbr*

CNR-IMOF, Research Institute for Vegetable and Ornamental Plant Breeding, Via Università 133, 80055 Portici, Italy

have an endosperm balance number (EBN) equal to 1 and 4, respectively (Hanneman and Bamberg 1986). Approaches based on ploidy and EBN manipulation made it possible to overcome sexual barriers in some instances (Ehlenfeldt and Hanneman 1988; Novy and Hanneman 1991; Carputo et al. 1997a). Alternatively, hybridization carried out at the somatic level by fusing protoplasts isolated from di(ha)ploid *Solanum* parents, allowed not only interspecific gene transfer across sexual borders in several species, but also in a single step to combine intact parental nuclear genomes, preserving allelic and nonallelic interactions, to restore normal tetraploidy, and to generate genetic variability at the cytoplasmic level as a consequence of sorting-out and/or rearrangement of parental plastidial and mitochondrial genomes (Waara and Glimelius 1995).

Protoplast fusion has been successfully used to produce symmetric somatic hybrids between various *S. commersonii* and *S. tuberosum* genotypes (Cardi et al. 1993b; Kim et al. 1993; Waara et al. 1998; Bastia et al. 2000a). The production of asymmetric hybrids and cybrids between the two species has also been reported (Perl et al. 1991; Nyman and Waara 1997; Waara et al. 1998). The symmetric somatic hybrids have been characterized for the transfer of resistance genes from the wild to the cultivated species and for cytological, morphological, and fertility-related aspects (Cardi et al. 1993a; Carputo et al. 1997b; Conicella et al. 1997; Nyman and Waara 1997; Cardi 1998; Chen et al. 1999a,c; Laferriere et al. 1999; Parrella and Cardi 1999). In this chapter, the results obtained on somatic hybridization including *S. commersonii* and *S. tuberosum* are reviewed, focusing on the potential utilization of the somatic hybrids in genetics and breeding of potato.

2 Somatic Hybridization

2.1 Isolation of Protoplasts

In vitro-grown plantlets have been used for protoplast isolation in all studies published so far (Table 1). Generally, single seed-derived clones have been isolated in four *S. commersonii* accessions, micropropagated and eventually used for protoplast isolation (Fig. 1a). Eight dihaploid and two tetraploid *S. tuberosum* clones have been used as fusion partners. In one series of experiments, *tbr* shoot cultures were bleached by adding the herbicide SAN 9789 (Sandoz) to the culture medium in order to stain protoplasts with fluorescein diacetate and estimate heterologous fusion frequency (Fig. 1b; Cardi et al. 1993b). Shoot cultures were generally maintained and protoplasts isolated according to protocols previously established for parental species (Haberlach et al. 1985; Cardi et al. 1990; Waara et al. 1991). Since bleached protoplasts tended to float, it was difficult to pellet them. Therefore, the isolation protocol was modified by including a flotation step on Percoll (Cardi et al. 1993b). Using the regular or a modified procedure, 3.0–8.5×10^6 and 0.7–2.6×10^6 protoplasts g^{-1} of fresh

Table 1. Fusion partners and conditions, and heterofusion frequencies in various experiments with *S. commersonii* (*cmm*) and *S. tuberosum* (*tbr*)

Fusion partners		Source of protoplasts		Pretreatment of protoplasts	Protoplast fusion		Cell density (prot.s ml^{-1})	Medium (mM)	Heterofusion frequency (%)	Reference
cmm	*tbr*[a]	*cmm*	*tbr*		Electrofusion parameters					
PI 243503	DH 81-7-1463, Desiree	Leaves from in vitro-grown plantlets	Bleached in vitro plantlets[b]	Stained with FDA (*tbr*)	AC: 73 V cm^{-1}, 1 MHz; DC: 2 kV cm^{-1}, 3 pulses, 50 µs each		4.8×10^5	Mannitol (440), CaCl$_2$ (0.8)	3–10	Cardi et al. (1993b)
PI 320266	PT56, R4	"	Leaves from in vitro-grown plantlets	None	AC: 90 V cm^{-1}, 0.5 MHz; DC: 1.25 kV cm^{-1}, 1 pulse, 100 µs		4.0×10^5	Mannitol (393), sucrose (59) CaCl$_2$ (0.5)	nd[c]	Kim et al. (1993); Laferriere et al. (1999)
PI 472834	At19	"	"	"	AC: 100 V cm^{-1}, 1 MHz; DC: 2 kV cm^{-1}, 3 pulses, 50 µs each		4.8×10^5	Mannitol (440), CaCl$_2$ (0.8)	"	Bastia et al. (2000a)
PI 243503	W730, SS543	"	"	"	"		"	"	"	N. Carotenuto and T. Cardi (unpubl.)
PI 472834, PI 473412	161:14, 67:9, 536:2	"	"	Stained with FDA (*cmm*) and scopoletin (*tbr*)	AC: 62 V cm^{-1}, 1 Mhz; DC: 1.7 kV cm^{-1}, 3 pulses, 40 µs each		5.0×10^5	Mannitol (600), CaCl$_2$ (1)	"	Waara et al. (1998)

[a] DH 81-7-1463, PT56, At19, 161:14, 67:9, 536:2, dihaploid clones ($2n = 2x = 24$) from breeding line W 72-22-492, cv. Superior, cv. Atlantic, cv. Stina, breeding line Y 67-20-40, cv. Matilda, respectively; Desiree and R4, tetraploid ($2n = 4x = 48$) variety and breeding line PI 203900, respectively.
[b] In vitro plantlets were grown in the presence of the herbicide SAN 9789.
[c] Not determined.

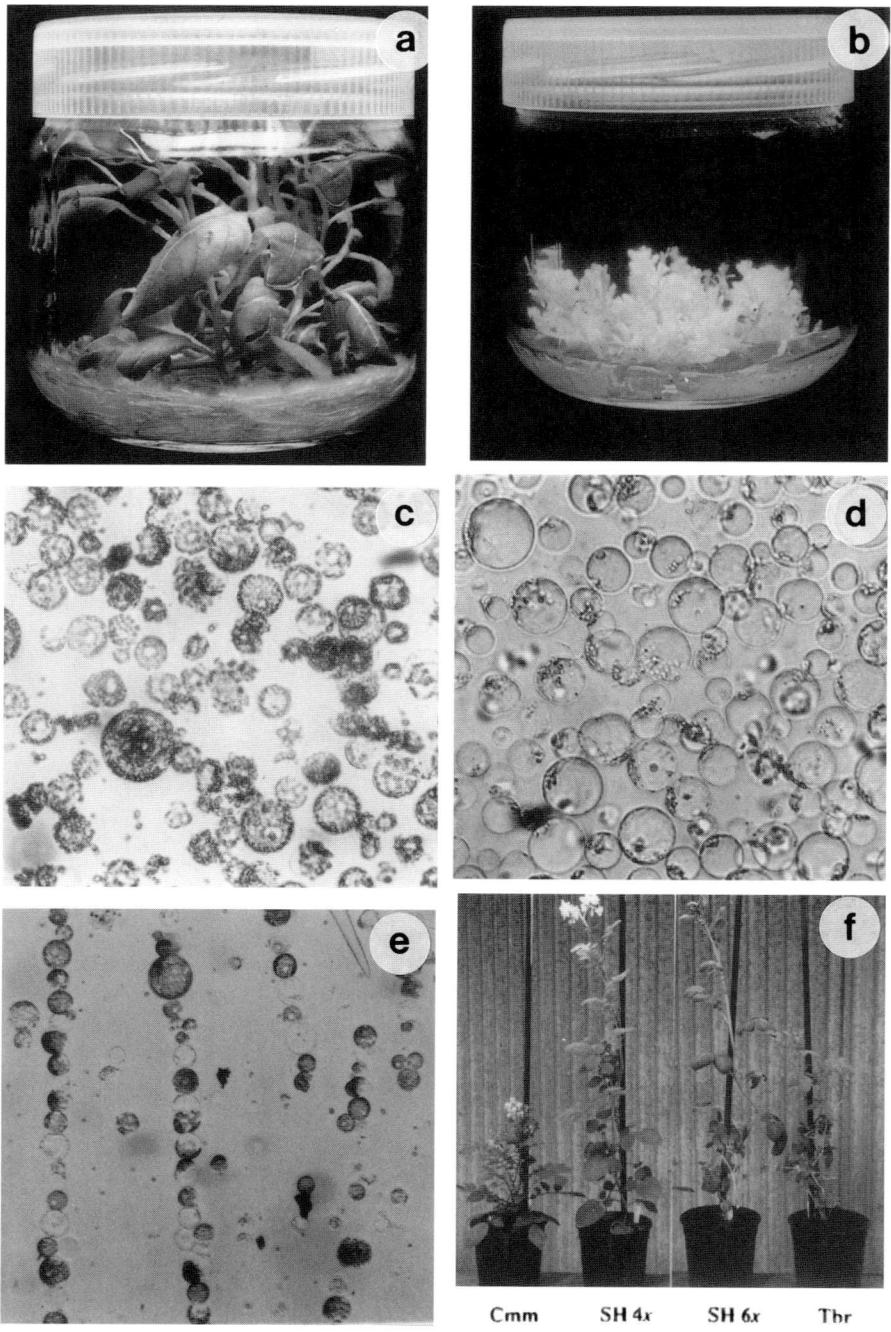

Fig. 1a–f. Parental and hybrid plants and protoplasts of *Solanum commersonii* (*cmm*) and *S. tuberosum* (*tbr*). **a** In vitro-grown plants of *cmm* PI 243503. **b** In vitro-grown bleached plants of *tbr* DH81-7-1463. **c** Mesophyll protoplasts isolated from *cmm*. **d** Protoplasts isolated from *tbr*. **e** Protoplasts aligned in an AC field and fused with a DC field. **f** Parental and hybrid plants regenerated after electrofusion experiments: from *left to right*. *Cmm*; tetraploid somatic hybrid (*SH4x*); hexaploid somatic hybrid (*SH6x*); *Tbr*. (**a** and **c** from Cardi et al. 1990)

tissue could be isolated from green and bleached material, respectively (Fig. 1c,d; Cardi et al. 1990, 1993b). The shoot cultures used for protoplast isolation or the excised leaves were sometimes preconditioned in the dark and/or at low temperature to improve protoplast yield and division ability (Kim et al. 1993; Waara et al. 1998).

2.2 Fusion of Protoplasts

Successful fusion of parental protoplasts was achieved by electrofusion (Table 1). An AC field between 62 and $100\,V\,cm^{-1}$ at 0.5–1 MHz was used to align protoplasts. These were resuspended at a density of 4–5×10^5 in a medium containing 0.44–0.66 M mannitol as osmoticum and 0.8–1 mM $CaCl_2$ as membrane stabilizer. Generally, one or three consecutive DC pulses (1.25–$2.0\,V\,cm^{-1}$), 40–100 μs each, were given to induce membrane fusion between adjacent protoplasts (Fig. 1e). Higher DC field strength determined a lower viability of treated protoplasts, the heterologous fusion frequency remaining almost the same. On the other hand, key factors for the improvement of fusion frequency were the increase of the density of protoplasts and the addition of Ca^{2+} to the fusion medium; varying with the genotype, heterologous fusion frequencies between 3 and 10% were obtained (Cardi et al. 1993b). Based on staining with ethidium bromide (Ye and Earle 1991), about 70% of all fusions were binary (N. Carotenuto and T. Cardi, unpubl.).

An alternative chemo-electrofusion method based on the use of PEG to aggregate protoplasts in place of the AC field was also tested in some experiments (Cardi et al. 1993b). However, the fusion frequency and the plating efficiency of treated protoplasts were considerably lower than with the usual AC/DC fusions, and no plants could be regenerated. Although PEG has been employed in chemical fusions in *Solanum* spp., it was found harmful for potato protoplasts in some cases (Zuba and Binding 1989; Cardi et al. 1993b).

In some experiments (Waara et al. 1998), parental mesophyll protoplasts were stained either with scopoletin or FDA, and fusion products showing dual fluorescence were selected by flow cytometric sorting. After adjustment of staining and sorting conditions to specific requirements of potato protoplasts, approximately 90% of sorted protoplasts were putative heterofusion products. It was calculated that about 55000 heterokaryons could be isolated per day using automatic cell sorting, in comparison with $500\,day^{-1}$ isolated by micromanipulation in other fusion combinations.

2.3 Culture of Fused Protoplasts

Established procedures for the culture of parental protoplasts were used for raising fusion products (Haberlach et al. 1985; Hunt and Helgeson 1989; Cardi et al. 1990; Waara et al. 1991). Mass culture of fused and unfused protoplasts was carried out in most experiments and putative hybrid calli were selected on the basis of growth characteristics in vitro (Table 2).

Table 2. Methods used for the selection of heterofusion products and confirmation of nuclear hybridity, and frequency of hybrid plants regenerated in various fusion experiments between S. commersonii and S. tuberosum

Fusion partners		Selection of heterofusion products	Confirmation of nuclear hybridity	Hybrid/regenerated plants (%)	Reference
cmm	tbr[a]				
PI 243503	DH 81-7-1463	Differential response of fusion partners and expression of hybrid vigor in vitro	POX and MDH isozymes, SSCP, RAPD	100	Cardi et al. (1993b); Carotenuto et al. (1996); Carotenuto (1997); T. Bastia and T. Cardi (unpubl.)
PI 320266	PT56	"	MDH and MNR isozymes, RAPD	97	Kim et al. (1993)
PI 472834	Atl9	"	RAPD, SSCP	100	Carotenuto et al. (1996); Carotenuto (1997); Bastia et al. (2000a)
PI 472834, PI 473412	161:14, 67:9, 536:2	Flow cytometric cell sorting	RAPD	98	Waara et al. (1998)

[a] For origin of genotypes see Table 1.

Preselected heterokaryons were cultured by Waara et al. (1998). Depending on the fusion combination, plating efficiencies ranging from 0.02 to 4.0% or from 0.1 to 5.3% were obtained with mass-cultured or flow cytometrically sorted protoplasts, respectively (Cardi et al. 1993b; Waara et al. 1998; Bastia et al. 2000a).

2.4 Regeneration of Somatic Hybrids

A number of procedures were used for shoot regeneration from putative somatic hybrid calli; generally, best results were obtained with protocols that had been previously set up with either one or both parental protoplasts (Cardi et al. 1990, 1993b; Kim et al. 1993; Waara et al. 1998). In experiments in which parental and fused protoplasts were mass-cultured after fusion, regeneration frequency ranged from 1 to 32% (Cardi et al. 1993b; Kim et al. 1993; Carotenuto 1997), whereas significantly higher figures were obtained when selected heterokaryons were cultured (Waara et al. 1998). This is somewhat expected considering the different response of parental and hybrid protoplasts in vitro and the heterotic vigor generally shown by the latter (Cardi et al. 1993b; Kim et al. 1993; Bastia et al. 2000a). As a matter of fact, significantly higher regeneration frequencies (32–60%), comparable to those obtained with sorted *cmm* (+) *tbr* fusion products, were obtained when protoplasts were isolated from PI 243503 (+) DH 81-7-1463 somatic hybrid plants (I. Florio and T. Cardi, unpubl.).

Nuclear hybridity of regenerated plants was confirmed with isozyme and/or DNA analyses (Table 2). The dimeric enzyme MDH proved to be particularly useful for recognizing true interspecific hybrids (Cardi et al. 1993b), and was also used by Kim et al. (1993). PCR-based molecular markers, such as RAPD, were also commonly used. Because of the reduced amount of tissue needed, the speed of the analysis, and the lack of necessity to use radioactive substances, they are a quick and convenient method to test the hybridity of regenerated material (Kim et al. 1993; Waara et al. 1998; Bastia et al. 2000a). In some experiments, SSCP analysis of specific regions within target genes was carried out to have more detailed data on molecular and chromosomal composition of somatic hybrids (Carotenuto et al. 1996; Carotenuto 1997).

In all cases, regenerated plants were almost exclusively hybrids (97–100%), indicating the high efficiency of methods used for the selection of the hybrids (Table 2). The differential response of fusion partners and the expression of hybrid vigor in vitro have been successfully applied as selecting criteria also in other intra- and interspecific fusion combinations in *Solanum* spp. (Cardi et al. 1993b; Polgár et al. 1993). Due to the possibility of sorting out nonhybrid protoplasts very early and quickly, thus reducing the cumbersomeness of in vitro subcultures and plant transfers, flow cytometric cell sorting of fusion products is an attractive alternative (Waara et al. 1998). However, the cost of the machine and the relatively high skill required for its use could limit a more general utilization.

2.5 Characterization of Somatic Hybrids

2.5.1 Cytology

Although plants with $2n = 4x = 48$ chromosomes are expected after the fusion of protoplasts isolated from di(ha)ploid parents ($2n = 2x = 24$), aneuploids and/or plants with higher ploidy level are commonly found among somatic hybrid regenerants (Ramulu et al. 1989). They probably derive from multiple protoplast fusions and/or fusions involving cells in different phases of the cell cycle, and to chromosome elimination during callus growth. As a matter of fact, 61–74% of regenerated plants were tetraploid or near-tetraploid in different *cmm* (+) *tbr* fusion combinations. The remaining plants mostly showed a 6x ploidy level (Fig. 2; Cardi et al. 1993a; Kim et al. 1993; Waara et al. 1998;

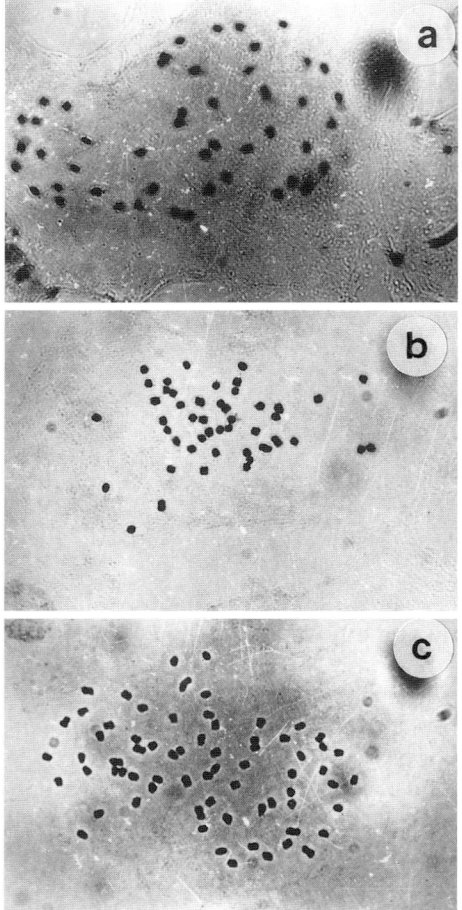

Fig. 2a–c. Mitotic chromosomes of **a** eutetraploid ($2n = 48$), **b** hypotetraploid ($2n = 47$), and **c** hypo-hexaploid ($2n = 68$) somatic hybrids

Bastia et al. 2000a). In a few cases, pentaploid, octoploid, mixoploid, or cytochimeric plants showing different ploidy in L_1 and L_3 histogenetic layers were also obtained (Cardi et al. 1993a; Waara et al. 1998; Bastia et al. 2000a). In earlier experiments, chromosome counts were performed and thus it was established that 33–50% of $4x$ plants were eutetraploids ($2n = 4x = 48$) (Cardi et al. 1993a; Kim et al. 1993). Flow cytometry was adopted in later experiments (Waara et al. 1998; Bastia et al. 2000a), while chloroplast count was also used as a quick method to select plants with the desired ploidy level by Cardi et al. (1993a) and Bastia et al. (2000a).

Although the average figure reported by Waara et al. (1998) varied between different fusion combinations, the frequency of $4x$ somatic hybrids ranged from 44 to 77%, and so it was remarkably higher than that obtained in previous experiments in which micromanipulation was used to select heterokaryons (Waara et al. 1998 and references cited therein).

At meiosis, all hybrids showed the formation of multivalents, suggesting the possibility of having gene exchange between the parental genomes and introgression of useful traits from *S. commersonii* in *S. tuberosum* (Conicella et al. 1997). Hybrids, however, differed in male fertility and in most cases exhibited abnormalities during meiosis in PMCs (see Sect. 2.5.3, Fertility). The microsporogenesis of a fertile somatic hybrid was characterized in detail to assess the abnormalities in meiotic cytoskeletal elements involved in altered cytokinesis and $2n$ pollen formation in *Solanum* spp. (Genualdo et al. 1998).

2.5.2 Morphology

Generally, somatic hybrids were much more vigorous than parental genotypes (Fig. 1f) (Cardi et al. 1993a; Nyman and Waara 1997; Cardi 1998; Bastia et al. 2000a). Similar results were reported also in other intra- or interspecific somatic hybridizations in *Solanum* spp. (Waara and Glimelius 1995). With respect to parental genotypes, both the increase in ploidy and heterozygosity could play a role in determining the heterotic vigor of the hybrid plants. However, no similar change in vigor was found in parents with somatically doubled ploidy level (Bastia et al. 1995). Therefore, the degree of nonadditive gene interactions and intralocus allelic diversity are more important causes for the expression of hybrid vigor in *Solanum* spp. (Cardi 1998 and references cited therein).

In comparison with parents, hybrids showed intermediate phenotypes for several characteristics, but they were generally more similar to the cultivated fusion partner (Cardi et al. 1993a; Nyman and Waara 1997; Bastia et al. 2000a). Similar results were obtained when parental and hybrid genotypes, grown either in a greenhouse or in the open field, were analyzed by means of multivariate statistical tests (Cardi 1998; Mazzei et al. 1999). Interestingly, the cultivated phenotype was easily recovered also in progenies derived from selfing or backcrossing the somatic hybrids (Basile et al. 1998; Mazzei et al. 1999) as well as in *cmm* × *tbr* sexual hybrids and derived progenies (Carputo et al. 1997a). These results suggest a "general dominance" of the cultivated phenotype and probably depend on the relative closeness of the two parental species on an evolutionary scale.

Some variations were observed among somatic hybrids. Their ploidy level was the most important factor for determining differences. In comparison with tetraploids, hexaploids were generally less vigorous and had lower numbers of stems and leaflets, larger leaflets and a reduced degree of flowering. However, some residual differences, probably due to aneuploidy or nuclear and cytoplasmic mutations, were also present among somatic hybrids (Cardi et al. 1993a; Cardi 1998).

Most somatic hybrids tuberized under long-day conditions as the cultivated parent, but generally on long stolons as the wild parent (Cardi et al. 1993a; Nyman and Waara 1997; Chen et al. 1996b; Bastia et al. 2000a).

2.5.3 Fertility

In order to introgress genes into the cultivated gene pool from wild species, the fertility of interspecific hybrids is an important aspect to consider in somatic hybridization experiments. Somatic hybrids between *cmm* and *tbr* generally resulted in male sterility (Cardi et al. 1993a; Nyman and Waara 1997; Bastia et al. 2000a). Male fertility was found, however, in one single regenerant by Cardi et al. (1993a), in all plants derived by one fusion combination by Nyman and Waara (1997), and in somatic hybrids derived by two fusions, including one line of *S. commersonii* PI 320266 (Chen et al. 1999b; Laferriere et al. 1999).

Male sterile hybrids usually flowered profusely and had either normal-looking or slightly deformed flowers and anthers; the latter contained no pollen grains, but only a few degenerated structures (Cardi et al. 1993a; Nyman and Waara 1997; Bastia et al. 2000a). They showed precocious breakdown of tapetum, and a number of meiotic abnormalities, such as complete omission of meiosis-II, partial division, and irregular cytokinesis. Defects in the organization of the cytoskeleton during meiosis were associated with the formation of monads, dyads, or triads, which eventually degenerated. In fact, only one male fertile hybrid derived from the same fusion combination produced functional tetrads (Alfano et al. 1997; Conicella et al. 1997; Alfano et al. 1998).

Male sterility of interspecific hybrids is a common phenomenon in various species combinations in *Solanum* spp., due to interactions between nuclear and cytoplasmic genes derived from different parental genotypes (Kaul 1988). Accordingly, sexual hybrids obtained by ploidy manipulation of *cmm* and *tbr* were male sterile only when the wild species was used as pollen donor (Novy and Hanneman 1991; Carputo et al. 1995). Hence, it could be hypothesized that male sterility in *cmm* (+) *tbr* hybrids could be due to interactions between nuclear and cytoplasmic factors of the wild and the cultivated species, respectively, and that the reversion to male fertility in a single regenerated plant depended on the sorting-out of the *tbr* factors (Cardi et al. 1993a, 1999a). This hypothesis was strengthened by the fact that the *tbr* genotype employed by

Nyman and Waara (1997) in the single fusion combination in which male fertile plants were recovered did not have a *tbr* cytoplasm. On the other hand, the male fertility of somatic hybrids reported by Laferriere et al. (1999) and by Chen et al. (1999b) could be due to the lack, in the *S. commersonii* genotype used, of the nuclear genes involved in such nuclear-cytoplasmic interactions.

Since genetic variation can be induced in organellar genomes by somatic hybridization and in vitro culture (Earle 1995), male fertile and male sterile hybrids were used in genetic and molecular analyses aimed to characterize the inheritance of the fertility phenotype and the composition of cytoplasmic organelle genomes (Scotti et al. 1998; Bastia et al. 1999a,b, 2000a,b; Cardi et al. 1999a). Maternal inheritance of the fertility phenotype was evidenced in the progenies derived by crossing or selfing male sterile and male fertile hybrids, respectively. Large genetic variability in chloroplast and mitochondrial DNA was evidenced within the somatic hybrid populations analyzed, and either random or preferential transmission of chloroplasts and specific mitochondrial genome regions was found in somatic hybrids. Although no clear association between any organelle gene and male sterility has been obtained so far, it could be demonstrated that male sterile hybrids preferentially inherited mitochondrial genes from the *tbr* fusion partner, whereas the fertile hybrid had a mitochondrial genome almost exclusively derived from the wild species.

Somatic hybrids were generally female fertile. Progenies could be derived by crossing somatic hybrids among them as well as with *tbr* varieties (Fig. 3), indicating the possibility of using somatic hybrids in genetic studies and in

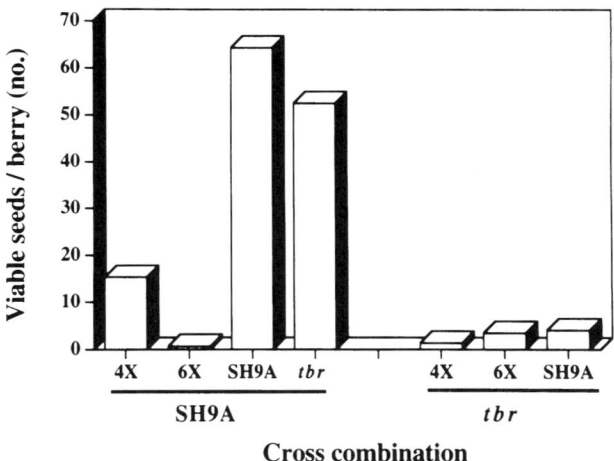

Fig. 3. Number of viable seeds per berry in different cross-combinations between SH somatic hybrids (Cardi et al. 1993b) and various *S. tuberosum* varieties. Genotypes *above* and *below horizontal lines* were used as female and male parents, respectively. *4X* eu- and hypotetraploid male-sterile somatic hybrids; *6X* eu- and hypohexaploid male sterile somatic hybrids; *SH9A* tetraploid male fertile somatic hybrid; *tbr* tetraploid varieties of *S. tuberosum*

Table 3. Resistance traits found in somatic hybrids produced by various *S. commersonii* (+) *S. tuberosum* fusion combinations

Fusion partners		Traits	Reference
cmm	tbr[a]		
PI 243503	DH 81-7-1463	Frost tolerance, cold acclimation capacity, tolerance to oxidative stress, resistance to blackleg, tuber soft rot, PVX	Cardi et al. (1993a); Carotenuto (1997); Carputo et al. (1997b); Parrella and Cardi (1999); Seppänen (2000)
PI 320266	PT56	Cold acclimation capacity, resistance to brown rot	Chen et al. (1999a,c) Laferriere et al. (1999)
PI 472834	Atl9	Frost tolerance, resistance to *Verticillium* wilt	Bastia at al. (2000a)
PI 472834	161:14, 67:9, 536:2	Frost tolerance, cold acclimation capacity, resistance to common scab	Nyman and Waara (1997)

[a] For origin of genotypes see Table 1.

breeding programmes aimed to introduce resistance and other useful genes from the wild into the cultivated species (Cardi et al. 1993a; Nyman and Waara 1997; Carputo et al. 1998; Chen et al. 1999b; Laferriere et al. 1999). Female fertility, however, was influenced by the EBN value and ploidy of the somatic hybrids used in crosses (Carputo et al. 1998; Laferriere et al. 1999). As a general rule, genotypes with the same EBN value can be crossed in *Solanum* spp., but it could be demonstrated that viable seeds could be obtained from reciprocal crosses between tetraploid somatic hybrids ($2n = 4x = 48$, 3EBN) and *Solanum tuberosum* varieties ($2n = 4x = 48$, 4EBN). The difference in parental EBN, however, was better tolerated when there was a maternal excess and therefore a significantly higher number of viable seeds per berry was obtained in *tbr* × somatic hybrids crosses than in the opposite direction (Carputo et al. 1998; Laferriere et al. 1999).

2.5.4 Stress Tolerance and Breeding

The transfer into the somatic hybrids of genes for resistance to various abiotic and biotic stresses has been demonstrated (Table 3). Frost tolerance and ability to acquire cold hardiness were generally analyzed by the ion leakage test conducted on excised leaflets (Cardi et al. 1993a; Nyman and Waara 1997; Chen et al. 1999a,c; Bastia et al. 2000a; Seppänen 2000). Results were confirmed with whole plants stressed at freezing temperatures in a growth chamber or in the field (Carotenuto 1997; Chen et al. 1999a). Hybrids generally showed a higher freezing tolerance and/or acclimation capacity than the *tbr* fusion partner. Similarly to what is found in *S. commersonii* × *S. cardiophyllum* sexual hybrids (Stone et al. 1993), both traits behaved as partially recessive, and in most cases the ability to cold acclimate was inherited to a greater extent than the frost

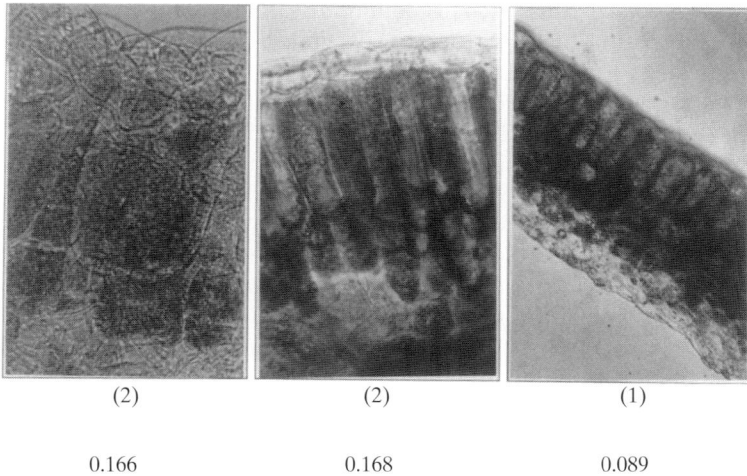

(2)   (2)   (1)

0.166   0.168   0.089

Fig. 4. Number of palisade layers (*upper part of the panel*) and stomatal index (*bottom line*) in *S. commersonii* (*left*), *S. tuberosum* (*right*), a somatic hybrid (*center*)

tolerance without acclimation (Cardi et al. 1993a, 1996; Chen et al. 1999a,c). However, LT_{50} values closer to that of the wild species for both acclimated and nonacclimated hybrid plants have been recently reported (Nyman and Waara 1997). Cold-tolerant hybrids showed also some physiological and morphological traits associated to cold tolerance, such as specific changes in phospholipid composition of membranes as well as higher stomatal index and number of palisade layers (Fig. 4; Cardi et al. 1996; Carotenuto and Cardi 1996; Kaur et al. 1996; Carotenuto 1997). A somatic hybrid between PI 243503 (*cmm*) and DH 81-7-1463 (*tbr*), and a population of progenies derived by selfing it and segregating for freezing tolerance, were also investigated for several molecular traits putatively involved in the plant response to freezing and oxidative stress (Seppänen et al. 1998, 2000; Seppänen 2000; Seppänen and Fagerstedt 2000).

Resistance to various diseases induced by bacteria, fungi, and viruses were also found in the somatic hybrids, indicating multiple transfer of resistance genes in most fusion combinations (Table 3). Thus, besides being cold-tolerant, hybrids between PI 243503 and DH 81-7-1463 proved to be resistant to PVX, and to blackleg and tuber soft rot induced by *Erwinia carotovora* (Carputo et al. 1997b; Parrella and Cardi 1999). However, the resistance to blackleg was probably derived by the dihaploid fusion partner (Carputo et al. 1997b). Similarly, cold-tolerant hybrids obtained in other fusion combinations were also resistant to brown rot by *Ralstonia solanacearum* (Laferriere et al. 1999), common scab by *Streptomyces scabiae* (Nyman and Waara 1997), and *Verticillium* wilt by *Verticillium dahliae* (Bastia et al. 2000a).

Progenies were obtained by selfing or intercrossing somatic hybrids and by backcrossing them with potato varieties. They were used for genetic studies as well as for breeding purposes, and genotypes with good agronomic characteristics and tolerance to abiotic and biotic stresses could be selected (Frusciante et al. 1993; Cardi et al. 1996; Carotenuto 1997; Nyman and Waara 1997; Basile et al. 1998; Cardi et al. 1999b; Chen et al. 1999b; Mazzei et al. 1999; Parrella and Cardi 1999; Carputo et al. 2000; Consiglio 2000). Variation among backcross progenies was also found for glycoalkaloid content, suggesting the possibility of selecting with relative ease improved genotypes showing acceptable nutritional characteristics (Esposito 1999; Esposito et al. 2000). In order to facilitate genetic analysis and the use of somatic hybrids in breeding, the production of dihaploids was attempted by interspecific crosses with parthenogenesis-inducer genotypes of *S. phureja* or by anther culture. In both cases, however, the frequency of dihaploids recovered was rather low (Bastia et al. 1996) and needs to be increased for a practical application of this technology.

By contrast with symmetric somatic hybrids, asymmetric hybrids produced between cv. Matilda and X-irradiated *S. commersonii* were sensitive to common scab and developed symptoms of chilling stress during cold acclimation (Nyman and Waara 1997). Thus, although more experiments with other fusion combinations should be performed, they did not show much agronomic interest so far.

3 Summary and Conclusion

Somatic hybridization has been successfully used to overcome sexual incongruity barriers between the frost-tolerant wild species *S. commersonii* and the cultivated potato *S. tuberosum*. Various genotypes of both species were fused by AC/DC electrofusion procedures. The latter gave significantly better results than chemophysical methods based on combined effects of PEG and DC field. Either fused protoplasts were mass-cultured and heterokaryons selected on the basis of growth characteristics in vitro or fusion products were preselected by flow cytometric cell sorting before culture. Hybrid plants were regenerated at a high frequency using established procedures for parental protoplasts. Confirmation of nuclear hybridity was achieved by isozyme, RAPD, and SSCP analyses.

Regenerated somatic hybrids were characterized for various aspects related to cytology, morphology, fertility, stress tolerance, and breeding. Most of the regenerated plants had the expected $4x$ ploidy level. However, (hypo)-hexaploid and a few (hypo)-octoploid or cytochimeric plants were also regenerated. Ploidy level was ascertained by chloroplast and/or chromosome counts and determination of nuclear DNA content by flow cytometry. Somatic hybrids were usually male sterile, probably as a consequence of nuclear-cytoplasmic interactions between the *cmm* and *tbr* genomes. Some male fertile plants, however, were regenerated in some cases, and they were used in genetic and molecular analyses. Large variability was evidenced in chloroplast and mitochondrial genomes of somatic hybrids. Progenies were obtained by selfing or crossing male sterile hybrids either with fertile somatic hybrids or with potato varieties. In both primary hybrids and in their progenies, several resistances to abiotic and biotic stresses were recovered. Generally, multiple gene transfer was achieved and thus genotypes were resistant to more than one type of stress or disease.

Hence, somatic hybridization proved to be not only a useful method to combine genomes of two incongruent species, such as *S. commersonii* and *S. tuberosum*, but also to generate, at both the nuclear and cytoplasmic level, useful variability for genetic analyses and breeding of common potato.

4 Protocol

Different protocols for protoplast culture and regeneration can be used depending on the genotypes employed. However, the following procedures can be adopted with a fairly large number of genotypes.

Shoot cultures for protoplast isolation are usually grown on MS medium (Murashige and Skoog 1962) plus 1–3% sucrose. Low sucrose concentrations, low temperatures (20 °C) and medium light intensity (50–100 $\mu E\,m^{-2}\,s^{-1}$ with 14–16 h photoperiod) are generally optimal for growth of shoots as well as for protoplast yield and plating efficiency. In some cases, the addition of an ethylene antagonist like silver nitrate at $2\,mg\,l^{-1}$ can increase leaf size and protoplast yield. A preincubation of shoot cultures for 24–48 h at 4–8 °C and/or in the dark can be beneficial for plating efficiency of protoplasts. In order to bleach shoot cultures, $3\,mg\,l^{-1}$ of the herbicide SAN 9789 (Sandoz) are added to the culture medium. In this case, the sucrose concentration is increased to 3%.

Protoplasts from leaves of green in vitro plants or from whole bleached shoot cultures are isolated with an overnight digestion with $10\,g\,l^{-1}$ Cellulase Onozuka-R10 and $2\,g\,l^{-1}$ Macerozyme R10 in half-strength (HS) V-KM medium (Bokelmann and Roest 1983), 0.205 M glucose, and 0.205 M mannitol. After enzyme digestion, protoplasts are washed twice in HS V-KM salts plus 0.24 M NaCl. The pellet is resuspended in 0.43 M sucrose, on the top of which 0.5 ml of fusion medium (0.44 M mannitol, 0.8 mM $CaCl_2$) are layered. Protoplasts collected at the interface are washed once in fusion medium and finally resuspended at $4.8 \times 10^5\,ml^{-1}$. Bleached protoplasts are stained by adding $3-5\,mg\,l^{-1}$ FDA to the enzyme solution. To improve their yield, the enzyme mixture with digested protoplasts is mixed 1:1 with a solution of Percoll and 0.44 M mannitol. After centrifugation, protoplasts are floated again on sucrose solution as usual and finally resuspended in fusion medium. Parental protoplasts are mixed 1:1 and kept on ice prior to fusion.

Electrofusion is carried out in a four-compartment chamber, in which 800 µl of protoplasts can be processed at once. Protoplasts are aligned with an AC field of $73-100\,V\,cm^{-1}$ at 1 MHz. When short chains are formed, they are fused with usually 3 DC pulses (50 µs each) at $2\,kV\,cm^{-1}$. After the DC pulses, the AC field is slowly lowered to zero, protoplasts are diluted with culture medium (HS V-KM medium, 0.17 M glucose, 0.17 mannitol, $0.1\,mg\,l^{-1}$ NAA, $0.5\,mg\,l^{-1}$ ZEA, $250\,mg\,l^{-1}$ cefotaxime), and left to rest for 30–60 min. Afterwards, the heterologous fusion frequency is estimated by counting the proportion of protoplasts showing the dual fluorescence of fluorescein and chlorophyll. Agarose is added to the cultures at a final concentration of 0.2% and plates are incubated in dim light at 26 °C. After a few days, cultures are checked for division and, if needed, diluted 1:3 with fresh culture medium.

After 3–4 weeks, microcalli are transferred onto solid callus growth medium (MS salts and organics, 1% sucrose, 0.2 mg l^{-1} NAA, 1 mg l^{-1} BAP). After 2 more weeks, calli are subcultured on shoot-induction medium (MS salts and organics, 1% sucrose, 0.01 mg l^{-1} NAA, 1 mg l^{-1} ZEA), and after 2.5 weeks on shoot-elongation medium (MS salts and organic constituents, 1% sucrose, 0.25 mg l^{-1} BAP, 0.1 mg l^{-1} GA$_3$). For some genotypes, better results are obtained with a modified regeneration procedure based on Cul M and DH media (Haberlach et al. 1985) in place of the callus-growth and shoot-induction media reported above, followed by SP medium (Austin and Cassels 1983) in place of the shoot-elongation medium. Elongated shoots are transferred on MS medium plus 1–3% sucrose for rooting.

Acknowledgments. Thanks are due to Dr. C. Allen (USA), Dr. J.P. Helgeson (USA), Dr. M. Nyman (Sweden), and Dr. S. Waara (Sweden) for providing preprints, and to Dr. K.S. Ramulu (The Netherlands), Dr. D. Carputo (Italy), Dr. C. Conicella (Italy), and Prof. L. Monti (Italy) for critically reading the manuscript. The financial support of the Italian Ministry of Agriculture, Project Biotecnologie Vegetali is acknowledged. CONTRIBUTION NO. 211 FROM CNR-IMOF

References

Alfano F, Bastia T, Cardi T, Errico A, Conicella C (1997) Cytological events involved in the cytoplasmic male sterility of interspecific somatic hybrids in *Solanum* spp. Atti Accad Fisiocrit Siena Ser XV – Suppl XVI:61–62
Alfano F, Cammareri M, Carputo D, Errico A, Conicella C (1998) The role of the cytoskeleton in potato meiosis. Int Symp Breeding Research on Potatoes, Groß Lüsewitz, June 23–26, pp 1–2
Austin S, Cassels AC (1983) Variation between plants regenerated from individual calli produced from separated potato stem callus cells. Plant Sci Lett 31:107–114
Bajaj YPS (1987) Biotechnology and 21st century potato. In: Bajaj YPS (ed) Biotechnology in agriculture and forestry, vol 3. Potato. Springer, Berlin Heidelberg New York, pp 3–22
Basile B, Carputo D, Zoina A, Monti L, Cardi T (1998) *Solanum commersonii* (+) *Solanum tuberosum* somatic hybrids: fertility in inter-EBN backcrosses and evaluation of progenies. Int Symp Breeding Research on Potatoes, Groß Lüsewitz, June 23–26, pp 14–15
Bastia T, Hasani A, Carotenuto N, Pelosi M (1995) Raddoppiamento somatico in vitro del livello di ploidia in specie selvatiche diploidi (2n = 2x = 24, EBN = 1) nel genere *Solanum*. XXXIX Annu Meet Italian Society of Agricultural Genetics, Vasto Marina, Italy, September 27–30, pp 167–168
Bastia T, Cardi T, Frusciante L (1996) Production of haploids from *Solanum commersonii* (+) *S. tuberosum* somatic hybrids by interspecific crosses or anther culture. XL Annu Meet Italian Society of Agricultural Genetics, Perugia, Italy, September 18–21, 177 pp
Bastia T, Scotti N, Cardi T (1999a) Universal primers for organelle DNA analysis in *Solanum* and *Brassica* spp. somatic hybrids. XLIII Annu Meet Italian Society of Agricultural Genetics, Molveno, Trento, Italy, September 22–25, 148 pp
Bastia T, Scotti N, Monti L, Earle ED, Cardi T (1999b) Genetic and molecular analysis of male fertility and cytoplasmic DNA variation in interspecific *Solanum* spp. somatic hybrids. In: Altman A, Ziv M, Izhar S (eds) Plant Biotechnology and in vitro biology in the 21st Century. Kluwer, Dordrecht, pp 105–108
Bastia T, Carotenuto N, Basile B, Zoina A, Cardi T (2000a) Induction of novel organelle DNA variation and transfer of resistance to frost and *Verticillium* wilt in *Solanum tuberosum* through somatic hybridization with 1EBN *S. commersonii*. Euphytica 116:1–10
Bastia T, Scotti N, Cardi T (2000b) Organelle DNA analysis of *Solanum* and *Brassica* somatic hybrids by PCR with "Universal Primers". Theor Appl Genet (in press)
Bokelmann GS, Roest S (1983) Plant regeneration from protoplasts of potato (*Solanum tuberosum* cv. Bintje). Z Pflanzenphysiol 109:259–265
Bradshaw JE, Mackay GR (eds) (1994) Potato genetics. CAB International, Wallingford

Cardi T (1998) Multivariate analysis of variation among *Solanum commersonii* (+) *S. tuberosum* somatic hybrids with different ploidy levels. Euphytica 99:35–41

Cardi T (1999) Genetic transformation of *Solanum commersonii* Dun. In: Bajaj YPS (ed) Biotechnology in agriculture and forestry, vol 45. Transgenic medicinal plants. Springer, Berlin Heidelberg New York, pp 312–326

Cardi T, Puite KS, Ramulu KS (1990) Plant regeneration from mesophyll protoplasts of *Solanum commersonii* Dun. Plant Sci 70:215–221

Cardi T, D'Ambrosio F, Consoli D, Puite KJ, Ramulu KS (1993a) Production of somatic hybrids between frost-tolerant *Solanum commersonii* and *S. tuberosum*: characterization of hybrid plants. Theor Appl Genet 87:193–200

Cardi T, Puite KJ, Ramulu KS, D'Ambrosio F, Frusciante L (1993b) Production of somatic hybrids between frost-tolerant *Solanum commersonii* and *S. tuberosum*: protoplast fusion, regeneration and isozyme analysis. Am Potato J 70:753–764

Cardi T, Carotenuto N, Garreffa P, Bastia T, Carputo D, Frusciante L, Monti L (1996) Somatic hybridization between 1EBN *Solanum commersonii* and 2EBN *S. tuberosum*: production, characterization, and utilization of hybrid plants. 13th Triennial Conf European Association for Potato Research, Veldhoven, The Netherlands, July 14–19, pp 126–127

Cardi T, Bastia T, Monti L, Earle ED (1999a) Organelle DNA and male fertility variation in *Solanum* spp. and interspecific somatic hybrids. Theor Appl Genet 99:819–828

Cardi T, Bastia T, Scotti N, Earle ED, Monti L (1999b) Interspecific somatic hybridization in *Solanum* spp.: a means to induce novel variability in the cytoplasmic genomes for genetic studies and breeding. 14th Triennial Conf European Association for Potato Research, Sorrento, Italy, May 2–7. Assessorato Agricoltura Regione Campania, Naples, pp 124–125

Carotenuto N (1997) Valutazione agronomica e caratterizzazione di cloni derivanti da ibridazione somatica tra *Solanum commersonii* e *S. tuberosum*. PhD Thesis, Dept of Agronomy and Plant Genetics, Univ of Naples Federico II, Portici, 96 pp

Carotenuto N, Cardi T (1996) Evaluation of frost tolerance of somatic hybrids *Solanum commersonii* (+) *Solanum tuberosum*. XL Annu Meet Italian Society of Agricultural Genetics, Perugia, Italy, September 18–21, pp 218–219

Carotenuto N, Martin R, Mok MC, Mok DWS (1996) SSCP analysis for chromosomal characterization of somatic hybrids in the *Solanum* genus. XL Annu Meet Italian Society of Agricultural Genetics, Perugia, Italy, September 18–21, 58 pp

Carputo D, Cardi T, Frusciante L, Peloquin SJ (1995) Male fertility and cytology of triploid hybrids between tetraploid *Solanum commersonii* ($2n = 4x = 48$, 2EBN) and *phureja-tuberosum* haploid hybrids ($2n = 2x = 24$, 2EBN). Euphytica 80:123–129

Carputo D, Barone A, Cardi T, Sebastiano A, Frusciante L, Peloquin SJ (1997a) Endosperm balance number manipulation for direct in vivo germplasm introgression to potato from a sexually isolated relative (*Solanum commersonii* Dun.). Proc Natl Acad Sci USA 94:12013–12017

Carputo D, Cardi T, Speggiorin M, Zoina ALF (1997b) Resistance to blackleg and tuber soft rot in sexual and somatic interspecific hybrids with different genetic background. Am Potato J 74:161–172

Carputo D, Garreffa P, Mazzei M, Monti L, Cardi T (1998) Fertility of somatic hybrids *Solanum commersonii* ($2x$, 1BN) (+) *S. tuberosum* haploid ($2x$, 2EBN) in intra- and inter-EBN crosses. Genome 41:776–781

Carputo D, Basile B, Cardi T, Frusciante L (2000) *Erwinia* resistance in backcross progenies of *Solanum tuberosum* × *S. tarijense* and *S. tuberosum* (+) *S. commersonii* hybrids. Potato Res 43:135–142

Chen Y-KH, Bamberg JB, Palta JP (1999a) Expression of freezing tolerance in the interspecific F_1 and somatic hybrids of potatoes. Theor Appl Genet 98:995–1004

Chen Y-KH, Palta JP, Bamberg JB (1999b) Freezing tolerance and tuber production in selfed and backcross progenies derived from somatic hybrids between *Solanum tuberosum* L. and *S. commersonii* Dun. Theor Appl Genet 99:100–107

Chen Y-KH, Palta JP, Bamberg JB, Kim H, Haberlach GT, Helgeson JP (1999c) Expression of nonacclimated freezing tolerance and cold acclimation capacity in somatic hybrids between hardy wild *Solanum* species and cultivated potatoes. Euphytica 107:1–8

Conicella C, Genualdo G, Lucia R, Ramulu KS, Cardi T (1997) Early tapetal degeneration and meiotic defects are involved in the male sterility of *Solanum commersonii* (+) *S. tuberosum* somatic hybrids. Theor Appl Genet 95:609–617

Consiglio F (2000) Introgressione mediante tecniche innovative di geni utili per caratteristiche agronomiche e tolleranza al freddo dalla specie selvatica *Solanum commersonii* Dun. nella patata coltivata (*Solanum tuberosum* L.). PhD Thesis, Dept of Agronomy and Plant Genetics, Univ of Naples Federico II, Portici, 62 pp

Earle ED (1995) Mitochondrial DNA in somatic hybrids and cybrids. In: Levings CS III, Vasil I (eds) The molecular biology of plant mitochondria. Kluwer, Dordrecht, pp 557–584

Ehlenfeldt MK, Hanneman RE Jr (1988) The transfer of the synaptic gene (*sy-2*) from 1EBN *Solanum commersonii* Dun. Euphytica 37:181–187

Esposito F (1999) Caratterizzazione chimico-fisica di genotipi di patata ottenuti con tecniche innovative di miglioramento genetico. MSc Thesis, Dept of Food Science, Univ of Naples Federico II, Portici, 59 pp

Esposito F, Fogliano V, Cardi T, Carputo D, Filippone E (2000) Chemical composition of genetically improved potatoes. J Agric Food Chem (submitted)

Frusciante L, Mazzei M, Carputo D, Cardi T (1993) Phenotypical analysis of a selfed progeny from tetraploid *S. commersonii* – *S. tuberosum* somatic hybrid. 12th Triennial Conf European Association for Potato Research, Paris, France, July 18–23, pp 97–98

Genualdo G, Errico A, Tiezzi A, Conicella C (1998) α-Tubulin and F-actin distribution during microsporogenesis in 2*n* pollen producer of *Solanum*. Genome 41:636–641

Haberlach GT, Cohen B, Reichert N, Baer M, Towill L, Helgeson JP (1985) Isolation, culture and regeneration of protoplasts of potato and several related species. Plant Sci 39:67–74

Hanneman RE Jr, Bamberg JB (1986) Inventory of tuber-bearing *Solanum* species, vol 533. Wis Agric Exp Stn Bull, 216 pp

Hawkes JG (1994) Origins of cultivated potatoes and species relationships. In: Bradshaw JE, Mackay GR (eds) Potato genetics. CAB International, Wallingford, pp 3–42

Hosaka K, Ogihara Y, Matsubayashi M, Tsunewaki K (1984) Phylogenetic relationship between the tuberous *Solanum* species as revealed by restriction endonuclease analysis of chloroplast DNA. Jpn J Genet 59:349–369

Hunt GJ, Helgeson JP (1989) A medium and simplified procedure for growing single cells from *Solanum* species. Plant Sci 60:251–257

Kaul MLH (1988) Male sterility in higher plants. Springer, Berlin Heidelberg New York

Kaur N, Palta JP, Greene D, Wykle RL, Workmaster BA, Karlsson B (1996) Plasma membrane phospholipid molecular species associated with genetic variations in cold acclimation ability in frost-sensitive and -tolerant *Solanum* species and their somatic hybrids. Annu Meet American Society of Plant Physiologists, San Antonio, Texas, July 27–31. Plant Physiol 111 (2 Suppl), 133 pp

Kim H, Choi SU, Chae MS, Wielgus SM, Helgeson JP (1993) Identification of somatic hybrids produced by protoplast fusion between *Solanum commersonii* and *S. tuberosum* haploid. Korean J Plant Tissue Cult 20:337–344

Laferriere LT, Helgeson JP, Allen C (1999) Fertile *Solanum tuberosum* + *S. commersonii* somatic hybrids as sources of resistance to bacterial wilt caused by *Ralstonia solanacearum*. Theor Appl Genet 98:1272–1278

Matsubayashi M (1991) Phylogenetic relationships in the potato and its related species. In: Tsuchiya T, Gupta PK (eds) Chromosome engineering in plants: genetics, breeding, evolution, part B. Elsevier, Amsterdam, pp 93–118

Mazzei M, Frusciante L, Cardi T (1999) Multivariate analysis of variation in field-grown *Solanum commersonii* (+) *S. tuberosum* somatic hybrids and progenies derived by selfing. 14th Triennial Conf European Association for Potato Research, Sorrento, Italy, May 2–7. Assessorato Agricoltura Regione Campania, Naples, pp 600–601

Murashige T, Skoog F (1962) A revised medium for rapid growth and bioassay with tobacco tissue cultures. Physiol Plant 15:473–497

Novy RG, Hanneman RE Jr (1991) Hybridization between Gp. *tuberosum* haploids and 1EBN wild potato species. Am Potato J 68:151–169

Nyman M, Waara S (1997) Characterisation of somatic hybrids between *Solanum tuberosum* and its frost-tolerant relative *Solanum commersonii*. Theor Appl Genet 95:1127–1132

Parrella G, Cardi T (1999) Transfer of a new PVX resistance gene from *Solanum commersonii* to *S. tuberosum* through somatic hybridization. J Genet Breed 53:359–362

Perl A, Aviv D, Galun E (1991) Nuclear-organelle interaction in *Solanum*: interspecific cybridizations and their correlation with a plastome dendrogram. Mol Gen Genet 228:193–200

Polgár Z, Preiszner J, Dudits D, Fehér A (1993) Vigorous growth of fusion products allows highly efficient selection of interspecifc potato somatic hybrids: molecular proofs. Plant Cell Rep 12:399–402

Ramulu KS, Dijkhuis P, Roest S (1989) Patterns of phenotypic and chromosome variation in plants derived from protoplast cultures of monohaploid, dihaploid and diploid genotypes and in somatic hybrids of potato. Plant Sci 60:101–110

Scotti N, Bastia T, Monti L, Earle ED, Cardi T (1998) Composition and expression of the mtDNA in male-fertile and male-sterile interspecific *Solanum* spp. somatic hybrids. In: Møller IM, Gardeström P, Glimelius K, Glaser E (eds) Plant mitochondria: from gene to function. Backhuys, Leiden, pp 95–98

Seppänen MM (2000) Characterization of freezing tolerance in *Solanum commersonii* (Dun.) with special reference to the relationship between freezing and oxidative stress. PhD Thesis, Dept of Plant Production, Univ of Helsinki, Helsinki, 55 pp

Seppänen MM, Fagerstedt K (2000) The role of superoxide dismutase activity in response to cold acclimation in potato. Physiol Plant 108:279–285

Seppänen MM, Majaharju M, Somersalo S, Pehu E (1998) Freezing tolerance, cold acclimation and oxidative stress in potato. Paraquat tolerance is related to acclimation but is a poor indicator of freezing tolerance. Physiol Plant 102:454–460

Seppänen MM, Cardi T, Borg Hyökki M, Pehu E (2000) Characterization and expression of cold-induced glutathione *S*-transferase in freezing-tolerant *Solanum commersonii*, sensitive *S. tuberosum* and their interspecific somatic hybrids. Plant Sci 153:125–133

Stone JM, Palta JP, Bamberg JB, Weiss LS, Harbage JF (1993) Inheritance of freezing resistance in tuber-bearing *Solanum* species: evidence for independent genetic control of nonacclimated freezing tolerance and cold acclimation capacity. Proc Natl Acad Sci USA 90:7869–7873

Waara S, Glimelius K (1995) The potential of somatic hybridization in crop breeding. Euphytica 85:217–233

Waara S, Wallin A, Ottosson A, Eriksson T (1991) Factors promoting sustained divisions of mesophyll protoplasts isolated from dihaploid clones of potato (*Solanum tuberosum* L.) and a cytological analysis of regenerated plants. Plant Cell Tissue Organ Cult 27:257–265

Waara S, Nyman M, Johannisson A (1998) Efficient selection of potato heterokaryons by flow cytometric sorting and the regeneration of hybrid plants. Euphytica 101:293–299

Ye GN, Earle ED (1991) Effect of cellulases on spontaneous fusion of maize protoplasts. Plant Cell Rep 10:213–216

Zuba M, Binding H (1989) Isolation and culture of potato protoplasts. In: Bajaj YPS (ed) Biotechnology in agriculture and forestry, vol 8. Plant protoplasts and genetic engineering. Springer, Berlin Heidelberg New York, pp 124–146

II.13 Somatic Hybridization Between *Solanum tuberosum* L. (Potato) and *Solanum phureja*

S. MILLAM and P. DAVIE

1 Introduction

1.1 Morphology, Importance and Distribution of These Plants

Solanum tuberosum is the fourth most important food crop in the world, after rice, maize and wheat. The common potato is of ancient origin, originally domesticated as a range of diploid types in South America. Spontaneous tetraploid lines arose in the Lake Titicaca area and these were brought to Europe during the latter part of the 16th century. The modern-day cultivated potato is *Solanum tuberosum* L, a tetraploid species with a chromosome number of 48. Hawkes (1990) recognised 235 species belonging to the tuber-bearing section of the genus *Solanum* (section petota). Seven of these are cultivated, and the remaining species divided into a diverse hierarchy of sections and series. The group is widely distributed from Southwest USA through Mexico, Central and South America. *Solanum phureja* Juz.et. Buk. is a member of the series Tuberosa, chromosome number 24, EBN number 2. A wide range of accessions of this species exists, with ca. 200 held in the Commonwealth Potato Collection at the Scottish Crop Research Institute (*http://www.scri.sari.ac.uk/*). These mainly originate from the Narino, Merida and Sogomoso provinces of Colombia, the Merida province of Venezuela, Crachi and Pichina in Ecuador and La Paz in Bolivia. Within these accessions there are lines with known resistance to a range of pathogens, including *Phytopthora infestans* (late blight), *Pseudomonas solanacearum* (bacterial wilt), *Erwinia carotovera* (soft rot), potato virus Y and *Meloidgyne incognita* (root-knot nematode). *S. phureja* is also utilised in potato-breeding programmes and genetic studies as an inducer for dihaploid production. There have been recent reports, however, that such an approach may cause problems, as DNA originating from *S. phureja* can affect the morphology of the potato dihaploids (Allainguillaume et al. 1997). There has also been recent interest in the commercialisation of selected *S. phureja* lines as a crop in the UK. Despite problems with dormancy and yield, the excellent cooking characteristics of the tubers are considered highly marketable.

Scottish Crop Research Institute Invergowrie, Dundee DD2-5DA, United Kingdom

1.2 Rationale for Somatic Hybridization

The cultivated potato is a tetraploid species ($2n = 4x = 48$) and, due to this, the segregation ratios of many important characters in the progeny of sexual crosses can be very complex. Conventional breeding strategies can involve initial selections from over 100000 seedlings and take 11–15 years to produce a new cultivar. Most of the important agronomic characters of a potato cultivar such as yield, tuber quality, starch and dry matter content, and also many of the pathogen-resistance traits are of a polygenic nature, thus precluding most existing *Agrobacterium*-mediated transformation systems as a means of rapid cultivar improvement. Furthermore, the public acceptance of such gene transfer strategies is a matter of current debate and controversy.

1.3 Review of Previous Work

Potato has proved an amenable species for somatic hybridization studies, and a wide range of germplasm has been utilised for a number of fundamental and applied studies (for a review see Millam et al. 1995). The main strategies for potato protoplast fusion fall into three main categories. Fusion of the tetraploid *tuberosum* with wild relatives (diploid or tetraploid), dihaploid:dihaploid fusion to recreate the tetraploid status, and thirdly, asymmetric approaches. Specific references to the use of *S. phureja* as a fusion partner have been related to all three approaches.

Protoplasts from diploid *S. tuberosum* and *S. phureja* were electrofused followed by selection of the heterokaryons using a micromanipulator, and 18 putative hybrid plantlets obtained according to the report of Puite et al. (1986). A more detailed study by Pijnacker et al. (1987) revealed the elimination of *S. phureja* nucleolar chromosomes in *S. tuberosum* + *S. phureja* somatic hybrids. Preferential elimination of nucleolar chromosomes of *S. phureja* and in somatic hybrids involving diploid *S. tuberosum* was further reported by Pijnacker et al. (1989). Patterns of phenotypic and chromosome variation in plants derived from protoplast cultures of monohaploid, dihaploid and diploid genotypes and in somatic hybrids of potatoes and *S. phureja* were reported by Ramulu et al. (1989). Using cytological and DNA cytophotometric analyses, the occurrence of instability in nuclear processes during the early stages of dedifferentiation of protoplasts and callis was detected. Protoplast culturability (defined by the authors as the ability of a genotype to develop callis from cultured protoplasts) was studied in *S. phureja* by Cheng and Veilleux (1991). Tetraploid potato hybrids derived from the electrofusion of *S. tuberosum* and *S. phureja* were field-trialled and the results reported by Mattheij and Puite (1992). Ten different fusion combinations were used and six resultant hybrids field-tested for a range of agronomic traits. One of the hybrids had a tuber yield three times that of the potato variety Bintje parent, and the authors added that the technique of somatic hybridization was included in commercial breeding programmes in The Netherlands. An interesting practical application of somatic hybridization was described by Craig et al. (1994), who reported a study of the expression of reducing sugar accumulation in interspecific somatic hybrids of *S. tuberosum* cv. Record (the most important crisping cultivar in Europe at the time of this work) and selected accessions of *S. phureja*. Interspecific hybrids were evaluated at the morphological and molecular level, indicating that the regenerant population were mainly asymmetric hybrids which had undergone elimination of most of the *S. phureja* chromosomes. Despite this apparent elimination, the somatic hybrids resembled the *S. phureja* parent in terms of reducing sugar content. The isolation, culture and regeneration of protoplasts of *S. tuberosum* and *S. phureja* were also reported by Carrasco et al. (1994), but regenerants only reached a callus stage.

2 Somatic Hybridization

2.1 Isolation of Protoplasts

2.1.1 Source Material

Probably the simplest method of maintaining material for protoplast isolation is in the form of in vitro shoot cultures. This enables a constant supply of uniform, sterile material. From our own findings, stock plant material was maintained on either medium HelMS (Haberlach et al. 1985), nodal medium (NM) of Coleman et al. (1991) or MS2BA as described by Foulger and Jones (1986), and the effects attributable to each media type on mean protoplast yield assessed. No significant differences in protoplast yield were found from any of the lines tested.

However, it was found that for *S. tuberosum* cv. Pentland Squire the initial medium did have an effect on regeneration potential, with media MS2BA being favoured. Mattheij and Puite (1992) also maintained source material on MS medium with 3% sucrose, but reduced sucrose to 1% prior to protoplast isolation. These authors also bleached some of the source material by treatment with norflurazon, and treated protoplasts with FDA in order to facilitate heterokaryon differentiation. Craig et al. (1994) also used shoot material on a half-strength Murashige and Skoog plus sucrose medium for their *S. tuberosum* cv. Record parent. However, these authors initiated a cell suspension culture of *S. phureja* from stem and leaf sections which was used for the iso-

Table 1. Effect of shoot culture medium on mean viable (assessed using FDA) protoplast yield ($\times 10^5 \text{g}^{-1}$ FW) of four *S. tuberosum* cultivars and *S. phureja* accession 4471

Medium	Brodick	P. Squire	Desiree	GL 76B2	*S. phureja*
HelMS	6.7 ± 1.3	5.2 ± 0.8	7.4 ± 0.2	6.4 ± 1.2	7.2 ± 1.3
NM	4.1 ± 1.1	7.0 ± 0.7	5.8 ± 0.5	6.6 ± 2.0	6.7 ± 1.2
MS2BA	8.6 ± 2.1	4.8 ± 0.8	3.8 ± 0.2	6.4 ± 0.8	8.6 ± 0.9

lation of protoplasts for fusion. Their procedure consisted of taking 5-mm leaf strips and 5–10-mm stem internode explants. These were cultured on a callus-inducing medium, composed of MS30 supplemented with 27.0 µM NAA and 1.0 µM kinetin, solidified with 8.0 g l^{-1} agar. The plates were cultured in the dark at 22 °C for 6 weeks. The resultant friable callus was excised, finely macerated and cultured in 50 ml of the same callus induction medium, lacking agar, in 250-ml Erlenmeyer conical flasks. These were placed on an orbital shaker at 150 rpm in a 16-h light:8-h dark regime of 80 µE m^2 s^{-1}. To enable the maintenance of cultures in a logarithmic phase of division for protoplast isolation, the cultures were subcultured every 7 days.

2.1.2 Isolation Conditions

In the work of Mattheij and Puite (1992), leaf protoplasts were isolated according to the methods of Puite et al. (1988). From our own work, we recommend that leaf strips (2 mm) of *S. tuberosum* should be plasmolysed in the solution of Nelson et al. (1983) supplemented with the major salts of medium A (Sheperd and Totten 1977), pH 5.8. These strips should, after 30-min plasmolysis, be transferred to an enzyme solution according to the methods of Foulger and Jones (1986) modified to employ 1% Onozuka R10 Cellulase and 0.3% Onozuka Macerozyme R10. We recommend that this solution be adjusted to 575 mOsm kg^{-1} H$_2$O. The enzymes were obtained from Yakult Pharmaceutical Company Ltd., Tokyo, Japan. Under optimum isolation conditions (22–25 °C, dark, and shaken at 80 rpm), a working population of protoplasts is released after 3–4 h. For suspension cultures of *S. phureja* a 3-day subculture is centrifuged at 800 rpm for 5 min and the pellet resuspended in 2.5% Cellulase R10, 1.0% Macerozyme R10, 0.5% Driselase in 0.5 M mannitol, solution filter-sterilised. The suspension is incubated overnight at 22 °C, dark, and shaken at 50 rpm. The resultant protoplasts are shown in Fig. 1.

2.2 Purification of Protoplasts

Interesting findings from our own work indicate that the method of purification of *Solanum* protoplasts has great significance for the efficiency of later stages of fusion, culture and regeneration. Empirically comparing (using the *S. tuberosum* cultivar Maris Piper) the mannitol and calcium chloride sedimentation method of Foulger and Jones (1986), the sucrose method of Shepard (1979) and the Percoll cushion method of Nelson et al. (1983) produced some interesting findings. The highest mean recovery of protoplasts was achieved using mannitol sedimentation and the least from the Percoll cushion method. However, there was also considerable debris from the sedimentation treatment alone. The sucrose treatment provided a particularly pure sample of intact protoplasts, similar to the methods employed by Mattheij and Puite (1992), where protoplasts were collected in 0.38 M mannitol layered over an 0.43 M sucrose solution. The Percoll sample had problems with chloroplast

Fig. 1. Protoplasts derived from cell suspension cultures of *S. phureja*. Derived from layer resultant from centrifugation as described. Size range of protoplasts 20–65 µM

contamination and a high incidence of misshapen cells. However, an FDA (fluoroscein diacetate) viability test revealed that the incidence of viable cells was actually highest in the Percoll sample (78%) and lowest in the sucrose (32%) sample. Purification of the cell suspension protoplasts was by a derived method. After filtration through a 100-µm stainless steel mesh, the sample was centrifuged at $80g$ for 8 min and the resultant pellet layered over 0.45 M sucrose, respun at $100g$ and the subsequent band collected and washed in 0.4 M mannitol.

2.3 Fusion of Protoplasts

The methods employed in our study were those reported by Craig et al. (1994), who mixed green mesophyll protoplasts derived from *S. tuberosum* and clear cell suspension-derived *S. phureja* protoplasts at a ratio of 1:1 in 0.5 M mannitol. The apparatus used was a Braun Biotech (Aylesbury, UK) multilamellar fusion chamber with parallel stainless steel electrodes spaced 1 mm apart. The chamber was connected to a Zimmerman Cell Fusion Instrument (GCA Corp, Chicago, USA) and protoplast aligned by the application of an alternating field of 1 Mhz and $40\,\text{V}\,\text{cm}^{-1}$, increasing to $120\,\text{V}\,\text{cm}^{-1}$ over 20s. Fusion was induced by two direct current pulses of 110 V and 35 µs duration, and directly after the fusion pulses the alignment field strength was returned

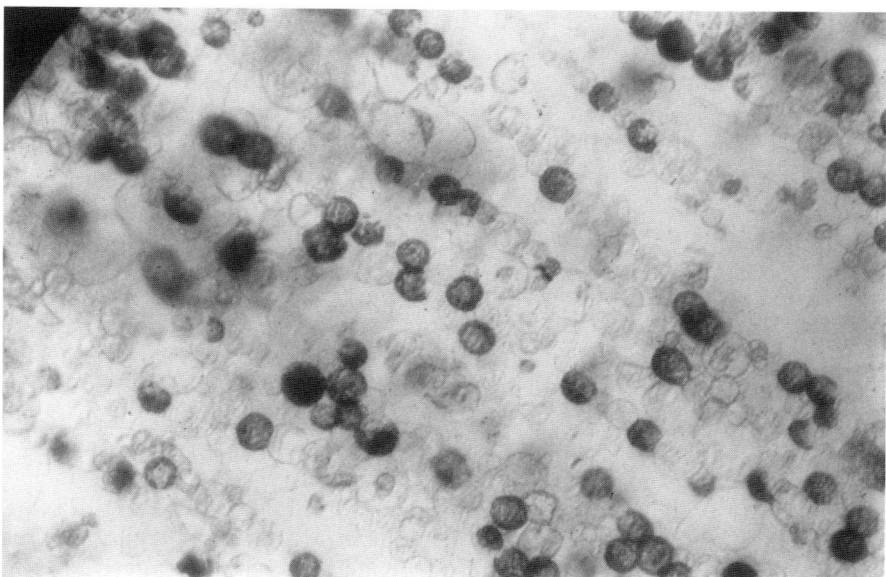

Fig. 2. Cell suspension protoplasts of *S. phureja* and mesophyll protoplasts of *S. tuberosum* cv. Pentland Squire aligned by the application of an alternating field of 1 Mhz and 40 V cm^{-1} over 20 s. Fusion was induced by two direct current pulses of 100 V and 35 µS duration. Directly after the fusion pulses, the alignment field strength was returned to 40 V cm^{-1} for 1 min to facilitate stability of fusion products

to 40 V cm^{-1} for 1 min to facilitate stability of fusion products (Fig. 2). Mattheij and Puite (1992) mixed green and bleached protoplasts (FDA stained) in a 1:1 or 2:1 ratio (due to the fragility of the green protoplasts) in 0.38 M mannitol, and performed electrofusion using a 0.8-ml four-compartment fusion chamber with an electrode distance of 3 mm. Protoplasts were aligned using an AC field of 100 V cm^{-1} at 1 MHz, DC pulses of 1300–2300 V cm^{-1} were applied, with 1–3 pulses of 50- to 100-µs duration.

Recent work (Millam et al. 1997) on limited chromosome transfer by asymmetrical fusion of commercial cultivars of potato with accessions of the Commonwealth Potato Collection, which produced several hundred fusion products for field trial within 1 year of commencing the project, utilised a chemical-based fusion approach. The fusion solution comprised 6 g PEG 1500, 1.79 g HEPES buffer in 20 ml of deionised water, pH 5.8, filter-sterilised.

2.4 Culture of Protoplasts

In our work the system of Craig et al. (1994) was used. This method was based on initial culture in the VKClg medium of Foulger and Jones (1986) followed by transfer to split-plates (Fish and Karp 1986), where the reservoir medium was supplemented with 1% activated charcoal. The antibiotic carbenicillin was originally added to the medium at 1 mg ml^{-1} to minimise bacterial contamination. The system was devised to preclude the regeneration of unfused *S. phureja* protoplasts. Mattheij and Puite (1992) used the V-KM culture medium (formulated by Bokelmann and Roest 1983 and derived from a combination of the V-47 medium of Binding 1974 and Kao and Michayluk 1975) at half-strength with 0.17 mannitol, 0.14M glucose, 0.1 mg l^{-1} NAA, 0.5 mg l^{-1} zeatin, 500 mg l^{-1} casein hydrolysate and 0.18 mg l^{-1} Cefotaxime. Fusion products could easily be identified in this work due to the simultaneous red and yellow fluorescence from chlorophyll and FDA. Fusion products were collected and transferred to a half-strength V-KM medium in the centre of a Falcon Dish (no. 3037), with the outer ring containing sterile water. Following a series of culture dilutions, callis were formed which were treated on callus-growth medium (2 weeks) shoot-induction medium (3 weeks) and shoot-elongation medium (Bokelmann and Roest 1983). Conversely, Cheng and Veilleux (1991) described methods for the regeneration of *S. phureja* protoplasts using a medium reported by Schumann and Koblitz (1983) modified to include 30.0 g l^{-1} glucose, 10.0 g l^{-1} sucrose, 50.0 g l^{-1} sorbitol, 1.25 mg l^{-1} NAA, 0.25 mg l^{-1} 2,4-D and 1.0 mg l^{-1} zeatin. The use of an agarose-embedding culture system was required in this protocol for reproducible protoplast derived-callus formation.

2.5 Assessments of the Hybrid Nature

2.5.1 Chromosome Counting

The basic cytological methods for root counts have been recently augmented from work in our laboratories, with particular reference to *S. phureja* (Braselton et al. 1997). The elimination of *S. phureja* nucleolar chromosomes in *S. tuberosum* + *S. phureja* somatic hybrids was studied in some detail by Pijnacker et al. (1987). These authors found that the elimination of two nucleolar chromosomes occurred preferentially, probably taking place during callus formation. Of the eight somatic hybrids selected for use in this study, the range of chromosomes present varied from 45–47 to 90–96, indicating degrees of hexaploidy and octoploidy.

2.5.2 Flow Cytometry

This method has proved valuable as a quick indicative guide to the gross cytological component of somatic hybrids, but has not, in our hands, proved discriminative enough for the identification of addition lines. Mattheij and Puite (1992) utilised such an approach for measurement of nuclear DNA content. They found in their somatic hybrid population of 28 plants that 15 were diploid, 54 tetraploid and 28 of a higher ploidy level. We have sourced a number of commercial companies that offer a flow cytometric service, analysing hybrid material for as little as $5 per sample.

2.5.3 Molecular Analysis

The range of molecular markers available for discriminating hybrids has increased enormously in recent years. RFLPs (restriction fragment length polymorphisms) have been utilised for characterising putative somatic hybrids. For example, Mattheij and Puite (1992) utilised such an approach, in order to provide further information on samples initially analysed by flow cytometry. The 54 tetraploid lines detected by flow cytometry were found, using RFLPs, to only contain 9 actual hybrids. However, RFLP technology can be relatively laborious and requires substantial amounts of plant material for analysis. The advent of PCR-based markers, which use smaller amounts of plant material, enables a more rapid and early test to be made. RAPD (randomly amplified polymorphic DNA) markers have been used, but several primers would have to be applied for unequivocal analysis. Our work and that of Craig et al. (1994) replicated the published methods of Baird et al. (1992). Waugh et al. (1992) utilised a similar system in the analysis of potato dihaploids generated after interspecific pollination of a tetraploid *S. tuberosum* cultivar by *S. phureja* inducer clones and concluded that the methodology was precise enough to allow the discrimination of bands arising from *S. phureja* alone. The major drawbacks of using RAPDs include problems with reproducibility and the comigration of heterologous bands on agarose gels showing misleading results, and also the relatively low levels of polymorphism associated with these dominant markers. More sophisticated methodology has recently been employed at SCRI for the detailed analysis of a range of somatic hybrid populations. Matthews et al. (1999) used 5' anchored simple sequence repeat primers to discriminate hybrids and simple sequence repeat primers were also used by Provan et al. (1996) to unambiguously demonstrate hybridity. Attempts to quantify the actual degree of hybridity using a refined dot-blot system were reported by Matthews et al. (1997). More detailed molecular analysis of ribosomal RNA genes in somatic hybrids between wild and cultivated *Solanum* species was reported by Harding and Millam (1999). Analysis of chromatin, nuclear DNA and organelle composition in somatic hybrids between *S. tuberosum* and *S. sanctae-rosae* was also reported by Harding and Millam (2000).

2.5.4 Morphological Characterisation

Many workers undertake preliminary analysis on somatic hybrids based on the morphology of intact plants. In the case of *S. tuberosum* + *S. phureja* putative hybrids, considerable phenotypic differences exist between the two parental types (see Fig. 3) for growth habit, leaf shape and foliage colour. Notably, *S. phureja* has purple flowers. For a commercial purpose, reducing sugars were assayed in the tubers of a hybrid population by Craig et al. (1994) but no significant differences were detected between the *S. phureja* parent and the hybrids. It was postulated that low reducing sugar was dominant to high reducing sugar accumulation in these particular samples. Sucrose levels in the

Fig. 3A–C. Field-grown *S. tuberosum* (**A**), *S. phureja* (**B**) and a somatic hybrid between the two species (**C**) showing intermediate morphology

tubers of this putative hybrid population were also used to assess the degree of hybridity, and detected a trend for sucrose expression towards the *S. tuberosum* parent, implying that low expression was dominant.

3 Summary and Conclusion

Compared to other *Solanum* species (notably *S. brevidens*) the reports of *S. phureja* as a direct fusion partner with *S. tuberosum* are relatively limited. The use of *S. phureja*-derived dihaploids for fusion, however, has been widely reported (Waara et al. 1991). The methodologies for successful isolation, fusion, culture and regeneration of somatic hybrids of *S. tuberosum* + *S. phureja* would appear to be well documented and reproducible. These species have been used to illustrate several approaches to the creation and identification of heterokaryons at early stages. To our knowledge, fusion material of this nature has also been extensively field-trialled and evaluated in the UK by commercial companies.

References

Allainguillaume J, Wilkinson MJ, Clulow S, Barr SNR (1997) Evidence that genes from the male parent may influence the morphology of potato dihaploids. Theor Appl Genet 94:241–248

Baird E, Cooper-Bland S, Waugh R, De Maine M, Powell W (1992) Molecular characterisation of inter- and intra-specific somatic hybrids of potato using randomly amplified polymorphic DNA (RAPD) markers. Mol Gen Genet 233:469–475

Binding H (1974) Regeneration of haploid and dihaploid plants from protoplasts of *Petunia hyrida*. Z Pflanzenphysiol 74:327–356

Bokelmann GS, Roest S (1983) Plant regeneration from protoplasts of potato (*Solanum tuberosum* cv. Bintje). Z Pflanzenphysiol 109:259–265

Braselton JP, Wilkinson MJ, Clulow SA (1997) Feulgen staining of intact plant tissue for confocal microscopy. Biotechnol Histochem 71:84–87

Carrasco A, Ruiz De Galarreta JI, Ritter E (1994) Isolation, culture and regeneration of protoplasts of *S. tuberosum* and *S. phureja*. Invest Agra Prod Prot Veg 9:347–357

Cheng J, Veilleux RJ (1991) Genetic analysis of protoplast culturability in *S. phureja*. Plant Sci 75:257–265

Coleman MC, Davie P, Vessey J, Powell W (1991) Intraclonal genetic variation for protoplast regeneration ability within *Solanum tuberosum* cv. Record. Ann Bot 67:459–461

Craig AL, Morrison I, Baird E, Waugh R, Coleman MC, Davie P, Powell W (1994) Expression of reducing sugar accumulation in interspecific somatic hybrids of potato. Plant Cell Rep 13:401–405

Fish N, Karp A (1986) Improvements in regeneration from protoplasts of potato and studies on chromosomal stability. Theor Appl Genet 72:405–412

Foulger D, Jones MGK (1986) Improved efficiency of genotype-dependent regeneration from protoplasts of important potato cultivars. Plant Cell Rep 5:72–76

Haberlach GT, Cohen BA, Reicher NA, Baer MA, Towill LE, Helgeson JP (1995) Isolation, culture and regeneration from protoplasts from potato and several related *Solanum* species. Plant Sci 39:67–74

Harding K, Millam S (1999) Analysis of ribosomal RNA genes in somatic hybrids between wild and cultivated *Solanum* species. Mol Breed 5:11–20

Harding K, Millam S (2000) Analysis of chromatin, nuclear DNA and organelle composition in somatic hybrids between *Solanun tuberosum* and *Solanum sanctae-rosae*. Theor Appl Genet 101:939–947

Hawkes JG (1990) The potato. Evolution, biodiversity and genetic resources. Belhaven Press, London and Smithsonian Institute Press, Washington, DC, 259 pp

Kao KN, Michayluk MR (1975) Nutritional requirements for the growth of *Vicia hajastana* cells and protoplasts at a very low population density in liquid media. Planta 126:105–110

Mattheij WM, Puite KJ (1992) Tetraploid potato hybrids through protoplasts fusions and analysis of their performance in the field. Theor Appl Genet 83:807–812

Matthews D, Harding K, Wilkinson MJ, Millam S (1997) A slot blot hybridization method for screening somatic hybrids. Plant Mol Biol Rep 15:62–70

Matthews D, McNicoll J, Harding K, Millam S (1999) 5′ anchored simple sequence repeat primers are useful for analysing potato somatic hybrids. Plant Cell Rep 19:210–212

Millam S, Davie PA, Harding K, Dale MF (1997) Non-transgenic applications of plant tissue culture in potato. Annu Rep Scott Crop Res Inst 1996(7):50–52

Millam S, Payne L, Mackay GR (1995) The integration of protoplast fusion-derived material into a potato breeding programme, review of progress and problems. Euphytica 85:451–455

Nelson R, Creissen GP, Bright SWJ (1983) Plant regeneration from protoplasts of *Solanum brevidens*. Plant Sci Lett 30:255–262

Pijnacker LP, Ferweda MA, Puite KJ, Roest S (1987) Elimination of *Solanum phureja* nucleolar chromosomes in *S. tuberosum* and *S. phureja* somatic hybrids. Theor Appl Genet 73:878–882

Pijnacker LP, Ferweda MA, Puite KJ, Schaart JG (1989) Chromosome elimination and mutation in tetraploid somatic hybrids of *Solanum tuberosum* and *Solanum phureja*. Plant Cell Rep 8:82–85

Provan J, Kumar A, Sheperd L, Powell W, Waugh R (1996) Analysis of intra-specific somatic hybrids of potato (*Solanum tuberosum*) using simple sequence repeats. Plant Cell Rep 16:196–199

Puite KJ, Roest S, Pijnacker LP (1986) Somatic hybrid plants after electrofusion of diploid *Solanum tuberosum* and *Solanum phureja*. Plant Cell Rep 5:262–265

Puite KJ, Broeke WT, Schaart J (1988) Inhibition of cell wall synthesis improves flow cytometric sorting of potato heterofusions resulting in hybrid plants. Plant Sci 56:61–68

Ramulu KS, Dijkhuis S, Roest S (1989) Patterns of phenotypic and chromosome variation in plants derived from protoplast cultures of monohaploid, dihaploid and diploid genotypes and in somatic hybrids of potatoes. Plant Sci 60:101–110

Schumann U, Koblitz H (1983) Studies on plant recovery from mesophyll protoplasts of *Solanum tuberosum* L. and *Solanum phureja* Juz. and Buk. Biol Plant 25:180–186

Shepard JF (1979) Mutant selection and plant regeneration from potato mesophyll protoplasts. In: Rubenstein I, Gengenbach B, Phillips RL, Green CE (eds) Genetic improvement of crop plants: emergent techniques. University of Minnesota Press, Minneapolis, pp 185–219

Shepard JF, Totten RE (1977) Mesophyll cell protoplasts of potato: isolation, proliferation and plant regeneration. Plant Physiol 60:313–316

Waara SL, Wallin A, Eriksonn T (1991) Production and analysis of intraspecific somatic hybrids of potato (*Solanum tuberosum* L.). Plant Sci 75:107–115

Waugh R, Baird E, Powell W (1992) The use of RAPD markers for the detection of gene introgression in potato. Plant Cell Rep 11:466–469

Section III
Medicinal and Aromatic Plants

III.1 Somatic Hybridization in *Dianthus* Species

M. Nakano[1,2], Y. Hoshino[1,3], and M. Mii[1]

1 Introduction

Flower crops may have higher feasibility in the application of somatic hybridization technology for their genetic improvement. For example, since flower crops are usually evaluated for their ornamental value, somatic hybrid plants with certain ornamental value can directly be used commercially without any further additional breeding. In addition, sterile plants occasionally obtained from somatic hybridization between distantly related species can also be used by clonal propagation using tissue culture methods. Although several tissue culture methods, such as a meristem culture and embryo rescue, have been routinely applied in some commercially important flower crops, including *Dianthus* spp., *Rosa* spp., *Lilium* spp., and *Dendranthema grandiflorum* (Mii et al. 1990; Sagawa and Kunisaki 1990; van Aartrijk et al. 1990), only a few studies had previously been reported on somatic hybridization in flower crops except for some solanaceous species such as *Nicotiana*, *Petunia*, and *Datura* (Sink 1991). This is largely due to their lesser importance in agriculture as compared with food crops. However, protoplast-to-plant systems have recently been developed in various flower crops other than solanaceous species (Roest and Gilissen 1989, 1993), indicating that somatic hybridization can be applicable to flower crops.

The genus *Dianthus* belongs to the family Caryophyllaceae and consists of more than 300 species. Several species in this genus, such as *D. caryophyllus* (carnation), *D. chinensis* (Chinese pink) and *D. barbatus* (sweet William), *D. plumarius* (cottage pink), *D. superbus*, and their interspecific hybrids, are widely cultivated as flower crops. The breeding of these species has so far been carried out mainly by means of intra- and interspecific hybridization and sport selection. Although somatic hybridization is expected to offer an alternative approach for the genetic improvement of floral and marketable qualities of

---

[1] Laboratory of Plant Cell Technology, Faculty of Horticulture, Chiba University, Matsudo 271-8520, Japan
[2] Present address: Laboratory of Horticulture, Faculty of Agriculture, Niigata University, 2-8050 Ikarashi, Niigata 950-2181, Japan
[3] Present address: Experimental Farms, Faculty of Agriculture, Hokkaido University, Kita 11, Nishi 10, Kita-Ku, Sapporo 060-0811, Japan

Dianthus species, no studies had been reported on the application of this technique. We previously made an attempt at somatic hybridization between *D. caryophyllus* and *Gypsophila paniculata* through fusion of cell suspension culture-derived protoplasts of both species (Nakano and Mii 1993a). Although roots were developed from an intergeneric somatic hybrid callus line, plant regeneration from this line could not be achieved, probably due to the lack of regeneration ability of the original cell suspension cultures of both parental species. However, systems for regenerating plants from leaf- and hypocotyl-derived protoplasts have recently been established in certain *Dianthus* species and cultivars (Nakano and Mii 1992, 1995). In this chapter, the results obtained in our laboratory on the production of somatic hybrid plants in *Dianthus* species are summarized.

2 Protoplast Isolation

Dianthus caryophyllus cv. Coral, *D. chinensis* cv. Gosun-sekichiku, *D. barbatus*, and *Gypsophila paniculata* cv. Bristol Fairy were used as plant materials. Protoplasts were isolated from leaves (Nakano and Mii 1992), hypocotyls (Nakano and Mii 1995) or cell suspension cultures (Nakano and Mii 1993a).

2.1 Leaf- and Hypocotyl-Derived Protoplasts

Fully expanded leaves of in vitro-grown plantlets of *D. chinensis*, *D. caryophyllus*, and *D. barbatus*, or hypocotyls of in vitro-sown seedlings of *D. chinensis* and *D. barbatus* were harvested and cut into pieces. The pieces were soaked into CPW solution (Frearson et al. 1973) containing 0.5 M mannitol and incubated for 1 h at 27 °C in the dark. Then, 1 g fresh weight (FW) of the pieces was transferred to 10 ml of an enzyme solution consisting of 2% Cellulase Onozuka RS (Yakult Pharmaceutical Co. Ltd., Japan), 0.1% Pectolyase Y-23 (Seishin Pharmaceutical Co. Ltd., Japan), 1% Driselase (Kyowa Hakko Kogyo Co. Ltd., Japan), 5 mM MES, and 0.5 M mannitol in CPW solution, pH 5.8. After 5 h incubation at 27 °C in the dark with gentle shaking, the mixture was passed through a nylon sieve and protoplasts were freed from debris by flotation on 0.5 M sucrose solution with centrifugation (120 g for 3 min). The protoplasts were washed twice in 0.5 M mannitol solution by resuspension and centrifugation (120 g for 3 min).

2.2 Cell Suspension Culture-Derived Protoplasts

Cell suspension cultures of *G. paniculata* were established from leaf-derived calli. They were subcultured weekly in an MS liquid medium containing 1 mg l^{-1} Picloram and 3% sucrose on an orbital shaker (100 cycles min^{-1}) at

27 °C under continuous illumination ($35\,\mu mol\,m^{-2}\,s^{-1}$). Suspension cultures were maintained for about 3 months prior to use in protoplast isolation.

One g FW of cells from 4-day-old cultures was transferred to 10 ml of a filter-sterilized enzyme solution containing 0.5% Macerozyme R-10, 1% Cellulase Onozuka RS, 0.05% Pectolyase Y-23, 0.5% Driselase, 5 mM MES, and 0.7 M sorbitol, pH 5.8, and incubated for 4 h with gentle shaking at 27 °C in the dark. Isolated protoplasts were purified by the same methods as those for leaf- and hypocotyl-derived protoplasts, but using 0.7 M sucrose and 0.7 M sorbitol solutions instead of 0.5 M sucrose and mannitol solutions, respectively.

3 Protoplast Culture

Purified protoplasts were cultured at a density of $1 \times 10^5\,ml^{-1}$ in MS- or KM8p (Kao and Michayluk 1975)-based medium. The former was supplemented with $5\,mg\,l^{-1}$ α-naphthaleneacetic acid (NAA), $1\,mg\,l^{-1}$ zeatin, and 0.5 M glucose, and the latter with $5\,mg\,l^{-1}$ NAA, $1\,mg\,l^{-1}$ zeatin, 2% sucrose, and 0.5 M mannitol. Both media were adjusted to pH 5.8 and filter-sterilized. Three-ml aliquots of protoplast suspensions were dispensed into 6-cm diameter petri dishes, which were then maintained at 27 °C in the dark. Protoplasts stated to divide 3 to 10 days after the onset of culture, and division frequencies of 5 to 25% were obtained after 14 days, depending on the plant species and protoplast source tissue. Generally, the frequency was higher in hypocotyl- and suspension culture-derived protoplasts and when a KM8p-based medium was used. In some experiments, 1 ml of the same fresh medium was added three times at 14-day intervals to promote colony growth. After 2 months of culture, visible colonies, ca. 1 mm in diameter, were transferred for callus proliferation to MS medium containing 1 or $5\,mg\,l^{-1}$ NAA, $1\,mg\,l^{-1}$ zeatin, 2% sucrose, and 0.2% gellan gum, pH 5.8. Cultures during and after callus proliferation were maintained at 27 °C under continuous illumination ($35\,\mu mol\,m^{-1}\,s^{-1}$).

One month after transfer to the callus proliferation medium, protoplast-derived calli were further transferred for inducing shoot regeneration onto MS medium containing $1\,mg\,l^{-1}$ NAA, $5\,mg\,l^{-1}$ zeatin, 2% sucrose, and 0.2% gellan gum, pH 5.8. After 1 to 4 months, shoot primordia of green nodular structure appeared on the surface of the calli and developed into adventitious shoots. Shoot regeneration frequency varied markedly among plant species and source tissues of protoplasts: regeneration frequencies of approximately 1 and 30% were obtained from leaf-derived protoplasts of *D. barbatus* and *D. chinensis*, respectively, while no shoots were regenerated from those of *D. caryophyllus*; shoots were regenerated from approximately 8% of calli derived from hypocotyl protoplasts of *D. barbatus*, while no shoot regeneration occurred in suspension culture-derived protoplasts of *G. paniculata*. Irrespective of the protoplast source, some shoots of *D. barbatus* produced flowers immediately after shoot regeneration.

Regenerated shoots, 1.5 to 2 cm in height, were detached from the callus and transferred to a half-strength MS medium containing 2% sucrose and 0.8% agar, pH 5.8, on which most of the shoots easily developed roots. Protoplast-derived plantlets of *D. chinensis* looked normal and were successfully transferred to the greenhouse after they completed acclimatization. In *D. barbatus*, however, almost all of the plantlets derived from both leaf and hypocotyl protoplasts were dwarf and flowered precociously under the in vitro conditions. These abnormal plantlets tended to die both during acclimatization and after transfer to the greenhouse. However, the abnormal plantlets occasionally recovered a normal morphology during prolonged culture on plant growth regulator-free medium or cultivation in the greenhouse.

4 Protoplast Fusion

Protoplast fusion was carried out by a polyethylene glycol (PEG) or electrofusion method. In some fusion experiments, protoplasts from one parent were treated with iodoacetamide (IOA) prior to fusion for the selection of somatic hybrids in combination with regeneration ability. Details are given in the Protocol (Sect. 7).

4.1 PEG-Mediated Fusion

In somatic hybridization experiments between *D. chinensis* and *D. barbatus* (Nakano and Mii 1993b) and between *D. chinensis* and *D. caryophyllus* (Nakano and Mii 1993c), protoplast fusion was carried out by the PEG-mediated method. In these experiments, leaf mesophyll protoplasts were used in both fusion parents, and therefore, the frequency of heterokaryon formation after PEG treatment could not be determined because of the morphological similarity of both parental protoplasts. The frequency was estimated to be 2 to 3% by using leaf-derived protoplasts of *D. chinensis* and suspension culture-derived protoplasts of *G. paniculata* (unpubl.).

4.2 Electrofusion

Electrofusion was carried out using a Somatic Hybridizer SSH-1 (Shimadzu Corp, Japan). In a fusion experiment between hypocotyl-derived protoplasts of *D. barbatus* and suspension culture-derived protoplasts of *G. paniculata*, the electrofusion method was applied, and the mean frequency of heterokaryon formation was estimated to be 5.2% (Nakano et al. 1996). This value is higher than that obtained from the PEG-mediated fusion described above (2 to 3%). In addition, electrofusion is less complicated and less time-consuming than PEG-mediated fusion. Therefore, it can be concluded that the electrofusion technique can be effectively used as a routine fusion procedure for somatic hybridization, not only in *Dianthus* but also in other caryophyllaceous species.

4.3 IOA Treatment

Selection of somatic hybrids by combining IOA inactivation and regeneration ability seems to be quite promising because it requires neither the production of mutants or transformants, nor special equipment such as a cell sorter or techniques such as manual isolation of heterokaryons. In *Dianthus* species and *G. paniculata*, in addition, shoot regeneration ability from protoplasts has been shown to be extremely dependent on the genotype and protoplast source tissue as described above.

The IOA treatment in combination with regeneration ability was applied for the selection of somatic hybrids in fusion experiments between *D. chinensis* and *D. caryophyllus* (Nakano and Mii 1993c) and between *D. barbatus* and *G. paniculata* (Nakano et al. 1996). Prior to PEG-mediated fusion or electrofusion, protoplasts from one fusion parent were treated with 10mM IOA. In these experiments, cell division of the IOA-treated protoplasts of *D. chinensis* (the former experiments) and *G. paniculata* (the latter experiments) was completely inhibited by the IOA treatment even when they were cocultured with non-treated protoplasts of the other fusion parent using a culture plate insert (Millicell-HA 12-mm diameter, 0.45-µm pore size, Millipore USA) placed in the center of the culture dish. The detailed data on the selection of somatic hybrids between *D. chinensis* and *D. caryophyllus* are given in the following section, Section 5.

5 Regeneration of Somatic Hybrids

We examined somatic hybridization between various interspecific and intergeneric combinations and successfully produced somatic hybrid plants or plantlets is several combinations.

5.1 Plant Regeneration from an Interspecific Somatic Hybrid Between *D. chinensis* and *D. barbatus*

We initially examined the possibility of somatic hybridization in *Dianthus* species and fused leaf-derived protoplasts of *D. chinensis* cv. Gosun-sekichiku and *D. barbatus* by the PEG method (Nakano and Mii 1993b). These two species were chosen for the following reasons: *D. chinensis* has been shown to have a high shoot regeneration ability from leaf mesophyll protoplasts, while calli derived from leaf protoplasts of *D. barbatus* regenerated shoots at a very low frequency under the same culture conditions; in addition, protoplast-derived shoots of *D. barbatus* have been shown to produce flowers immediately after shoot regeneration under in vitro conditions. Therefore, we expected that somatic hybrids could be detected, without applying any other selection methods, by shoot regeneration with precocious flowering and the expression of novel flower color and morphology.

The PEG-treated protoplasts started to divided 10 days after the onset of culture and formed visible colonies after 2 months. Upon transferring these colonies to the callus-proliferation medium, some calli showed a vigorous growth as compared to the other calli and those of control parental cultures. These vigorously growing calli were then picked up and further transferred to the regeneration medium. Among 30 calli transferred, two formed shoot primordia and one of them regenerated many shoots 2 months after transfer. In the control parental cultures, approximately 30% of protoplast-derived calli of *D. chinensis* regenerated shoots, whereas no shoot regeneration was observed from protoplast-derived calli of *D. barbatus*.

Shoots obtained from the PEG-treated protoplasts produced flowers immediately after shoot regeneration (Fig. 1a). The flower color of these shoots (light pink with red stripes) was distinctly different from that of both parental plants (*D. chinensis* cv. Gosun-sekichiku, deep pink; *D. barbatus*, red with white rim; Fig. 1b). Therefore, it was concluded that these shoots were derived from a fusion product between *D. chinensis* cv. Gosun-sekichiku and *D. barbatus*. The shoots, which grew up to 1 to 2 cm in height, were detached from the callus and transferred to a plant growth regulator-free medium, on which roots were developed within 2 weeks. The plantlets thus obtained were severely dwarf with short internodes and dark green leaves, as compared with parental plantlets growing in vitro, and continuously produced flowers under the in vitro condition (Fig. 1c). The addition of 1mg l^{-1} GA_3 to the medium induced elongation of shoots but did not stop flowering. Transfer of the cultures to a short-day condition (8-h photoperiod) did not affect flowering. Continuous flower production was still observed during acclimatization and after transfer to the greenhouse, and most plantlets ultimately died. However, plants with a normal morphology were developed from abnormal ones more than 6 months after transfer to the greenhouse (Fig. 1d).

To confirm the hybridity of the regenerated plants, esterase isozyme analysis was employed (Wetter and Dick 1983). Esterase zymograms were distinctly different between *D. chinensis* and *D. barbatus* and the regenerated plants had bands common to both parents (Fig. 1e). Restriction endonuclease analysis of rDNA (Honda and Hirai 1990) provided additional evidence for hybridization. Among the restriction enzymes tested, *Eco*RV was shown to be the best for discriminating between rDNA fragments of both parents (Fig. 1f). A clear and specific rDNA fragment originating from the nuclear DNA of *D. chinensis* was 6.5 kbp in length, whereas that of *D. barbatus* was 6.0 kbp. *Eco*RV-digested nuclear DNA of the regenerated plants contained both 6.5- and 6-kbp fragments. These results confirmed that the regenerated plants are indeed an interspecific somatic hybrid between *D. chinensis* cv. Gosun-sekichiku and *D. barbatus*.

Fig. 1a–g. Plant regeneration from an interspecific somatic hybrid between *Dianthus chinensis* cv. Gosun-sekichiku and *D. barbatus*. **a** Adventitious shoots derived from the PEG-treated protoplasts which produced flowers immediately after shoot regeneration. **b** Flowers of *D. barbatus*, somatic hybrid, and *D. chinensis* (*left to right*). **c** A somatic hybrid plantlet growing precociously in vitro. **d** Somatic hybrid plants with abnormal (*left*) and normal (*right*) appearance. **e** Isozyme patterns for esterase of *D. barbatus*, somatic hybrid, and *D. chinensis* (*left to right*). *Arrowheads* indicate bands specific to the parental species. **f** Blot hybridization of digoxigenin-labeled rDNA fragments to *Eco*RV-digested total DNAs of *D. barbatus*, somatic hybrid, and *D. chinensis* (*left to right*). *Arrowheads* indicate the 6.5- and 6.0-kbp fragments (*top to bottom*). **g** Chromosomes in a root-tip cell of a somatic hybrid plant ($2n = 52$). (**a–c**, **e–g**, Nakano and Mii 1993b; **d**, unpubl.)

The results from chromosome counts of the somatic hybrid revealed that it had $2n = 52$ (Fig. 1g), which was less than the sum of *D. chinensis* ($2n = 30$) and *D. barbatus* ($2n = 30$). Thus, the somatic hybrid obtained in this experiment was an asymmetric one. It would appear that a certain number of chromosomes were eliminated during protoplast culture and/or shoot regeneration. However, because of the difficulty in identifying the individual chromosomes of each parental species, it is still unclear whether specific elimination of certain parental chromosomes occurred. Although most flowers of the somatic hybrid plants growing both in vitro and in the greenhouse were male sterile with undeveloped stamens, some flowers developed stamens with mature pollen grains. Pollen fertilities of these flowers were up to 60%, as determined by acetocarmine staining.

In this experiment, an interspecific somatic hybrid was obtained without applying any selection. The somatic hybrid plants were regenerated from a callus, which showed a vigorous growth on the callus-proliferation medium. These results suggest that at least some hybrid vigor was expressed during the callus-proliferation stage.

5.2 Plant Regeneration from Somatic Hybrids Between *D. chinensis* and *D. caryophyllus*

In the somatic hybridization experiment mentioned above, no special method was employed to increase the efficiency for the selection of somatic hybrids. Therefore, we then employed a selection method by using IOA inactivation and regeneration ability of protoplasts. Leaf-derived protoplasts of *D. caryophyllus* cv. Coral were fused with IOA-treated leaf protoplasts of *D. chinensis* cv. Gosun-sekichiku (Nakano and Mii 1993c). Since protoplasts of *D. caryophyllus* have been shown to divide to from callus but not regenerate shoots, fusion-derived callus lines which regenerated shoots could be tentatively selected as somatic hybrids.

The PEG-treated protoplasts formed visible colonies 2 months after the onset of culture. As shown in Table 1, more than 1000 visible colonies were obtained from five independent fusion experiments. These colonies were transferred to the callus-proliferation medium, and calli obtained on this medium were further transferred to the shoot-regeneration medium. Among over 200 calli transferred, four regenerated many shoots by 2 months after transfer. In the control cultures, IOA-treated protoplasts of *D. chinensis* did not divide, and protoplast-derived calli of *D. caryophyllus* never regenerated shoots under the culture conditions used in this study.

Shoots derived from these four lines easily rooted and were successfully transferred to the greenhouse. No abnormalities, such as dwarf shoots and precocious flowering, were observed in them. Plant height (Fig. 1a) and flower morphology (Fig. 1b) of plants derived from three callus lines were intermediate between those of the parents. Flowers of *D. chinensis* cv. Gosun-sekichiku were deep pink, whereas *D. caryophyllus* cv. Coral had red flowers. Flower color of the regenerated plants which exhibited intermediate morphology was

Table 1. The data on colony formation and plant regeneration from five independent fusion experiments between IOA-treated leaf protoplasts of *Dianthus chinensis* cv. Gosun-sekichiku and leaf protoplasts of *D. caryophyllus* cv. Coral. (Nakano and Mii 1993c)

Fusion experiment (no.)	Total protoplasts plated[a]	No. of visible colonies formed	No. of calli transferred to the regeneration medium	No. of calli regenerating plants	No. of somatic hybrids
1	4×10^5	250	75	0	0
2	4×10^5	250	53	0	0
3	2×10^5	180	33	2	2
4	2×10^5	120	35	1	0
5	6×10^5	330	40	1	1
Total	18×10^5	1080	236	4	3

[a] Equal numbers of both parental protoplasts were mixed and treated with PEG.

red and resembled *D. caryophyllus* cv. Coral. The results of chromosome counts revealed that these plants had $2n = 60$, which was the sum of both parents ($2n = 30$ for each parent) (Fig. 2c). From these results, three out of four callus lines selected in this study were assumed to be interspecific somatic hybrids. All of the putative hybrid plants were completely male sterile with undeveloped stamen.

To confirm the hybrid nature of the putative hybrid plants, esterase isozyme analysis and RAPD analysis (Williams et al. 1990) were carried out. At least two plants from each callus line were subjected to each analysis. Esterase zymograms were distinctly different between *D. chinensis* and *D. caryophyllus* and all of the putative hybrid plants had bands from both parents (Fig. 2d). Additional evidence for hybridization was provided by RAPD analysis (Fig. 2e). The profiles of the amplified products from both parents by SSU-2F primer (Omura et al. 1991) were clearly different from each other. A specific amplified DNA fragment of *D. chinensis* was 1.1 kbp, while that of *D. caryophyllus* was 0.8 kbp. In all the putative hybrid plants, both 1.1- and 0.8-kbp fragments were amplified. These results indicated that the three selected lines contained the genetic materials of both parents.

However, plants similar to *D. chinensis* cv. Gosun-sekichiku were also regenerated from one of the four selected callus lines. They had 30 chromosomes and had the same esterase patterns as *D. chinensis*. These results indicated that this callus line is not a hybrid and possessed only the *D. chinensis* genome. Although no cell division of IOA-treated *D. chinensis* protoplasts was observed in this study even when they were cocultured with nontreated protoplasts of *D. caryophyllus*, it is possible that some escapes might have happened by chance due to a "nurse effect" of nontreated protoplasts. Alternatively, it is also possible that the callus line possessing only the *D. chinensis* genome might have been derived from the elimination of the *D. caryophyllus* genome during the initial stages of cell division of the heterokaryocyte. In this case, the plants derived from this callus line may be cybrids.

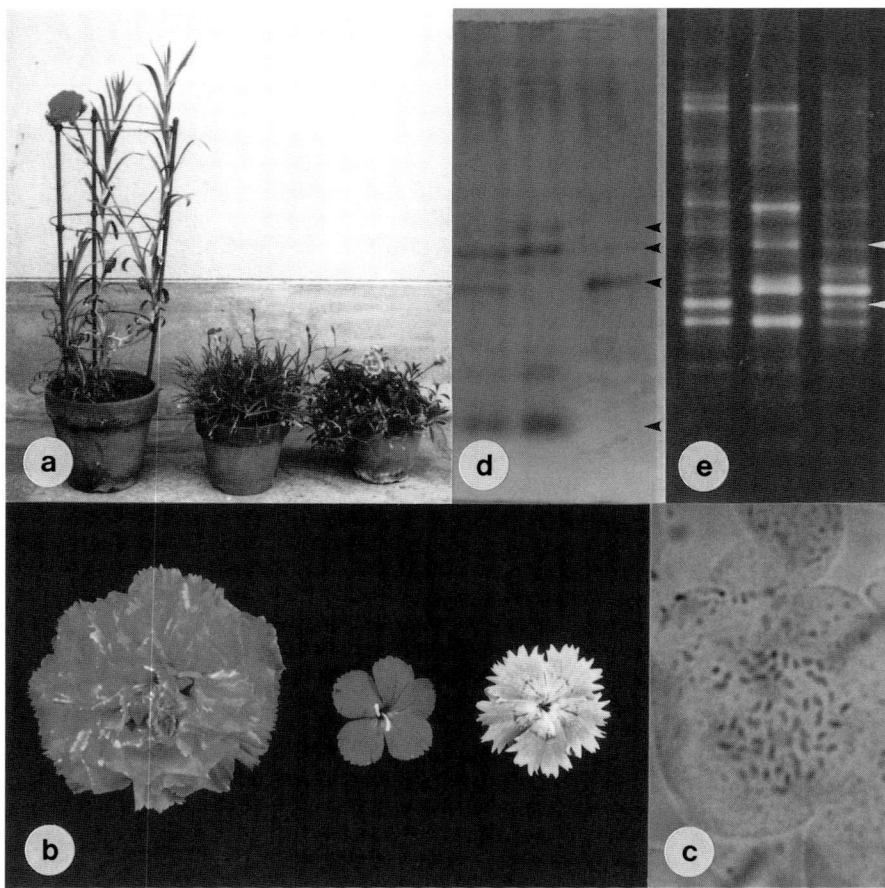

Fig. 2a–e. Plant regeneration from interspecific somatic hybrids between *Dianthus chinensis* cv. Gosun-sekichiku and *D. caryophyllus* cv. Coral. **a** Greenhouse-grown plants of *D. caryophyllus*, somatic hybrid, and *D. chinensis* (*left to right*). **b** Flowers of *D. caryophyllus*, somatic hybrid, and *D. chinensis* (*left to right*). **c** Chromosomes in a root-tip cell of a somatic hybrid plant ($2n = 60$). **d** Isozyme patterns for esterase of somatic hybrid, *D. chinensis* and *D. caryophyllus* (*left to right*). *Arrowheads* indicate bands specific to the parental species. **e** RAPD profiles of *D. caryophyllus*, *D. chinensis* and somatic hybrid (*left to right*). Total DNAs from leaves were used as a template for PCR amplification using the primer SSU-2F. *Arrowheads* indicate the 1.1- and 0.8-kbp fragments (*top to bottom*). (Nakano and Mii 1993c)

5.3 Plantlet Regeneration from Intergeneric Somatic Hybrids Between *D. barbatus* and *Gypsophila paniculata*

In the above two somatic hybridization experiments, we produced somatic hybrid plants by the PEG-mediated method between sexually compatible species. We recently produced sexually incompatible, intergeneric somatic hybrid plantlets between *D. barbatus* and *G. paniculata* cv. Bristol Fairy. Hypocotyl-derived protoplasts of *D. barbatus*, which were pretreated with

IOA, were fused electrically with suspension culture-derived protoplasts of *G. paniculata*, which could divide to form callus but regenerate no shoots (Nakano et al. 1996).

The electrofusion-treated protoplasts started to divide within 7 days in culture, and formed a number of visible colonies after 2 months. In the control cultures, suspension-derived protoplasts of *G. paniculata* also formed many visible colonies, whereas cell division was never observed in IOA-treated hypocotyl protoplasts of *D. barbatus*. Therefore, visible colonies formed after electrofusion seemed to be derived from either *G. paniculata* protoplasts or heterokaryons. From five independent fusion experiments, over 300 calli were obtained, and they were then transferred to the shoot-regeneration medium. Among these calli, eight formed green shoot primordia, and two of them further developed many shoots. These two calli were derived from independent fusion experiments. The remaining six calli turned brown without shoot development and ultimately died. In the control cultures, calli derived from protoplasts of *G. paniculata* cv. Bristol Fairy never regenerated shoots under the culture conditions used in this study, while the shoot regeneration frequency of about 8% were obtained from those derived from non-IOA-treated protoplasts of *D. barbatus*.

Regenerated shoots from two putative hybrid callus lines (SH-1 and SH-2) were rooted on plant growth regulator-free medium within 2 weeks. All plantlets on this medium were severely dwarf and flowered precociously in vitro (Fig. 3a). Flower color of plantlets of both SH-1 and SH-2 was red with a white center, while parental *D. barbatus* and *G. paniculata* cv. Bristol Fairy had red and double white flowers, respectively. All of these flowers were male sterile with undeveloped stamens. SH-1 and SH-2 plantlets also continued to produce flowers during acclimatization and ultimately died after transfer to the greenhouse.

To confirm the hybrid nature of the regenerated plantlets (SH-1 and SH-2), restriction endonuclease analysis of rDNA was performed (Fig. 3b). *Bam*HI digestion gave fragments of 6.7, 5.6, 5.1, and 4.6 kbp specific for *D. barbatus*, and of 7.6 and 5.5 kbp specific for *G. paniculata*. Both SH-1 and SH-2 had the fragments specific for both parents. This result indicated that they were novel hybrids between *D. barbatus* and *G. paniculata*, and could be called *Gypsodianthus*.

Chromosome counts of hybrid plantlets from SH-1 and SH-2 revealed that they had about $2n = 55$, which were less than the sum of *D. barbatus* ($2n = 30$) and *G. paniculata* ($2n = 34$), indicating that both SH-1 and SH-2 were asymmetric hybrids. Because of the similarity in size and morphology of chromosomes of *D. barbatus* and *G. paniculata*, parental chromosomes could not be distinguished in these hybrids. Analysis of nuclear DNA content also indicated the asymmetric nature of both hybrids (Fig. 3c). Nuclei were isolated from leaves of in vitro-grown plantlets and stained with 4′,6-diamidino-2-phenylindole (DAPI). Their DNA content was then measured by using a CA II flow cytometer (Partec, Germany), and the position of the G_0/G_1 peaks of DAN histograms were compared between the parental and somatic hybrid plantlets. Based on the position of the peaks, hybrids were identified as

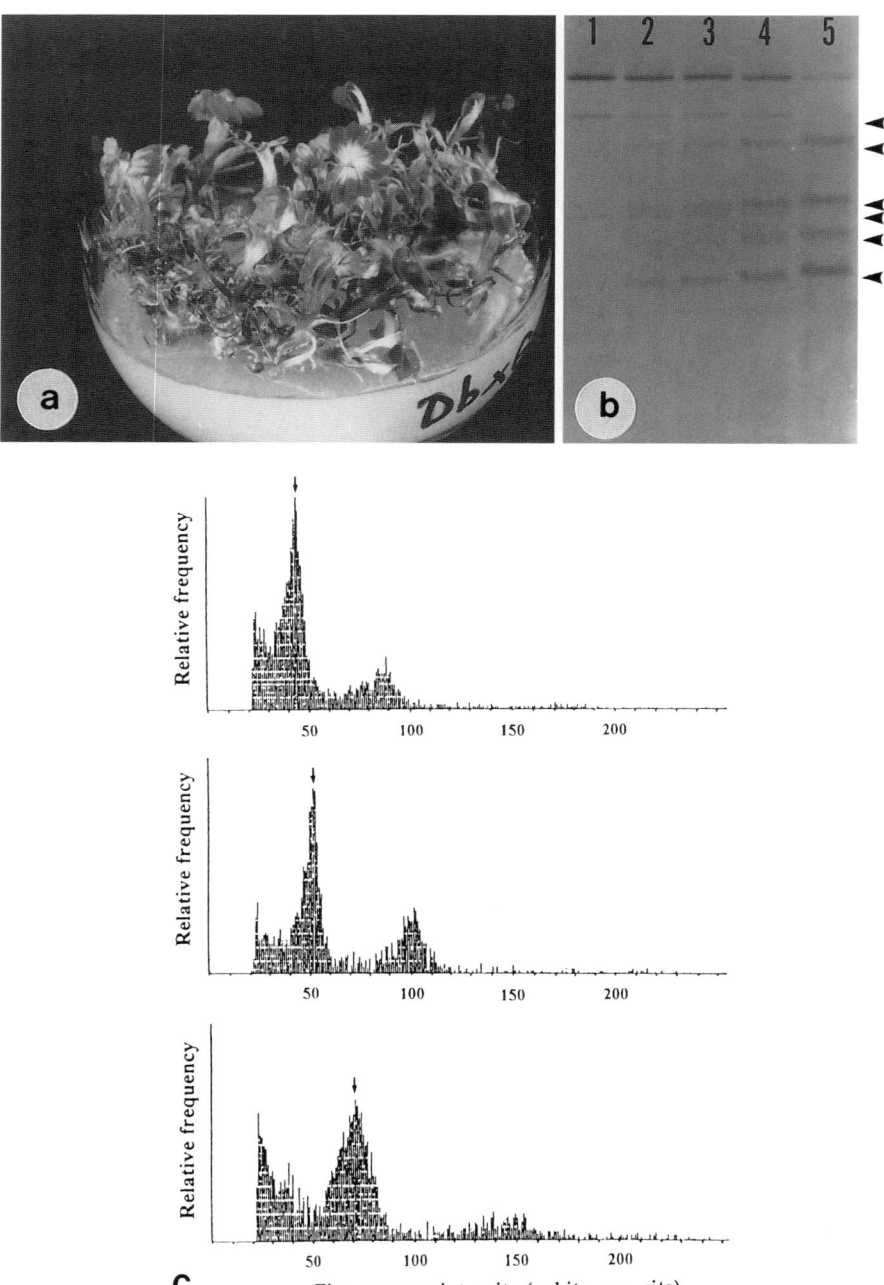

Fig. 3a–c. Plantlet regeneration from intergeneric somatic hybrids between *Dianthus barbatus* and *Gypsophila paniculata* cv. Bristol Fairy. **a** Somatic hybrid plantlets continuously growing in vitro. **b** Blot hybridization of digoxigenin-labeled rDNA fragments to *Bam*HI-digested total DNAs of the parental species and somatic hybrid. *Arrowheads* indicate the 7.6-, 6.7-, 5.6-, 5.5-, 5.1-, and 4.6-kbp fragments (*top to bottom*). Lane 1 *G. paniculata*; lanes 2 and 3 somatic hybrid SH-1; *lane 4* somatic hybrid SH-2; *lane 5 D. barbatus*. **c** Flow cytometric analysis of nuclear DNAs of in vitro-grown plantlets of *D. barbatus*, *G. paniculata*, and somatic hybrid SH-2 (*top to bottom*). *Arrows* indicate the G_0/G_1 peak. (**a**, **b** Nakano et al. 1996; **c**, unpubl.)

aneuploids which had a DNA content lower than the sum of *D. barbatus* and *G. paniculata*.

All of the intergeneric somatic hybrid plantlets obtained in this study have still retained the abnormal characters, such as precocious flowering, during more than 5 years of subculture on media without plant growth regulators or containing $1\,\text{mg}\,\text{l}^{-1}$ GA_3, and their transfer to the greenhouse has not yet been successful. An examination of the cultural conditions required for inducing plants with a normal growth from the abnormal hybrid plantlets is now in progress.

6 Summary and Conclusion

We have provided the possibility of applying somatic hybridization for genetic improvement in *Dianthus* species. Somatic hybrid plants were produced in two interspecific combinations between *D. chinensis* and *D. barbatus* and between *D. chinensis* and *D. caryophyllus* and, in addition, somatic hybrid plantlets were regenerated in one intergeneric combination between *D. barbatus* and *G. paniculata*. Although the interspecific combinations examined here were sexually compatible ones, direct production of amphidiploid or aneuploid plants will have some advantages in breeding these flower crops. Further evaluation of these somatic hybrid plants and their progenies is needed.

An efficient method for selecting somatic hybrids by using IOA inactivation and regeneration ability was also developed in *Dianthus* species. At the same time, it was confirmed that the regeneration ability is dominantly transmitted to interspecific as well as intergeneric somatic hybrids in *Dianthus* species. Therefore, it is expected that somatic hybrids could be easily obtained by utilizing the species combination with high regeneration ability in at least one parent.

For the practical application of somatic hybridization in *Dianthus* breeding, many problems still remain unsolved, among which the regeneration of shoots or plantlets with abnormal morphology is the most serious. Generally, abnormal shoots or plantlets were severely dwarf and showed precocious flowering. Although the mechanism involved in the occurrence of such abnormal characters has not been identified at present, these characters appear to be induced by some transient physiological disorders rather than to arise as a result of genetic variations, aneuploidy, or genetic incompatibilities, because plants not exhibiting these characters sometimes developed from abnormal ones during prolonged culture both in vitro and in the greenhouse in the interspecific somatic hybrid between *D. chinensis* cv. Gosun-sekichiku and *D. barbatus* (Nakano and Mii 1993b) as well as in protoplast-derived plantlets of *D. barbatus* (Nakano and Mii 1992, 1995). Breeding of flower crops in *Dianthus* species by somatic hybridization would be facilitated if this problem were overcome.

7 Protocol

7.1 IOA Treatment

1. Prepare protoplasts with an enzyme solution and purify them by resuspension and centrifugation.
2. Suspend the protoplasts in a CPW solution containing 10 mM IOA and 0.5 M mannitol, pH 5.8, and incubate them for 15 min at 4 °C in the dark.
3. Wash the protoplasts three times with a CPW solution containing 0.5 M mannitol, pH 5.8, by resuspension and centrifugation (120 g for 3 min).

7.2 PEG-Mediated Fusion

1. Suspend the purified protoplasts in 0.5 M mannitol solution at a density of 1×10^6 ml^{-1}.
2. Mix the protoplast suspensions of both parental species in equal volumes.
3. Place 1 ml of a PEG solution (40% PEG, MW 4000, and 50 mM $CaCl_2 \cdot 2H_2O$ in 50 mM HEPES buffer, pH 6.5) in the center of a petri dish, 6 cm in diameter.
4. Add 1 ml of the mixed protoplast suspension onto the center of the PEG solution. Then stir gently to mix the protoplast and PEG solution. Keep the petri dish at 27 °C for 30 min.
5. Confirm the aggregation of protoplasts under a light microscope.
6. Add 10 ml of a washing solution containing 50 mM $CaCl_2 \cdot 2H_2O$ in 0.5 M mannitol solution, pH 10.5 gradually, and gently stir the contents of the petri dish.
7. Transfer the protoplast-PEG mixture into a test tube, and wash the protoplasts twice with 0.5 M mannitol solution by resuspension and centrifugation (120 g for 3 min).

7.3 Electrofusion

1. Suspend the purified protoplasts in an electrofusion buffer (0.5 M mannitol and 1 mM $CaCl_2 \cdot 2H_2O$ in 5 mM MES buffer, pH 5.8) at a density of 0.5 to 1×10^6 ml^{-1}.
2. Mix the protoplast suspensions of both parental species in equal volumes.
3. Introduce a 1-ml aliquot of the mixture into a fusion chamber. Keep the fusion chamber for 5 min.
4. Subject the protoplasts to an alternating current (AC) field of 2 MHz, 50 V cm^{-1} for 20 s, followed by a 30-ms direct current (DC) pulse of 1 kV cm^{-1} twice.
5. Wash the protoplasts once with 0.5 M mannitol solution by resuspension and centrifugation (120 g for 3 min).

References

Frearson EM, Power JB, Cocking EC (1973) The isolation, culture and regeneration of *Petunia* leaf protoplasts. Dev Biol 33:130–137
Honda H, Hirai A (1990) A simple and efficient method for identification of hybrids using non-radioactive rDNA as probe. Jpn J Breed 40:339–348
Kao KN, Michayluk MR (1975) Nutritional requirements for growth of *Vicia hajastana* cells and protoplasts at a very low population density in liquid media. Planta 126:105–110
Mii M, Buiatti M, Gimelli F (1990) Carnation. In: Ammirato PV, Evans DR, Sharp WR, Bajaj YPS (eds) Handbook of plant cell culture, vol 5. McGraw-Hill, New York, pp 284–381

Murashige T, Skoog F (1962) A revised medium for rapid growth and bioassays with tobacco tissue cultures. Physiol Plant 15:473–497

Nakano M, Mii M (1992) Protoplast culture and plant regeneration of several species in the genus *Dianthus*. Plant Cell Rep 11:225–228

Nakano M, Mii M (1993a) Callus and root formation from an intergeneric somatic hybrid between *Dianthus caryophyllus* and *Gypsophila paniculata*. Sci Hortic 53:13–19

Nakano M, Mii M (1993b) Somatic hybridization between *Dianthus chinensis* and *D. barbatus* through protoplast fusion. Theor Appl Genet 86:1–5

Nakano M, Mii M (1993c) Interspecific somatic hybridization in *Dianthus*: selection of hybrids by the use of iodoacetamide inactivation and regeneration ability. Plant Sci 88:203–208

Nakano M, Mii M (1995) Plant regeneration from protoplasts in *Dianthus*: comparison of cultural behavior of different donor tissues. Plant Tissue Cult Lett 12:62–67

Nakano M, Hoshino Y, Mii M (1996) Intergeneric somatic hybrid plantlets between *Dianthus barbatus* and *Gypsophila paniculata* obtained by electrofusion. Theor Appl Genet 92:170–172

Omura M, Hidaka T, Nakamura I (1991) Variations in PCR patterns of Citrus DNA. Abst Jpn Soc Hortic Sci Spring Meet, pp 284–381 (in Japanese)

Roest S, Gilissen LJW (1989) Plant regeneration from protoplasts: a literature review. Sci Hortic 38:1–23

Roest S, Gilissen LJW (1993) Regeneration from protoplasts – a supplementary literature review. Sci Hortic 42:1–23

Sagawa Y, Kunisaki JT (1990) Micropropagation of floricultural crops. In: Ammirato PV, Evans DR, Sharp WR, Bajaj YPS (eds) Handbook of Plant Cell Culture, vol 5. McGraw-Hill, New York, pp 25–56

Sink KC (1991) Protoplast fusion and somatic hybridization. In: Harding J, Singh F, Mol JNM (eds) Genetics and breeding of ornamental species. Kluwer, Dordrecht, pp 53–68

van Aartrijk J, Blom-Barnhoon GJ, van der Linder PCG (1990) Lilies. In: Ammirato PV, Evans DR, Sharp WR, Bajaj YPS (eds) Handbook of plant cell culture, vol 5. McGraw-Hill, New York, pp 535–576

Wetter L, Dick J (1983) Isoenzyme analysis of cultured cells and somatic hybrids. In: Evans DA, Sharp WE, Ammirato PV, Yamada Y (eds) Handbook of plant cell culture, vol 1. MacMillan, New York, pp 607–628

Williams JGK, Kubelik AE, Levak KJ, Rafalski JA, Tingey SC (1990) DNA polymorphisms amplified by arbitrary primers are useful as genetic markers. Nucleic Acids Res 18:6531–6535

III.2 Somatic Hybridization Between *Nicotiana sylvestris* Speg. & Comes and *N. plumbaginifolia* Viv.

C.-C. Chen, Y.-Y. Kao, F.-M. Lee, and R.-F. Lin

1 Introduction

Although *Nicotiana sylvestris* and *N. plumbaginifolia* belong to the same section, viz., Alatae of the subgenus *Petunioides*, they differ markedly in morphology, geographic distribution, and karyotype (Goodspeed 1954). *N. sylvestris* is the maternal parent of cultivated allotetraploid *N. tabacum* (Gray et al. 1974), which is in the subgenus *Tabacum*, and possesses some morphological characters not present in the section Alatae. *N. sylvestris* occurs exclusively in northwestern Argentina, while *N. plumbaginifolia* has a much wider distribution, extending from northwestern Argentina to Brazil, Peru, Guatemala, Cuba, and Mexico. The chromosomes of *N. sylvestris* ($2n = 24$) are rather uniform in size and are clearly bi-armed, while those of *N. plumbaginifolia* ($2n = 20$) differ significantly in size and are telocentric or acrocentric (Fig. 1A,B).

    N. sylvestris and *N. plumbaginifolia* are sexually incompatible (Christoff 1928) but can be crossed asexually through protoplast fusion. The earliest work in somatic hybridization between these two species is that of Cséplö et al. (1984), who fused ^{60}Co-irradiated protoplasts from lincomycin-resistant *N. sylvestris* with untreated protoplasts from wild-type *N. plumbaginifolia*. The fusion products were cultured in lincomycin medium for selection of cybrids containing *N. sylvestris* chloroplasts and *N. plumbaginifolia* nucleus. Gamma-irradiated *N. sylvestris* protoplasts (wild-type) were also fused with protoplasts from *N. plumbaginifolia* nitrate reductase-deficient (NR$^-$) mutant. The fusion products were cultured in medium with nitrate as the sole nitrogen source for selection of asymmetric somatic hybrids (Famelaer et al. 1989, 1990). Gleba et al. (1987) studied the spatial relationships of parental chromosomes in somatic hybrids obtained from fusion of protoplasts from wild-type *N. sylvestris* with those from an NR$^-$, kanamycin-resistant double mutant of *N. plumbaginifolia*, and subsequent culture of the fusion products in medium containing kanamycin and with nitrate as the sole source of nitrogen.

    We describe here a method of somatic hybridization between *N. sylvestris* and *N. plumbaginifolia* employed in our laboratory and discuss applications of the somatic hybrids in basic research and plant breeding.

Department of Botany, National Taiwan University, Taipei, Taiwan, Republic of China

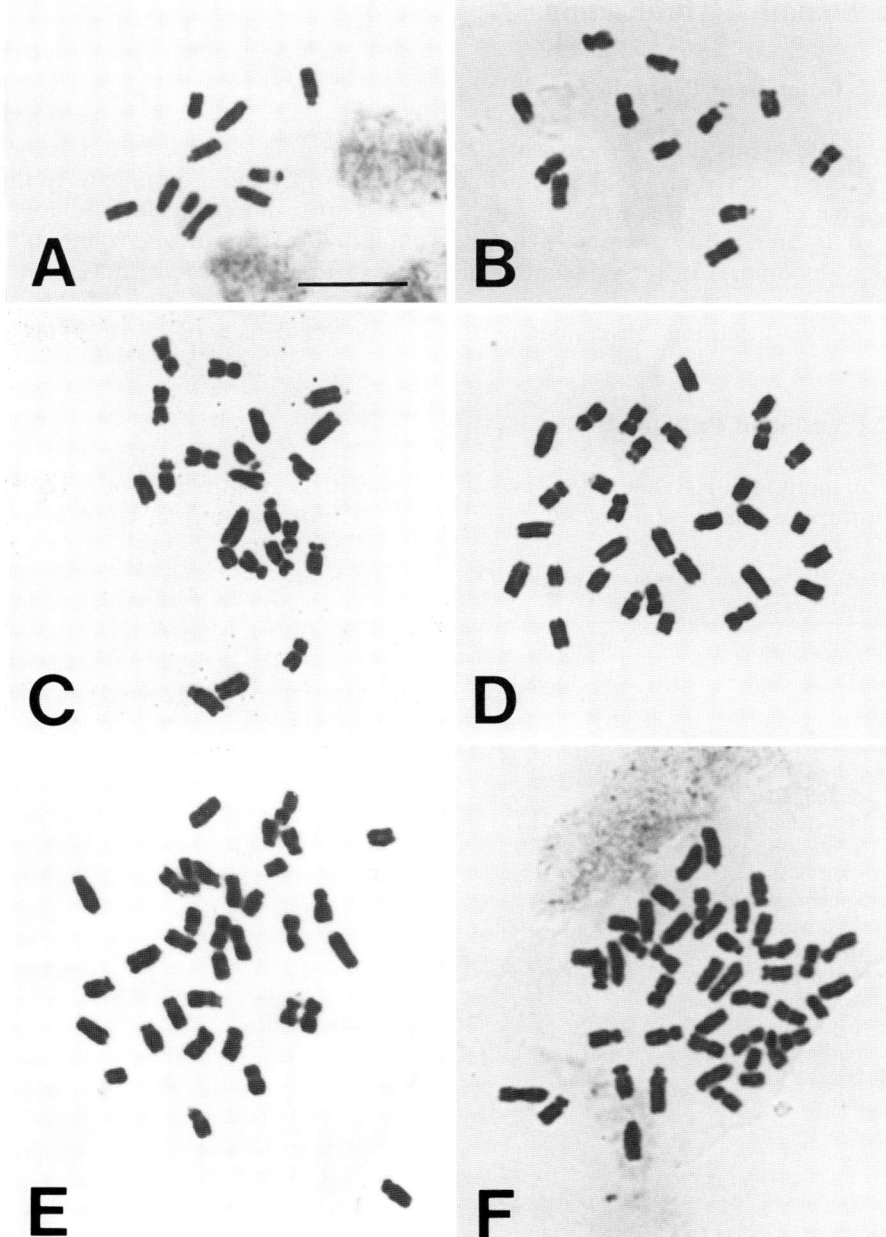

Fig. 1A–F. Somatic metaphase chromosomes of parental species and somatic hybrids. **A** Haploid *Nicotiana plumbaginifolia* (P) showing 10 chromosomes. **B** Haploid *N. sylvestris* (S) showing 12 chromosomes. **C** Somatic hybrid PS, $2n = 22$. **D** Somatic hybrid PPS, $2n = 32$. **E** Somatic hybrid PSS, $2n = 34$. **F** Somatic hybrid PPSS, $2n = 44$. *Bar* 10 µm

2 Somatic Hybridization

2.1 Isolation of Protoplasts

Haploid plants of *N. sylvestris* and *N. plumbaginifolia*, obtained from anther culture (Chen et al. 1985) and maintained in vitro by shoot cutting, were the source of protoplasts. Approximately 4 weeks after shoot transfer, the upper two to three fully expanded leaves were cut into fine strips in an enzyme solution (medium A in Table 1) and incubated for 16 h in the dark at 25 °C. Protoplasts released from the treatment were washed twice in W_5 solution (Medgyesy et al. 1980), and then suspended in this solution at a density of $2 \times 10^5 \, \text{ml}^{-1}$.

2.2 Fusion of Protoplasts

The method of Kao et al. (1974) was followed with some modifications. Briefly, protoplast suspensions of *N. sylvestris* and *N. plumbaginifolia* were mixed

Table 1. Media for isolation and culture of *Nicotiana* protoplasts. (After Huang and Chen 1988)

Component	Medium				
	A	B	C	D	E
Inorganic salts	K3[a]	K3	K3	MS[b]	MS
Vitamins (mg l^{-1})					
m-Inositol	100	100	100	100	
Thiamine-HCl	10	10	10	1	
Pyridoxine-HCl	1	1	1		
Nicotinic acid	1	1	1		
Organic acids (mg l^{-1})					
Sodium pyruvate			5	5	
Citric acid			10	10	
Malic acid			10	10	
Fumaric acid			10	10	
Growth regulators (mg l^{-1})					
Naphthaleneacetic acid	3	3	0.1		
Benzylaminopurine	1	1	0.2	0.2	
Sugars (g l^{-1})					
Sucrose	137	20	20	30	10
Glucose		62	40		
Enzyme (g l^{-1})					
Cellulase Onozuka R10	10				
Macerozyme R10	2				
Solidifying agents (g l^{-1})					
Sea Plaque agarose			6		
Phytagar				8	8
MES[c] (g l^{-1})	0.5	0.5	0.5		
pH	5.6	5.6	5.6	5.7	5.7

[a] Nagy and Maliga (1976).
[b] Murashige and Skoog (1962).
[c] 2-(N-morpholino)-ethanesulfonic acid.

at equal volumes. The protoplast mixture was treated with a 50% polyethylene glycol (PEG) solution for 12 min, eluted once with a high pH-high Ca^{2+} solution and then twice with protoplast culture medium B (Table 1). The protoplasts were washed thoroughly with medium B and cultured in the same medium in the dark at 25 °C. In the control experiment, protoplasts of the two species were mixed and cultured in medium B without treatment with PEG.

To test the efficiency of our method, protoplasts of *N. sylvestris* and *N. otophora* were fused and examined by the aceto-orcein staining method 20 h after culture (Lee and Chen 1990). *N. otophora* was chosen as a fusion partner because its interphase nuclei possess up to ten darkly stained heterochromatin blocks which were not present in *N. sylvestris* (Fig. 2A,B). This difference can be used as a basis to distinguish heterokaryotic from homokaryotic fusions. Our data showed that in the control culture 99.1% protoplasts contained a single nucleus of either *N. sylvestris* or *N. otophora*, and that all bi- and multi-nucleate protoplasts (0.9%) were homokaryotic (Fig. 2C,D). In the PEG-treated culture, the frequencies of bi- and multinucleate protoplasts increased to 9.7 and 1.2%, respectively, and among these, approximately two thirds resulted from heteroplasmic fusion (Fig. 2E). We assume that similar frequencies of fusions may occur in somatic hybridization between *N. sylvestris*

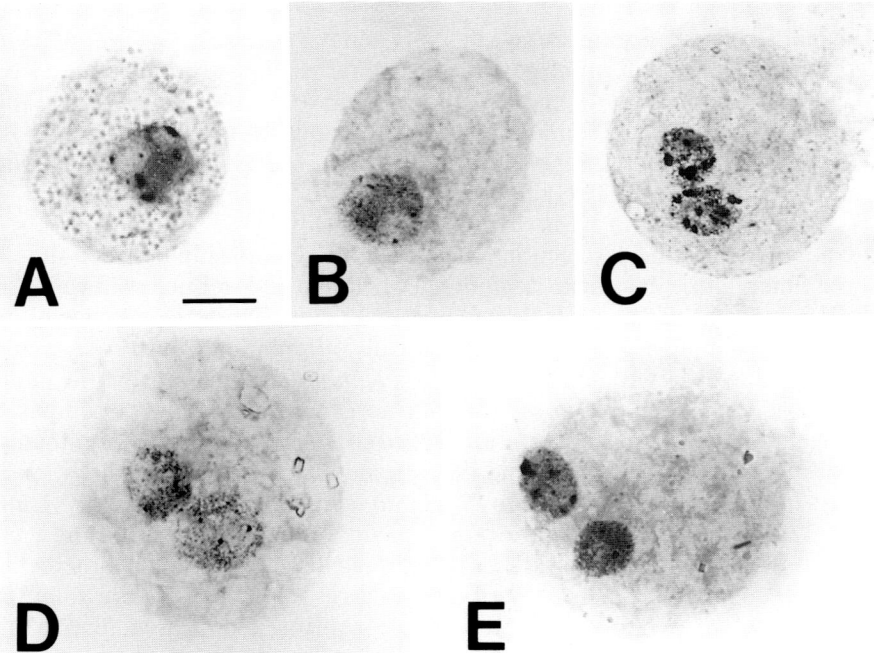

Fig. 2A–E. Aceto-orcein staining of protoplasts 20 h after treatment with PEG. **A** *N. otophora*. **B** *N. sylvestris*. **C** A protoplast with two *N. otophora* nuclei. **D** A protoplast with two *N. sylvestris* nuclei. **E** A protoplast with one *N. otophora* and one *N. sylvestris* nucleus. *Bar* 10 µm

and *N. plumbaginifolia*, in which the two parental nuclei cannot be distinguished by conventional staining methods.

2.3 Culture of Fused Protoplasts

After 10 days of culture in medium B, protoplast-derived colonies were plated at a density of $1.5 \times 10^2 \, \text{ml}^{-1}$ in medium C (Table 1) in which no selective agents were added. The cultures were kept at 25 °C in the dark for 1 week and then at 25 °C under 16-h daily illumination with 1000-lx light. Under these conditions, colonies became visible 1 week after plating, and after another 10 days the most advanced colonies had reached the size of 2 mm in diameter. The plating efficiency in control and PEG-treated experiments was approximately 12 and 6%, respectively.

Calli of 2 mm in diameter were transferred to auxin-free medium (medium D in Table 1) for plant regeneration. Shoots began to emerge about 10 days after the transfer, and they continued to emerge until approximately 50 days. Shoots were cut off and embedded in medium E (Table 1) for root formation.

2.4 Genome Constitutions of Regenerated Plants

The marked differences in karyotypes between *N. sylvestris* and *N. plumbaginifolia* enabled us to determine genome constitutions of plants regenerated from protoplast fusion. In two separate experiments we consistently observed that most regenerants were somatic hybrids of various ploidy levels such as PS, PPS, PSS, PPSS (Fig. 1C–F), etc.; parental species, especially *N. sylvestris*, were rare (Lin and Chen 1990; Hung et al. 1993).

To understand the basis of preferential recovery of somatic hybrids in the absence of artificial selection, we studied the relationship between time of shoot emergence (days from beginning of culture) and frequency of calli producing parental and hybrid plants (Lee and Chen 1992). In the control experiment (without PEG treatment), 50 calli differentiated into *N. sylvestris* plants and 8 differentiated into *N. plumbaginifolia* plants; no hybrid plants were observed. When the times of shoot emergence from calli were compared, it appeared that plant regeneration occurred earlier in *N. sylvestris* than in *N. plumbaginifolia* (Fig. 3). In the PEG-treated experiment, 45 calli differentiated into hybrid plants, 11 into *N. plumbaginifolia* plants, and only 2 into *N. sylvestris* plants. As shown in Fig. 3, plants regenerated during the early periods of culture (36–55 days) were almost exclusively somatic hybrids, while those regenerated in the last period of culture (66–75 days) were solely *N. plumbaginifolia*. When compared with the control, it appeared that hybrid plants regenerated earlier than both parents. These results suggest that *N. sylvestris* protoplasts are sensitive to PEG but develop rapidly, whereas those of *N. plumbaginifolia* are more resistant to PEG but develop slowly. The

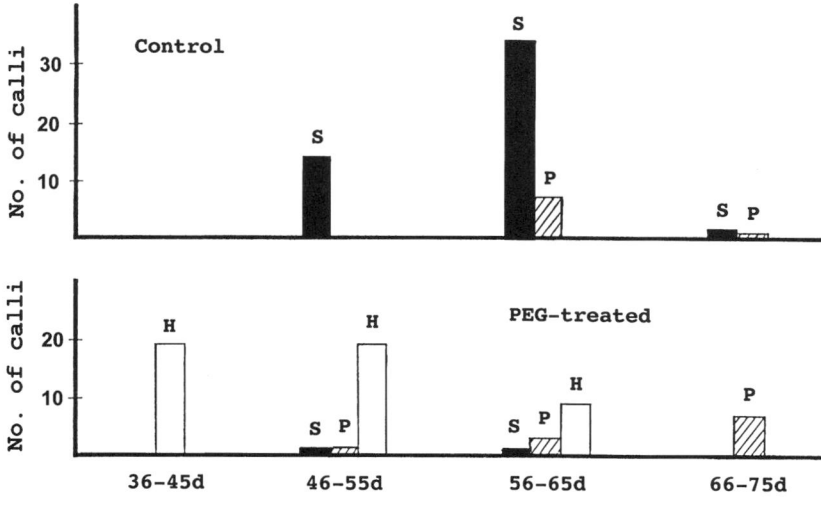

Fig. 3. Relationship between time of shoot emergence (days from beginning of culture) and frequency of calli producing parental and hybrid plants. *P N. plumbaginifolia*; *S N. sylvestris*; *H* somatic hybrid

preferential recovery of somatic hybrids may be the result of genetic complementation between the two parents.

A similar phenomenon has also been observed by other investigators in *Nicotiana* (Smith et al. 1976), *Datura* (Schieder 1978, 1980), and *Lycopersicon* (O'Connell and Hanson 1987). Thus, vigor in growth and development of hybrid cells appears to be a common phenomenon in plant tissue culture, and in many cases may be used as a basis for isolation of somatic hybrids when selectable markers are not available in the parents.

2.5 Identification of Somatic Hybrids

In addition to karyotype, other criteria that have been used for the identification of *N. sylvestris* + *N. plumbaginifolia* somatic hybrids are: plant morphology, isozyme patterns (Lee and Chen 1992), restriction fragment length polymorphism (RFLP) (Suen et al. 1997), and fluorescence of chromosomes revealed by genomic in situ hybridization (GISH) (Fig. 4A). GISH, in which total genomic DNA of a parent is used as a probe, can "paint" parental genomes in different colors (Schwarzacher et al. 1989), and is, therefore, a very useful technique for hybrid identification.

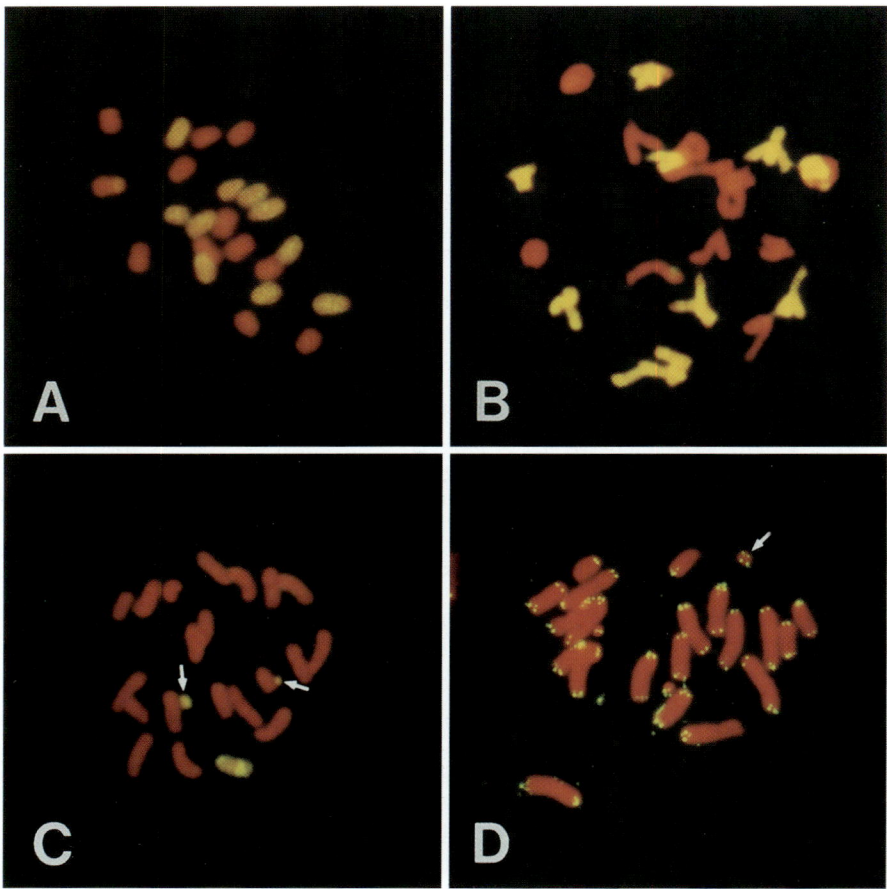

Fig. 4A–D. Fluorescence in situ hybridization of *N. sylvestris* + *N. plumbaginifolia* somatic hybrids and *N. plumbaginifolia-sylvestris* monosomic addition lines. **A** Somatic chromosomes of PS hybrid probed with *N. plumbaginifolia* total genomic DNA. **B** Meiotic chromosomes of PPSS hybrid probed with *N. plumbaginifolia* total genomic DNA, showing autosyndetic pairing. **C** Somatic chromosomes of a monosomic addition line probed with *N. sylvestris* total genomic DNA. In addition to the *N. sylvestris* chromosome, the two satellites of *N. plumbaginifolia* chromosomes (*arrows*, heteromorphic in size) were also labeled. **D** Somatic chromosomes of a monosomic addition line probed with the oligonucleotide (TTTAGGG)$_4$, showing both ends of the terminally deleted *N. sylvestris* chromosome (*arrow*) contain telomere sequences. The hybridization sites fluoresce yellow while the other parts of chromosomes fluoresce red

3 Applications of Somatic Hybrids

3.1 Genomic Relationships Between Parental Species

N. sylvestris and *N. plumbaginifolia* could not be crossed sexually (Christoff 1928). The successful production of somatic hybrids containing one set of chromosomes from each species provides an opportunity to investigate genome relationships between these two species through analysis of meiotic chromosome associations in the hybrids. Studies by the conventional aceto-carmine smear technique revealed predominant univalent and bivalent formation in PS diploid and PPSS tetraploid hybrids, respectively (Lin and Chen 1990). Analysis by GISH showed that all bivalents observed in the tetraploid hybrids were the result of autosyndesis (Fig. 4B). Thus, our results indicate that although *N. sylvestris* and *N. plumbaginifolia* were taxonomically classified into the same section (Goodspeed 1954), there is little homology between their chromosomes.

3.2 Nuclear-Cytoplasm Interaction

It has been reported that following protoplast fusion there is rapid segregation of parental chloroplasts. In some cases, segregation of chloroplasts seems to be random (Chen et al. 1977; Belliard et al. 1978), while in other cases it is biased (Kumar et al. 1982; Sundberg et al. 1991). The mechanism involved in chloroplast segregation is not fully understood; however, some studies suggest that the ratio of parental nuclear genomes in the fusion products may be an influencing factor (Sundberg and Glimelius 1991; Bonnema et al. 1992). To test this possibility, we analyzed the restriction patterns of chloroplast DNA (cpDNA) of somatic hybrids PS, PPS, PSS, and PPSS which differ in the ratio of parental nuclear genomes (Hung et al. 1993). We found that among the 118 plants analyzed, two had a pattern corresponding to a mixture of parental cpDNA while all the others had the pattern of either *N. sylvestris* or *N. plumbaginifolia*. In the latter case, the ratio of the two parental types fits 1:1 in all the four genome constitutions (PS, PPS, PSS, and PPSS) studied. It should be mentioned that protoplasts used in the fusion experiment were isolated from plants grown under identical conditions and, therefore, they should be physiologically similar. Also, during culture, the hybrid cells were not deliberately selected. Thus, our results suggest that chloroplast segregation in the somatic hybrids is independent of the nuclear background of the fusion products (Hung et al. 1993).

3.3 Establishment, Characterization and Utilization of Monosomic Alien Addition Lines

Monosomic alien addition lines, in which a single chromosome from one species is added to the genome of another, are useful materials for gene transfer (Jena and Khush 1989), gene/DNA marker assignment (Struss et al. 1992), isolation of chromosome- and region-specific DNA sequences (Chen et al. 1998; Ananiev et al. 1998a,b,c), chromosome isolation by flow cytometry (Doležel et al. 1994), and for high-resolution physical mapping as discussed by Ananiev et al. (1997).

In order to establish chromosome addition lines for such studies, we backcrossed somatic hybrids PPS to *N. plumbaginifolia* (Suen et al. 1997). A total of 102 monosomic addition plants, each containing two sets of *N. plumbaginifolia* chromosomes and a single *N. sylvestris* chromosome, were obtained in the BC_1 and BC_2 generations. Based on chromosome morphology and synteny relationships of RFLP markers, we classified these plants into 9 normal and 15 aberrant lines. The *N. sylvestris* chromosomes in the normal lines were chromosomes 2, 4, 5, 6, 7, 8, 9, 10, and 12, respectively. Four aberrant lines were the result of chromosome translocations, three were terminal deletions, and eight possessed telocentric chromosomes arising from centromere misdivision of normal chromosomes.

All the monosomic addition lines were characterized by molecular cytogenetic techniques. GISH revealed that there was no DNA introgression between the donor and recipient genomes during the process of establishment of monosomic addition lines (Fig. 4C). Fluorescence in situ hybridization using oligonucleotide $(TTTAGGG)_4$ as a probe showed that both ends of deficient and telocentric *N. sylvestris* chromosomes in the addition lines contained telomere sequences (Fig. 4D). According to the estimation of Narayan (1987), the nuclear DNA content of *N. sylvestris* (5.74 pg/2C) is about 1.5 times smaller than that of *N. plumbaginifolia* (8.29 pg/2C). Therefore, the monosomic addition lines, especially those containing deficient and telocentric chromosomes, would be ideal materials for the isolation of *N. sylvestris* chromosomes by flow cytometry and for the construction of chromosome- and arm-specific libraries.

Some of the RFLP markers localized on *N. sylvestris* chromosomes using monosomic addition lines (Suen et al. 1997) were mapped to the genetic map of *N. plumbaginifolia*. Comparison of the distribution of markers common to the synteny groups of *N. sylvestris* and the linkage groups of *N. plumbaginifolia* suggests that extensive chromosome reorganization has taken place during the evolution of these two species.

All the chromosome addition lines could be propagated and maintained vegetatively by in vitro culture and sexually through backcross to *N. plumbaginifolia* and self-fertilization. If more DNA markers, preferably cDNA and cloned genes, become available, these lines could be used for high-resolution physical mapping in *N. sylvestris*, using the irradiation technique similar to that practiced in mammalian hybrid-cell systems (Cox et al. 1990).

4 Summary and Conclusion

We present here our experiences for somatic hybridization between *N. sylvestris* and *N. plumbaginifolia*. The uniqueness of this method is that the protoplasts participating in fusion were from in vitro-cultured haploid plants, and that no deliberate selection was applied against parental cells during culture. One advantage of using in vitro-cultured plants as the source of protoplasts is that more consistent results can be obtained among experiments owing to uniformity of the growth conditions of donor plants. Because protoplasts of *N. sylvestris* and *N. plumbaginifolia* differ in response to PEG treatment and culture, and complementation of these characteristics makes the fusion products more viable and makes them grow faster than parental protoplasts, artificial selection is not necessarily required. Utilization of haploid protoplasts for fusion enables us to obtain hybrid plants with one set of chromosomes from each parent. These plants are invaluable for genome analysis when parental species cannot be crossed sexually. Because of the frequent occurrence of multiple fusions and somaclonal variation, hybrids with certain polyploidy can be produced from somatic hybridization using haploid protoplasts. These hybrids are useful materials for studying the effect of parental nuclear genomes on chloroplast segregation in somatic hybrids and for establishing chromosome addition lines for basic research as well as plant breeding.

Although the method described was developed primarily for somatic hybridization in *Nicotiana*, we believe that with some modifications it could be applicable to other plant genera.

5 Protocol

5.1 Protoplast Isolation

Haploid plants of *N. sylvestris* and *N. plumbaginifolia* obtained from anther culture are maintained as shoot cultures on MS salts (Murashige and Skoog 1962) with 1% sucrose and 0.8% Difco agar at 25 °C under 16-h illumination of 2000 lx. About 4 weeks after shoot transfer, the upper two to three fully expanded leaves of in vitro-cultured plants are excised and cut into 1-mm-wide strips in an enzyme solution containing 1% Cellulase Onozuka R10 (Yakult Honsha, Japan) and 0.2% Macerozyme R10 (Yakult Honsha, Japan) (medium A in Table 1). After incubation in the dark at 25 °C for 16 h, the protoplast suspension is gently shaken, poured through nylon sieves (75-μm aperture), and then centrifuged at $100\,g$ for 3–5 min. The floating protoplasts are picked up with a pipette, washed twice with W_5 solution (154 mM NaCl, 125 mM $CaCl_2 \cdot 2H_2O$, 5 mM KCl, 5 mM glucose), and then resuspended in this solution at a density of $2 \times 10^5\,ml^{-1}$.

5.2 Protoplast Fusion

Protoplast suspensions of the two *Nicotiana* species are mixed at equal volumes. Approximately 0.2 ml of the protoplast mixture is pipetted onto a 22×22-mm coverglass, which is held in a 60×15-mm Falcon 1007 petri dish by silicone fluid. The protoplasts are allowed to sediment for 5 min, and then 0.5 ml of a solution containing 50% PEG 1500, 0.2 M glucose, 10 mM $CaCl_2 \cdot 2H_2O$, and 0.7 mM KH_2PO_4, pH 5.8, is added dropwise. After 12 min of incubation, the PEG solution is eluted once with a high pH-high Ca^{2+} solution (50 mM glycine, 0.3 M glucose, 50 mM $CaCl_2 \cdot 2H_2O$, pH 10.5) and then twice with culture medium B (Table 1). The protoplasts are washed thoroughly with medium B, and then 0.7 ml of the same medium is added. The petri dishes are sealed with parafilm and incubated in the dark at 25 °C.

5.3 Protoplast Culture and Plant Regeneration

After 10 days of incubation, protoplast-derived colonies are collected on a nylon screen (pore size 20μm) and plated at a density of $1.5 \times 10^2 \, ml^{-1}$ in medium C (Table 1). The plates are kept at 25°C in the dark for 1 week and then at 25°C in the light (1000 lx, 16 h day^{-1}). When calli reach the size of 2 mm in diameter, they are transferred to medium D (Table 1) for shoot induction. Shoot cuttings are embedded in medium E (Table 1) for root formation.

Acknowledgments. This work was financially supported by the National Science Council, Republic of China. We thank Miss M.C. Liu for providing Fig. 4C,D.

References

Ananiev EV, Riera-Lizarazu O, Rines HW, Phillips RL (1997) Oat-maize chromosome addition lines: a new system for mapping the maize genome. Proc Natl Acad Sci USA 94:3524–3529
Ananiev EV, Phillips RL, Rines HW (1998a) Complex structure of knob DNA on maize chromosome 9: retrotransposon invasion into heterochromatin. Genetics 149:2025–2037
Ananiev EV, Phillips RL, Rines HW (1998b) A knob-associated tandem repeat in maize capable of forming fold-back DNA segments: are chromosome knobs megatransposons? Proc Natl Acad Sci USA 95:10785–10790
Ananiev EV, Phillips RL, Rines HW (1998c) Chromosome-specific molecular organization of maize (*Zea mays* L.) centromeric regions. Proc Natl Acad Sci USA 95:13073–13078
Belliard G, Pelletier G, Vedel F, Quetier F (1978) Morphological characteristics and chloroplast DNA distribution in different cytoplasmic parasexual hybrids of *Nicotiana tabacum*. Mol Gen Genet 165:231–238
Bonnema AB, Melzer JM, Murray LW, O'Connel MA (1992) Nonrandom inheritance of organellar genomes in symmetric and asymmetric somatic hybrids between *Lycopersicon esculentum* and *L. pennellii*. Theor Appl Genet 84:435–442
Chen CC, Huang CR, To KY (1985) Anther cultures of four diploid *Nicotiana* species and chromosome numbers of regenerated plants. Bot Bull Acad Sin 26:147–153
Chen K, Wildman SG, Smith HH (1977) Chloroplast DNA distribution in parasexual hybrids as shown by polypeptide composition of fraction I protein. Proc Natl Acad Sci USA 74:5109–5112
Chen ZJ, Phillips RL, Rines HW (1998) Maize DNA enrichment by representational difference analysis in a maize chromosome addition line of oat. Theor Appl Genet 97:337–344
Christoff M (1928) Cytological studies in the genus *Nicotiana*. Genetics 13:233–277
Cox DR, Burmeister M, Price ER, Kim S, Myers RM (1990) Radiation hybrid mapping: a somatic cell genetics method for constructing high-resolution maps of mammalian chromosomes. Science 250:245–250
Cséplö A, Nagy F, Maliga P (1984) Interspecific protoplast fusion to rescue a cytoplasmic lincomycin resistance mutation into fertile *Nicotiana plumbaginifolia* plants. Mol Gen Genet 198:7–11
Doležel J, Lucretti S, Schubert I (1994) Plant chromosome analysis and sorting by flow cytometry. Crit Rev Plant Sci 13:275–309
Famelaer I, Gleba YY, Sidorov VA, Kaleda VA, Parakonny AS, Boryshuk NV, Chereb NN, Negrutiu I, Jacobs M (1989) Intrageneric asymmetric hybrids between *Nicotiana plumbaginifolia* and *Nicotiana sylvestris* obtained by gamma-fusion. Plant Sci 61:105–117
Famelaer I, Negrutiu I, Mouras A, Vaucheret H, Jacobs M (1990) Asymmetric hybridization in *Nicotiana* by gamma fusion and progeny analysis of self-fertile hybrids. Theor Appl Genet 79:513–520
Gleba YY, Parokonny A, Kotov I, Negrutiu I, Momot V (1987) Spatial separation of parental genomes in hybrids of somatic plant cells. Proc Natl Acad Sci USA 84:3709–3713
Goodspeed TH (1954) The genus *Nicotiana*. Chronica Botanica, Waltham, Massachusetts

Gray JC, Kung SD, Wildman SG (1974) Origin of *Nicotiana tabacum* L. detected by polypeptide composition of fraction I protein. Nature 252:226–227
Huang HC, Chen CC (1988) Genome multiplication in cultured protoplasts of two *Nicotiana* species. J Hered 79:28–32
Hung CY, Lai YK, Feng TY, Chen CC (1993) Chloroplast segregation in somatic hybrids of *Nicotiana plumbaginifolia* and *N. sylvestris* having different ratios of parental nuclear genomes. Plant Cell Rep 13:83–86
Jena KK, Khush GS (1989) Monosomic alien addition lines of rice: production, morphology, cytology, and breeding behavior. Genome 32:449–455
Kao KN, Constabel F, Michayluk MR, Gamborg OL (1974) Plant protoplast fusion and growth of intergeneric hybrid cells. Planta 120:215–227
Kumar A, Cocking EC, Bovenberg WA, Kool AJ (1982) Restriction endonuclease analysis of chloroplast DNA in interspecies somatic hybrids of *Petunia*. Theor Appl Genet 62:377–383
Lee FM, Chen CC (1990) Somatic hybridization between *Nicotiana sylvestris* and *N. otophora* without application of selection. J Hered 81:313–317
Lee FM, Chen CC (1992) Preferential recovery of somatic hybrids from protoplast fusion of two *Nicotiana* species in the absence of artificial selection. Taiwania 37:11–18
Lin RF, Chen CC (1990) Cytological studies of interspecific somatic hybrids in *Nicotiana*. Bot Bull Acad Sin 31:179–187
Medgyesy P, Menczel L, Maliga P (1980) The use of cytoplasmic streptomycin resistance: chloroplast transfer from *Nicotiana tabacum* into *Nicotiana sylvestris*, and isolation of their somatic hybrids. Mol Gen Genet 179:693–698
Murashige T, Skoog F (1962) A revised medium for rapid growth and bioassays with tobacco tissue cultures. Physiol Plant 15:473–497
Nagy JI, Maliga P (1976) Callus induction and plant regeneration from mesophyll protoplast of *Nicotiana sylvestris*. Z Pflanzenphysiol 78:453–455
Narayan RKJ (1987) Nuclear DNA changes, genome differentiation and evolution in *Nicotiana* (Solanaceae). Plant Syst Evol 157:161–180
O'Connell MA, Hanson MR (1987) Regeneration of somatic hybrid plants formed between *Lycopersicon esculentum* and *L. pennellii*. Theor Appl Genet 75:83–89
Schieder O (1978) Somatic hybrids of *Datura innoxia* Mill + *Datura stramonium* L. var. *tatula* L. I. Selection and characterization. Mol Gen Genet 162:113–119
Schieder O (1980) Somatic hybrids between a herbaceous and two tree *Datura* species. Z Pflanzenphysiol 98:119–127
Schwarzacher T, Leitch AR, Bennett MD, Heslop-Harrison JS (1989) In situ localization of parental genomes in a wide hybrid. Ann Bot 64:315–324
Smith HH, Kao KN, Combatti NC (1976) Interspecific hybridization by protoplast fusion in *Nicotiana*. J Hered 67:123–128
Struss D, Quiros CF, Robbelen G (1992) Mapping of molecular markers on *Brassica* B-genome chromosomes added to *Brassica napus*. Plant Breed 108:320–323
Suen DF, Wang CK, Lin RF, Kao YY, Lee FM, Chen CC (1997) Assignment of DNA markers to *Nicotiana sylvestris* chromosomes using monosomic alien addition lines. Theor Appl Genet 94:331–337
Sundberg E, Glimelius K (1991) Effects of parental ploidy level and genetic divergence on chromosome elimination and chloroplast segregation in somatic hybrids within Brassicaceae. Theor Appl Genet 83:81–88
Sundberg E, Lagercrantz U, Glimelius K (1991) Effects of cell type used for fusion on chromosome elimination and chloroplast segregation in *Brassica oleracea* (+) *Brassica napus* hybrids. Plant Sci 78:89–98

III.3 Somatic Hybridization Between *Nicotiana tabacum* L. (Tobacco) and *Atropa belladonna* L. (Deadly Nightshade)

M.K. ZUBKO[1,3], E.I. ZUBKO[1], O.A. KHVEDYNICH[1], S.V. LOPATO[1], S.A. LATIPOV[2], and Yu. Yu. GLEBA[1]

1 Introduction

Somatic hybridization based on protoplast fusion appeared as an alternative to sexual hybridization of higher plants. The term was proposed by Melchers and Labib (1974). Numerous experiments on somatic hybridization have been carried out using *Nicotiana* species (Gleba and Sytnik 1984; Negrutiu et al. 1989; Bates 1992; Puite 1992). The genus *Nicotiana* belongs to the Solanaceae family and includes about 70 species showing significant genetic and morphological variability (Goodspeed 1954). The most important species is *N. tabacum* L., or tobacco, cultivated widely as a source for the tobacco industry. *N. tabacum* ($2n = 48$) originated from Central and South America (Darlington and Wylie 1955) but is now unknown in the wild. Different varieties of tobacco display strong vigor of stems possessing simply organized large leaves covered by abundant trichomes. Determinate inflorescences contain about 60–100 pink flowers, which after self- or cross-pollination develop capsules full of small brown seeds. Due to these characteristics, which allow easy propagation and manipulation with cells and tissues in vitro, tobacco became a convenient model species for plant cell- and genetic engineering. The success of whole-plant regeneration from cultivated tobacco protoplasts (Takebe et al. 1971) further provided basis for the generation of hybrids by protoplast fusions. First somatic hybrids were produced in interspecific *Nicotiana* combinations (Carlson et al. 1972; Melchers and Labib 1974; Gleba et al. 1975; Kameya 1975). In these experiments, basic foundations for parasexual hybridization of higher plants were established. Since the possibilities of overcoming distant sexual cross-incompatibility were soon demonstrated (Gleba and Hoffman 1978, 1979; Melchers et al. 1978; Krumbiegel and Schieder 1979, 1981), subsequent efforts focused on the development of reproducible technologies for the creation of true hybrid plants combining traits from both parental species. *Nicotiana* species were further involved in the construction of intergeneric somatic hybrids with different genera (Table 1).

---

[1] Institute of Cell Biology and Genetic Engineering, National Academy of Sciences of Ukraine, Zabolotnogo Str. 148, 252022 Kiev, Ukraine
[2] Institute of Genetics, Moldovian Academy of Sciences, Lesnaya Str. 20, 277018 Kishinev, Moldova
[3] Present address: School of Biological Sciences, Manchester University, 3.614 Stopford Building, Oxford Road, Manchester M13 9PT, United Kingdom

The integration of complete parental genomes (symmetric hybridization) is often prevented by incompatibility (Krumbiegel and Schieder 1979, 1981; Endo et al. 1988; Lazar 1989; Wolters et al. 1993). Asymmetric (partial) transfer of genetic material has been shown to be the most promising way to solve this problem (see reviews of Gleba and Sytnik 1984; Negrutiu et al. 1989; Bates 1992; Puite 1992). Asymmetrization of somatic hybrids often occurs spontaneously due to elimination of parental chromosomes, and irradiation of donor protoplasts provides a reliable approach for increasing asymmetry in hybrids (Piastuch and Bates 1990; Hinnisdaels et al 1991; Babiychuk et al. 1992; Bates 1992; Puite 1992; Wolters et al. 1993; Kisaka and Kameya 1994).

Many aspects of symmetry versus asymmetry in phylogenetically distant somatic hybrids were investigated in different *Nicotiana* spp. (+) *Atropa belladonna* combinations. *A. belladonna* (deadly nightshade, $2n = 72$) is a tropane alkaloid-bearing plant (tribe Lycinae of the Solanaceae family) widely distributed in Europe, SW Asia, and India (Darlington and Wylie 1955; D'Arcy 1979). Nightshade is a perennial herb with sympodial shoots and entire leaves having a clammy glandular surface. The fruit is a black poisonous berry. The main alkaloid of the plant, atropine, is used in medicine. The usage of nightshade stimulated the involvement of this plant in distant hybridization with other alkaloid-producing plants including species of *Datura, Hyoscyamus, and Nicotiana* (Table 1). Hybridization between *Nicotiana* and *Atropa* was extensively carried out. Being quite different in developmental characteristics and secondary metabolism, tobacco and nightshade serve as a valid model system for somatic hybridization to investigate the potential of integration between distant genomes. In fact, almost all phenomenology of somatic hybridization between distantly related species could be observed in the *Nicotiana* (+) *Atropa* combination. The stable presence of both parental nuclear genomes was shown in cell lines *A. belladonna* (+) *N. chinensis* after 6 months of cultivation in vitro (Gleba et al. 1982). Some of these lines regenerated imperfect shoots. After 12 months of cultivation, most cell lines retained chromosomes of both parents, but in some other clones species-specific elimination of nearly all *Atropa* chromosomes had occurred (Gleba et al. 1983). Polyploidization and chromosomal changes (reconstituted and ring chromosomes) were observed in *A. belladonna* (+) *N. chinensis* hybrids (Gleba et al. 1982, 1983). In these studies, the first evidence for non-random arrangement of parental chromosomes in somatic hybrids was obtained (Gleba et al. 1983). Investigation of a cell line of *N. tabacum* (+) *A. belladonna* revealed a new class of ribosome DNA repeat which was suggested to be a result of interspecific recombination between two rDNA units or amplification of preexisting units (Borisjuk et al. 1988). Studies on the transmission of T-DNA crown gall characters from *A. belladonna* to tobacco resulted in the production of abnormal hybrid shoots (Gleba et al. 1986; Kanevsky and Gleba 1986). Generation of cell lines or abnormal shoots indicated the limitation of regeneration capacity in most studies on intergeneric hybridization where model *Nicotiana* species were used as one of the parents (Kao 1977; Wetter 1977; Chien et al. 1982; Skarzhynskaya et al. 1983; Imamura et al. 1987; Kishinami and Widholm 1987; Vries et al. 1987; Ye et al. 1987; Endo et al. 1988; Gilissen et al. 1992a,b;

Table 1. Distant somatic hybridization in genera of *Nicotiana* and *Atropa*

Combination of parental species	Hybrid material obtained	Reference
Hybrids		
Glycine max (+) *N. glauca*	Cell lines	Kao (1977); Wetter (1977); Wetter and Kao (1980)
N. tabacum (+) *G. max*	Cell lines	Chien et al. (1982)
Datura innoxia (+) *A. belladonna*	Shoots; plants (asymmetric hybrids)	Krumbiegel and Schieder (1979); Krumbiegel and Schieder (1981)
N. tabacum (+) *Solanum tuberosum*	Cell lines; plants (asymmetric hybrids)	Skarzhynskaya et al. (1982; 1983)
N. tabacum (+) *S. sucrense*	Cell lines (polyploid symmetric hybrids)	Skarzhynskaya et al. (1983)
N. plumbaginifolia (+) *S. tuberosum*	Shoots (asymmetric hybrids)	Vries et al. (1987)
S. tuberosum (+) *N. plumbaginifolia*[a]	Cell lines (asymmetric hybrids)	Gilissen et al. (1992a,b)
Hyoscyamus muticus (+) *N. tabacum*	Plants (asymmetric hybrids)	Potrykus et al. (1984); Lazar (1989)
N. tabacum (+) *H. muticus*	Cell lines (asymmetric hybrids)	Imamura et al. (1987)
N. plumbaginifolia (+) *H. muticus*	Cell lines	Kishinami and Widholm (1987)
N. tabacum (+) *Hordeum vulgare*	Plants (highly asymmetric hybrids)	Somers et al. (1986)
N. tabacum (+) *Daucus carota*	Plants (asymmetric hybrids)	Dudits et al. (1987)
N. plumbaginifolia (+) *D. carota*	Cell lines (symmetric hybrids)	Ye et al. (1987)
D. carota (+) *N. tabacum*	Plants (highly asymmetric hybrids)	Kisaka and Kameya (1994)
A. belladonna (+) *N. chinensis*	Cell lines; shoots	Gleba et al. (1982; 1983)
N. tabacum (+) *A. belladonna*	Cell lines; shoots; plants (symmetric hybrids)	Gleba et al. (1986); Kanevsky and Gleba (1986); Borisjuk et al. (1988)
N. tabacum (+) *A. belladonna*	Plants (highly asymmetric hybrids)	Babiychuk et al. (1990)
N. plumbaginifolia (+) *A. belladonna*	Plants (asymmetric hybrids)	Babiychuk et al. (1992)
N. plumbaginifolia (+) *A. belladonna*	Cell lines, shoots; plants (asymmetric hybrids)	Gleba et al. (1988); Yemets et al. (2000)
Duboisia hopwoodi (+) *N. tabacum*	Shoots (asymmetric hybrids)	Endo et al. (1988)
N. tabacum (+) *Physochlaine officinalis*	Plants (symmetric hybrids)	Babiychuk et al. (1990)
Lycopersicon esculentum (+) *N. tabacum*	Plants (asymmetric hybrids)	Wolters et al. (1993)

Somatic Hybridization Between *Nicotiana tabacum* L. and *Atropa belladonna* L.

<u>*L. peruvianum*</u> (+) *N. plumbaginifolia*	Plants (asymmetric hybrids)	Wolters et al. (1993)
<u>*N. plumbaginifolia*</u> (+) *L. esculentum*	Plants (asymmetric hybrids)	Vlahova et al. (1997)
A. belladonna (+) <u>*H. muticus*</u>	Plants (asymmetric hybrids)	Ahuja et al. (1993); Giri and Ahuja (1995)
Cybrids		
<u>*P. hybrida*</u> (+*N. tabacum*)[b]	Plants	Pental et al. (1986)
<u>*N. tabacum*</u> (+*Petunia hybrida*)	Plants	Glimelius and Bonnett (1986); Dragoeva et al. (1999)
<u>*N. tabacum*</u> (+*A. belladonna*)	Plants	Kushnir et al. (1987)
<u>*N. tabacum*</u> (+*Salpiglossis sinuata*)	Plants	Thanh et al. (1988)
<u>*N. tabacum*</u> (+) *S. tuberosum*	Plants (recombinant plastid DNA)	Thanh and Medgyesy (1989)
<u>*N. tabacum*</u> (+*Scopolia carniolica*)	Plants	Kushnir et al. (1991); Babiychuk et al. (1995); Zubko et al. (1996)
A. belladonna (+*N. tabacum*)	Plants (chlorophyll-deficient)	Kushnir et al. (1991)
<u>*S. tuberosum*</u> (+*N. sylvestris*)	Plants (recombinant mitochondrial DNA)	Perl et al. (1991)
<u>*N. tabacum*</u> (+*Lycium barbarum*)	Plants	Babiychuk et al. (1995)
<u>*N. tabacum*</u> (+*P. officinalis*)	Plants	Babiychuk et al. (1995)
<u>*N. tabacum*</u> (+*Nolana paradoxa*)	Plants	Babiychuk et al. (1995)
<u>*N. plumbaginifolia*</u> (+*A. belladonna*)	Plants	Babiychuk et al. (1995)
<u>*N. tabacum*</u> (+*Hyoscyamus niger*)	Plants (recombinant mitochondrial DNA)	Zubko et al. (1996)

Underlined species represents a complete or a greater part in hybrid or cybrid nuclear genomes.
[a] Different hybrid lines in the fusion combination are represented by predominant proportions of nuclear genomes from different parental species.
[b] Species in parentheses are donors of plastome.

Wolters et al. 1993; Yemets et al. 2000). Morphologically normal plants have been produced only for highly asymmetric hybrids (Negrutiu et al. 1989). Hybrid plants *N. tabacum* (+) *D. carota* showed morphological variability (Dudits et al. 1987), whereas other intergeneric hybrids were uniform and similar to one of the parental species (Somers et al. 1986; Lazar 1989; Hinnisdaels et al. 1991; Wolters et al. 1993; Kisaka and Kameya 1994). In this respect, fertile asymmetric hybrids, produced after fusions of *N. plumbaginifolia* protoplasts and γ-irradiated *A. belladonna* protoplasts, contained only few chromosomes of *A. belladonna* (Gleba et al. 1988). No irradiation dose effect on chromosome elimination degree was found in these experiments. Spontaneous extensive asymmetrization leading to morphologically normal plants that retained only single *A. belladonna* chromosomes under kanamycin-selective background was observed with low frequency in *N. tabacum* (+) *A. belladonna* hybrids (Babiychuk et al. 1992). Spontaneous and γ-irradiation-induced asymmetrization was also achieved in reciprocal somatic hybrids *N. plumbaginifolia* (+) *A. belladonna*, retaining a whole set of chromosomes from one parent and few chromosomes from another, after selection for kanamycin and amiprophosmethyl resistance (Yemets et al. 2000). UV irradiation was used as an alternative tool for the asymmetric transfer of two to four chromosomes from tomato to *N. plumbaginifolia* using a selection in the presence of kanamycin (Vlahova et al. 1997). A unique example of regeneration of morphologically normal plants is known for completely symmetric hybrids *N. tabacum* (+) *A. belladonna*. Crossed with tobacco, they nevertheless produced asymmetric progeny with significant loss of *A. belladonna* chromosomes (Babiychuk et al. 1990). Uniformity and normal *Nicotiana* morphology have been also reported to be feature of *Nicotiana* cybrids possessing plastids from *A. belladonna* (Kushnir et al. 1987; Babiychuk et al. 1995) as well as from other phylogenetically distant donors (Glimelius and Bonnett 1986; Thanh et al. 1988; Babiychuk et al. 1995; Dragoeva et al. 1999).

2 Somatic Hybridization

The above data show that in most cases the construction of morphologically balanced somatic hybrids between phylogenetically remote species is still not routine. Further approaches in this area are required to provide success in generating hybrid plants exhibiting broad spectra of genetic variability as a source for alteration of developmental and metabolic characteristics. In this respect, we describe here the production of asymmetric somatic hybrids between *N. tabacum* and *A. belladonna* using different schemes. In the first experiment, "gamma-plants" of *N. tabacum* and *A. belladonna*, each obtained by γ-irradiation of seeds (1000 Gy) and subsequent rescue in vitro, were used for protoplast fusions. In the second experiment, fusions were carried out between protoplasts from a gamma-plant of *N. tabacum* and γ-irradiated protoplasts (200 Gy) from *A. belladonna*-A4 transformed by *Agrobacterium rhizogenes*.

The third experiment was based on genetic complementation between albino plants of N. tabacum (plastome mutant) and white cybrid A. belladonna (+N. tabacum).

The main aims of this work were: (1) to generate somatic hybrids N. tabacum (+) A. belladonna to examine the approaches undertaken for inducing asymmetry and genetic variability; (2) to perform morphological, biochemical, and karyological analyses to estimate the extent of asymmetry in resultant hybrids; (3) to study meiosis and histological peculiarities during flowering in sterile hybrid plants.

2.1 Hybrids Between "Gamma-Plants" of N. tabacum and A. belladonna (Experiment 1)

It has previously been shown that seeds of higher plants γ-irradiated at high doses germinate and only reach the two-cotyledon stage in soil, forming so-called gamma seedlings which die very soon (Grodzynsky 1989). We reported that gamma seedlings may be restored in vitro to functional gamma-plants or cell lines which are presumably less genetically conserved than normal plants, and they therefore could be a subject of genetic and cell manipulations (Zubko et al. 1990a,b, 1993).

Fusions between protoplasts of gamma-plants N. tabacum R100a and A. belladonna 100-01 resulted in the isolation of 14 colonies capable of greening on C medium. Hybridity of these isolates was predicted based upon chlorophyll deficiency of line R100a and the inability of A. belladonna protoplasts to divide on SW medium. Three clones did not regenerate. Regenerants from the remaining 11 clones could be divided into two groups according to their morphology.

The first group included hybrid lines Rab_1, Rab_3, Rab_5, Rab_6, Rab_7, Rab_8, Rab_9, Rab_{10}. The general characteristic of this group is that they grew only in the presence of cytokinin and expressed a stable phenotype. These hybrids had poorly developed leaves (Fig. 1A,B) and they were not capable of rooting and development on hormone-free medium. All lines had trichomes on their leaves, which is a typical feature of tobacco, but not Atropa. All lines differed in plant phenotype in vitro by size and shape of shoots, leaf morphology, and degree of greening. The leaves of hybrid Rab_5 synthesized anthocyanins (an Atropa trait). Figure 2 shows hsp patterns for tobacco, nightshade, and randomly selected hybrids Rab_1, Rab_6, Rab_8, and Rab_9. The low molecular mass (18–25 kDa) hsp patterns of all hybrids reflected the main bands from both parents. Hybrid Rab_1 expressed two high molecular mass hsp of approximately 94 and 65 kDa, and they are absent in both parents. Hybrids from the first group contained isozyme bands of aspartate-aminotransferase from both tobacco and nightshade (data not shown). Esterase isozymes appear as a basic set of bands from Nicotiana, some bands from Atropa, in addition to new bands (Fig. 3A). Tobacco amylase bands were clearly resolved in hybrids; but no Atropa bands were detected (Fig. 3B). Thus, morphological and biochemical analyses confirm that regenerants of the first group express different parts

Fig. 1A–F. Morphological diversity of asymmetric hybrids resulting after fusions of protoplasts from gamma-plants of *N. tabacum* and *A. belladonna*. **A, B** Different morphology of shoot-like hybrids, lines Rab_6 and Rab_9, growing on the medium MS containing $1 mg l^{-1}$ of kinetin. **C–F** Hybrid plants growing in soil, lines Rab_{2-1}, Rab_{4a}, Rab_{11-2}, and Rab_{11-21g}, respectively

of the nuclear genome from *Atropa*, and these must represent slightly asymmetric hybrids.

The second group included three lines, Rab_2, Rab_4, and Rab_{11}. They were morphologically normal plants resembling wild-type tobacco, but differed in their phenotypic characters such as leaf and flower composition and growth

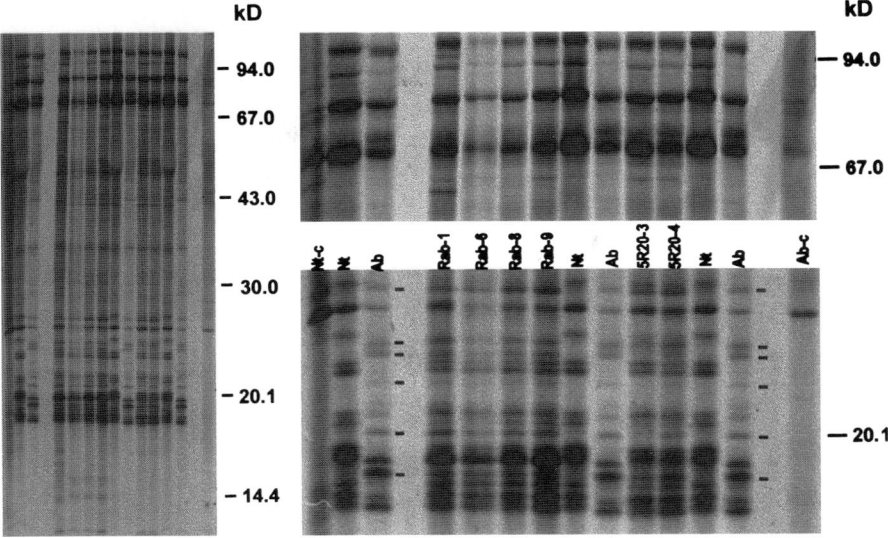

Fig. 2. Autoradiograms of hsp of *A. belladonna, N. tabacum* and their somatic hybrids: whole hsp profile (*left panel*), high molecular mass hsp (*top right*) and low molecular weight hsp (*bottom right panel*; *bars* indicate species-specific proteins of *A. belladonna*)

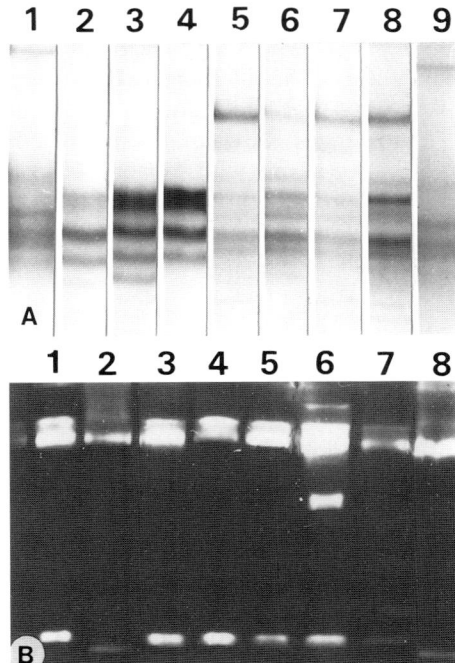

Fig. 3A,B. Isozyme analysis of somatic hybrids. **A** Esterase isozymes of *A. belladonna* (*1,9*) *N. tabacum*, line R100a1 (*2*), and their somatic hybrids: Rab_{4a} (*3*); Rab_{2-1} (*4*); Rab_5 (*5*); Rab_6 (*6*); Rab_8 (*7*); Rab_9 (*8*). **B** Amylase isozymes of *N. tabacum*, line R100a1 (*1*); *A. belladonna* (*2,8*) and somatic hybrids: Rab_1 (*3*); Rab_{4a} (*4*); Rab_{11} (*5*); Rab_{2-1} (*6*); Rab_6 (*7*)

habit under field conditions. Line Rab_2 developed morphologically normal shoots with no ability to form roots on hormone-free medium. Line Rab_2 was thought to be unstable in the phenotype since the derived line Rab_{2-1} (Fig. 1C) could form normal roots. Hybrid Rab_{4a} was morphologically regular and stable in phenotype at least during 1-year observation in the subculture (Fig. 1D).

Plants of the line Rab_{11} (Fig. 1E) were capable of rooting but were not completely normal and stable in their morphology. First, a derived line Rab_{11-2} with dark green leaves segregated out from Rab_{11}. From this line, we isolated a secondary lateral shoot with light-green leaves (Rab_{11-21g}, Fig. 1F). Imperfect offspring such as Rab_{11m} had wrinkled leaf blades, deviations in apical dominance, and abnormal roots. Leaves in Rab_{11m} plants usually had light green sectors. Regenerants obtained from light green segments of leaves were abnormal green plants, unable to root, and with slow growth. Regenerants from darker green segments of the same leaf grew rapidly and were similar to the normal phenotype of line Rab_{11}. Because of their morphological regularity and capacity to grow in the soil, these plants were studied more extensively.

Plants from the second group displayed aspartate-aminotransferase isozyme patterns identical to those in tobacco (data not shown). The esterase patterns were similar to tobacco type, but differed quantitatively (Fig. 3A). Two hybrids, sublines Rab_{2-1} and Rab_{11-2}, had novel bands in the amylase pattern (Fig. 3B). It is interesting to note that the initial line Rab_{11} was identical to tobacco for the amylase isozyme set (Fig. 3B). The plants possessed plastids of *A. belladonna* (Fig. 4A) and mitochondria of tobacco type (Fig. 4B). Hsp patterns in two lines, Rab_{2-1} and Rab_{4a}, resembled those in tobacco (data not shown). Cytological studies showed there were different chromosome numbers in the hybrids. Karyotypes of lines Rab_{2-1}, Rab_{4a}, and Rab_{11} consisted of 64–66, 46–48, and 72–76 chromosomes, respectively.

Flowers of Rab_{2-1} plants possessed petalloid anthers and produced seeds after backcrossing with tobacco; 6% of these seeds were capable of germination and production of plants. Rab_{4a} hybrids possess normal anthers, which produce a small quantity of pollen, but no seeds were produced after self-pollination.

To further investigate causes of self-sterility, we carried out a cytological study of flower composition and meiosis of the hybrid line Rab_{4a}. Their flowers were morphologically similar to those in wild-type tobacco (Fig. 5A,C). The stamens of Rab_{4a} were predominantly quadrilocular, but sometimes trilocular due to the absence of a partition between two loculi (Fig. 5D). The loculi of anthers appeared to be narrow, because the microsporocytes were degenerate. At the same time, the basal tissue and the anther wall were enlarged in comparison with the wild type. In the wall, the fibrous layer was developed most intensely (Fig. 5E).

An unusual organization of chromosome apparatus was observed in Rab_{4a} (Fig. 6). In meiotic metaphase I, all chromosomes were involved in the formation of rings. We found two rings per cell (Fig. 6B) In control experi-

Fig. 4. RFLP analyses of organelle DNA in asymmetric hybrids. **A** cp-DNA (*Hind*III) restriction patterns of *N. tabacum*, line R100a1 (*1*), *A. belladonna* (*2*), hybrids Rab$_{4a}$ (*3*), and Rab$_{11}$ (*4*); (*5*) Lambda *Hind*III markers. **B** DNA gel blot hybridization of mtDNA from *N. tabacum* (*1*), *A. belladonna* (*2*), hybrids Rab$_{4a}$ (*3*), and Rab$_{11}$ (*4*) with *coxI* probe. mtDNAs were digested with *Bam*HI

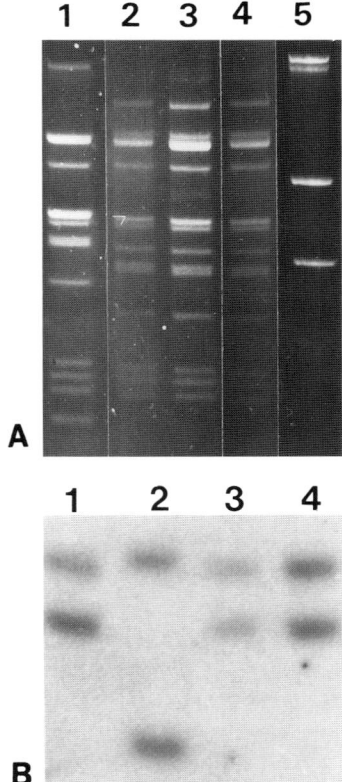

ments, no ring-like structures were observed in meiotic divisions of tobacco gamma-plants (data not shown). Meiotic figures were also characterized by multivalent formation in diakinesis and metaphase I, lagging (delay) of chromosomes, multiple chromosomal and chromatid bridges with fragments in anaphase (Fig. 6A,C,D).

The gynoecium of Rab$_{4a}$ hybrid consisted of a pistil with a biolcular ovary (Fig. 5F). On the placenta, seed buds were formed, but their development was changed. Some seed buds (in Fig. 5F, they are small) consisted of nucellus and integument. In the nucelli, megasporocytes were differentiated and proceeded to meiosis. As a tendency, meiosis was not completed by megaspore formation, and megasporocytes degenerated. Megasporocytes were not differentiated in a number of seed buds. In such cases, meristematic embryos proliferated. Embryos were at first small hillocks, which then grew and changed into oval shapes and then, after elongation, into cylinders (Fig. 5F,D, large seed buds). After pollination of flowers with wild-type tobacco, the pollen tubes reached ovaries, but grew between seed buds and on their surface, preventing fertilization.

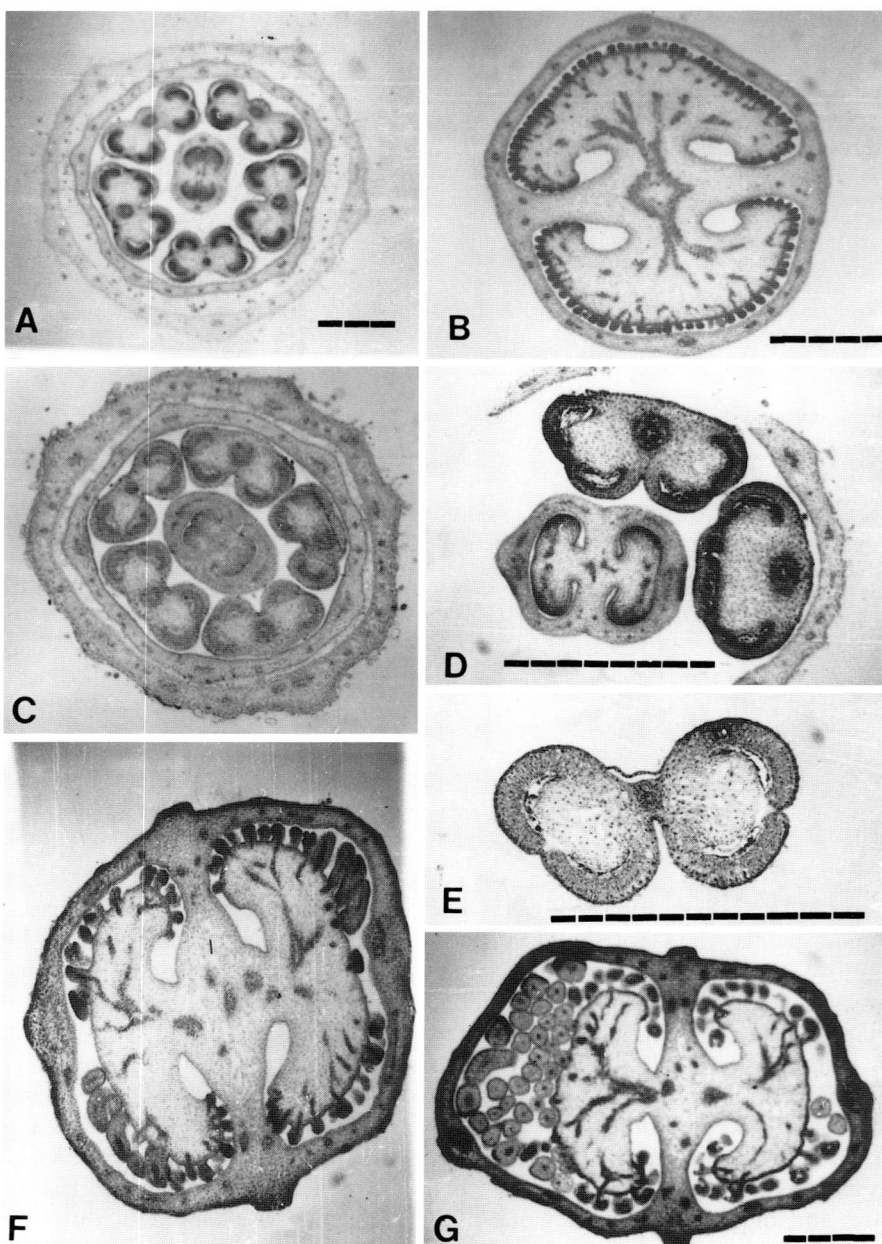

Fig. 5A–G. Cross-sections through flower organs of wild-type tobacco and asymmetric tobacco-like hybrid Rab$_{4a}$. **A** Mature tobacco flower composition. **B** Normal ovary of tobacco flower. **C, D** Cross sections through flower of hybrid Rab$_{4a}$. **E** Stamens of hybrid Rab$_{4a}$. **F, G** Ovaries of Rab$_{4a}$ hybrid with abnormal seed buds. *Bar* 1 mm

Fig. 6A–D. Anomalies of meiosis in asymmetric hybrid Rab$_{4a}$. **A** Formation of uni-, three- and tetravalents in methaphase I. **B** Formation of chromosomal rings in methaphase I. **C** Origin of multiple chromosomal and chromatidic bridges with fragments in anaphase I. **D** The lagging of chromosomes in anaphase

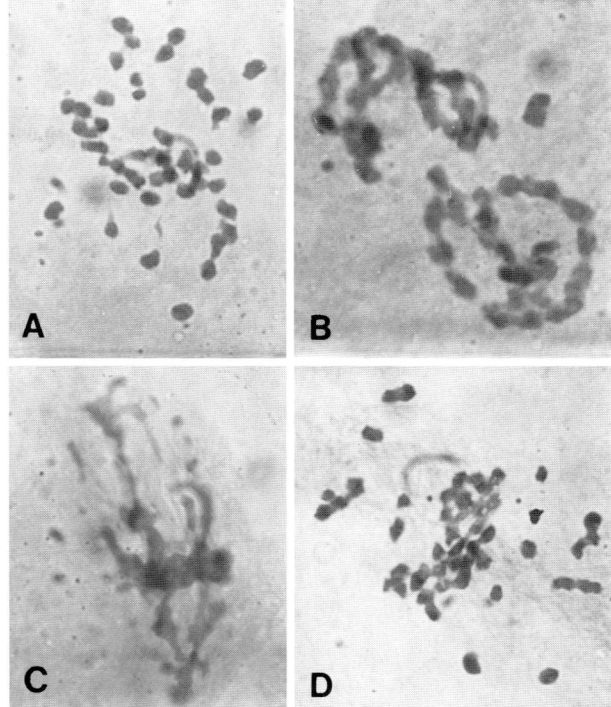

2.2 Hybrids Between a Gamma-Plant of *N. tabacum* and *A. belladonna* Transformed by *A. rhizogenes* (Experiment 2)

Protoplasts of *A. belladonna* A4 were subjected to γ-irradiation at 200 Gy before fusions. Sixteen hybrid clones with different capacity to morphogenesis resulted after the fusion experiment. Six lines produced shoots on regenerating medium, two of which, 5R20-1 and 5R20-2, were able to grow in soil. Their stamens were often converted to petals, but some anthers had normal morphology and contained small amounts of pollen. These plants produced capsules containing a few dozen of seeds. Four other lines grew slowly both in vitro and in vivo. Two of these hybrid lines, 5R20-3 and 5R20-4, produced roots on regeneration medium with cytokinin (Fig. 7A,B). The dissected roots were able to grow autonomously on the hormone-free medium (Fig. 7C).

Secondary regenerants of line 5R20-4 were different in their morphology and ability to root. Among the plants capable of rooting, there were sublines characterized by the hairy roots syndrome, a typical feature of the donor (nightshade) phenotype (Fig. 7D,E). Some leaves of those plants formed roots similar in phenotype to the hairy roots of *A. belladonna*-A4. Roots of lines 5R20-3 and 5R20-4 synthesized chlorophyll. Leaves exhibited purple zones of anthocyanins (Fig. 8). Aseptic 5R20-3 plants had light green, rather than green, pigmentation.

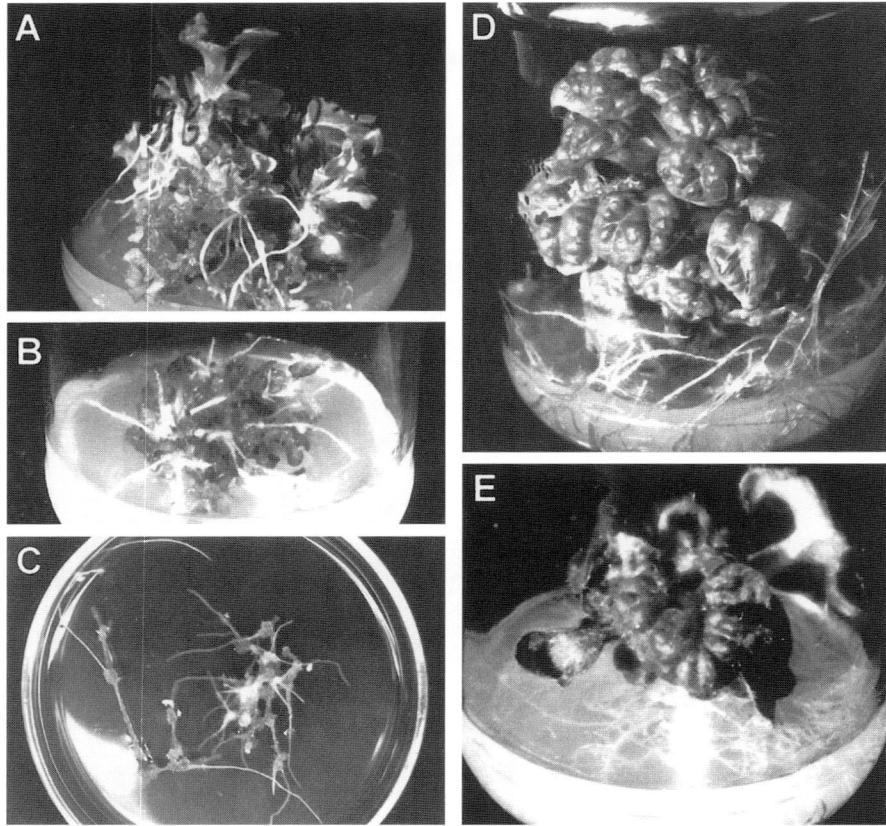

Fig. 7A–E. Expression of hairy roots syndrome of *A. belladonna*-A4 in somatic hybrids *N. tabacum* (+) *A. belladonna*-A4. **A, B** Formation of hairy roots in hybrids 5R20-3 and 5R20-4 on MS medium with 1 mg l^{-1} of BAP. **C** Autonomous growth of hybrid hairy roots on hormone-free medium. **D, E** Wrinkled leaves in *A. belladonna*-A4 and hybrid 5R20-4

Results of isoenzyme analyses revealed the hybrid status of the lines produced. The patterns of esterases in hybrids in particular were different from those in tobacco and contained some bands characteristic of nightshade (Fig. 9). At the same time, all analyzed hybrids manifested spectra of aspartate-aminotransferases identical to tobacco type (data not shown).

Hsp patterns of hybrids 5R20-3 and 5R20-4 were different from those of the Rab hybrid group (Fig. 2). In particular, hybrids 5R20-3 and 5R20-4 did not express (or displayed low expression) of *Atropa*-specific hsp with molecular mass of about 23 kDa which is highly expressed in Rab hybrids. In addition, 5R20-3 and 5R20-4 plants highly expressed a new hsp (about 26 kDa), which was almost undetectable in Rab plants.

Fig. 8A–C. Synthesis of anthocyanins in callus (**A**), regenerating shoots (**B**) and rooted plants (**C**) of the hybrid 5R20-4

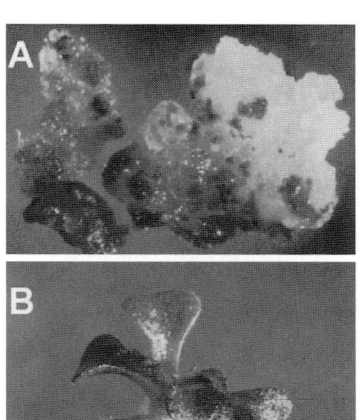
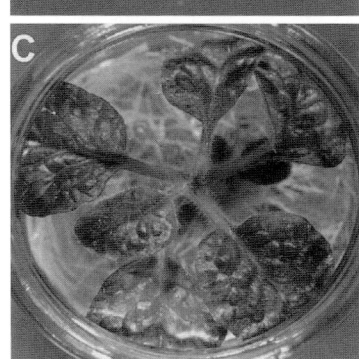

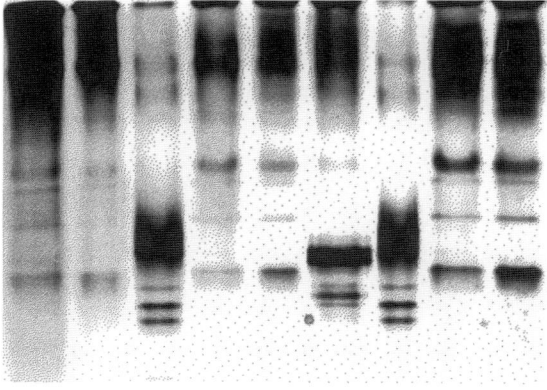

Fig. 9. Esterase isozymes of *N. tabacum* (*Nt*), *A. belladonna*-A4 (*Ab*) and their five somatic hybrids (*1–5*)

2.3 Complementation of Chlorophyll Deficiency by Hybridization Between *N. tabacum* and the Cybrid *A. belladonna* (+*N. tabacum*) (Experiment 3)

This experiment was designed to provide a possibility of generating nuclear hybrids *N. tabacum* (+) *A. belladonna* with only plastid genome of *N. tabacum*. For this purpose, fusions were carried out between protoplasts of tobacco chlorophyll-deficient plastome mutant DSR-A15 and an albino cybrid *A. belladonna* (+*N. tabacum*) containing a nucleus from nightshade and plastids from tobacco. It was previously shown that plastids of *N. tabacum* are not functional in the nuclear background of *A. belladonna*, and resultant cybrids *A. belladonna* (+*N. tabacum*) have a chlorophyll-deficient phenotype due to this incompatibility (Kushnir et al. 1991), whereas tobacco cybrids with plastids of nightshade produce functional green plants (Kushnir et al. 1987).

We produced albino cybrids *A. belladonna* (+*N. tabacum*) by fusion of protoplasts from transgenic *A. belladonna* resistant to kanamycin (Kushnir et al. 1991) and a plastome mutant *N. tabacum* SR1 resistant to streptomycin (Maliga et al. 1975). Chlorophyll-deficient cybrids were selected as described (Kushnir et al. 1991). In a further complementation experiment, protoplasts from albino cybrid line A59 (Fig. 10C) and the chlorophyll-deficient plastome tobacco mutant DSR-A15 (Fig. 10D; Svab and Maliga 1986) were fused. A number of green colonies which resulted from such a fusion were previously described as pure tobacco types with functional plastome (Kushnir et al. 1991). We also found that a predominant amount of green regenerants from the experiment were tobacco plants (Fig. 10E). Nevertheless, several regenerants with altered morphology were found to be similar in their phenotypes to nuclear hybrids *N. tabacum* (+) *A. belladonna* (Fig. 10F). At the later stage of regeneration, their leaves manifested anthocyanin pigmentation (Fig. 10G), which is a nuclear coded trait of *A. belladonna*. On the other hand, leaves possessed trichomes encoded by the nuclear genome of *N. tabacum*. Therefore, nuclear hybrids *N. tabacum* (+) *A. belladonna* with tobacco plastid genome could be constructed via fusions between plastome albino mutants of tobacco and albino cybrids *A. belladonna* (+*N. tabacum*). Their genetic constitution and extent of symmetry versus asymmetry needs to be investigated further.

Fig. 10A–G. Phenotypes of aseptic plants: **A** *N. tabacum* SR1. **B** *A. belladonna*. **C** Albino cybrid line A59 resulting after fusion of protoplasts from green plants of *N. tabacum* SR1 and *A. belladonna*. **D** Mutant *N. tabacum* DSR-A15; green tobacco-like plant (**E**) and nuclear hybrid (**F, G**) both resulted after protoplast fusion between albino plants of *N. tabacum* DSR-A15 and A59

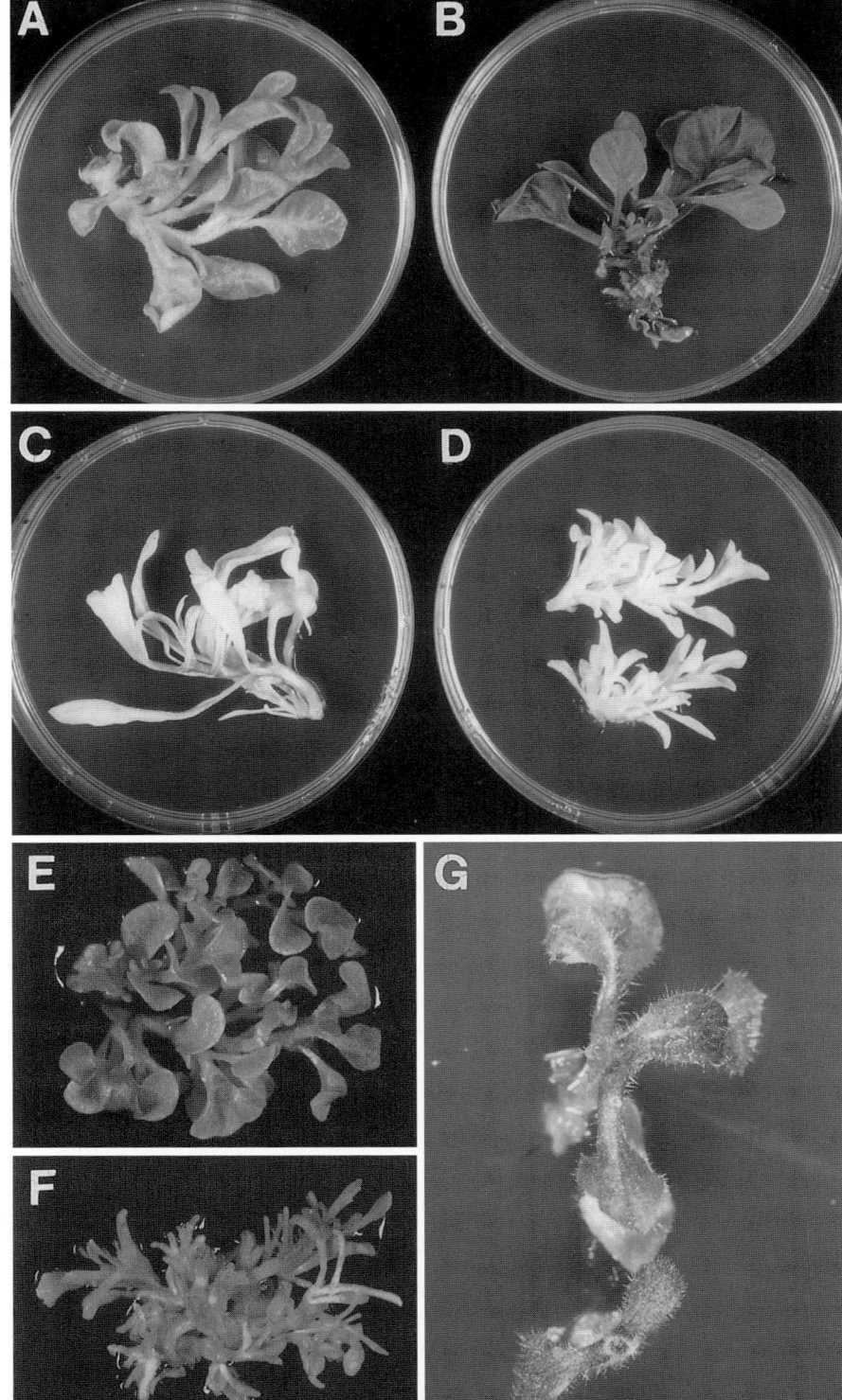

3 Conclusions and Prospects

We have produced intertribal somatic hybrids using fusions between: (1) protoplasts from gamma plants of *N. tabacum* R100a and *A. belladonna* 100-01 derived in vitro from highly irradiated seeds; (2) protoplasts of gamma plant *N. tabacum* R100a and irradiated protoplasts of *A. belladonna*-A4 transformed with *A. rhizogenes*; (3) protoplasts from chlorophyll-deficient plastome mutant *N. tabacum* DSR-A15 and albino cybrid *A. belladonna* (+*N. tabacum*).

Previous experiments on fusions of nonirradiated protoplasts from normal plants of *N. tabacum* and *A. belladonna* resulted in either cybrids (Kushnir et al. 1987) or symmetric hybrids which showed stable nuclear hybridity for a long time (Babiychuk et al. 1990). In both cases, no morphological variations were observed. Extremely asymmetric hybrids in this combination were obtained with low frequency due to an extensive elimination of *Atropa* chromosomes during regeneration of plants in the presence of kanamycin as a selective agent (Babiychuk et al. 1992). Plants regenerated in that experiment were also uniform in their morphology and identical to those of tobacco.

In experiment 1, hybrids with different levels of asymmetry were obtained without γ-inactivation of protoplasts before fusion. The high frequency of asymmetry may reflect radiation damage of the gamma-plant genome. Genetic changes occur at the inception of gamma plants (Zubko et al. 1990a,b). With the aim of possible selection, we used plastome chlorophyll-deficient *N. tabacum* gamma-plants (line R100a). In the previous experiments (Zubko et al. 1991), no reversions to the green phenotype were observed at least among 10^6 protoclones of mutant line R100a. This confirms the stability of chlorophyll deficiency in R100a, which prevents the occurance of green parental tobacco plants after protoplast fusions.

Combining parental phenotypic features and biochemical analyses confirmed the great extent of nuclear hybridity among plants of the first group. Nevertheless, their morphological and biochemical diversity may be explained by the differing extent of asymmetry in these hybrids. With respect to other plant groups, we regard them as being extremely asymmetric hybrids possessing the limited part both of the genome and donor plastids. The following characters of plants from the second group provide evidence for their nuclear hybridity: (1) phenotypic variations between plants of different lines under field conditions; (2) the segregation of normally rooting plants among hybrids Rab_2 initially incapable of rooting; (3) instability of the morphogenetically normal line Rab_{11}, the resulting subline being incapable of rooting and efficient growth on hormone-free medium; (4) the different chromosome numbers in karyotypes. Gamma plants never display such a broad range of phenotypes as these hybrid plants do (unpubl. data).

The extreme asymmetry in one quarter of the hybrids may be a consequence of the γ-irradiation applied before the generation of parental gamma plants. In addition, asymmetrization could have occurred as a result of spontaneous chromosome elimination, which has been reported for somatic hybrids obtained without irradiation (Krumbiegel and Schieder 1979, 1981;

Gleba et al. 1983; Vries et al. 1987; Endo et al. 1988; Lazar 1989; Babiychuk et al. 1992; Gilissen et al. 1992a,b; Wolters et al. 1993; Kisaka and Kameya 1994).

Heat shock proteins (hsp) were previously shown to be valuable markers to identify expression of hsp genes from both parents in somatic hybrids *A. belladonna* (+) *N. chinensis* (Lopato and Gleba 1985). The appearance of new protein bands in the hybrids described here might presumably reflect DNA recombination events or altered regulation (for instance, derepression of genes, deviations in splicing mechanisms, etc.).

The formation of chromosomal rings during meiosis is a mechanism usually related to reciprocal translocations. Previously, this phenomenon was described for a limited number of species (e.g., *Oenothera* spp., *Rhoeo discolor*) in which translocational complexes are balanced evolutionarily (Swanson et al. 1967). The chromosomal rings observed in our hybrids (Fig. 6B) implies an interesting model system to study chromosomal rearrangements in *Nicotiana*. The appearance of chromosomal rings could be explained by extensive translocational activities in hybrid cells. Chromosome translocations have been well documented for somatic hybrids (Vries et al. 1987; Piastuch and Bates 1990; Hinnisdaels et al. 1991).

Hybrids between *N. tabacum* R100a and *A. belladonna* A4 (experiment 2) revealed the possibility of transferring specific *A. belladonna* traits to hybrids plants. The hairy roots syndrome, a characteristic trait coded by the integrated Ri-plasmid (Tepfer 1984), was expressed in an original transformant of *A. belladonna* A4 as wrinkled leaves and abundant roots capable of independent growth on hormone-free medium (Fig. 7). These traits were expressed in two hybrid lines. The formation of abundant roots during regeneration of those lines in the presence of cytokinin could reflect a significant alteration in hormonal status of hybrid shoots due to expression of the hairy roots syndrome. The transfer of the parental hairy roots trait to somatic hybrids was discussed previously as a potential application to biotechnological purposes (Ahuja et al. 1993).

The synthesis of anthocyanins in hybrid leaves confirms the possibility of altering developmental regulation of secondary metabolic pathways as a result of somatic hybridization. One of the most valuable applications of this aspect is thought to be the generation of somatic hybrid lines and plants with modified patterns of medicinally important alkaloids (Ahuja et al. 1993).

Analyses of isoenzymes and hsp revealed an asymmetry of hybrids *N. tabacum* (+) *A. belladonna* A4. The differences in hsp patterns of hybrids from groups 5R20 and Rab could be interpreted in terms of different representation of donor (nightshade) genome in the hybrids of these two combinations of fusions. These data additionally reveal an asymmetric status of the generated hybrids.

The recovery of hybrid plants combining nuclear encoded phenotypic traits of *N. tabacum* and *A. belladonna* in experiment 3 regarding complementation of chlorophyll deficiency showed that the use of parents with different genomes and the same plastome could be advantageous for distant somatic hybridization restricted to transfer of nuclear-coded traits. Previously, this system was shown to be useful for substitution of mutated plastome by wild-type plastome in tobacco (Kushnir et al. 1991). Our present results show that such a transmission of incompatible plastome to a compatible nuclear background could be a tool for constructing true nuclear hybrids. This approach potentially provides possibilities for somatic hybridization schemes which operate stably with components of genetic systems of more than two species. In the concrete case, for instance, somatic hybridization between highly irradiated protoplasts of *N. tabacum* and albino protoplasts of *A. belladonna* (+*N. tabacum*) cybrids might allow the study of genetic mechanisms of nucleo-plastome incompatibility on the basis of complementation of the cybrids by a limited amount of tobacco nuclear material sufficient for recovery of green plants of *A. belladonna* (+*N. tabacum*).

The somatic hybrids described in this work demonstrate different levels of asymmetry and genetic heterogeneity, which result in morphological and biochemical diversity. They represent an interesting resource for investigating genetic processes and metabolic modifications in genome-engineered plants. In general, we suggest that the use of gamma-plants for somatic hybridization could provide an additional source of genetic variability. Gamma-plants, once obtained, may be maintained in an aseptic culture for a long time. Their protoplasts can be used for production of asymmetric hybrids without irradiation before fusion if a selective system for the evaluation of hybrids is available.

4 Protocol

4.1 Generation of Gamma-Plants

Gamma-plants of *N. tabacum* L. (chlorophyll-deficient albino mutant, line R100a) and *A. belladonna* L. (line 100-01) were recovered in vitro from gamma seedlings using a 1000-Gy dose to irradiate seeds (Zubko et al. 1990a,b). The chlorophyll deficiency of the mutant *N. tabacum* R100a is controlled by the plastome (Zubko et al. 1991). Plants were aseptically propagated on MS medium (Murashige and Skoog 1962) with $30\,g\,l^{-1}$ sucrose.

4.2 Genetic Transformation of *A. belladonna* with *A. rhizogenes*

Leaf segments ($0.5 \times 0.25\,cm$) of *A. belladonna* aseptic plants were placed for 10–15 min on the surface of overnight culture of *A. rhizogenes*, strain A4 (Tepfer 1984) in LB medium (Maniatis et al. 1982). Tissues were blotted slightly on filter paper and transferred on solidified MS medium containing Cefataxim and carbenicillin at a concentration of $400\,mg\,l^{-1}$. Every 2 weeks, explants were transferred to the fresh medium with a gradual decrease of antibiotic concentration to $100\,mg\,l^{-1}$. Transformants selected as autonomously growing hairy roots on hormone-free medium were then regenerated into plants on MS medium with $1\,mg\,l^{-1}$ of 6-benzylaminopurine (BAP). Occasionally, hairy roots formed plants on hormone-free medium. Plants were propagated on MS medium.

4.3 Fusion of Protoplasts, Selection and Regeneration of Hybrids

Protoplasts were isolated from leaves of aseptic plants using an enzyme mixture containing 0.3% Onozuka R10, 0.2% Macerozyme R10 (Yakult Pharmaceutical, Japan), and 0.1% Driselase (Sigma, USA) dissolved in W_5 solution (Medgyesy et al. 1980). The fusion procedure of Menczel et al. (1981) was slightly modified. PEG solution (400 µl) was pipetted into a flow centrifuge tube (15 ml, 1 cm in diameter). About 2×10^5 parental protoplasts (mixed in the ratio 1:1) were washed and resuspended in 0.5 ml of W_5 medium. The protoplast mixture was carefully layered on top of the PEG solution with a Pasteur pipette. After centrifugation at 600 rpm for 40 s, a fraction of the protoplasts formed a compact band at the interphase between PEG and W_5 solutions; 15 min later, 0.5 ml of high-pH buffer (pH 10.5) was added carefully to the upper W_5 phase and allowed to settle for 15–20 min. The protoplast mixture was then washed with 10 ml of W_5 solution using gentle centrifugation (300 rpm, 2 min). Protoplasts were resuspended gently in SW medium (Sidorov et al. 1987) and cultivated in this medium at a density of 10^4 cells ml^{-1}.

Protoplasts were incubated in the dark at 26 °C for the first week after fusion. After adding fresh medium (about 20% of the initial volume), dishes were transferred to a reduced illumination (500 lx). Two weeks later, microcolonies were diluted three times with liquid C medium (Shepard and Totten 1977) and plated onto the surface of C medium with 0.8% Difco agar. Microcolonies were cultivated at 2000 lx. After 4 weeks, greenish colonies appeared, and these were transferred onto MS medium with 1 mg l^{-1} kinetin and 10 g l^{-1} sucrose to promote plant regeneration. Induced shoots were rooted on the hormone-free medium. Shoots that did not root were maintained on MS medium with 1 mg l^{-1} of BAP or kinetin. Rooted in vitro plants were transferred to pots in a greenhouse and then to field conditions.

4.4 Cytological Analyses

Root tips of aseptic plants were treated with 0.02% colchicine for 3–4 h, fixed in 3:1 ethanol: glacial acetic acid overnight, stained with 1% aceto-orcein, and used to make chromosome squashes. Meiosis was studied in three plants of hybrid Rab$_{4a}$ growing in the field. Flower buds were fixed in a mixture of 1% acetohematoxylin and iron acetate. Chromosomes of pollen mother cells were examined in preparations from the anthers crushed in 40% chloral hydrate.

For light microscopy, buds and flowers of tobacco (cv. Lechija) and hybrid (Rab$_{4a}$) plants at different stages of development were fixed by Navashin's method, embedded, sectioned, and stained with Schiff's reagent and gallocyanin-chrome alum according standard protocols (Pausheva 1974).

4.5 Biochemical and Molecular Analyses

Multiple molecular forms of esterase, amylase, and aspartate-aminotransferase were analyzed using extracts from leaf tissues. Electrophoresis was carried out as described by Maurer (1976). The staining was performed according to Brewer (1970).

Chloroplast (cp) and mitochondrial (mt) DNAs were prepared from leaves of aseptic plants according to standard procedures (Bookjans et al. 1984; Wilson and Chourey 1984). DNAs were digested with *Hind*III and *Bam*HI enzymes, then separated by electrophoresis on 0.8% agarose. Restriction cpDNA patterns were visualized by staining gels with ethidium bromide. Digested mtDNAs were probed with mitochondrial gene *coxI* (cytochrome oxidase, subunit I) from *Oenothera* (a gift from Prof. A. Brennicke, Berlin).

For analyses of heat shock proteins (hsp), shoots or leaves of aseptic plants (1 g) were used. Stress conditions, labeling, protein extraction, and gel electrophoresis were as described in a previous work (Lopato and Gleba 1985). Before autoradiography, polyacrylamide gels were fixed with 60% isopropanol and stained with Coomassie Blue R-250. Then gels were washed in 7% acetic acid, dried onto Whatman 3 MM filter paper and exposed at −80 °C for 3–14 days.

Acknowledgments. We thank L. Lalakina and Y. Patskovsky for technical assistance, W. Waterworth and C. West for grammar suggestions. This work was partially granted by Körber-Stiftung (Hamburg, FRG).

References

Ahuja PS, Rahman L, Bhargava SC, Banerjee S (1993) Regeneration of intergeneric somatic hybrid plants between *Atropa belladonna* L. and *Hyoscyamus muticus* L. Plant Sci 92:91–98

Babiychuk EL, Kushnir SG, Gleba YuYu (1990) Fertile intertribal asymmetric somatic hybrids in the Solanaceae. Biopolymers Cell 6:86–90

Babiychuk E, Kushnir S, Gleba YY (1992) Spontaneous extensive chromosome elimination in somatic hybrids between somatically congruent species *Nicotiana tabacum* L. and *Atropa belladonna* L. Theor Appl Genet 84:87–91

Babiychuk E, Schantz R, Cherep N, Weil J-H, Gleba Y, Kushnir S (1995) Alterations in chlorophyll a/b binding proteins in Solanaceae cybrids. Mol Gen Genet 249:648–654

Bates GW (1992) Molecular analysis of nuclear genes in somatic hybrids. Physiol Plant 85:308–314

Bookjans G, Stummann BM, Henningsen KW (1984) Preparation of chloroplast DNA from pea plastids isolated in a medium of high ionic strength. Anal Biochem 141:244–247

Borisjuk NV, Momot VP, Gleba Y (1988) Novel class of rDNA repeat units in somatic hybrids between *Nicotiana* and *Atropa*. Theor Appl Genet 76:108–112

Brewer GJ (1970) An introduction to isozyme techniques. Academic Press New York; 250 pp

Carlson PS, Smith HH, Dearing RD (1972) Parasexual interspecific plant hybridization. Proc Natl Acad Sci USA 69:2292–2294

Chien YC, Kao KN, Wetter LR (1982) Chromosomal and isozyme studies of *Nicotiana tabacum-Glycine max* hybrid cell lines. Theor Appl Genet 62:301–304

D'Arcy WG (1979) The classification of the Solanaceae. In: Hawkes JG, Lester RN, Skelding AD (eds) The biology and taxonomy of the Solanaceae, no. 7. Academic Press, London, pp 3–49

Darlington CD, Wylie AP (1955) Chromosome atlas. George Allen & Unwin, London

Dragoeva A, Atanassov I, Atanassov A (1999) CMS due to tapetal failure in cybrids between *Nicotiana tabacum* and *Petunia hybrida*. Plant Cell Tissue Organ Cult 55:67–70

Dudits D, Maroy E, Praznovszky T, Olah, Z, Gyorgyey J, Cella R (1987) Transfer of resistance traits from carrot into tobacco by asymmetric somatic hybridization. Regeneration of fertile plants. Proc Natl Acad Sci USA 84:8434–8438

Endo T, Komiya T, Mino M, Nakanishi K, Fujita S, Yamada Y (1988) Genetic diversity among sublines originating from a single somatic hybrid cell of *Duboisia hopwoodii* + *Nicotiana tabacum*. Theor Appl Genet 76:641–646

Gilissen LJW, van Staveren MJ, Verhoeven HA, Sree Ramulu K (1992a) Somatic hybridization between potato and *Nicotiana plumbaginifolia*. 1. Spontaneous biparental chromosome elimination and production of asymmetric hybrids. Theor Appl Genet 84:73–80

Gilissen LJW, van Staveren MJ, Ennik E, Verhoeven HA, Sree Ramulu K (1992b) Somatic hybridization between potato and *Nicotiana plumbaginifolia*. 2. Karyotypic modification and segregation of genetic markers in hybrid suspension cultures and sublines. Theor Appl Genet 84:81–86

Giri CC, Ahuja PS (1995) Characterizing intergeneric regenerants of protoplast fusion between *Hyoscyamus muticus* (Egyptian henbane) and *Atropa belladonna* (Indian sagangur). Curr Sci 69:458–461

Gleba YY, Hoffmann F (1978) Hybrid cell lines *Arabidopsis thaliana* + *Brassica campestris*: no evidence for specific chromosome elimination. Mol Gen Genet 165:257–264

Gleba YY, Hoffmann F (1979) "Arabidobrassica": plant-genome engineering by protoplast fusion. Naturwissenschaften 66:547–554

Gleba Yu, Sytnik K (1984) Protoplast fusion and hybridization of distantly related plant species. In: Schoerman R (ed) Protoplast fusion, genetic engineering in higher plants. Monogr Theor Appl Genet 8:115–161

Gleba YY, Butenko RG, Sytnik KM (1975) Protoplast fusion and somatic hybridisation of *Nicotiana tabacum* L. Dokl Acad Sci USSR 221:1196–1198

Gleba YY, Momot VP, Cherep NN, Skarzynskaya MV (1982) Intertribal hybrid cell lines of *Atropa belladonna* (×) *Nicotiana chinensis* obtained by cloning individual protoplast fusion products. Theor Appl Genet 62:75–79

Gleba YuYu, Momot VP, Okolot AN, Cherep NN, Skarzynskaya MV, Kotov V (1983) Genetic processes in intergeneric cell hybrids *Atropa* + *Nicotiana*. I. Genetic constitution of cells of different clonal origin grown in vitro. Theor Appl Genet 65:269–276

Gleba YY, Kanevsky IF, Skarzynskaya MV, Komarnitsky IK, Cherep NN (1986) Hybrids between tobacco crown gall cells and normal somatic cells of *Atropa belladonna* – isolation and characterization of cell lines. Plant Cell Rep 5:394–397

Cleba YY, Hinnisdaels S, Sidorov VA, Kaleda VA, Parokonny AS, Boryshuk NV, Cherep NN, Negrutiu I, Jacobs M (1988) Intergeneric asymmetric hybrids between *Nicotiana plumbaginifolia* and *Atropa belladonna* obtained by "gamma-fusion". Theor Appl Genet 76:760–766

Glimelius K, Bonnett HT (1986) *Nicotiana* cybrids with *Petunia* chloroplasts. Theor Appl Genet 72:794–798

Goodspeed TH (1954) The genus *Nicotiana*. Chronica Botanica, Waltham, MA, 536 pp

Grodzynsky DM (1989) Plant radiobiology. Kiev. Naukova Dumka.

Hinnisdaels S, Bariller L, Mouras A, Sidorov V, Del-Favero J, Veuskens J, Negrutiu I, Jacobs M (1991) Highly asymmetric intergeneric nuclear hybrids between *Nicotiana* and *Petunia*: evidence for recombinogenic and translocation events in somatic hybrids plants after "gamma-fusion". Theor Appl Genet 82:609–614

Imamura J, Saul MW, Potrykus I (1987) X-ray irradiation promoted asymmetric somatic hybridization and molecular analysis of the products. Theor Appl Genet 74:45–450

Kameya T (1975) Induction of hybrids through somatic cell fusion with dextran sulfate and gelatin. Jpn J Genet 50:236–246

Kanevsky IF, Gleba YuYu (1986) Analysis of characters coded for by T-DNA using hybridization between tumorous and normal plant cells. Plant Cell Rep 5:352–355

Kao KN (1977) Chromosomal behaviour in somatic hybrids of soybean + *Nicotiana glauca*. Mol Gen Genet 150:225–230

Kisaka H, Kameya T (1994) Production of somatic hybrids between *Daucus carota* L. and *Nicotiana tabacum*. Theor Appl Genet 88:75–80

Kishinami I, Widholm JM (1987) Auxotrophic complementation in intergeneric hybrid cells obtained by electrical and dextran-induced protoplast fusion. Plant Cell Physiol 28:211-218

Krumbiegel G, Schieder O (1979) Selection of somatic hybrids after fusion of protoplasts from *Datura innoxia* Mill. and *Atropa belladonna* L. Planta 145:371–375

Krumbiegel G, Schieder O (1981) Comparison of somatic and sexual incompatibility between *Datura innoxia* and *Atropa belladonna*. Planta 153:466–470

Kushnir SG, Shlumukov LR, Pogrebnyak NJ, Berger S, Gleba Y (1987) Functional cybrid plants possessing a *Nicotiana* genome and an *Atropa* plastome. Mol Gen Genet 209:159–163

Kushnir SG, Babiychuk E, Bannikova M, Momot V, Komarnitsky I, Cherep N, Gleba Y (1991) Nucleo-cytoplasmic incompatibility in cybrid plants possessing an *Atropa* genome and a *Nicotiana* plastome. Mol Gen Genet 225:225–230

Lazar G (1989) Somatic hybridization in *Hyoscyamus* × *Nicotiana*. In: Bajaj YPS (ed) Biotechnology in agriculture and forestry, vol 8. Plant protoplasts and genetic engineering I. Springer, Berlin Heidelberg New York, pp 356–369

Lopato SV, Gleba YuYu (1985) Heat shock proteins from cell cultures of higher plants and their somatic hybrids. Plant Cell Rep 4:19–22

Maliga P, Breznovits AS, Marton L, Joo F (1975) Non-Mendelian streptomycin-resistant tobacco mutant with altered chloroplasts and mitochondria. Nature 255:401–402

Maniatis T, Fritsch EF, Sambrook J (1982) Molecular cloning: a laboratory manual. Cold Spring Harbor Laboratory Press, Cold Spring Harbor, NY

Maurer HR (1976) Disk electrophoresis and related techniques of polyacrylamide gel electrophoresis. Walter de Gruyter, Berlin, 187 pp

Medgyesy P, Menczel L, Maliga P (1980) The use of cytoplasmic streptomycin resistance: chloroplast transfer from *Nicotiana tabacum* into *Nicotiana sylvestris* and isolation of their somatic hybrids. Mol Gen Genet 179:693–698

Melchers G, Labib G (1974) Somatic hybridization of plants by fusion of protoplasts. Selection of light-resistant hybrids of "haploid" sensitive varieties of tobacco. Mol Gen Genet 135:277–294

Melchers G, Sacristan MD, Holder AA (1978) Somatic hybrid plants of potato and tomato regenerated from fused protoplasts. Carlsberg Res Commun 43:203–218

Menczel L, Nagy F, Kiss ZR, Maliga M (1981) Streptomycin resistant and sensitive somatic hybrids of *Nicotiana tabacum* + *Nicotiana knightiana*: correlation of resistance to *N. tabacum* plastids. Theor Appl Genet 59:191–195

Murashige T, Skoog F (1962) A revised medium for rapid growth and bioassays with tobacco tissue cultures. Physiol Plant 15:473–497

Negrutiu I, Mouras A, Gleba YY, Sidorov V, Hinnisdaels S, Famelaer Y, Jacobs M (1989) Symmetric versus asymmetric fusion combinations in higher plants. In: Bajaj YPS (ed) Biotechnology in agriculture and forestry, vol 8. Plant protoplasts and genetic engineering I. Springer, Berlin Heidelberg New York, pp 305–319

Pausheva ZP (1974) Practicum on cytology of plants. Kolos, Moscow

Pental D, Hamill JD, Pirrie A, Cocking EC (1986) Somatic hybridization of *Nicotiana tabacum* and *Petunia hybrida*. Recovery of plants with *P. hybrida* nuclear genome and *N. tabacum* chloroplast genome. Mol Gen Genet 202:342–347

Perl A, Aviv D, Galun E (1991) Protoplast fusion mediated transfer of oligomycin resistance from *Nicotiana sylvestris* to *Solanum tuberosum* by intergeneric cybridization. Mol Gen Genet 225:-11–16

Piastuch WC, Bates GW (1990) Chromosomal analysis of *Nicotiana* asymmetric somatic hybrids by dot blotting and in situ hybridization. Mol Gen Genet 222:97–103

Potrykus I, Jia J, Lazar GB, Saul M (1984) *Hyoscyamus muticus* + *Nicotiana tabacum* fusion hybrids selected via auxotroph complementation. Plant Cell Rep 3:68–71

Puite KJ (1992) Progress in plant protoplast research. Physiol Plant 85:403–410

Shepard JF, Totten RE (1977) Mesophyll cell protoplasts of potato (*Solanum tuberosum* L.). Isolation, proliferation, and plant regeneration. Plant Physiol 60:313–316

Sidorov VA, Zubko MK, Kuchko AA, Komarnitsky IK, Gleba YY (1987) Somatic hybridization in potato: use of γ-irradiated protoplasts of *Solanum pinnatisectum* in genetic reconstruction. Theor Appl Genet 74:364–368

Skarzhynskaya MV, Cherep NN, Gleba YY (1982) Somatic hybridization and obtaining of cell lines and plants potato + tobacco. Cytol Genet 6:42–48

Skarzhynskaya MV, Cherep NN, Gleba YY (1983) Generation and studies on hybrid cell lines *Nicotiana tabacum* + *Solanum tuberosum* and *N. tabacum* + *S. sucrense*. Ukr Bot J 34(1):64–68

Somers DA, Narayanan KR, Kleinhofs A, Cooper-Blaud S, Cocking EC (1986) Immunological evidence for transfer of the barley nitrate reductase structural gene to *Nicotiana tabacum* by protoplast fusion. Mol Gen Genet 204:296–301

Svab Z, Maliga P (1986) *Nicotiana tabacum* mutants with chloroplast encoded streptomycin resistance and pigment deficiency. Theor Appl Genet 72:637–643

Swanson CP, Merz T, Young WJ (1967) Cytogenetics. Prentice-Hall, Englewood Cliffs

Takebe I, Labib G, Melchers G (1971) Regeneration of whole plants from isolated mesophyll protoplasts of tobacco. Naturwissenschaften 58:318–320

Tepfer D (1984) Transformation of several species of higher plants by *Agrobacterium rhizogenes*: sexual transmission of the transformed genotype and phenotype. Cell 37:959–967

Thanh ND, Medgyesy P (1989) Limited chloroplast gene transfer via recombination overcomes plastome-genome incompatibility between *Nicotiana tabacum* and *Solanum tuberosum*. Plant Mol Biol 12:87–93

Thanh ND, Pay A, Smith MA, Medgyesy P, Marton L (1988) Intertribal chloroplast transfer by protoplast fusion between *Nicotiana tabacum* and *Salpiglossis sinuata*. Mol Gen Genet 213:186–190

Vlahova M, Hinnisdaels S, Frulleux F, Claeys M, Atanassov A, Jacobs M (1997) UV irradiation as a tool for obtaining asymmetric somatic hybrids between *Nicotiana plumbaginifolia* and *Lycopersicon esculentum*. Theor Appl Genet 94:184–191

Vries SE de, Ferwerda MA, Loonen AEHM, Pijnacker LP, Feenstra WJ (1987) Chromosomes in somatic hybrids between *Nicotiana plumbaginifolia* and a monoploid potato. Theor Appl Genet 75:170–176

Wetter LR (1977) Isoenzyme patterns in soybean-*Nicotiana* hybrid cell lines. Mol Gen Genet 150:231–235

Wetter LR, Kao KN (1980) Chromosome and isoenzyme studies on cells derived from protoplast fusion of *Nicotiana glauca* with *Glycine max-Nicotiana glauca* cell hybrids. Theor Appl Genet 57:273–276

Wilson AJ, Chourey PS (1984) A rapid unexpensive method for the isolation of restrictable mitochondrial DNA from various plant sources. Plant Cell Rep 3:237–239

Wolters AMA, Vergunst AC, van der Werff F, Koornneef M (1993) Analysis of nuclear and organellar DNA of somatic hybrid calli and plants between *Lycopersicon* spp. and *Nicotiana* spp. Mol Gen Genet 241:707–718

Ye J, Hauptmann RM, Smith AG, Widholm JM (1987) Selection of a *Nicotiana plumbaginifolia* universal hybridizer and its use in intergeneric somatic hybrid formation. Mol Gen Genet 208:474–480

Yemets AI, Kundel'chuk OP, Smertenko AP, Solodushko VG, Rudas VA, Gleba YY, Blume YB (2000) Transfer of amiprophosmethyl resistance from a *Nicotiana plunbaginifolia* mutant by somatic hybridization. Theor Appl Genet 100:847–857

Zubko M, Zubko E, Gleba Y (1990a) Restoration of gamma seedlings viability in vitro: potentials for mutagenesis. VII Int Congr on Plant Tissue and Cell Culture. Amsterdam, June 24–29. Abstracts, p 170

Zubko MK, Zubko EI, Gleba YuYu (1990b) Restoration of gamma seedlings viability in vitro. Dokl Akad Nauk SSSR 313:453–457

Zubko MK, Zubko EI, Kapranov PhV (1991) Induction of chlorophyll-deficient tobacco mutants as markers for cell engineering. Biopolymers Cell 7:72–79

Zubko MK, Schmeer K, Glaβgen WE, Bayer E, Seitz HU (1993) Selection of anthocyanin-accumulating potato (*Solanum tuberosum* L.) cell lines from calli derived from seedlings produced by gamma-irradiated seeds. Plant Cell Rep 12:555–558

Zubko MK, Zubko EI, Patskovsky YV, Khvedynich OA, Fisahn J, Gleba YY, Schieder O (1996) Novel "homeotic" CMS patterns generated in *Nicotiana* via cybridization with *Hyoscyamus* and *Scopolia*. J Exp Bot 47:1101–1110

III.4 Somatic Hybridization and Cell Grafting in *Senecio*

G. Wang[1,2] and H. Binding[1]

1 Introduction

Interspecific combinations in the genus *Senecio* have been investigated in order to explore somatic hybridization and cell grafting in the daisyflower family (Wang and Binding 1993a,b). During the present studies, reports were published on somatic hybrids in this family with *Rudbeckia* (Al-Atabee et al. 1990) and *Lactuca* (Matsumoto 1991). Screening experiments on regeneration efficiencies from isolated protoplasts with flowering plant species included 24 species of 13 genera of the Asteraceae (Binding and Nehls 1980; Binding et al. 1981). Highest plating and regeneration efficiencies were obtained with members of the genera *Senecio* and *Cichorium*. After further screening of five species of these genera, finally two *Senecio* species were selected. The experimental details, including the terminology, are comprised after presentation and discussion of the results.

Senecio L. is a genus of the subfamily Asteroideae. The selected species belong to the subgenus Eusenecio, *S. fuchsii* C. Gmel. (*S.f*; *S. nemorensis* L. *subsp. fuchsii* Celac.) to the section Sarracenii, *S. jacobaea* L. (*S.j*) to the Jacobaeae. They are perennial herbs, possessing 20 chromosomes in the haploid state. Both species are native in temperate Europe; *S. jacobaea* is naturalized in North America. They have been used as medicinal plants and are supposed to be slightly poisonous to human and cattle due to senecio-alkaloids.

2 Results and Discussion

2.1 Screening of *Cichorium* and *Senecio* Species for Protoplast Regeneration Efficiency

Protoplasts were cultured, isolated from single seed-derived shoot clones of *C. endivia* L., *C. intybus* L., *S. jacobaea* L., *S. fuchsii* C. Gmel., *S. viscosus* L.,

---

[1] Botanical Institute, Christian-Albrechts-University, 24098 Kiel, Germany
[2] Present address: Department of Cellular and Molecular Physiology, The Pennsylvania State University, Hershey, PA 17033, USA

Table 1. Highest regeneration efficiencies and respective plating efficiencies in selected experiments with protoplasts of *Cichorium* and *Senecio* species. (1 = 100%)

Species	Plating efficiency	Regeneration efficiency
C. endivia L.	0.6	0.24
C. intybus L.	0.4	0.11
S. fuchsii C. Gmel.	0.4	0.06
S. jacobaea L.	0.6	0.53
S. viscosus L.	0.6	0.24
S. vulgaris L.	0.75	0.20

and *S. vulgaris* L. The data shown in Table 1 reflect the highest regeneration efficiencies obtained. These were most reliably approximated in repeated experiments with *S. jacobaea*; henceforth, further experiments were concentrated on this material. *S. fuchsii* (*S.j*) was chosen as the fusion partner because of its significantly different leaf shapes and relatively close relationship. Cell division activity started on the 2nd day of culture in *S.j* and on the 3rd to 4th day in *S.f*. First shoot primordia were visible after 18 and 32 days in the respective species.

2.2 Selection of Chlorophyll-Deficient Mutants of *Senecio jacobaea*

Chlorophyll deficiencies are useful traits in somatic hybridization, providing selective maker systems for the genomes and plastomes. *S. jacobaea* was chosen for the selection of chlorophyll-deficient mutants because of its high regeneration rate. Among 2534 regenerant lines, 14 lines exhibited chlorophyll deficiency about 7 weeks after a 4-day incubation of plastocytes with nitrosomethylurea (NMU) at concentrations of 50 to 100 µM; a single mutant grew in a population of protoplast-derived callis, which had been cultured on agar medium containing 1 mM NMU for a period of 30 days and on drug-free medium for another 14 days. The following mutant clones were selected for further studies.

S.j-Wa1 appeared as a pure albino primary shoot on a mutagenized callus. It was separated and subcultured. The purity of the albino primary shoot suggested a nuclear genome mutation. The mutant clone had the complete diploid chromosome set of $2n = 40$. It is hence most likely that the mutant allele is dominant.

S.j-Wa2 occurred in a variegated shoot after mutagenesis of plastocytes. The pattern of variegation suggested a plastome mutation in the chlorophyll-deficient somatic subclone. An albino subline was established from the adventitious shoot population. No green tissue was found in 12 151 protoclones of *S.j*-Wa1 and in 10 397 clones of *S.j*-Wa2. This proved the clonal purity and stability of the mutant allele.

2.3 Growth in Biparental Cultures of Protoplasts

The cocultures of the mutants with a green wild-type clone of *S. fuchsii*, treated with the fusogen (Fig. 1A) or untreated, were grown to investigate the conditions for somatic hybridization and cell grafting in the family of Asteraceae, to study the fates of nuclei, plastids, and mitochondria in the fusant lines, and to establish the localization and phenotypic expression of the mutated genes. Vigorous callus growth was obtained in any type of culture (Fig. 1B). The organization of albino shoots commenced within 18 to 30 days after protoplast plating. Green and variegated shoots grew from the 30th day on. Repeated protoplast regeneration was used for the isolation of pure fusant clones and subclones, as well as for discrimination between hybrids and graft chimeras.

2.4 Selection of Biparental Regenerant Lines and Clones

The selection was based on the pigmentation, hair, and shape of the leaves, being denticulate in *S.f*, pinnately insected in *S.j*, and intermediate or irregular in the fusants and cell graft chimeras. All the biparental lines (Tables 2 and 3) stemmed from separate culture lenses and hence were derived from individual protoplast pairing.

2.5 The Graft Chimeras

The graft chimera of lines formed chimeral primary shoots, from which monotypic clones resembling either parents segregated in form of adventitious shoots and protoplast regenerants. All chimeral primary shoots showed periclinal organization. *S.j*-Wa1 was found in the L1 or core, while *S.j*-Wa2 tissue was covered with an *S.f* epidermis. Graft chimeras were detected in mixed cultures, to which no fusion stimulus was applied. That no cell grafts were found in the fusion cultures was most likely a mere accident. The occurrence of graft chimeras indicated the possibility of multicellular origin of shoot primordia in Asteraceae. A discrimination between fusants and cell grafts was hence

Table 2. Results of coculture (+) and hybridization (×) experiments

Fusion and coculture experiment	No. of regenerant lines	No. of lines with shoots of the following phenotypes					Lines with biparental traits (%)	
		S.f	*S.j*	Graft	Hybrid	Cybrid	Graft	Fusant
S.f (×) *S.j*-Wa1	3390	125	3251	0	9	5	–	0.41
S.f (+) *S.j*-Wa1	2944	97	2845	2	0	0	0.07	–
S.f (×) *S.j*-Wa2	5049	187	4843	0	16	3	–	0.26
S.f (+) *S.j*-Wa2	2845	93	2747	5	0	0	0.18	–
Total	14228	502	13686	7	25	8	0.12	0.39

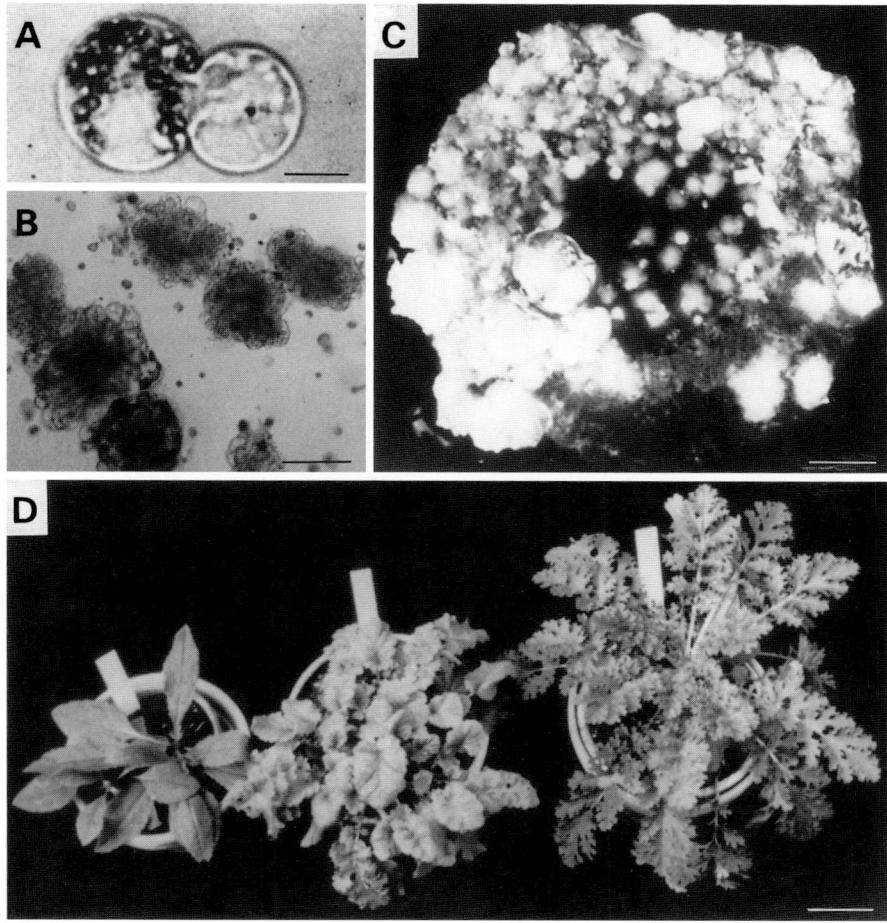

Fig. 1A–D. Somatic hybridization between *Senecio fuchsii* and *S. jacobaea*-Wa2. **A** Fusion body. *Bar* 20 μm. **B** Cell clusters 7 days after plating. *Bar* 320 μm. **C** Calli in an agarose gel lense after 1 week in V-KM and 1 week on B5BC; 0.5 mm. **D** Greenhouse-grown plants of the cybrid *S.f(x)j*-C7 (*middle*), *S.f* (*left*), and *S.j.* (*right*); *bar* 4 cm

necessary, which was only possible by the investigation of cloned material from the biparental lines.

2.6 The Hybrid Lines

The development of the hybrids was retarded. No plant could be grown in the greenhouse. The viability of 13 lines was even insufficient for further investigations. The other hybrid lines were characterized by the traits used for their selection, the transmission of these traits to the protoplast-derived clones, chromosome numbers, isoenzyme patterns, and RFLP. The lack of cybrid sub-

Table 3. Characterization of hybrid and cybrid lines

Fusant lines	Genome		Plastome		Chondriome
	2n	Morphology and isoenzymes	Chlorophyll	RFLP	RFLP
S.f(x)S.j-Wa1-H6	80	fxj	f	f	f
H7	80	fxj	f/(j)	f/j	j
H8	80	fxj	f	f/j	j
H9	80	fxj	f	j	f
H10	80	fxj	f	j	j
H11	80	fxj	f	f	j
H12	65–76	fxj	f	j	f
C1, 2	40	f	f	j	j
C3, 4, 5	40	f	f	f	j
S.f(x)S.j-Wa2-H1	80	fxj	f/j	f/j	j
H2	60–70	fxj	j	j	j
H3	68–75	fxj	f/j	f/j	j
H4, 5	80	fxj	f/j	f/j	f
C6	40	j[a]	f	f	j
C7, 8	40	j[a]	f	f	j

[a] Malformed leaves.

lines or clones suggests a high probability of early hybrid nucleus formation in the heterokaryons.

The interspecific genome compatibility was sufficient to allow organized growth of the hybrids. However, aneuploidy was detected in 3 of the 12 hybrids and indicated by small white spots in another line. The failure of growth in the greenhouse as well as the chromosomal instability indicates some limitation in the compatibility. It may be argued that the limited survival of the other hybrids was due to more restrictive aneuploidy, and that a certain number of hybrids was even not detected by their inability of organized growth. The hybrids showed an entirely additive pattern of the four investigated types of isozymes (Fig. 2A).

Fig. 2A–H. Determination of the genetic constitution of the hybrid and cybrid lines *Senecio fuchsii*(x)*jacobaea* through isoenzyme analysis and RFLP probing. **A,B** Peroxidase isoenzyme patterns. **A** The hybrids and parents; *lanes 1–12* lines H1–H12; *lane 13* S.j-Wa1; *lane 14* S.f; *lane 15* S.j-Wa2. **B** The cybrids and parents; *lanes 1–8* lines C1–8; *lane 9* S.j; *lane 10* S.j-Wa1; *lane 11* S.j-Wa2. **C–F** RFLP patterns of total DNA *Eco*RI + *Bam*HI fragments marked with a barley ptDNA probe of the genes ndh E + G; the kb values are concluded from the positions of bacteriophage λ *Hin*dIII fragments. **C** Hybrid lines and parents; *lanes 1–12* lines H1–H12; *lane 13* S.j-Wa1; *lane 14* S.j. **D** Cybrid lines and parents; *lanes 1–8* lines C1–8; *lane 9* S.j-Wa1; *lane 10* S.j. **E** Protoplast-derived clones of the hybrid line S.f(x)j-Wa2-H1 and the parents; *lanes 1–7* chlorophyll-deficient clones; *lanes 8–11* green clones; *lane 12* S.j-Wa2; *lane 13* S.j. **F** Protoplast-derived clones of the hybrid line S.f(x)j-Wa1-H7 and the parents; *lanes 1–11* clones (all green); *lane 12* S.j-Wa1; *lane 13* S.j. **G,H** RFLP patterns of total DNA *Eco*RI fragments marked with a maize mtDNA probe of the gene *cox* II; the kb values are concluded from the positions of bacteriophage λ *Hin*dIII fragments. **G** Single clones of the hybrids and parents; lanes ordered as in **C**. **H** Clones of the cybrids and parents; lanes ordered as in **D**

The plastome constitution of the hybrids was traced with the pigmentation [in the lines S.f(x)S.j-Wa2-H1-H5] and the ptDNA restriction fragments hybridized with four probes (illustrated in Fig. 2C). Plastid segregation in the hybrid lines was investigated with protoclones grown after 1 year. All clones

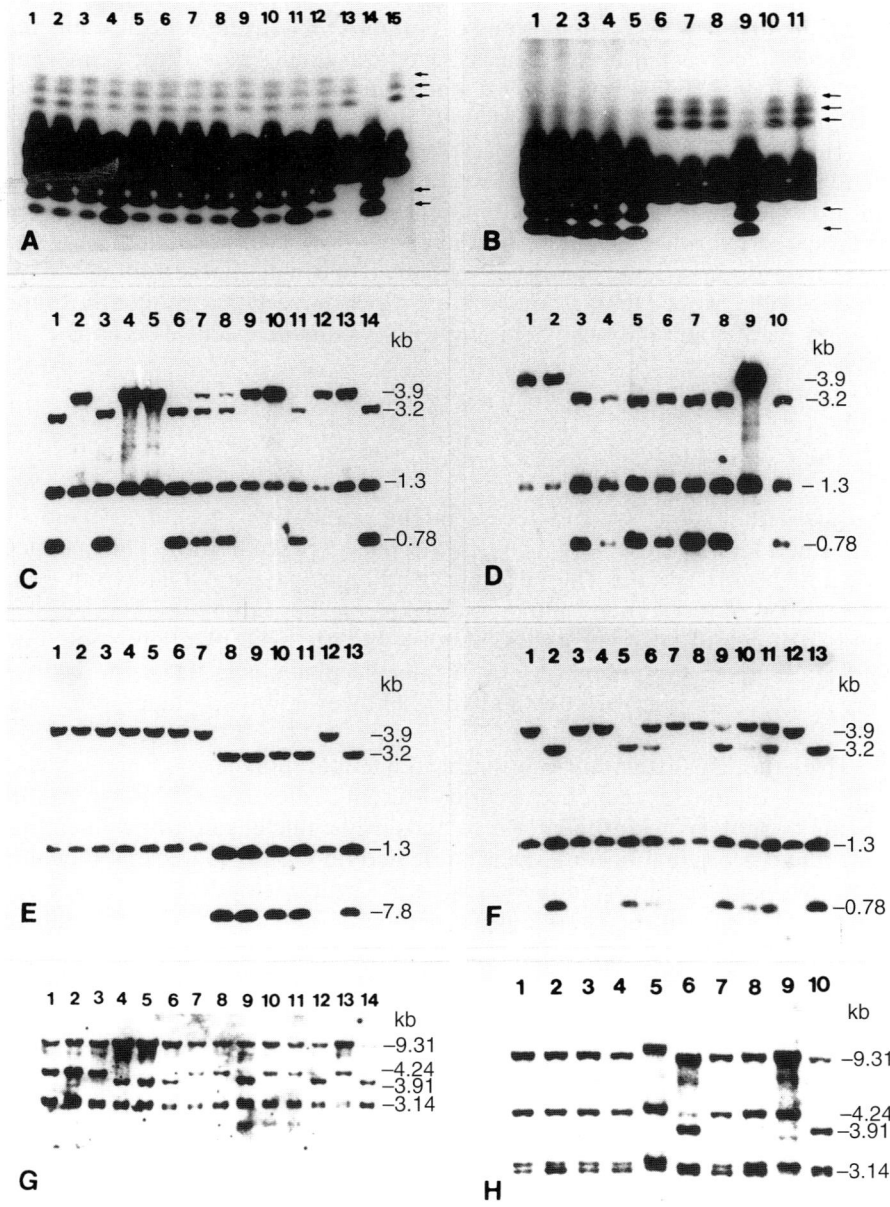

of *S.f(x)S.j*-Wa1-H7 were green; out of 11 clones, 6 clones only showed the *S.j* RFLP pattern, 2 clones the *S.f* pattern and 3 a mixed pattern (Fig. 2F). In the case of *S.f(x)S.j*-Wa2-H1, no variegated protoclones appeared. The RFLP analysis of seven chlorophyll-deficient clones and four green clones (Fig. 2E) revealed patterns corresponding to the pigmentation. It could be concluded that mixed cells were still present in *S.f(x)S.j*-Wa1-H7, while the plastid segregation appeared completed in *S.f(s)S.j*-Wa2. The number of investigated genetic traits of the plastids was too low to exclude the presence of ptDNA recombinants. The recombination of ptDNA has rarely been detected after protoplast fusion in Solanaceae (Medgyesy and Fejes 1985; Thanh and Medgyesy 1989; Fejes et al. 1990) and after mixed protoplast culture of *Solanum* and *Potentilla* (Wang and Binding 1994).

The heritage of the mitochondria was investigated in each single clone of the hybrid lines. It could be traced with the mitochondrial DNA probe of the maize *Cox* II gene (Fig. 2G). No biparental RFLP pattern was found in the investigated hybrid clones. Four hybrid clones showed *S.f*-specific bands and seven hybrids *S.j*-specific bands, respectively. The single marker did not allow the detection of mt-DNA recombinants, which have been frequently found in other plant combinations beginning with *Nicotiana* species (Belliard et al. 1979).

2.7 The Cybrid Lines

Cybrids with *S.j*-Wa1 were detected in the population of lines selected by malformed leaves, those with *S.j*-Wa2 selected by pigmentation. They formed only a few cybrid shoots. The missing cybrid types have most likely not been detected because of their similarity to one of the parents and/or the low number of cybrid lines and cybrid shoots obtained. The plantlets, rooted on MSI, grew up in the greenhouse into infertile plants (Fig. 1D). The cybrids were analyzed as described for the hybrids; however, only a single protoclone of each line.

Even though all genomic traits indicated uniparental heritage of the nuclei (Fig. 2B), diploid asymmetric hybrids cannot be excluded. However, the missing of sublines with the hybrid or allospecific nucleus in cybrid lines suggests that the original fusion bodies contained only a single nucleus, meaning that a cytoplast was involved in the fusion. No cybrid with the *S.j*-Wa1 nucleus was found, while all the cybrids in the combinations with *S.j*-Wa2 contained the *S.j* nucleus.

Three types of genome/plastome combinations have been detected (Fig. 2D). Cybrids with the *S.j* nucleus and *S.j*-Wa2 plastids as well as with both plastomes were missing. Both types of nuclei were associated to the *S.j* mitochondrial trait, and only one cybrid also contained the *S.j* mtDNA trait (Fig. 2H). The malformation of the green cybrids with the *S.f* nuclei was strictly correlated to the *S.j* mitochondria.

2.8 The Localization of the Mutant Genes

The genome location of the mutated gene of S.j-Wa1 is confirmed by the green pigmentation of all respective hybrids and cybrids, independently of the types of plastomes. The mutant character is masked by a gene product of S.f. The rare white spots in S.f(x)S.j-Wa1-H7 may be due to demasking caused by genomic instability, even though not reflected in the chromosome number. While the mutant allele was supposed to be dominant within S.j, expression in the hybrid is interpreted rather as hypostasis than recessivity.

The plastome location of the mutation of S.j-Wa2 was proved by strict correlation of the pigmentation type with the ptDNA RFLP pattern. Hence, the mutant character was a valid trait for the plastome-based selection and further analysis of the fusants.

3 Summary and Conclusion

Hybrids, cybrids, and cell graft chimeras of members of the Asteraceae have been produced by procedures which had been already successful in other higher plant families. Also, the developmental phenomena, namely high regeneration rates via adventitious shoot formation, multicellular shoot initiation, and variable chromosomal stability, were features which have been described for other taxa. The combination of marker systems of morphology, pigmentation, isoenzymes, DNA restriction fragments, and subcloning through protoplasts was appropriate for the selection, identification, and analysis of biparental lines. The various heterospecific combinations of genomes, plastomes, and mitochondrial traits indicate heterospecific tolerance. For the detection and analysis of extremely asymmetric hybrids and plastome DNA recombinants, the application of more molecular markers would be needed. The aptitude of monocot pt- and mtDNA species for RFLP probing in the dicots was a new experience at the time of experimentation.

It is clearly demonstrated with the present investigations that somatic hybridization, cybridization, and cell grafting are open for application to members of the Asteraceae, provided that sufficient regeneration is obtained.

4 Protocol

Seeds were plated on phytohormone-free B5 medium (Gamborg et al. 1968). Shoots of the seedlings were propagated on MS (Murashige and Skoog 1962) with $2.5\,\mu M$ 6-bynzylaminopurine (BAP) in 10-cm petri dishes in a tissue culture room at $25 \pm 1\,°C$ and a 16-h day by white fluorescent light photon fluxes of $30–60\,\mu mol\,m^{-2}s^{-1}$. Shoots were rooted on MSI (MS with $30\,\mu M$ indole-3-acetic acid).

For protoplast isolation, each 100 mg sliced shoot tip material was incubated in a 10-cm petri dish with an enzyme broth which contained 3% Rohament PC (Roehm), 0.5 M sorbitol, 5 mM

Can(NO$_3$)$_2$, pH 5.8, at 25 °C and continuous photon flux of 5 µmol m^{-2} s^{-1}. The protoplasts were purified by sieving through a steel sieve of 40-µm mesh, spinning for 5 min at 100 g, and a second sedimentation at 80 g with V-NaCl.3 (salts of V-KM 10^{-1} in 0.3 M NaCl). The protoplast culture medium was V-KM (Binding and Nehls 1977; containing the organic nutrients of the 8p medium, Kao and Michayluk 1975). For untreated culture, 1–4 × 10^3 protoplasts were plated in 3-cm petri dishes, distributed to four to five 15-µl streaky lenses using V-KM with 2% agarose. The lenses were covered with 1 ml liquid V-KM (Binding et al. 1988a).

For protoplast fusion, roughly 5 × 10^4 protoplasts of both genotypes were embedded in four sandwich lenses using 0.2 M Ca(NO$_3$)$_2$ (pH 6.2) with 2% agarose at 37 °C. The lenses were covered with 0.2 M Ca(NO$_3$)$_2$ (adjusted with NH$_4$OH to pH 10.3) for fusion induction. The fusogen was washed off with liquid V-KM after 20 min. After 5 min, the liquid was replaced with 1 ml of fresh V-KM (Binding et al. 1988b).

The dishes were sealed with household polyethylene foil strips and placed in a growth chamber at 25 ± 1 °C and permanent fluorescent light photon flux of 10 µmol m^{-2} s^{-1}. The liquid V-KM was exchanged on the 2nd, 4th, and 6th day. The lenses were transferred to B5BC (B5 with 2.5 µM BAP and 5% liquid coconut endosperm) on the 7th day. Seven days later, the plating efficiencies were calculated from the relation of multicellular structures and plated protoplasts, and the microcalli were separated for the initiation of callus lines.

The regenerated shoots of a callus line were subcultured on MSB as a regenerant line. The relation of regenerant lines to plated protoplasts was used as the measure of regeneration efficiency. However, lines which formed pure primary shoots of both partner species or a fusant and a parental type were treated as two separate lines.

Lines with variegated shoots, composed of uniparental cells of both partners, were classified as cell graft chimeras (Binding et al. 1987). The fusant lines were assigned with (x) (see Gleba and Sytnik 1984; Binding et al. 1986). The hybrid lines were labeled with H-numbers and the cybrid lines with C-numbers. Cybrids (Cocking 1977) are characterized by monotypic nuclei and foreign plastome traits (Binding et al. 1986).

Isoenzyme patterns were studied by standard procedures. The included extraction and purification of proteins from shoots of 30-day-old subcultures, polyacrylamid gel electrophoresis, and detection of proteins by specific activities of peroxidase, esterase, malate dehydrogenase, and glutamate-oxaloacetate transaminase according to protocols of Allen et al. (1984), Brewer (1970), and Gabriel (1971).

For RFLP probing of ptDNA, mtDNA, and total DNA was isolated by CATB protocol (Rogers and Bendich 1988), cut with *Eco*RI and *Bam*H1 following the manufacturer's instruction (Boehringer, Mannheim), separated by agarose gel electrophoresis, blotted onto a nitronylon membrane (Sambrook et al. 1989) and hybridized with the following probes: ptDNA probes from barley *ndh*E + G (Krupinska 1992), *psb*A and *rbc*L (Falk et al. 1993), and *psb*F + E (Krupinska and Berry-Lowe 1988); mtDNA probes from maize HindIII segment of *cob* and Pst/EcoRI segment of *cox*II (Pratje, Hamburg). The ptDNA probes, inserted in the plasmid vectors pUC (provided by K. Krupinska), and the mtDNA probes in pBR322 (provided by E. Pratje) were cloned in the *Escherichia coli* strain DH5α. The inserts were excised with the suitable restriction enzymes and nick-labeled with ^{32}P-ATP. For more details on the RFLP investigations see Wang and Binding (1994).

References

Al-Atabee JS, Mulligan BJ, Power BJ (1990) Interspecific somatic hybrids of *Rudbeckia hirta* and *R. laciniata* (Compositae). Plant Cell Rep 8:517–520

Allen RC, Saravis CA, Maurer HR (1984) Gel electrophoresis and isoelectric focusing of proteins, selected techniques. Walter de Gruyter Berlin

Belliard G, Vedel F, Pelletier G (1979) Mitochondrial recombination in cytoplasmic hybrids of *Nicotiana tabacum* by protoplast fusion. Nature 281:401–403

Binding H, Nehls R (1977) Regeneration of isolated protoplasts to plants in *Solanum dulcamara* L. Z Pflanzenphysiol 85:279–280
Binding H, Nehls R (1980) protoplast regeneration to plants in *Senecio vulgaris* L. Z Pflanzenphysiol 99:183–185
Binding H, Nehls R, Kock R, Finger J, Mordhorst G (1981) Comparative studies on protoplast regeneration in herbaceous species of the Dicotyledoneae class. Z Pflanzenphysiol 101:119–130
Binding H, Krumbiegel-Schroeren G, Nehls R (1986) Protoplast fusion and early development of fusants. In: Reinert J, Binding H (eds) Differentiation of protoplasts and transformed cells. Springer, Berlin Heidelberg New York, pp 37–66
Binding H, Witt D, Monzer J, Mordhorst G, Kollmann R (1987) Plant cell graft chimeras obtained by co-culture of isolated protoplasts. Protoplasma 141:64–73
Binding H, Görschen E, Jörgensen J, Krumbiegel-Schroeren G, Ling HQ, Rudnick J, Sauer A, Zuba M, Mordhorst G (1988a) Protoplast culture in agarose media with particular emphasis to streaky culture lenses. Bot Acta 101:233–239
Binding H, Zuba M, Rudnick J, Mordhorst G (1988b) Protoplast gel fusion. J Plant Physiol 133:409–413
Brewer GJ (1970) An introduction to isozyme techniques. Academic Press, New York
Cocking EC (1977) Uptake of foreign material by plant protoplasts. In Bourne GH, Danielli JF (eds) International review of cytology 48. Academic Press, New York, pp 323–343
Falk J, Schmidt A, Krupinska K (1993) Characterization of plastid DNA transcription in ribosome-deficient plastids of heat-bleached barley leaves. J Plant Physiol 141:176–181
Fejes E, Engler D, Maliga P (1990) Extensive homologous chloroplast DNA recombination in the pt14 *Nicotiana* somatic hybrid. Theor Appl Genet 79:28–32
Gabriel O (1971) Locating enzymes on gels. In: Jakoby WB (ed) Enzyme purification and related techniques, vol 22. Academic Press, New York, pp 578–604
Gamborg OL, Miller RA, Ojima K (1968) Nutrient requirements of suspension cultures of soybean root cells. Exp Cell Res 50:151–158
Gelba YY, Sytnik KM (1984) Protoplast fusion. Springer, Berlin Heidelberg New York
Kao KN, Michayluk MR (1975) Nutritional requirements for growth of *Vicia hajastana* cells and protoplasts at a very low population density in liquid media. Planta 126:105–110
Krupinska K (1992) Transcriptional control of plastid gene expression during development of primary foliage leaves of barley grown under a daily light-dark regime. Planta 186:294–303
Krupinska K, Berry-Lowe S (1988) Characterization and in vitro expression of cytochrome b-559 genes of barley. I. Localization and sequence of the genes. Carlsberg Res Commun 53:43–55
Matsumoto E (1991) Interspecific somatic hybridization between lettuce (*Lactuca sativa*) and wild species *L. virosa*. Plant Cell Rep 9:531–534
Medgyesy P, Fejes G (1985) Interspecific chloroplast recombination in a *Nicotiana* somatic hybrid. Proc Natl Acad Sci 82:6960–6964
Murashige T, Skoog F (1962) A revised medium for rapid growth and bioassay with tobacco tissue cultures. Physiol Plant 15:473–497
Rogers SO, Bendich AJ (1988) Extraction of DNA from plant tissue. In: Gelvin SB, Schilperoort RA, Verma DPS (eds) Plant molecular biology manual. Kluwer, Dordrecht, A6, pp 1-10
Sambrook J, Fritsch EF, Maniatis T (1989) Molecular cloning, a laboratory manual. Cold Spring Harbor Laboratory Press, Cold Spring Harbor, New York
Thanh ND, Medgyesy P (1989) Limited chloroplast gene transfer via recombination overcomes plastome-genome incompatibility between *Nicotiana tabacum* and *Solanum tuberosum*. Plant Mol Biol 12:87–93
Wang G, Binding H (1993a) Somatic hybridization between *Senecio fuchsii* Gmel. and *S. jacobaea* L. In vitro culture. Acta Hortic 336:315–320
Wang G. Binding H (1993b) Somatic hybrids and cybrids of *Senecio fuchsii* Gmel. (x) *S. jacobaea* L. Theor Appl Genet 87:561–567
Wang G, Binding H (1994) A *Solanum(x)Potentilla* interfamily protoplast regenerant with *Solanum* characteristics but with a *Potentilla* plastome fraction. Physiol Plant 91:155–160

Section IV
Legumes/Pasture Crops

IV.1 Somatic Hybridization Between *Medicago sativa* L. (Alfalfa) and *Lotus corniculatus* L. (Birdsfoot Trefoil)

M. Niizeki

1 Introduction

Alfalfa (*Medicago sativa* L.) is the most important forage crop in terms of high protein content and good palatability. However, it does not grow well in acid soils and under humid conditions. On the other hand, birdsfoot trefoil (*Lotus corniculatus* L.) is a forage crop which grows well under adverse soil and climate conditions, especially under wet conditions. Mutual combinations of the good traits from these leguminous species would be desirable, but cannot be achieved due to the lack of sexual hybridization between them. Notably, somatic cell hybridization has been useful as a means of bypassing barriers to sexual hybridization in order to introduce unavailable but desirable germplasm for crop improvement (Bajaj 1994). Thus, it is worthwhile to combine favorable genetic traits between alfalfa and birdsfoot trefoil.

Nonetheless, agriculturally useful somatic hybrid production has been very difficult, since there is an imbalance in the genomes of the parents in most cases, which results in the rearrangement or partial elimination of the chromosomes of one parent and an inability to achieve morphogenesis (Kao 1977; Chien et al. 1982; Sala et al. 1985). Thus, attempts have been made to fuse X-ray, γ-ray, or UV-irradiated donor protoplasts with recipient protoplasts (Sidorov et al. 1981; Tanno-Suenaga et al. 1988; Samoylov and Sink 1996; Sakai et al. 1996; Oberwalder et al. 1997; Hansen and Earle 1997; Vlahova et al. 1997; Atanassov et al. 1998; Forsberg et al. 1998; Liu et al. 1999). In this case, the retention or integration of chromosome segments from the donor protoplasts can be expected in the recipient protoplasts, and plant regeneration from the fused cells may be expected.

To date, there have been a limited number of reports on successful hybridization between leguminous species (Sano et al. 1988; Niizeki and Saito 1989; Kihara et al. 1992; Kaimori et al. 1998). This report shows a successful case for the production of calli and regenerated plants, within the leguminous species, through asymmetric protoplast fusion of X-irradiated alfalfa with iodoacetamide (IOA)-treated birdsfoot trefoil. Further, the significance of this trial is interpreted from the view point of plant breeding.

Laboratory of Plant Breeding and Genetics, Faculty of Agriculture and Life Science, Hirosaki University, Hirosaki, Aomori-ken 036-8561, Japan

2 Symmetric Somatic Cell Hybridization

2.1 Callus Induction and Protoplast Isolation

The materials used were alfalfa, cv. DuPuits, Rangelander, Vernal, and Rambler, and birdsfoot trefoil, cv. Viking. Seeds of these materials were sterilized by a sodium hypochlorite solution containing 1.5% active chlorite for 15 min.

Sterilized seeds of alfalfa were germinated on the medium of Nitsch and Nitsch (1969) without growth regulators. For callus induction, hypocotyls of 2-week-old seedlings excised about 1 cm were cultured on B5 basal medium (Gamborg et al. 1968) supplemented with $1.0\,mg\,l^{-1}$ 2,4-dichlorophenoxyacetic acid (2,4-D), and $0.2\,mg\,l^{-1}$ kinetin (B5h medium). In the case of birdsfoot trefoil, about 1-cm segments of hypocotyls from 10-day-old seedlings were transferred onto the medium of Murashige and Skoog (1962) containing $4\,mg\,l^{-1}$ l-naphthaleneacetic acid (NAA) and $2.5\,mg\,l^{-1}$ kinetin (M-1N medium). All cultures were maintained under $38\,\mu E\,m^{-2}\,s^{-1}$ illumination at 25 °C. An enzyme solution containing 4% Cellulase Onozuka RS (Yakult Co. Ltd., Tokyo), 1% Macerozyme R10 (Yakult Co. Ltd., Tokyo), 0.2% Pectolyase Y23 (Seishin Co. Ltd., Tokyo), and 0.7 M mannitol (pH 5.8) was used for the isolation of protoplasts from calli of both species. The isolated protoplasts were separated from undigested cell clumps by passage through eight layers of cotton gauze (Iwatsuki Co. Ltd.). The enzyme was removed by four successive washings with 0.6–0.7 M mannitol and 50 mM $CaCl_2$ (pH 5.8) by centrifugation at $80\,g$ for 5 min each.

2.2 Protoplast Fusion and Culture of Parental and Fused Protoplasts

First, it was confirmed that 5 mM IOA treatment inhibited the division of callus protoplasts of birdsfoot trefoil, while without IOA treatment colony formation was not inhibited at all. It has been reported that the colony formation of alfalfa depends upon the genotype of the cultivars and individual plants in the same cultivar (Niizeki and Saito 1987). Indeed, most alfalfa protoplasts divided at the early stage of culture, but in a certain genotype, continuous cell division ceased and colonies were not formed (Table 1). The limited colony formation of alfalfa protoplasts made it possible to select hybrid cells resulting from the fusion with birdsfoot trefoil protoplasts whose cell division was inhibited by IOA.

In the fusion of protoplasts between the two species, protoplasts of birdsfoot trefoil were treated with 5 mM IOA for 15 min at 4 °C to inactivate cell division, while alfalfa protoplasts from five selected callus lines were used. Alfalfa protoplasts showed limited cell division only during the early stage of culture. The protoplast fusion treatment with polyethyleneglycol (PEG) was performed by applying a slightly modified method of Melchers et al. (1978). The detailed procedure has been shown in Niizeki et al. (1985). After fusion treatment, the protoplasts were cultured in 0.5 ml liquid medium of a modi-

Table 1. Frequencies of genotypes which formed colonies from callus protoplasts of alfalfa, *Medicago sativa*, in modified 8p medium

Cultivar	No. of individuals with different genotypes used	No. of individuals with different genotypes which formed colonies from protoplasts
DuPuit	6	4
Rangelander	9	2
Vernal	3	1
Rambler	2	1
Total	20	8

fied KM8p medium (Kao and Michayluk 1975), containing $0.5\,mg\,l^{-1}$ benzyladenine instead of zeatin and lacking coconut milk, at 25 °C. The protoplast population density ranged from 10^4 to $10^5\,ml^{-1}$. After the protoplasts divided a few times, they were exposed to light ($38\,\mu E\,m^{-2}\,s^{-1}$) and subsequently colony formation was observed. For the induction of calli, the colonies were transferred onto M-1N medium.

By this procedure 128 calli were obtained. Only 2 out of 128 calli were brownish green and notably soft. These calli were most likely derived from the fusion between cv. DuPuits in alfalfa and birdsfoot trefoil, since they showed intermediate characteristics between the two parents. However, the remaining calli did not appear to be hybrid calli because they were green or dark green and were very hard, as in the case of birdsfoot trefoil. Alfalfa calli were usually yellow or light green and very soft. This result indicates that the selection system was incomplete and that 5 mM IOA could not completely suppress the cell division of birdsfoot trefoil protoplasts. This may be due to the use of calli of different genotypes arising from individuals resulting from outcrossing of birdsfoot trefoil in the process of protoplast fusion, although only the callus of one genotype was examined for the response to IOA treatment. Inhibition of cell division in birdsfoot trefoil protoplasts by IOA could vary among the genotypes.

2.3 Isozyme and Chromosome Analysis

The protocols of isozyme and chromosome analysis of the calli were previously described by Niizeki et al. (1985) and Niizeki and Grant (1971), respectively.

The analysis of peroxidase isozyme by the disk electrophoresis method was performed for the two brownish green calli. The electrophoretic pattern of one callus indicated mixed banding pattern of both parents, suggesting that it was a hybrid (Fig. 1). The other callus did not exhibit a unique band of alfalfa. The isozyme analysis for the identification of hybrids may be incomplete, because it does not rule out the possibility that the examined callus could consist of a chimera of parent cells, and not be in a hybrid state. Therefore, as

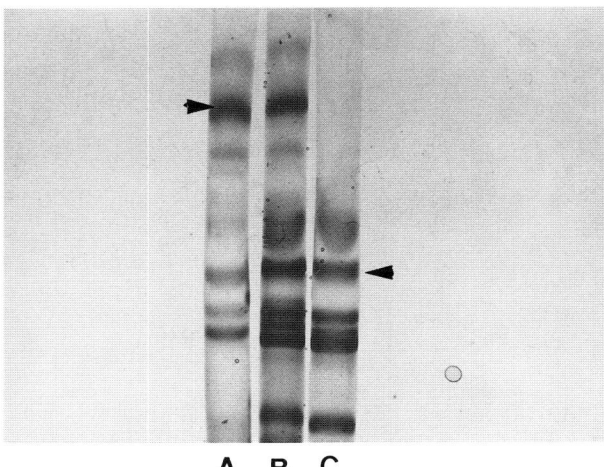

Fig. 1. Electrophoretic banding pattern of peroxidase isozyme of alfalfa, birdsfoot trefoil, and hybrid calli. In the hybrid calli, unique bands of parents indicated by *arrowheads* were observed. *A* Alfalfa callus; *B* hybrid callus; *C* birdsfoot trefoil callus. (Niizeki and Saito 1989)

further evidence, we analyzed the chromosomes of the calli after 7 months (Fig. 2). The chromosome number of alfalfa is 32 and that of birdsfoot trefoil 24. The callus which showed the peroxidase isozyme bands of both parents was examined for chromosome analysis. The chromosome number varied among the cells and was generally less than 56 (32 + 24), which is the theoretical chromosome number of an amphidiploid. The number of chromosomes of alfalfa, which are longer than those of birdsfoot trefoil, seemed to be lower than the expected number. Gradual chromosome elimination of one parent or chromosome segment elimination has been reported in some cases (Kao 1977; Wetter 1977; Krumbiegel and Schieder 1981; Chien et al. 1982; Yamada et al. 1998; Polgar et al. 1999). In this study, it may be considered that some of the alfalfa chromosomes might have been eliminated after protoplast fusion. These results also support the supposition that the other brownish green callus, which did not show the mixed isozyme banding pattern of the parents, might originate from fusion products.

In our system, the calli of birdsfoot trefoil produced adventitious buds which could develop into shoots and maintained their totipotency for long periods of 1 to 2 years on the M-1N medium. On the other hand, the calli of alfalfa lost their totipotency after several subcultures on the B5h medium. The hybrid calli, however, produced no adventitious buds or embryos on either media. Some fusion combinations between distantly related species grow as calli but cannot be regenerated into plants (Zenkteler and Melchers 1978; Binding and Nehls 1978). These facts may be attributed to the imbalance of parental chromosomes or incomplete genome of one parent. It may, however, be possible to obtain plants if the genetic information of one fusion parent is reduced to an extent so that it would not interfere with the predominating genome any more. An approach to this issue may be represented by the hybrid callus between alfalfa and birdsfoot trefoil, in which most of the alfalfa chromosomes would be lost after a long period of culture.

Fig. 2A–C. Chromosomes of birdsfoot trefoil, alfalfa, and their somatic hybrid callus. **A** Birdsfoot trefoil callus. **B** Alfalfa callus. **C** Hybrid callus. (Niizeki and Saito 1989)

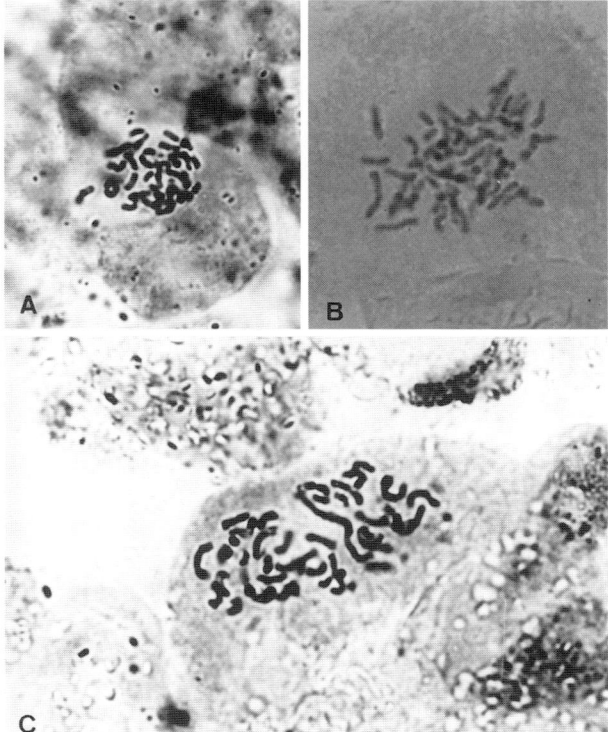

3 Asymmetric Somatic Cell Hybridization

3.1 Protoplast Isolation and Culture

Callus induction and protoplast isolation of alfalfa, cv. Rangelander and birdsfoot trefoil, cv. Viking were carried out by the methods described above.

Isolated protoplasts were settled in the modified KM8p medium and solidified with 1.25% agarose (Sea Plaque agarose). The agarose was cut into blocks and transferred to a 60-mm plastic dish containing 5 ml of the modified KM8p medium with the nurse cells of alfalfa or birdsfoot trefoil.

Isolated alfalfa protoplasts were X-ray-irradiated at four levels of 100, 200, 300, and 400 Gy, and then washed once before giving protoplast fusion treatment. The results are indicated in Fig. 3. Protoplast division in the culture with nurse cells of alfalfa ceased at an X-ray irradiation level of 300 Gy. Thus, in this study, protoplasts X-ray-irradiated at a level of 400 Gy were predominantly used for protoplast fusion, unless otherwise specified.

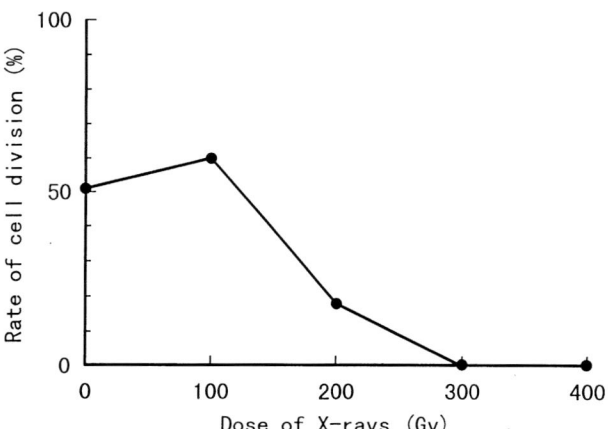

Fig. 3. Division rate of the cells derived from alfalfa protoplast after irradiation by X-rays. This investigation was carried out after 2 weeks of culture. (Kaimori et al. 1998)

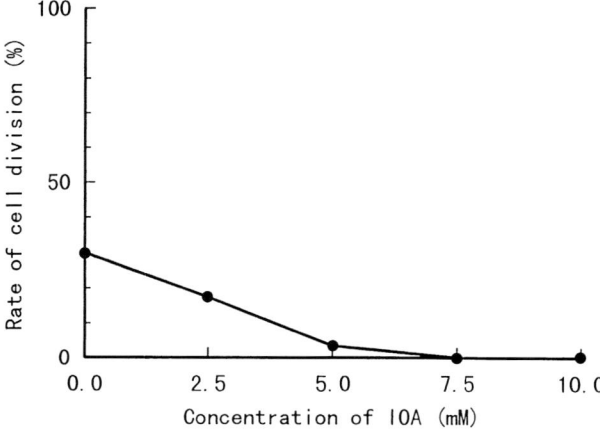

Fig. 4. Division rate of the cells derived from birdsfoot trefoil protoplasts after treatment with IOA. This investigation was carried out after 2 weeks of culture. (Kaimori et al. 1998)

Using the agarose-bead method, protoplasts of birdsfoot trefoil treated with IOA were cultured in the modified KM8p medium with nurse cells of birdsfoot trefoil. The results are shown in Fig. 4. Protoplast division was completely inhibited after treatment with 7.5 mM IOA, but protoplasts treated with 10 mM IOA were used in protoplast fusion.

3.2 Protoplast Fusion and Culture of Fused Protoplasts

Protoplast fusion by PEG was carried out using the method of Niizeki et al. (1985).

The fused protoplasts were cultured using the same method as was used for alfalfa and birdsfoot trefoil protoplast when cultured separately. After 2 weeks the protoplasts began to divide and give rise to cell clusters. The agarose blocks were then washed with 0.5M sucrose solution in order to remove the nurse cells of birdsfoot trefoil, and were again transferred to 5ml of the modified KM8p medium. When the colonies became visible, they were transferred to the M-1N medium containing $3.5\,g\,l^{-1}$ agar; 155 calli obtained were subcultured at intervals of about 1 month.

Selection of the fused hybrid protoplasts was performed by means of metabolic and physical complementation of IOA-treated and X-ray-irradiated protoplasts. Nurse cells were very effective for cell division in both the parent protoplasts themselves and the fused protoplasts.

3.3 Regenerated Plants

When vigorous shoot formation occurred, excised shoots were transferred to the basal medium of Nitsch and Nitsch (1969) without plant growth regulators, solidified with 0.2% Gelrite (Scott Laboratory, Inc., CA, USA). After roots were formed, the plantlets were transferred to sterilized soil in pots and 1/2 MS basal medium was added to the soil.

Shoots regenerated from the calli resembled the deep green leaf of birdsfoot trefoil. The shoots regenerated at an early stage of callus culture were frequently teratogenic (Fig. 5). However, the number of teratogenic

Fig. 5A–C. Regenerated shoots excised from the calli were cultured in the medium of Nitsch and Nitsch (1969) without growth regulators.
A A teratogenic shoot regenerated from an early stage of cultured callus. The growth of the shoot was very slow. **B, C** Shoots regenerated from the calli cultured for a longer term. The shoots are normal and morphologically resembled birdsfoot trefoil; one of them proliferated a root. (Kaimori et al. 1998)

regenerated shoots decreased as time went on and the morphology of regenerated plants growing in soil resembled those of birdsfoot trefoil.

3.4 Isozyme Analysis

Four kinds of isozymes in the calli and regenerated plants were analyzed using the starch gel system described by Ishikawa (1994). They were, namely aminopeptidase (AMP), esterase (EST), glutamate dehydrogenase (GDH), and catalase (CAT).

Out of 155 calli, 17 were identified as being somatic cell hybrids through the use of isozyme analysis (Table 2; Fig. 6). After the culture of hybrid calli for 1 month, the banding pattern of both parents was seen for four isozymes, which were analyzed. In addition, an isozyme banding pattern of one parent was seen in a certain callus, while in the other isozyme, the banding pattern of the other parent was also found in the same callus. After another month of subculture on the medium M-1N, most of the banding patterns had altered to those of birdsfoot trefoil in most of the calli, although only one callus indicated the banding patern of both parents. In another experiment at an early stage of cultured hybrid calli, seven hybrid calli had a

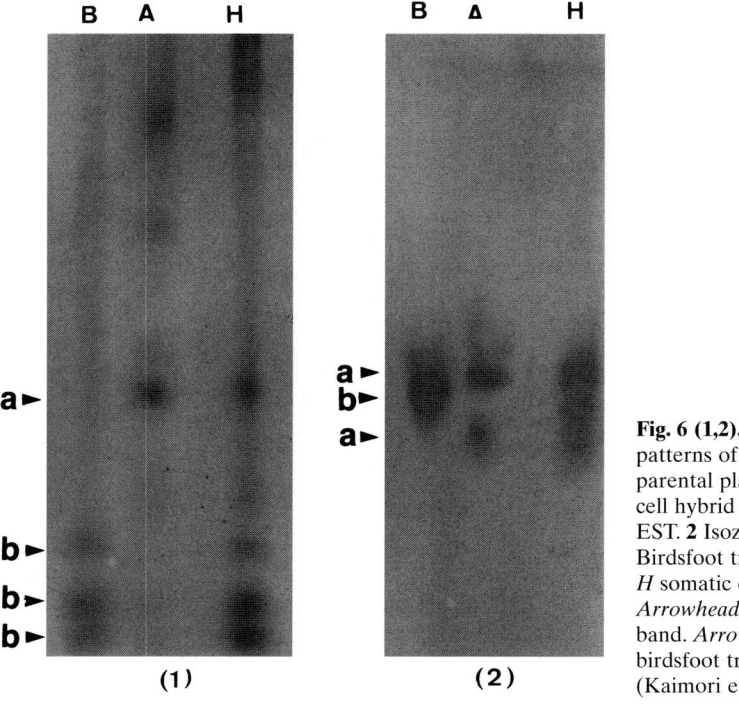

Fig. 6 (1,2). Banding patterns of isozymes for parental plants and somatic cell hybrid calli. **1** Isozyme EST. **2** Isozyme GDH. *B* Birdsfoot trefoil. *A* alfalfa; *H* somatic cell hybrids. *Arrowhead a* is the alfalfa band. *Arrowhead b* is the birdsfoot trefoil band. (Kaimori et al. 1998)

Table 2. Isozyme analysis of asymmetric somatic cell hybrids of alfalfa and birdsfoot trefoil

Hybrid calli	Plant regeneration[a]	Isozyme pattern I[b]				Isozyme pattern II[b]				Isozyme pattern III[b]			
		AMP	EST	GDH	CAT	AMP	EST	GDH	CAT	AMP	EST	GDH	CAT
I-1	–	A+B	–	–	A	B	B	B	B				
I-3	–	A+B	–	–	A	B	B	B	B				
I-7	+	A	–	–	B	B	B	B	B	B	B	B	B
I-16	–	A+B	–	–	A	–	–	–	–				
I-20	–	B	B	–	A	B	B	B	B				
I-27	+	A+B	B	–	A+B	A+B	B	B	B	–	–	–	–
I-29	+	A+B	A+B	–	A+B	B	B	B	B	B	B	B	B
I-31	–	A+B	A+B	–	A+B	B	B	B	B				
I-78	–	A+B	A+B	B	A+B	–	–	–	–				
I-82	–	A+B	A	A	A+B	–	–	–	–				
II-14	–	A+B	B	B	A+B	B	B	B	B				
II-34	+	A+B	B	B	A+B	B	B	B	B	–	–	–	–
II-44	–	A	A+B	A+B	A+B	–	–	–	–				
II-45	+	A+B	A	A+B	A	B	B	B	B	B	B	B	B
II-46	–	A	A	A+B	A	–	–	–	–				
II-47	–	A+B	–	A	B	B	B	B	B				
II-63	+	A+B	A+B	B	A+B	B	B	B	B	–	–	–	–

[a] +, Plant regeneration; –, no plant regeneration.
[b] Isozyme pattern I, investigation of the calli cultured for 1 month, Isozyme pattern II, investigation of the calli cultured for 2 months, Isozyme pattern III, investigation of the regenerated plants. A: Isozyme band pattern of alfalfa. B: Isozyme band pattern of bridsfoot trefoil, A + B: Isozyme band pattern of both parents. –: Indistinct band pattern or no investigation, AMP: Aminopeptidase, EST: Esterase, GDH: Glutamate dehydrogenase, CAT: Catalase.

catalase with the banding pattern of alfalfa and five were found to have the banding pattern of the other parent, birdsfoot trefoil (Fig. 7, no. 1). Only one callus exhibited the banding pattern of both parents. However, all the isozyme banding patterns found in regenerated shoots were the same as those of birdsfoot trefoil (Fig. 7, no. 2).

3.5 Southern Blot Analysis of Chloroplast Genome

The preparation of total DNA from calli was performed according to the method of Varadarajan and Prakash (1991): 5μg total DNA were digested with 20 units of HindIII. Electrophoresis was carried out on 0.8% agarose gel in TAE buffer (40mM Tris-acetate, 1mM EDTA, pH 8). The Southern blots were prepared by transferring the DNA to a nylon membrane (Duralon-UV membrane: Stratagene) and covalently binding the DNA to the membrane by ultraviolet irradiation. The P2 fragment, which is one of the fragments derived from rice chloroplast DNA, was used as a probe. The probe was radioactively labeled using Prime-It II random primer labeling kit (Stratagene). The blots were hybridized with the labeled probe in 20ml of hybridization buffer (6 × SSC, 2mM EDTA, pH 8, 10mM Tris-HCl, pH 7.5, 5 × Denhardts, 0.2 mg ml^{-1}

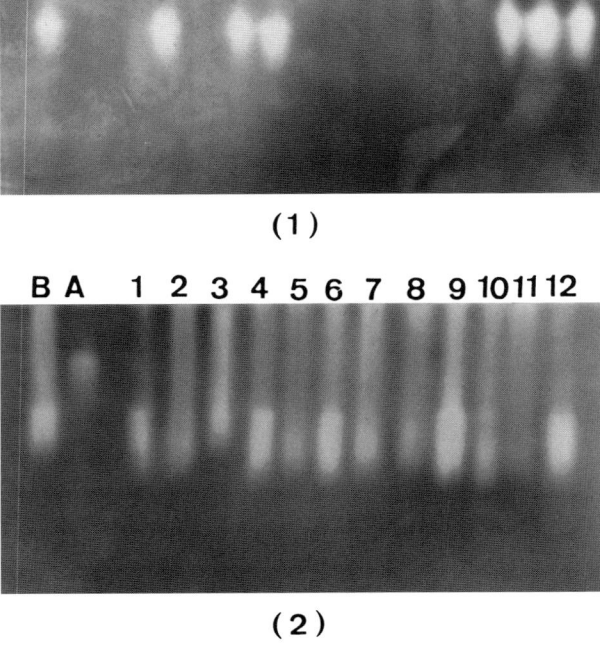

Fig. 7. 1 Banding patterns of isozyme CAT for calli. *A* Alfalfa; *B* birdsfoot trefoil; *1–13* calli cultured for 1 month. No. 1 of hybrid calli has the bands of both parents and the other hybrid calli have the bands of one of the parents. 2 Banding patterns of isozyme CAT of shoots. *A* Alfalfa; *B* birdsfoot trefoil; *1–12* regenerated shoots from hybrid calli. All shoots have the bands of birdsfoot trefoil. (Kaimori et al. 1998)

salmon spermary DNA, 10mM Na₃PO₄, 1% Na N-lauroyl salcosinate), and autoradiographed using a Kodak XJB-1 X-ray film.

The analyses of 14 hybrid calli indicated that two hybrid calli, II-44 and II-46, have the chloroplast genomes of alfalfa and the other 12 have those of birdsfoot trefoil (Fig. 8). According to isozyme analysis, it was confirmed that the callus of II-44 has nuclei of birdsfoot trefoil (Table 2). Therefore, the II-44 callus seems to have the nuclei of birdsfoot trefoil and the genomes of alfalfa chloroplast. The color of the two hybrid calli with the alfalfa chloroplast genomes was green, which was intermediate between the dark green calli of birdsfoot trefoil and the yellow of alfalfa. No shoot, however, was regenerated from these hybrid calli.

4 Summary and Conclusion

Some successes in asymmetric protoplast fusion based on the complementation of X-ray- or γ-ray-irradiated and IOA-treated protoplasts have been reported (Sidorov et al. 1981; Tanno-Suenaga et al. 1988; Sakai et al. 1996; Hansen and Earle 1997; Liu et al. 1999). Our results also support these results.

Many hybrid calli cultured for 1 month were found to have isozymes indicating the banding patterns of both parents. Certain calli were found to have the banding pattern of one parent in some isozymes and that of another parent in others. These facts may suggest that chromosomes or chromosome segments

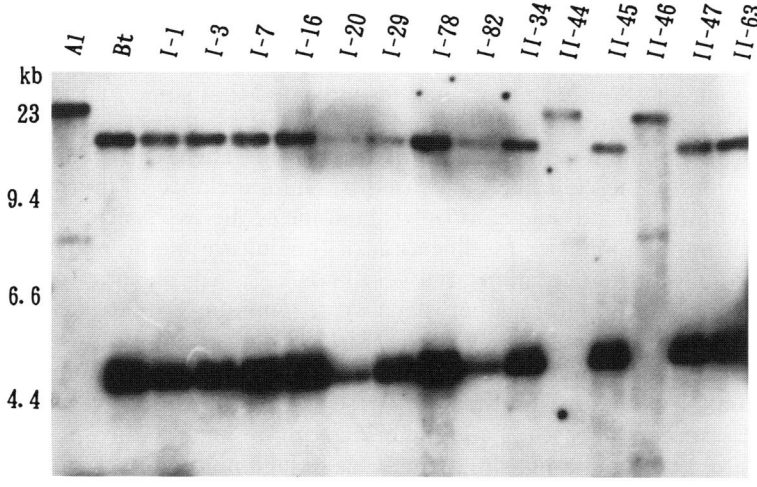

Fig. 8. Southern blot analyses of chloroplast genomes in the 14 hybrid calli and their parents. Total DNA was digested with *Hin*dIII and P2 fragment of rice chloroplast DNA was used as a probe. *Al* Alfalfa; *Bt* birdsfoot trefoil; *I-1~II-63* hybrid calli. The II-44 and II-46 hybrid calli show a banding pattern identical to alfalfa and the other to birdsfoot trefoil. (Kaimori et al. 1998)

of both parents may be eliminated randomly at the early stage of callus culture. On the other hand, most of the banding patterns had altered to those of the birdsfoot trefoil in 2 months of culture. Therefore, the number of calli having the chromosomes of both parents would seem to decrease and most of the calli had only birdsfoot trefoil chromosomes. This may indicate that callus cells with birdsfoot trefoil genomes rapidly came to be selected as dominant.

Shoot regeneration did not occur from the symmetric somatic hybrid calli of birdsfoot trefoil and alfalfa. This fact might be attributed to the imbalance of birdsfoot trefoil and alfalfa genomes or the incomplete genome of alfalfa, some of whose chromosomes were eliminated after protoplast fusion. On the other hand, normal shoot regeneration occurred from asymmetric somatic cell hybrid calli. All the isozyme banding patterns in the shoots investigated so far were the same as those of birdsfoot trefoil. This result indicates that most of the alfalfa chromosomes irradiated by X-rays degenerated during the subcultures. Accordingly, it may be possible to regenerate the shoots from the calli derived from asymmetric hybrids with X-ray-irradiated donor protoplasts, while regeneration of novel shoots is not likely from symmetrical somatic hybrids carrying complete chromosomes of both parents or some chromosomes of one parent.

Niizeki et al. (1990) and Kihara et al. (1992) reported that X-ray-irradiated soybean protoplasts and IOA-treated birdsfoot trefoil protoplasts were fused in order to transfer the desirable germplasm from soybean into birdsfoot trefoil. The morphology of the regenerated plants resembled that of birdsfoot trefoil. However, five regenerated plants were teratological with an erect and short plant height even though all plants had no soybean chromosomes. In this study, the morphology of all plant regenerated at the late stage of asymmetric hybrid callus culture resembled that of normal birdsfoot trefoil. This discrepancy may be caused by the difference in phylogenetic distances for the genomes of the organelle. The distance between soybean and birdsfoot trefoil may be greater than that between alfalfa and birdsfoot trefoil.

Isozyme and Southern blot analyses indicated that some hybrid calli had nuclei of birdsfoot trefoil and chloroplast genomes of alfalfa. This fact proved that the calli obtained were real asymmetric hybrids. The recalcitrance in shoot regeneration by the calli may be caused by the imbalance in morphogenetic potential of the nucleus and chloroplast genome of the two species concerned. However, improvement or modification of the culture media may provide a breakthrough for regeneration of novel plants.

Somatic hybrid cell lines between two leguminous species, soybean and hyacinth bean, were obtained by Sano et al. (1988). In attempts to use the leguminous species for protoplast fusion, I also succeeded in obtaining hybrids between alfalfa and birdsfoot trefoil. These results, which may represent one of the few successful cases of wide hybridization of leguminous species, show the possibility of obtaining somatic hybrids in leguminous species in addition to those reported in the Solanaceae and Cruciferae species. However, even in species in the Cruciferae, male sterility or low fertility has often been found in interspecific somatic hybridizations (Kirti et al. 1992; Lelivelt and Krens 1992). Hansen and Earle (1997) suggested that one reason for low fertility may be alloplasmic male sterility caused by incompatibility between the nuclear and the cytoplasmic genomes, whereas Nothnagel et al. (1997) indicated that backcrosses with one parent may become a useful tool in overcoming the male sterility or low fertility in some cases of Cruciferae. Somatic hybrids of birdsfoot trefoil and alfalfa also showed male sterility, so that no seed was obtained. Thus, from the point of view of plant breeding it may be also worthwhile to attempt to backcross somatic hybrids with one parent in these leguminous species.

5 Protocol

5.1 Protoplast Isolation

Alfalfa
1. Use 1–2-g calli subcultured for 10–15 days on B5 basal medium (Gamborg et al. 1968) supplemented with 1.0 mg l^{-1} 2,4-D and 0.2 mg l^{-1} kinetin (B5h medium).
2. Incubate the calli in a solution containing 4% Cellulase Onozuka RS, 1% Macerozyme R10, 0.2% Pectolyase Y23, and 0.7 M mannitol (pH 5.3) for 1–4 h on a reciprocal shaker (60 strokes min^{-1}) at 24 °C.
3. Separate the protoplasts from undigested cell clumps with eight layers of cotton gauze and remove the enzyme solution in four successive washings with 0.6–0.7 M mannitol and 50 mM CaCl$_2$ (pH 5.8) by centrifugation at 80 g for 5 min.
Birdsfoot Trefoil
1. Use 1–2 g calli subcultured for 10–15 days on MS basal medium (Murashige and Skoog 1962) containing 4 mg l^{-1} NAA and 2.5 mg l^{-1} kinetin (M-1N medium) for protoplast isolation.
2. The procedure for protoplast isolation is the same as for alfalfa.

5.2 Fusion of Protoplasts

1. For the selection of symmetric hybrid protoplasts, treat the protoplasts of birdsfoot trefoil with 5 mM IOA solution containing 0.56 M mannitol and 0.05 M CaCl$_2$ (pH 5.8) for 15 min at 4 °C and use five selected callus lines of alfalfa, from which isolated protoplasts divide only during the early stage of culture. In the case of an asymmetric somatic hybrid, subject the protoplasts of alfalfa to 400 Gy of X-ray irradiation.
2. Adjust the density of the two species of protoplast suspension to 10^4–10^5 ml^{-1} in 0.7 M mannitol solution.
3. Fuse the two species of protoplasts with a solution of PEG 1540: 10 mM CaCl$_2$, 0.7 mM KH$_2$PO$_4$, and 0.1 M glucose for 15–20 min at 25 °C.
4. Wash the protoplasts four times with a solution containing 0.6 M mannitol and 50 mM CaCl$_2$.

5.3 Culture of Heterokaryons

1. Settle the fused protoplasts in modified KM8p medium (add 0.5 mg l^{-1} BA instead of zeatin, and omit coconut milk) containing 1.25% Sea Plaque agarose, and cut the solidified agarose into

blocks. Transfer the blocks into a 60-mm plastic dish containing 5 ml modified KM8p medium and add callus cell clumps of birdsfoot trefoil as nurse cells.
2. Incubate at 25 °C in the dark and expose to light (38 μE m^{-2} s^{-1}) after the protoplast starts to divide.
3. Remove the nurse cells with 0.5 M sucrose solution after the protoplasts give rise to cell clusters, and transfer them to 5 ml of modified KM8p medium.
4. Transfer the colonies to the M-1N medium containing 3.5 g l^{-1} agar when the colonies become visible to the naked eye.
5. After 1–2 weeks, transfer the colonies to modified KM8p medium solidified with 5.5 g l^{-1} agar.
6. Cut the shoots formed from the calli and transfer them to the medium of Nitsch and Nitsch (1969) without growth regulator for root development.

References

Atanassov II, Atanassova SA, Dragoeva AI, Atanassov AI (1998) A new CMS source in *Nicotiana* developed via somatic cybridization between *N. tabacum* and *N. alata*. Theor Appl Genet 97:982–985

Bajaj YPS (1994) Somatic hybridization – a rich source of genetic variability. In: Bajaj YPS (ed) Biotechnology in agriculture and forestry, vol 27. Somatic hybridization in crop improvement I. Springer, Berlin Heidelberg New York, pp 3–32

Binding H, Nehls R (1978) Somatic cell hybridization of *Vicia faba* + *Petunia hybrida*. Mol Gen Genet 164:137–143

Chien YC, Kao KN, Wetter LR (1982) Chromosome and isozyme studies of *Nicotiana tabacum-Glycine max* hybrid cell lines. Theor Appl Genet 62:301–304

Forsberg J, Lagercrantz U, Glimelius K (1998) Comparison of UV light, X-ray and restriction enzyme treatment as tools in production of asymmetric somatic hybrids between *Brassica napus* and *Arabidopsis thaliana*. Theor Appl Genet 96:1178–1185

Gamborg OL, Miller RA, Ojima K (1968) Nutritional requirement of suspension cultures of soybean root cells. Exp Cell Res 50:151–158

Hansen LN, Earle ED (1997) Somatic hybrids between *Brassica oleracea* L. and *Sinapis alba* L. with resistance to *Alternaria brassicae* (Berk.) Sacc. Theor Appl Genet 94:1078–1085

Ishikawa R (1994) Genetical studies on isozyme genes in rice. Bull Fac Agric Hirosaki Univ 57:105–180 (in Japanese)

Kaimori N, Senda M, Ishikawa R, Akada S, Harada T, Niizeki M (1998) Asymmetric somatic cell hybrids between alfalfa and birdsfoot trefoil. Breed Sci 48:29–34

Kao KN (1977) Chromosomal behavior in somatic hybrids of soybean-*Nicotiana glauca*. Mol Gen Genet 150:225–230

Kao KN, Michayluk MR (1975) Nutritional requirements for growth of *Vicia hajastana* cells and protoplasts at a very low population density in liquid media. Planta 126:105–110

Kihara M, Cai K-N, Ishikawa R, Harada T, Niizeki M (1992) Asymmetric somatic hybrid calli between leguminous species of *Lotus corniculatus* and *Glycine max* and regenerated plant from the calli. Jpn J Breed 42:55–64

Kirti PB, Narasimuhulu SB, Prakash S, Chopra VL (1992) Production and characterization of intergeneric somatic hybrids of *Trachystoma ballii* and *Brassica juncea*. Plant Cell Rep 11:-90–92

Krumbiegel G, Schieder O (1981) Comparison of somatic and sexual incompatibility between *Datura innoxia* and *Atropa belladonna*. Planta 153:466–470

Lelivelt CLC, Krens FA (1992) Transfer of resistance to the beet cyst nematode (*Heterodera schachtii* Schm.) into the *Brassica napus* L. gene pool through intergeneric somatic hybridization with *Raphanus sativus* L. Theor Appl Genet 83:887–894

Liu B, Liu ZL, Li XW (1999) Production of a highly asymmetric somatic hybrid between rice and *Zizania latifolia* (Griseb): evidence for inter-genomic exchange. Theor Appl Genet 98:1099–1103

Melchers G, Sacristàn MD, Holder AA (1978) Somatic hybrid plants of potato and tomato regenerated from fused protoplasts. Carlsberg Res Commun 43:203–218

Murashige T, Skoog F (1962) A revised medium for rapid growth and bioassays with tobacco tissue culture. Physiol Plant 15:473–497

Niizeki M, Grant F (1971) Callus, plantlet formation, and polyploidy from cultured anthers of *Lotus* and *Nicotiana*. Can J Bot 49:2041–2051

Niizeki M, Saito K (1987) Genotypic variation in plant regeneration from calli and protoplasts of alfalfa, *Medicago sativa* L. Plant Tissue Cult Lett 4:27–31

Niizeki M, Saito K (1989) Callus formation from protoplast fusion between leguminous species of *Medicago sativa* and *Lotus corniculatus*. Jpn J Breed 39:373–377

Niizeki M, Tanaka M, Akada S, Hirai A, Saito K (1985) Callus formation of somatic hybridization of rice and soybean and characteristics of the hybrid callus. Jpn J Genet 60:81–92

Niizeki M, Cai K, Kihara M, Nakajo S, Harada T (1990) Somatic cell hybrids between birdsfoot trefoil and soybean. Lotus Newsl 21:14–17

Nitsch JP, Nitsch C (1969) Haploid plants from pollen grains. Science 163:85–87

Nothnagel T, Budahn H, Straka P, Schrader O (1997) Successful backcrosses of somatic hybrids between *Sinapis alba* and *Brassica oleracea* with the *Brassica oleracea* parent. Plant Breed 116:89–97

Oberwalder B, Ruoβ B, Schilde-Rentschler L, Hemeleben V, Ninnemann H (1997) Asymmetric fusion between wild and cultivated species of potato (*Solanum* spp.)-detection of asymmetric hybrids and genome elimination. Theor Appl Genet 94:1104–1112

Polgar Z, Susan M, Wielgus SM, Horvath S, Helgeson JP (1999) DNA analysis of potato + *Solanum brevidens* somatic hybrid lines. Euphytica 105:103–107

Sakai T, Liu HJ, Iwabuchi M, Kohno-Murase J, Imamura J (1996) Introduction of a gene from fertility restored radish (*Raphanus sativa*) into *Brassica napus* by fusion of X-irradiated protoplasts from a radish restorer line and iodacetoamide-treated protoplasts from a cytoplasmic male-sterile cybrid of *B. napus*. Theor Appl Genet 93:373–379

Sala C, Morandi MG, Nielsen C, Parisi E, Sala F (1985) Selection and nuclear DNA analysis of cell hybrids between *Daucus carota* and *Oryza sativa*. J Plant Physiol 118:409–419

Samoylov VM, Sink KC (1996) The role of irradiation dose and DNA content of somatic hybrid calli in producing asymmetric plants between in interspecific tomato hybrid and eggplant. Theor Appl Genet 92:850–857

Sano H, Suzuki Y, Oono K (1988) Somatic cell hybridization of hyacinth bean and soybean. Plant Tissue Cult Lett 5:11–14

Sidorov VA, Menczel L, Nagy F, Maliga P (1981) Chloroplast transfer in *Nicotiana* based on metabolic complementation between irradiated and iodoacetate-treated protoplasts. Planta 152:341–345

Tanno-Suenaga L, Ichikawa H, Imamura J (1988) Transfer of CMS trait in *Daucus carota* L. by donor-recipient protoplast fusion. Theor Appl Genet 76:855–860

Varadarajan GS, Prakash CS (1991) A rapid and efficient methods for the extraction of total DNA from the sweet potato and its related species. Plant Mol Biol Rep 9:6–12

Vlahova M, Hinnisdaels S, Frulleux F, Claeys M, Atanassov A, Jacobs M (1997) UV irradiation as a tool for obtaining asymmetric somatic hybrids between *Nicotiana plumbaginifolia* and *Lycopersicon esculentum*. Theor Appl Genet 94:184–191

Wetter LR (1977) Isozyme patterns in soybean-*Nicotiana* somatic hybrid cell lines. Mol Gen Genet 150:231–235

Yamada T, Hosaka K, Nakagawa K, Kaide N, Misoo S, Kamijima O (1998) Nuclear genome constitution and other characteristics of somatic hybrids between dihaploid *Solanum acaule* and tetraploid *S. tuberosum*. Euphytica 102:239–246

Zenkteler M, Melchers G (1978) In vitro hybridization by sexual methods and by fusion of somatic protoplasts. Theor Appl Genet 52:81–90

IV.2 Somatic Hybridization Between *Medicago sativa* L. (Alfalfa) and *Medicago falcata* L.

S. Arcioni, F. Damiani, F. Paolocci, and F. Pupilli

1 Introduction

According to the more recent classification of the *Medicago* genus (Quiros and Bauchan 1988), alfalfa is a unique large species named *Medicago sativa-falcata* complex, reassembling many subspecies, *sativa*, *falcata*, *coerulea*, *glutinosa*, and *glomerata*, that were previously considered as separate species.

The difficulty in producing sexual hybrids and the presence of subspecific distinctive traits are the main reasons for the classification of subspecies.

Chromosome number, flower color, and pod shape are more distinctive traits among subspecies, but the application of molecular techniques discloses new opportunities to evaluate variability among subspecies and new criteria to measure their similarity.

M. sativa subsp. *sativa* is tetraploid ($2n = 4x = 32$); *M. falcata* subsp. *falcata* includes diploid populations ($2n = 16$) and tetraploid ones. The possibility of producing sexual hybrids between these two subspecies is strictly related to the ploidy level; in fact, tetraploid accessions of *falcata* easily intercross with *sativa*, while the diploid ones can do that only if they are able to produce unreduced gametes. This latter fact makes it more complex. A distinctive marker between the two subspecies is the color of the flower: yellow for *falcata* and violet for *sativa*. The pod shape is also distinctive; *sativa* has pods coiled in two to five tight coils, while *falcata* has straight or sickle-shaped pods. *M. falcata* is well adapted to cold regions and is distributed over a wide geographical range from southern Germany to Siberia and from the Black Sea coasts to St. Petersburg; consequently, it has been utilized as a gene source for the winter-hardy alfalfa varieties currently grown in the northern USA and Canada. Recent investigations on variability at molecular levels, measured through RFLP and RAPD analysis, have shown a larger polymorphism between *falcata* and *sativa* than in comparisons between *sativa* and other subspecies of the *Medicago* complex (Osborn et al. 1997). It remains to be understood what advantages the production of somatic hybrids between these two subspecies could provide. In fact, although it is possible to produce sexual hybrids, various experiments of parasexual hybridization among them have been performed;

Istituto di Ricerche sul Miglioramento Genetico delle Piante Foraggere del CNR, Via della Madonna Alta 130, 06128 Perugia, Italy

indubitably, one of the main reasons was the necessity of finding suitable parents for setting up conditions of fusion in the genus *Medicago*, and from both subspecies it was possible to obtain a suitable plant regeneration system for protoplasts. In addition, the practical application was also a reason for using this technique with these materials. The increase in ploidy level is much more easily achieved through this method and the utilization of different subspecies assures a higher level of heterozygosity that is determinant for producing high-yielding plants, which can better tolerate cycles of selfing when the environmental conditions do not allow bees to visit flowers (Bingham and Binek 1969). Through somatic hybridization it is also possible to exploit the novel variability induced by cytoplasm-nucleus or cytoplasm-cytoplasm interactions. Such interactions have been shown to be responsible for the induction of male sterility in potato somatic hybrids (Cardi et al. 1993). This technique can also be useful to introduce resistance to some pathogens such as *Leptotrochilla medicagins*, *Cylindrocarpon chrenbergi* and *Empoasca fabae* from *falcata* subspecies to cultivated alfalfa.

For *Medicago sativa* and *Medicago falcata*, efficient systems for protoplast culture and plant regeneration are available, which is a prerequisite for producing somatic hybrid plants. Both species are cross-fertilized and since plant regeneration is genotype-specific, extensive screening and selection are required for the identification of plants with high regeneration capacity (Mitten et al. 1984; Brown and Atanassov 1985; Arcioni et al. 1990). However, even if regenerative genotypes can be identified in almost all the cultivars (Chen and Marowitch 1987), the germplasm of *M. falcata* has been considered to be more regenerable than that of *M. sativa* and the cultivars with the highest percentage of plants with high regeneration capability contained *M. falcata* germplasm (Mitten et al. 1984; Brown and Atanassov 1985). In both *M. sativa* and *M. falcata*, tetraploid genotypes are generally more embryogenic than plants belonging to diploid accessions (S. Arcioni et al. unpubl.). In the genus *Medicago*, the first somatic hybrid plants were reported by Téoulé (1983) following PEG fusion of mesophyll protoplasts of cytoplasmic male sterile *M. sativa* with those of cytoplasmic male fertile *M. falcata*. These species are tetraploid. The fusion products were cultured and even without the application of selection systems, somatic hybrid plants were recovered. The exciting results of this pioneer work stimulated different laboratories, spread throughout the world, to attempt protoplast fusion between *M. sativa* and other *Medicago* species, but in spite of all the efforts, new hybrid plants were not recovered until the early 1990s. In the meantime, the hybrid plants recovered by Téoulé were intensively analyzed and the fusion process and regeneration capacity of parental protoplasts improved. Research was carried out on protoplast culture and somatic hybridization procedures for developing methods supporting genotype-independent plant regeneration (Téoulé and Dattée 1987). Gilmour et al. (1987a) fused *M. falcata* L. and *M. quasifalcata* protoplasts with those of *M. sativa* L., and heterokaryons grew into calli which were not regenerated into plants. The results of Téoulé were confirmed by Mendis et al. (1991) following PEG fusion of protoplasts isolated from embryogenic cell suspensions of tetraploid *M. sativa* and

M. falcata. The first somatic hybrid plants between *M. sativa* and the sexually incompatible diploid *M. falcata* were obtained by Crea et al. (1997) by electrofusion. The resulting plants were hyperaneuploid ($2n = 33$) and showed gross genomic alterations where five of six parental nucleolar organizing regions were retained and new r-DNA RFLPs were created and amplified differentially among the hybrid regenerants (Cluster et al. 1996). The hybrid plants were also investigated with a highly repeated sequence (C300) cloned from *M. coerulea* and present in *M. sativa-coerulea falcata* complex with a different copy number (Calderini et al. 1997).

2 Isolation and Fusion of Protoplasts

Protoplast fusion between these two species has been carried out only in three laboratories in the world: Orsay-Paris (France), Nottingham (UK) and Perugia (Italy), and they were all successful in obtaining hybrid plants by alternative routes (Table 1). For protoplast isolation, different explants were used. In the French laboratory, Téoulé (1983) utilized young leaves and the same protocol for protoplast isolation for both parents; the protocol consists of incubating peeled leaf in a mixture 1:1 of the protoplast culture medium K77 (Kao 1977) with the enzymatic solution: 1% Cellulase R10 (Yakult Honsha Co. Ltd. Tokyo, Japan), 0.5% Macerozyme R10 (Yakult Honsha Co. Ltd.), 1% Rhozyme HP (Pollock and Pool Ltd. Reading, UK), 5.5% mannitol, and 5.5%

Table 1. Comparison of materials and methods adopted in the three laboratories where somatic hybrid plants between *M. sativa* and *M. falcata* were obtained

	Téoulé (1983)	Mendis et al. (1991)	Crea et al. (1997)
Explant			
M. sativa	Leaves	Cell suspension	Leaves
M. falcata	Leaves	Cell suspension	Callus
Fusion method	Ca^{2+} at high pH	PEG	Electrofusion
Selection	None	Dual fluorescence	Dual fluorescence
Protoplast culture	Liquid	Liquid	Solid
Nurse culture	None	Albino protoplasts	Mesophyll protoplasts
Hybrid isolation	None	After a few days with micromanipulator	Isolation of minicalli at 70–80-cell stage
Regeneration induction	$4\,g\,l^{-1}$ starch in Kao (1977) liquid medium	Growth-regulator-free MS solid medium	MS solid medium plus $10\,mg\,l^{-1}$ 2,4-D
Development	Filter paper soaked in liquid medium A (Kao and Michayluk 1975)	MS solid medium overlaid with liquid UM medium plus $0.4\,mg\,l^{-1}$ 2,4-D and $0.05\,mg\,l^{-1}$ kinetin	MS solid medium plus $0.1\,mg\,l^{-1}$ IAA and $1\,mg\,l^{-1}$ 2 iP

sorbitol in 3mM MES, pH 5.6). After 16h of treatment at 26°C protoplasts are isolated through centrifugation, washed and resuspended in K77 at a density of 2.5×10^5 protoplasts ml^{-1}.

In Nottingham the source of protoplasts for *M. falcata* was cell suspension, while for *M. sativa* cotyledons and cell suspension were utilized in two subsequent experiments. In the first experiment (Gilmour et al. 1987a), where cell suspension protoplasts were fused with cotyledon protoplasts, the enzymatic mixture, common to both parents, consisted of a CPW salt solution (Frearson et al. 1973) containing 2% Meicelase (Meiji Seika Kaisha Ltd. Tokyo, Japan), 2% Rhozyme HP, 0.03% Macerozyme R10, and 13% mannitol (pH 5.8); incubation was performed for 16h at 25°C, protoplasts were released by gently squeezing the cotyledon strips, sieved in 64-µm mesh, repeatedly washed, and resuspended in K8p medium (Kao 1977). In the second experiment (Mendis et al. 1991), which is different from the first successful case in producing hybrid plants, cell suspension was utilized as protoplast source for both parents and the protocol of protoplast isolation was similar to the previous one with a few differences: the presence of 2% Driselase (Fluka Chemie AG Buchs, Switzerland) in the enzymatic solution, 1h reduction of enzymatic incubation, and a more accurate purification of protoplasts with flotation, on 21% sucrose solution.

Protoplast isolation in the experiment carried out at Perugia showed some differences for the explant utilized and for the method of protoplast isolation. In fact, it was observed that the donor explant could be determinant in obtaining hybrid calli resulting in regeneration; starting from a *M. sativa* cell suspension, derived from an embryogenic genotype, with high plating efficiency, no regenerants were obtained (Pupilli et al. 1991) and leaf mesophyll has been identified as the most suitable alfalfa explant for somatic hybridization experiments. Leaves are always available, and easily manipulated in vitro. They release large amounts of protoplasts with high plating efficiency and, if derived from an embryogenic genotype, the protoplasts maintain the regeneration capability. For protoplast isolation, peeled *M. sativa* leaves were incubated for about 5h at 28°C in the enzymatic mixture: 2% Cellulase R10, 0.5%, Macerozyme R10 0.1%, Pectolyase Y23 (Seishin Co. Ltd. Tokyo, Japan), 10% mannitol in CPW solution, pH 5.8) diluted 2:1 with the protoplast culture medium KM8p (Kao and Michayluk 1975), sieved through a 63-µm mesh and the enzymatic mixture washed off with repeated centrifugation.

Protoplasts of *M. falcata* were obtained from 3-month-old calli dispersed in callus culture liquid medium for 2 days under moderate shaking, incubated for 5h, at 28°C, with the enzymatic solution [0.5% Cellulase Onozuka YC (Seishin Co. LTD), 2% Driselase, 1.2% Pectinase (Sigma Aldrich St. Louis Mo-USA), 0.3% Pectolyase Y23, mannitol 10% in CPW solution], sieved through a 63-µm mesh and floated on 25% sucrose. The novelties of this method consist of the use of calli of *M. falcata* in place of the more time-consuming cell suspension and the lack of Rhozyme in the enzymatic mixture; Rhozyme is, in fact, no longer commercially available.

Protoplast fusion has been performed differently in all three labs. Téoulé (1983) mixed equal volumes of parental protoplasts and replaced K77 medium with fusogenic solution consisting of 0.05 M $CaCl_2$ dihydrated, 9.5% mannitol, 0.05 M glycine brought to pH 10.4, with NaOH. After 15 min at 32 °C, protoplasts were washed with a CPW solution containing 13% mannitol and 0.05 M $CaCl_2 \cdot 2H_2O$ at pH 5.7, a second time with K77 medium, and finally resuspended in the same medium at a density of 5×10^4 protoplasts ml^{-1}. Téoulé and Dattée (1987) claim that the use of a fusogenic solution containing PEG allowed the recovery of a higher yield of heterokaryons but, at the same time, PEG adversely affected the viability of protoplasts, and that the method based on the solution of Ca^{2+} at high pH is more suitable for producing somatic hybrids in the genus *Medicago*. This conclusion is contradicted by the fusion method applied by Mendis et al. (1991), who utilized PEG at a high concentration and succeeded in producing symmetric somatic hybrids. To 1 ml of CPW (10% mannitol) solution containing an equal amount of parental protoplasts at a final density of 1×10^5 protoplasts ml^{-1}, they added 1 ml of fusogenic solution containing PEG (50% PEG 1500, 75 mM HEPES, pH 8) and left it for 25 min; then the PEG solution was eluted, adding a Ca^{2+}-free solution containing 0.2% BSA and 7% mannitol at pH 5.8, elution was performed by adding five drops of this solution while withdrawing an equal volume of the fusion solution; the elution was repeated every 5 min over a period of 30 min. At Perugia, somatic hybrids were obtained through electrofusion, which has many advantages over the chemical methods previously discussed: the absence of any toxic compound for protoplast viability, efficiency, reproducibility, and versatility (Arcioni et al. 1997), but expensive equipment is required. The electrofusion method utilized is that of Damiani et al. (1988) with a few modifications (Pupilli et al. 1991). It consists of the application of an alternate current field (AC) to a mixture of parental protoplasts, resuspended in the electrofusion solution (11% mannitol, 0.6 mM histidine, 0.2 mM $CaCl_2$), which induces protoplasts to stick together, followed by a strong direct current pulse which causes reversible membrane breakdown with the consequent fusion of the held-together protoplasts. The experimental parameters are adjusted to each parental combination in order to maximize the percentage of binary, heterokaryotic, fused protoplasts. For hybridization between *M. sativa* and *M. falcata*, equal amounts of parental protoplasts, at the final density of 3×10^5 protoplasts ml^{-1}, an AC field of $170 V cm^{-1}$ was applied followed after 1 min by three pulses of $1700 V cm^{-1}$, $35 \mu s$ long (Crea et al. 1997).

3 Culture and Selection of Fused Protoplasts

Neither antibiotic-resistant, nor auxotrophic mutants have been employed to date in somatic hybridization of *M. sativa* with *M. falcata*. Consequently, selection of heterokaryons at cell level was not performed at all (Téoulé 1983) or consisted of the identification of hybrid cells on the basis of their dual color (Gilmour et al. 1987a; Mendis et al. 1991; Crea et al. 1997). Following fusion,

Téoulé (1983) cocultivated heterokaryons and parental protoplasts in Kao77 liquid medium (Kao 1977) at a density of 5×10^4 protoplasts ml^{-1} in the dark at 28 °C. Gilmour et al. (1987a) were the first to use fluorescent dyes to identify and isolate putative hybrid calli: they labeled colorless protoplasts of *M. falcata* with fluoresceine diacetate (FDA) and fused them with green protoplasts of *M. sativa*. The fusion products were cultured in Kao77 culture medium with some modifications (glucose 100 g l^{-1}, sucrose 250 g l^{-1} Thiamine HCl 1 mg l^{-1}) at a density of 1×10^4 protoplasts ml^{-1} in the dark at 28 °C. Single heterokaryons, identified by their dual florescence, green for FDA and red for chlorophyll, were physically isolated with a micromanipulator 24 h after fusion and mixed with albino *M. sativa* nurse protoplasts isolated 2–3 days previously. The osmolality of the medium was reduced weekly by diluting the protoplast culture medium with the respective cell culture medium K8 (Kao 1977) modified as follows: thiamine-HCl 1 mg l^{-1}, hydrolysed casein 125 mg l^{-1}, coconut milk 1% (v/v), sucrose 20 g l^{-1}, 2,4-D 0.1 mg l^{-1} and zeatin 0.2 mg l^{-1}. After 1 month of culture, the developing calli were transferred to agar-solidified K8 medium in which green putative hybrid calli appeared 3 weeks later. These calli were then selected manually and transferred onto agar-solidified UM-based medium (Uchimiya and Murashige 1974) for further growth. Mendis et al. (1991) labeled the parental protoplasts with FDA (yellow fluorescence) and with rhodamine isothiocyanate (red fluorescence, RITC), alternatively, so that heterokaryons were double-stained under fluorescent light. The same micromanipulator device as in Gilmour et al. (1987a) was utilized to isolate heterokaryons, which were not mixed with nurse cells but physically separated from them either by a filter membrane (Gilmour et al. 1987b) or by embedding them in agarose droplets of 0.2 ml volume (Mendis et al. 1991). As in Gilmour et al. (1987a), the osmolality of the nurse medium was gradually reduced by adding an equal volume of cell culture medium every 10 days and the total volume was halved prior the every subculture in order to maintain the culture volume around 5 ml. After 70 days of culture, tissues developed from heterokaryons were transferred to agar-solidified and growth-regulator-free MS (Murashige and Skoog 1962) medium for embryo formation. Dual fluorescent selection followed by nurse culture of heterokaryons was also used by Crea et al. (1997): they labeled the colourless protoplasts from callus of *M. falcata* with fluorescein isothiocyanate (yellow-green florescence, FITC) and fused them with mesophyll protoplasts of *M. sativa*. The fusion products were cultured for 2 days with an excess of mesophyll *M. sativa* parental protoplasts, as nurse, in KM8p liquid medium, and afterwards embedded in 0.4% (w/v) agarose-solidified KM8p:KM8 (3:1) (Kao and Michayluk 1975) medium. The agarose bed was then cut into slices which were immobilized in petri dishes with melted agarose (2%) and surrounded by frequently renewed culture medium consisting of a mixture of KM8p:KM8 liquid media in which the proportion of KM8 was gradually increased (Pupilli et al. 1991). The hybrid cells were identified under the fluorescent microscope, their positions recorded, and their development was followed over time. When the putative hybrid calli reached the stage of around 70–80 cells, they were manually picked up with a needle, cultured onto KM8 solidified with 0.5% (w/v) agarose, and surrounded by actively dividing nurse calli. The cultures were incubated in the dark at 28 °C for 3 weeks and then transferred to the light ($27\,\mu E\,m^{-2}\,s^{-1}$) onto UM medium supplemented with $2\,mg\,l^{-1}$ 2,4-D for further growth. Dual fluorescence allows the selection of hybrid tissues between virtually any parents and does not need the use of mutants or very expensive devices such as the fluorescence activated cell sorter (FACS).

4 Regeneration of Somatic Hybrids

The main obstacle encountered while attempting somatic hybridization in *Medicago* is the difficulty in regenerating plants from somatic hybrid calli (Arcioni et al. 1994); nevertheless *M. sativa* + *M. falcata* somatic hybrid plants were among the first reported in literature (Téoulé 1983). This author fused embryogenic protoplasts isolated from mesophyll of both species. The addition of starch ($4\,g\,l^{-1}$) to the culture medium 1 month after fusion enhanced the formation of embryos that were individually picked up and cultured in a filter paper soaked in starchless medium A (Kao and Michayluk 1975). Well-developed shoots were then rooted in growth-regulator-free MS medium. Arcioni et al. (1994) stressed the necessity of using embryogenic protoplasts from both parents or, alternatively, at least one parent should contribute highly dividing protoplasts and the other almost dormant ones. Under these conditions, heterokaryons follow the developmental pathway of the morphogenetic parent. In fact, Gilmour et al. (1987a) failed to regenerate plants from hybrid calli derived from *sativa* cotyledon protoplasts and cell suspension protoplasts of *falcata*, probably as a consequence of the inability of the parents to regenerate. A further confirmation of the critical role played by the embryogenic capacity of parental protoplasts in the regeneration of hybrid plants comes from the work of Mendis et al. (1991), who obtained somatic hybrid plants between highly regenerable cell lines of both *M. sativa* and *M. falcata*. These authors stressed the use of plant hormone-free liquid media for the maintenance of the embryogenic capacity of parental cell lines that can be further enhanced if such media are alternated with a callus-induction medium containing a high concentration of auxins. Somatic embryos obtained from hybrid calli and developed in MS plant hormone-free solid medium were transferred to the same solid medium overlaid with a film of liquid UM medium supplemented with $0.4\,mg\,l^{-1}$ 2,4-D and $0.05\,mg\,l^{-1}$ kinetin. From these cultures, well-developed plants were generated that, when they became 6–8 cm in size, were transferred to a soil-less compost for further growth.

Crea et al. (1997) first reported obtaining somatic hybrid plants between *M. sativa* and *M. falcata* at different ploidy levels. Putative hybrid microcalli were cultured in several regeneration media all based on MS medium supplemented with various combinations of plant hormones, but only calli cultured in the presence of $10\,mg\,l^{-1}$ 2,4-D developed embryo-like structures that were transferred onto CDM1 medium (MS salts plus 0.1 mg l^{-1} IAA and 1 mg l^{-1} 2 iP). In 3 weeks time these embryos evolved into well-developed shoots which were transferred in Magenta vessels containing plant hormone-free RL (Phillips and Collins 1979) or plant hormone-free MS medium for root induction. Unlike the results with another *Medicago* somatic hybrid (Nenz et al. 1996), the addition of a synthetic auxin, BOA (1,2 benzisoaxazole-3-acetic acid), to the regeneration medium was not effective. In all three successful attempts to produce *M. sativa* + *M. falcata* somatic hybrid plants, 2,4-D (an auxin) was used either to maintain the regenerative capability of parental cell lines (Mendis et al. 1991) or in callus-induction media (Téoulé 1983), or even to induce regeneration when used at high concentration (Crea et al. 1997). Following the 2,4-D exposure, regeneration is achieved on media with low levels or without auxins: this seems a common feature for species related to *Medicago sativa*, since also other species belonging to the *sativa-coerulea-falcata* complex follow this rule (Arcioni et al. 1990). Once again, even in experiments aimed at obtaining somatic hybrids between *M. sativa* and *M. falcata*, the principal role of the genotype in the regeneration of hybrid plants is evident, but the regeneration efficiency can be strongly influenced by the experimental conditions. Potentially morphogenetic protoplasts can lose their regeneration ability when derived from long-time cultured cell suspension, and, on the other hand, the regeneration frequency of the hybrid calli can be enhanced by either varying the plant hormone composition or when these calli are cultured in close proximity to highly regenerating tissues. The last two observations may indicate that regeneration efficiency can be stimulated by diffusible factors in a manner similar to that already observed for enhancing cell division in protoplast nurse cultures.

5 Indentification and Characterization of Somatic Hybrid Plants

Even though selection has been performed at the cell level, the hybrid nature of the regenerated plants has been assessed to avoid the risk of misclassification of selected calli. The general criterion followed for this purpose is that polymorphic traits of the parents should be simultaneously present in the hybrids and, to discard the effect of somaclonal variation, chromosome counts should be considered as complementary evidence. The presence of both parent-specific peroxidase and β-amylase isozymes in the hybrid zymograms and the chromosome count, which was the sum (64) of parental complements, confirmed the hybrid nature of the regenerated plants of Téoulé (1983). Similarly, Mendis et al. (1991) used esterase and leucine-amino-peptidase to confirm the hybridity of selected plants showing intermediate phenotypes; the zymograms of hybrid plants showed additional bands not present in the parents (Table 2). The same hybrids showed wide variation in chromosome

Table 2. Cytological, molecular, and morphological traits of *M. sativa* + *M. falcata* somatic hybrid plants[a]

	Téoulé (1983)	Mendis et al. (1991)	Crea et al. (1997)
Chromosome number	33, 48, 64	34–58	33
Isoenzyme phenotype	F and I	H	nd
Morphological traits:			
Plant habit	nd	F	S
Leaf shape	F and H	I	S
Inflorescence length	nd	F	nd
Flowers/			
inflorescence	nd	I	nd
Flower size	nd	H	nd
Flower color	F and H	I	S
Pod shape	nd	I	nd

[a] S = *M. sativa*-like; F = *M. falcata-like*; I = intermediate; H = heterotic; nd = not determined.

complements ranging from 34 to 58, and there were no cases of plants exhibiting the exact sum of the chromosomes of both parents. Crea et al. (1997) encountered many more difficulties in providing evidence of hybridity because of the strong asymmetric nature of their hybrids. In fact, while almost all the genome of *M. sativa* parent was incorporated into the hybrid genome, only few or at least only one chromosome of *M. falcata* was retained among the 33 chromosomes of these hybrids. Therefore, more decisive markers such as isozymes are required to detect polymorphic loci located in the few introgressed chromosomes. RAPD (random amplified polymorphic DNA) analysis proved effective in revealing *M. falcata*-specific fragments that were present in the amplification profile of hybrid plants. A detailed characterization at molecular, cytological, and morphological levels exists for each of the three examples of somatic hybrids *M. sativa* + *M. falcata* reported in the literature. A study of the meiotic behavior of Téoulé's hybrids was reported by Coulaud et al. (1990). Although these hybrids produced a large array of abnormal tetrads as a consequence of meiotic abnormalities such as lagging chromosomes, the pollen viability was still relatively high. Curiously enough, one of the somatic hybrids lost a set of chromosomes during vegetative propagation, becoming hexaploid ($2n = 48$). This plant showed the highest percentage (26.15%) of abnormal tetrads in comparison with other fusion products, sexual hybrids *M. sativa* × *M. falcata* and parental lines, but still the percentage of pollen viability was high. For the high morphological similarity of the chromosomes of the two species, it was not possible to establish if the chromosome elimination originated prevalently in one or both parents. Since the percentage of gametic abnormalities was remarkably higher in somatic hybrids than in normal tetraploids, the authors hypothesized that in these latter plants a mechanism exists that can counterbalance the errors of meiosis. Such a mechanism is less effective in somatic hybrids. However, the authors concluded that the wide assortments of ploidy levels presumably present in the gametes of the somatic hybrid could generate useful variability to be utilized in breeding programs. The same somatic hybrid plants were studied for cytoplasmic composition by D'Hont et al. (1987), the chloroplast DNA of these plants was identical to that of *M. sativa* parent, while the mitochondrial DNA showed various extents of rearrangement such as the appearance of new bands, not present in both parents, and different assortments of parent-specific bands. The morphological traits of these plants were also investigated; the young leaves were more indented, rigid, and hairy than those of the parents, and their disposition on the rachis was intermediate between the parents. The flowers showed an array of variegation ranging from yellow (*M. falcata*) to violet (*M. sativa*), similar to that observed in the populations of *M. sativa* characterized by various extents of *M. falcata* introgression (Téoulé 1983). Yellow-violet and larger flowers were also noted by Mendis et al. (1991). Many phenotypic traits of the somatic hybrids, such as plant habit, internode distance, leaf morphology, and length, were intermediate to those of the parental species. An interesting correlation was noted between the chromosome complements and phenotypic variation: the higher the chromosome number, the larger the extent of phenotypic abnormality detected.

All the somatic hybrids analyzed showed a chromosome number lower than the sum of the parental chromosomes. The somatic hybrids obtained by Crea et al. (1997) were intensively studied, because they represented an interesting case of gene amplification at the ribosomal DNA (rDNA) level. Ten hybrids, deriving from the same fusion event, had the same chromosome number ($2n = 33$), indicating that 15 parental chromosomes were missing. In all plants one extra-long and one extra-short chromosome were observed, probably as a consequence of a translocation (Fig. 1e). The long chromosome showed two secondary constrictions, but only one of them contained the nucleolar organising region (NOR) and therefore the ribosomal genes. Fluorescent in situ hybridisation (FISH), using a homologous 18S probe, was performed to locate the NORs loci on parental chromosomes and to follow their evolution in hybrid plants. Five out of six parental NORs, four from *M. sativa* and two from *M. falcata* were retained in the somatic hybrid, but the origin of the missing NOR was not determined (Fig. 2). Other than these macroscopic rearrangements, the ribosomal genes of these plants showed the creation and amplification of new IGS (intergenic spacer of the rDNA gene, Fig. 1c) variants (Cluster et al. 1996). The molecular mechanism proposed by these authors as most likely responsible for the creation of these new variants, characterized by the insertion or elimination of one or more 340-bp subrepeating elements (Fig. 1d), is the unequal recombination between homologous and nonhomologous chromosomes, which it is known to create copy-number variation in tandem-repeated gene families (Smith 1976), which occurred throughout callus growth, as demonstrated by the variability reported among plants regenerated from the same callus.

Differences between parental lines were also detected in the amplification level of another tandem-repeated sequence that was cloned in *M. coerulea* and found to be interspersed in the genome of the *sativa-falcata-coerulea* complex. The copy number of this sequence, called C300, was remarkably lower in *falcata* than in *sativa*, as is noticeable on comparing the band

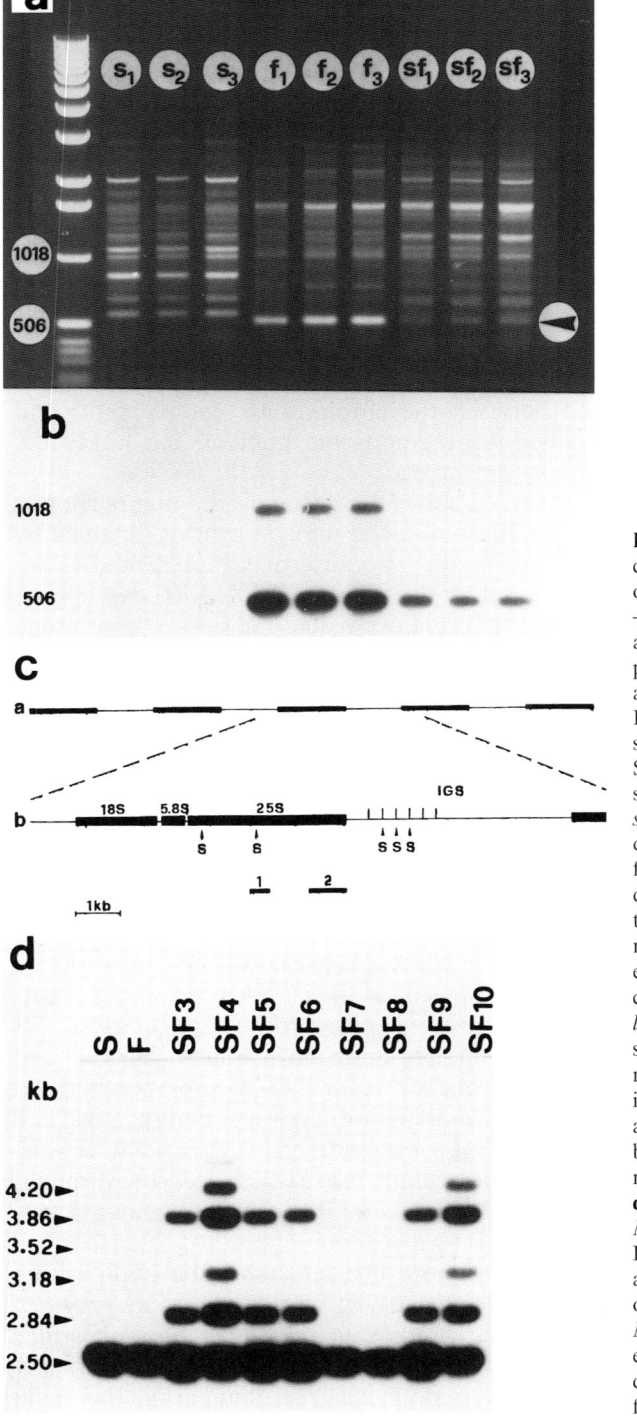

Fig. 1a–g. Molecular and cytological characterization of somatic hybrids *M. sativa* + *M. falcata*. **a** RAPD analysis obtained with the primer RP7. **b** Southern analysis of the RAPD from RP7 (probe: 530-bp *falcata*-specific RP7 fragment. s_{1-3} Somaclones of *M. sativa*,; f_{1-3} somaclones of *M. falcata*; sf_{1-3} three somatic hybrids deriving from the same fusion event. **c** rDNA diagram. *a* Indicates the tandem organization of rDNA genes and *b* is an expanded map of one complete gene unit. *Thick lines* represent coding sequences and *thin lines* non-coding IGS regions; *S* indicates *Ssp*I sites; *1* and *2* are two sequences belonging to the 25S coding region used as probes. **d** RFLP patterns from the *M. sativa* + *M. falcata* hybrid DNAs digested with *Ssp*I and probed with sequence 2 of **c**. S *M. sativa* and F *M. falcata* parents; SF_{3-10} eight somatic hybrids deriving from the same fusion event

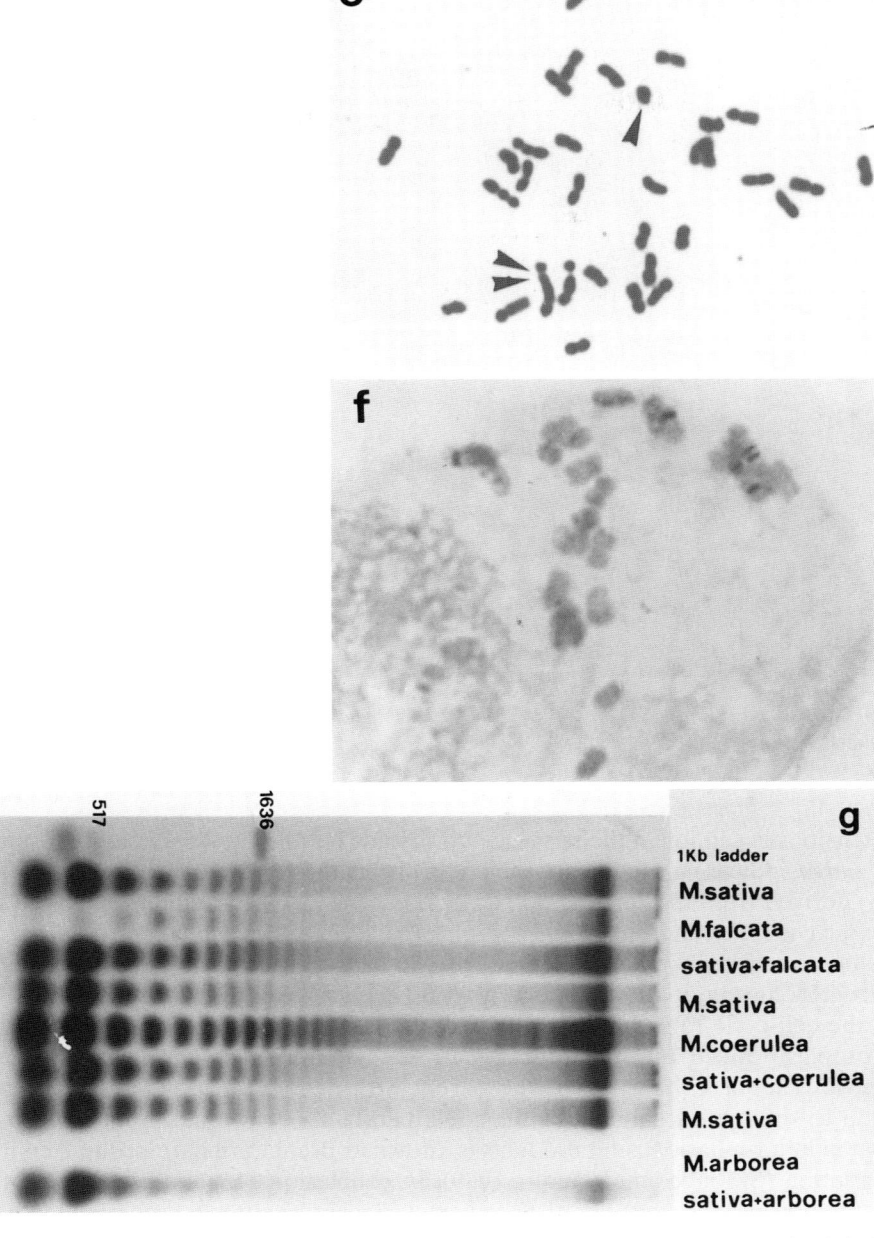

Fig. 1a–g. (*Continued*) **e** Metaphase spread of a somatic hybrid plant: the *two arrowheads* indicate two chromosomes of unexpected length, probably a consequence of a translocation phenomenon. **f** A metaphase of a somatic hybrid stained with silver nitrate. **g** Southern analysis of four *Medicago* spp. and three somatic hybrid DNAs digested with *Dra*I and probed with C300. Molecular weights are expressed in bp

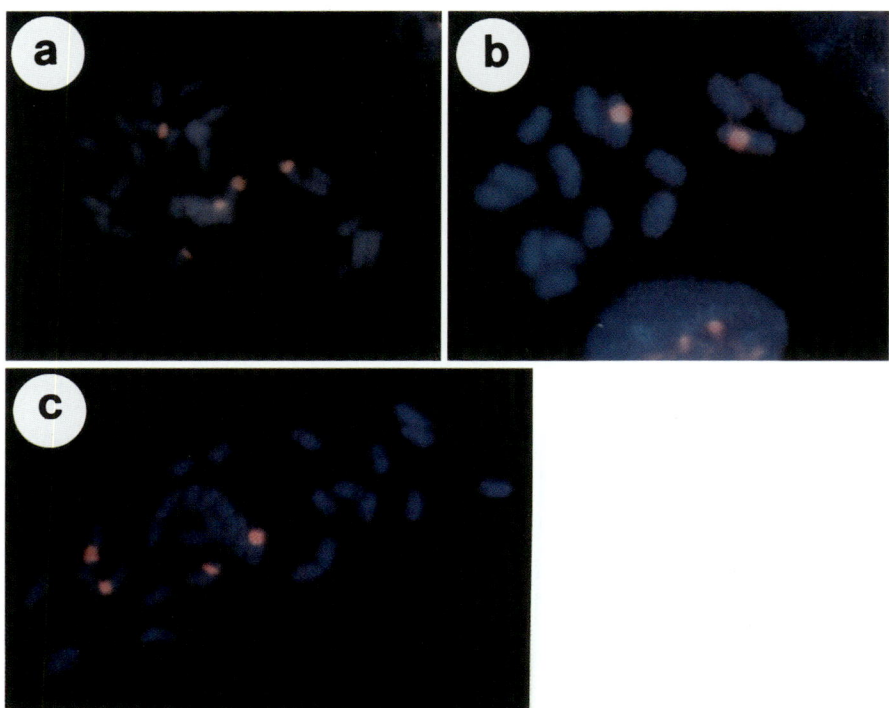

Fig. 2a–c. 18S FISH to mitotic chromosomes of **a** *Medicago sativa*, **b** *Medicago falcata*, and **c** somatic hybrid *M. sativa* + *M. falcata*

intensities in Fig. 1g: in the somatic hybrids this number was very close to that of *sativa*, indicating that most of the hybrid chromosomes (probably 32 out of 33) derived from this last parent (Calderini et al. 1997). Morphological analyses provided an indirect confirmation of this hypothesis, most of the traits being considered similar to *sativa*. One possible cause of the *falcata*-specific chromosome elimination in somatic hybrids, could be the significant difference in the cell-cycle rates between the parental cell lines, being mesophyll *sativa* protoplasts much more mitotically active than those of *falcata* cell lines. Previous reports showed that the parent with a slow cell cycle contributed little to the hybrid genome (Stutz 1962; Gupta 1969) through a mechanism that in heterophasic fusion products is known as premature chromosome condensation (Szabados and Dudits 1980). In conclusion, somatic hybridization created formidable stress that caused extensive and randomly occurring DNA rearrangements involving any portion of the genome. Differences on the synchrony of parental protoplasts can affect chromosome elimination more than genetic distance or ploidy levels. Nonetheless, regenerated plants appeared quite normal and fertile because the regeneration process eliminates unsuccessful genomic combinations arising from rearrangements at the level of the coding regions.

6 Summary and Conclusion

Experiments in somatic hybridization between the two subspecies of the *Medicago sativa* complex, *falcata* and *sativa*, have provided basic information about the problems encountered on protoplast fusion in the genus *Medicago* and the strategies to adopt to better overcome these problems. After the production of somatic hybrids between these two subspecies, other hybrids have been obtained between other subspecies of the group: *sativa* and *coerulea* (Pupilli et al. 1992), between incompatible and phylogenetic distant *Medicago* such as *M. arborea* and *M. sativa* (Nenz et al. 1996), and between cultivated alfalfa and other leguminous species such as *Onobrychis viciifolia* (Li et al. 1993). This last experiment, which was performed to introduce condensed tannins into alfalfa, resulted in the successful production of asymmetric somatic hybrids but did not succeed in improving alfalfa forage quality; nevertheless, all these experiments have indicated a novel route for alfalfa breeding.

The technical problems of the selection of best protoplast source, of somatic hybridization procedures, and of the identification and selection of hybrids have been solved by adopting different strategies. The most relevant acquisition from these experiments is the key role played by the parental protoplasts in determining the embryogenetic ability of hybrid calli; therefore, it has been concluded that to produce hybrid plants it is necessary to combine in one parental protoplast population embryogenetic ability and high plating efficiency. The synchrony of the cell cycle and the homogeneity of the chromosome complement of the two parents seem more important than genetic distance in determining equal representation of the two parental genomes in the hybrids, as parents with slow cell cycle or with low chromosome number do not contribute proportionally to the hybrid genome (Arcioni et al. 1997). The application of these techniques to plant breeding has been limited in all the experiments performed by the low number of hybrid plants produced: this means that an effort to optimize large-scale production of somatic hybrids still needs to be made to select within a large hybrid population the individuals carrying the most useful gene combination. In fact, through this technique it is possible to transfer multigenic traits simultaneously, but for this purpose it is necessary to select for the right allele combination within a large population of recombinants. Other than for plant breeding purposes, somatic hybridization showed in these experiments its relevance for studying the mechanisms of molecular evolution; in fact putting together alien genomes allowed the observation of infrequent phenomena such as chromosome deletion and duplication, translocation, and gene loss that could be referred to the phenomenon of the species-specific chromosome elimination observed in barley hybrids (Kasha and Kao 1970), or concerted evolution (Arnheim 1983), events that are rarely observed in eukaryotic organisms.

7 Protocol

All solutions used must be sterile; protoplast culture medium and enzymatic solution should be sterilized through filtration, all the others through autoclaving (121 °C, 1.2 atm, 20 min).

7.1 Protoplast Isolation

M. sativa, leaves: 30 heart-shape, hair-free, fully expanded leaflets grown axenically are necessary.
1. Remove lower epidermis with the aid of forceps and lay leaflets on plasmolyzing solution (CPW salts with 11% mannitol).
2. Place peeled leaves in a petri dish (Ø = cm 5) containing 6 ml of the enzymatic mixture (2% Cellulase R10, 0.5% Macerozyme, 0.1% Pectolyase Y23, 10% mannitol in CPW solution, pH 5.8) and 1 ml of the protoplast culture medium KM8p.
3. Incubate for 5–6 h at 28 °C in a rotary shaker (20 rpm).
4. Sieve all the contents of the petri dish through a 63-µm mesh to remove debris and then centrifuge (40 rpm, 5 min).
5. Remove supernatant and resuspend pellet in 8 ml CPW 11% mannitol solution.
6. Centrifuge again as previously and resuspend pellet in 2 ml of fusogenic solution (0.6 mM Histidine, 0.02 mM $CaCl_2$, 0.6% mannitol).
7. Count protoplasts in a hemocitometer and adjust the concentration of protoplasts with fusogenic solution at a density of 3×10^5 protoplasts ml^{-1}.

M. falcata, callus
1. Transfer 3 g of friable callus in an Erlenmeyer flask with 40 ml of UM liquid medium with 2 mg l^{-1} 2,4-D and 0.25 mg l^{-1} kinetin; incubate 4 days on a rotary shaker (150 rpm, 22 °C, 20 µE s^{-1} m^{-2}).
2. Centrifuge (300 g, 5 min) the content of one flask.
3. Resuspend the pellet in 21 ml enzymatic solution: 0.5% Cellulase YC, 2% Driselase, 1.2% Pectinase, 0.3%, Pectolyase, 10%, mannitol, and 110 µg FITC, fluoresceine isothiocianate, in CPW salt solution).
4. Incubate in a rotary shaker (50 rpm, 6 h, 28 °C) in the dark.
5. Sieve the suspension in a 63-µm mesh and centrifuge (40 g, 5 min).
6. Resuspend the pellet in CPW 25S solution (CPW salt solution containing 25% w/v of sucrose) and centrifuge (80 rpm, 15 min).
7. Add gently a few drops of fusogenic solution in the top of the tube and recover this solution where floated protoplasts are diffused in a new tube; if necessary, add a few more drops of fusogenic solution and repeat protoplast recovery.
8. Add 8 ml of fusogenic solution and centrifuge (40 g, 5 min).
9. Resuspend pellet in fusogenic solution, count protoplasts, and adjust the density at 3×10^5 protoplasts ml^{-1}.

7.2 Electrofusion

1. Mix equal volumes of parental protoplasts in a tube.
2. Sterilize the fusion chamber with immersion in 70% ethanol for 10 min, and leave them in the air flow of a sterile hood to allow ethanol to evaporate.
3. Load the fusion chamber with an aliquot of mixed protoplasts.
4. Apply and alternate current (AC) field of 170 V cm^{-1} for 1 min, apply three pulses of direct current (1700 V cm^{-1}, 35 µs long), let the AC field progressively reduce to 0.
5. Transfer the treated protoplasts to a petri dish and add 2 ml of mesophyll protoplasts resuspended in KM8p medium at the density of 2.5×10^5 protoplasts ml^{-1}.
6. After 2 days collect the contents of three petri dishes and spin (90 g, 10 min).

7. Remove 6 ml of the supernatant and add 0.8 ml of KM8p, 0.75 ml of 2× KM8p and 0.25 ml of 2× KM8, gently resuspend protoplasts.
8. Transfer into a petri dish (Ø = 5 cm) and add 1 ml of warm (30 °C) 2% low gelling agarose solution, mix gently and let solidify in the sterile hood.
9. After 1 day cut the contents of each petri dish into slices and split the contents of each dish into two, fix slices on the bottom of the dish with a few drops of warm agarose and fill dishes with 1 ml of liquid KM8p : KM8 (3 : 1) mixture.
10. Incubate in the dark (28 °C) and replace culture medium every 5 days, gradually increasing the percentage of KM8, after 1 month KM8p is totally replaced by KM8.
As soon as slices are fixed to the dish, it is possible to start heterokaryon identification

7.3 Heterokaryon Identification

1. For this purpose, dishes are marked and placed under an inverted microscope in a fixed position.
2. Observation under UV illumination allows identification of heterokaryons which show dual fluorescence induced by FITC for callus protoplasts and by chlorophyll for mesophyll protoplasts; position is recorded through coordinates.
3. Follow the development of hybrids and when they reach the stage of 70–80 cells, transfer with a needle to a new petri dish containing solid KM8 medium and place around five to six mesophyll-derived calli.
4. Incubate for 3 weeks under fluorescent light (27 $\mu E\ s^{-1}\ m^{-2}$; 12-h photoperiod, 24 °C).
5. Transfer to UM medium supplemented with 2 mg l^{-1} of 2,4-D.

7.4 Regeneration of Hybrid Calli

1. Transfer to MS solid medium containing 10 mg l^{-1} of 2,4-D.
2. After 10 days transfer embryo-like structures to MS medium supplemented with 0.1 mg l^{-1} IAA and 1 mg l^{-1} 2 iP.
3. When embryos develop into shoots, move to a magenta vessel containing growth regulator-free MS medium.

References

Arcioni S, Damiani F, Pezzotti M, Lupotto E (1990) Alfalfa, lucerne (*Medicago* spp.) In: Bajaj YPS (ed) Biotechnology in agriculture and forestry, vol 10. Legumes and oilseed crops I. Springer, Berlin Heidelberg New York, pp 197–235
Arcioni S, Damiani F, Pupilli F (1994) Somatic hybridization in the genus *Medicago*. In: Bajaj YPS (ed) Biotechnology in agriculture and forestry, vol 27. Somatic hybridization in crop improvement I. Springer, Berlin Heidelberg New York, pp 145–164
Arcioni S, Damiani F, Mariani A, Pupilli F (1997) Somatic hybridization and embryo rescue for the introduction of wild germplasms. In: McKersie BD, Brown DCW (eds) Biotechnology and the improvement of forage legumes. CAB International, Oxon, New York, pp 61–89
Arnheim N (1983) Concerted evolution and multigene families. In: Nei M, Koehn RK (eds) Evolution of genes and proteins. Sinauer, Sunderland, pp 38–61
Bingham ET, Binek A (1969) Hexaploid alfalfa, *Medicago sativa* L.: origin, fertility and cytology. Can J Genet Cytol 11:359–366

Brown DCW, Atanassov A (1985) role of genetic background in somatic embryogenesis in *Medicago*. Plant Cell Tissue Organ Cult 4:111–122

Calderini O, Pupilli F, Paolocci F, Arcioni S (1997) A repetitive and species-specific sequence as a tool for detecting the genome contribution in somatic hybrids of the genus *Medicago*. Theor Appl Genet 95:734–740

Cardi T, D'Ambrosio F, Consoli D, Puite KJ, Ramulu KS (1993) Production of somatic hybrids form frost-tolerant *Solanum commersonii* and *S. tuberosum*: Characterization of hybrid plants. Theor Appl Genet 87:193–200

Chen THH, Marowitch J (1987) Screening of *Medicago falcata* germplasm for in vitro regeneration. J Plant Physiol 128:271–277

Cluster PD, Calderini O, Pupilli F, Crea F, Damiani F, Arcioni S (1996) The fate of ribosomal genes in three interspecific somatic hybrids of *Medicago sativa*: three different outcomes including the rapid amplification of new spacer-length variants. Theor Appl Genet 93:801–808

Coulaud J, Siliak-Yakovlev S, Téoulé E (1990) Comportement méiotique chez des plantes de luzerne (*Medicago sativa*) provenant d'une expérience de fusion somatique. Can J Bot 68:73–78

Crea F, Calderini O, Nenz E, Cluster PD, Damiani F, Arcioni S (1997) Chromosomal and molecular rearrangements in somatic hybrids between tetraploid *Medicago sativa* and diploid *Medicago falcata*. Theor Appl Genet 95:1112–1118

Damiani F, Pezzotti M, Arcioni S (1988) Electric field-mediated fusion of protoplasts of *Medicago sativa* L. and *Medicago arborea* L. J Plant Physiol 132:474–479

D'Hont A, Quétier F, Téoulé E, Dattée Y (1987) Mitochondrial and chloroplast DNA analysis of interspecific somatic hybrids of a Leguminosae: *Medicago* (alfalfa). Plant Sci 53:237–242

Frearson EM, Power JB, Cocking EC (1973) The isolation, culture and regeneration of *Petunia* leaf protoplasts. Dev Biol 33:130–137

Gilmour DM, Davey MR, Cocking EC (1987a) Isolation and culture of heterokaryons following fusion of protoplasts from sexually compatible and sexually incompatible *Medicago* species. Plant Sci 53:263–270

Gilmour DM, Davey MR, Cocking EC, Pental D (1987b) Culture of low-number forage legume protoplasts in membrane chambers. J Plant Physiol 126:457–465

Gupta SB (1969) Duration of mitotic cycle and regulation of DNA replication in *Nicotiana plumbaginifolia* and a hybrid derivative of *N. tabacum* showing chromosome instability. Can J Genet Cytol 11:133–142

Kao KN (1977) Chromosomal behaviour in somatic hybrids of soybean *Nicotiana glauca*. Mol Gen Genet 150:225–230

Kao KN, Michayluk MR (1975) Nutritional requirements for growth of *Vicia hajastana* cells and protoplasts at very low population density in liquid media. Planta 126:105–110

Kasha KJ, Kao KN (1970) High frequencies haploid production in barley (*Hordeum vulgare* L.). Nature 225:874–876

Li YG, Tanner GJ, Delves AC, Larkin PJ (1993) Asymmetric somatic hybrid plants between *Medicago sativa* L. (alfalfa, lucerne) and *Onobrychis viciifolia* Scop. (sainfoin). Theor Appl Genet 87:455–463

Mendis MH, Power JB, Davey MR (1991) Somatic hybrids of the forage legumes *Medicago sativa* L. and *M. falcata* L. J of Exp Bot 42, 245:1565–1573

Mitten DH, Sato SJ, Skokut TA (1984) In vitro regenerative potential of alfalfa germplasm sources. Crop Sci 24:943–945

Murashige T, Skoog F (1962) A revised medium for rapid growth and bioassay with tobacco tissue cultures. Physiol Plant 15:473–497

Nenz E, Pupilli F, Damiani F, Arcioni S (1996) Somatic hybrid plants between the forage legume *Medicago sativa* L. and *Medicago arborea* L. Theor Appl Genet 93:183–189

Osborn TC, Brouwer D, McCoy TJ (1997) Molecular marker analysis of alfalfa. In: McKersie BD, Brown DCW (eds) Biotechnology and the improvement of forage legumes. CAB International, Oxon, New York, pp 91–109

Phillips GC, Collins GB (1979) In vitro tissue culture of selected legumes and plant regeneration from callus culture of red clover. Crop Sci 19:59–64

Pupilli F, Arcioni S, Damiani F (1991) Protoplast fusion in the genus *Medicago* and isozyme analysis of parental and somatic hybrid cell lines. Plant Breed 106:122–131

Pupilli F, Scarpa MG, Damiani F, Arcioni S (1992) Production of interspecific somatic hybrid plants in the genus *Medicago* through protoplast fusion. Theor Appl Genet 84:792–797

Quiros CF, Bauchan GR (1988) The genus *Medicago* and the origin of the *Medicago sativa* complex. In: Hanson AA, Barnes DK, Hill RR, (eds) Alfalfa and alfalfa improvement. Agronomy Monograph 29. American Society of Agronomy, Madison, pp 93–124

Smith GP (1976) Evolution of repeated DNA sequences by unequal crossover. Science 191:528–535

Stutz HC (1962) Asynchronous meiotic chromosome rhythm as a cause of sterility in *Triticale*. Genetics 47:988

Szabados L, Dudits D (1980) Fusion between interphase and mitotic plant protoplasts. Exp Cell Res 127:442–446

Téoulé E (1983) Hybridization somatique entre *Medicago sativa* L. et *Medicago falcata* L. CR Acad Sci Paris 197:13–16

Téoulé E, Dattée Y (1987) Recherche d'une méthode fiable de culture de protoplasts, d'hybridation somatique et de régénération chez *Medicago*. Agronomie 7:575–584

Uchimiya H, Murashige T (1974) Evaluation of parameters in the isolation of viable protoplasts from cultured tobacco cells. Plant Physiol 54:936–944

Subject Index

acclimation capacity   257
African marigold   177
agarose-bead culture   154, 346
Agrobacterium rhizogenes   315
Agropyron elongatum   61
alfalfa   341, 356
alginate embedding   86
amphidiploid   344
amylase isozyme   311, 363
anthocyanin   317
Arabidopsis thaliana   81
asymmetric protoplast fusion   18, 146, 181, 351
asymmetric somatic hybrid   106, 199, 209, 212, 287, 306, 335, 352
asparagus   95
aspartate aminotransferase   312
Asparagus macowanii   95, 108
Asparagus officinalis   95, 108
Asteraceae   112, 328, 335
Atropa belladonna   304

barley   3, 37
betaine   9, 14, 42
Bialaphos resistance   38
birdfoot trefoil   341
blackleg disease   257
Brassica campestris   87
Brassica napus   81
B. napus + Arabidopsis thaliana   81
Brassica nigra   86

carnation   277
carrot   3
Caryophyllaceae   277
catalase isozyme   348
cell cycle   368
cell viability   13, 45
cell grafting   328
chicory   112
chlorophyll deficiency   318
chloroplast genome   53
chromosome aberration   67

chromosome counting   12, 76, 161
chromosome elimination   66, 106, 174, 211, 308, 344
chromosome fragmentation   49
chromosome number   40, 104, 293
Cichorium endivia   113
Cichorium intybus   112, 119
Citrus aurantifolia   125
Cirus limon   125
Citrus sinensis   124
C. sinensis + C. unshiu   124
Citrus sudachi   125
Citrus unshiu   124
cold tolerance   4, 11, 38
cold treatment   100
common scab disease   258
Convolvulaceae   164
ctDNA   5
cucumber   139
Cucumis hardwickii   141
Cucumis sativus   139
Cucurbitaceae   139
cybrid   18, 25, 27, 32, 34, 119, 129, 307, 318, 331, 334
cybridization   17, 113, 124
cytoplasmic male sterility (CMS)   17, 27, 32, 113, 120, 125, 136

Daucus carota   3
D. corota + Hordeum vulgare   3
deadly nightshade   304
dihaploid production   264
dimethyl sulfoxide (DMSO)   117, 191, 200
Dianthus barbatus   277, 286, 289
Dianthus caryophyllus   277, 289
Dianthus chinensis   277, 289
D. chinensis + D. caryophyllus   277
Diospyos glandulosa   160
Diospyros kaki   152
D. kaki + D. glandulosa   160
disease resistance   48, 183
donor-recipient protoplast fusion   18, 200
dot blot hybridization   205

eggplant 199
eggplant rootstock 243
electrofusion 12, 24, 38, 97, 106, 109, 143, 148, 173, 177, 242, 249, 259, 280, 290, 358, 360, 370
embryogenic callus 51
esterase isozyme 56, 70, 102, 192, 282, 311, 348
Erwinia carotovora 257
Etuberosa species 188

fertility restoration 27
field trial 273
flow cytometry 154, 161, 202, 270, 288
fluorescein diacetate (FDA) 268, 361
forage crop 341
freezing tolerance 257

gamma-irradiation 18, 66, 200, 308
gamma-ray 67
genetic complementation 297, 309
genomic in situ hybridization (GISH) 297
graft chimera 330
Gypsophila paniculata 278, 286

Haynaldia villosa 48, 49
heat shock protein 321
Helianthus annus 115, 119
heterokaryocyte 117
heterokaryon 23, 51, 179, 295, 360, 362
Hordeum vulgare 3, 37
hybrid rice 17, 30
hybrid vigor 160, 181, 284

indica rice 29
interfamililal somatic hybrid 3, 10
intergeneric somatic hybrid 3, 81, 165, 286, 305
interspecific somatic hybrid 155, 165, 172
intertribal somatic hybrid 81
intraspecific somatic hybrid 155, 281
introgression 91
iodoacetamide (IOA) 18, 21, 33, 50, 65, 99, 109, 164, 280, 341
Ipomoea batatas 164, 177
I. batatas + I. trifida 180
Ipomoea erecta 166
Ipomoea lacunosa 166
Ipomoea trifida 166
Ipomoea triloba 165

japonica rice 29

leaf miner resistance 217
Leymus chinensis 61, 65
Lotus corniculatus 341
low-temperature tolerance 4

Lycopersicon esculentum 188, 218
L. esculentum + Solanum etuberosaum 188
L. esculentum + Solanum melongena 199
L. esculentum + Solanum ochranthum 217
Lycopersicon pennellii 201

male sterility 5
melon 139
Medicago falcata 356
Medicago sativa 341, 356
M. sativa + Lotus corniculatus 341
metabolic complementation 24, 347
mitochondrial genome 28, 40
monosomic addition 298
mouse-ear cress 81
mtDNA recombination 125, 334
mt-DNA rearrangement 134

Nicotiana sylvestris 292
N. sylvestris + N. plumbaginifolia 292
Nicotiana plumbaginifolia 292
Nicotiana tabacum 304
N. tabacum + Atropa belladonna 304
nitrosomethylurea (NMU) 329
norflurazon 266
nuclear-cytoplasmic interaction 299

Oryza sativa 37
O. sativa + Hordeum vulgare 37
Onobrychis viciifolia 369

parasexual hybridization 304
peroxidase isozyme 174
persimmon 152
peroxidase isozyme 70, 176, 224, 332, 344, 363
Phomopsis asparagi 95
Phytophthora infestans 217
plastid segregation 333
plastome mutant 318
plating efficiency 87
polypoidization 211
polyploidy 164
pollen viability 228
Poncitrus trifoliata 130
polyethylene glycol (PEG) 24, 50, 69, 117, 121, 143, 173, 191, 221, 229, 200, 221, 290, 342
PEG-DMSO fusion 200
potato 245, 264
potato dihaploid 264
potato virus X (PVX) 257
protoplast fusion 4, 33, 44, 62, 85, 92, 121, 126, 143, 153, 160, 213, 249, 268, 342
protoplast isolation 32, 44, 75, 85, 91, 97, 108, 121, 126, 144, 153, 160, 166, 213, 246, 266, 342

Subject Index

Psathyrostachys juncea 61
Pseudomonas solanacearum 233, 241
P. solanacearum resistance 234
P. solanacearum wilt-inducing product 241
ptDNA recombination 334

Ralstonia solanacearum 257
random amplified polymorphic DNA (RAPD) 58, 63, 72, 76, 103, 146, 157, 161, 192, 224, 240, 251, 202, 271, 285, 364, 366
rape 81
restriction fragment length polymorphism (RFLP) 161, 206, 271, 313, 331, 358
root-knot nematode 181

salt tolerance 4, 11, 38
Satsuma mandarin 124
Senecio fuchsii 328
Senecio jacobaea 328
sexual incompatibility 164, 188
Solanaceae 304
Solanum aethiopicum 233
Solanum brevidens 188
Solanum commersonii 245
Solanum etuberosum 188
Solanum integrifolium 233, 240
Solanum khasianum 233
Solanum melongena 233, 241, 199
S. melongena + *S. sanitwongsei* 233
Solanum nigrum 233
Solanum ochranthum 217
Solanum phureja 264
Solanum sanitwongsei 233, 241
Solanum sisymbriifolium 233

Solanum torvum 233
Solanum tuberosum 245, 264
S. tuberosum + *S. commersonii* 245
S. tuberosum + *S. phureja* 264
somaclonal mutation 29
somaclonal variation 301
somatic embryogenesis 96
– asparagus 48, 95, 113, 141, 164, 217, 223, 245, 328, 341, 356
somatic hybridization 133, 182, 202
Southern blot hybridization 350
Streptomyces scabiae 258
sunflower 117
sweet orange 124
sweet potato 164, 177
symmetric somatic hybrid 306, 352

Tagetes erecta 167
telomere sequence 298
tomato 188, 218, 199
tomato late blight disease 217
Triticum aestivum 48, 65
T. aestivum + *Haynaldia villosa* 48
T. aestivum + *Leymus chinensis* 65
TTC reduction test 7, 9, 41, 45
tuber soft rot disease 257

Verticillium dahlia 258
Verticillium wilt 258
vitrification 146

wheat 48, 49, 65

X-irradiation 18, 143, 172, 341
X-ray 21, 33, 85

Printing (Computer to Film): Saladruck Berlin
Binding: Stürtz AG, Würzburg